Aerosol Pollution Impact on Precipitation

Zev Levin · William R. Cotton
Editors

Aerosol Pollution Impact on Precipitation

A Scientific Review

Springer

Editors

Zev Levin
69978 Tel Aviv
Dept. of Geophysics
Tel Aviv University
Israel
zevlev@post.tau.ac.il

William R. Cotton
Ft.Collins CO 80523
Colorado State University
USA
william.cotton@colostate.edu

ISBN 978-1-4020-8689-2 ISBN 978-1-4020-8690-8 (eBook)
DOI 10.1007/978-1-4020-8690-8

Library of Congress Control Number: 2008933938

Printed on acid-free paper

9 8 7 6 5 4 3 2 1

springer.com

Lead Authors

Name	Affiliation	Country
Andreae, Meinrat O.	Max Planck Institute for Chemistry, Mainz	Germany
Artaxo, Paulo	University of Sao Paulo, Sao Paulo	Brazil
Barrie, Leonard. A.	WMO, Geneva	Switzerland
Brenguier, Jean-Louis	Meteo France, Paris	France
Cotton, William R.	Colorado State University, Fort Collins, CO	USA
Feingold, Graham	NOAA, Earth System Research Laboratory, Boulder, CO	USA
Gong, Sunling L.	Environment Canada, Dufferin St., Toronto, Canada	Canada
Hegg, Dean A.	University of Washington, Seattle, WA	USA
Kaufman,Yoram (deceased)	NASA, Goddard, Greenbelt, MD	USA
Levin, Zev	Tel Aviv University, Tel Aviv	Israel
Lohmann, Ulrike	ETH, Zurich	Switzerland
Tanré, Didier	CNRS, University of Lille, Lille	France
Yuter, Sandra	North Carolina State University, Raleigh, NC	USA

Liaisons

L.A. Barrie (WMO)

R. List (IUGG)

Co-authors and Contributor Authors

Co-authors and Contributors	Chapters
Arimoto, R.	4
Baltensperger, U.	5
Barrie, L. A.	3
Brenguier, J. L.	2
Bruintjes, R. T.	8
Cotton, W. R.	6
Feichter, J.	3
Feingold, G.	2, 5, 6, 7
Guibert, S.	4
Hallett, J.	2, 6
Hobbs, P. V. *(deceased)*	2, 8
Kinne, S.	4
Kloster, S.	3
Kreidenweis, S. M.	3, 4
Levin, Z.	2, 3, 5, 8
Li, S.M.	4
Liousse, C.	3
Lohmann, U.	2
Radke, L. F.	5
Schulz, M.	4
Shepherd, J. M.	5
Stier, P.	3
Strawbridge, K.	4
Sugimoto, N.	4
Terblanche, D.	8
Textor, C.	3, 4
Zhang, L.M.	4

Scientific Reviewers

Chairperson: Dr. George Isaac, Environment Canada

Name	Affiliation	Country
Ayers, Greg	CSIRO Marine and Atmospheric Research	Australia
Barth, Mary	National Center of Atmospheric Research	USA
Bormann, Stephan	Johannes-Gutenberg-University	Germany
Choularton, Thomas	University of Manchester	UK
DeMott, Paul	Colorado State University	USA
Flossmann, Andrea	Laboratoire de Météorologie Physique/OPGC Université Blaise Pascal/CNRS	France
Kahn, Ralph	Jet Propulsion Laboratory	USA
Khain, Alexander	The Hebrew University of Jerusalem	Israel
Leaitch, Richard	Environment Canada	Canada
Pandis, Spyros	University of Patras	Greece
Rosenfeld, Daniel	The Hebrew University of Jerusalem	Israel
Ryan, Brian	CSIRO Marine and Atmospheric Research	Australia
Twohy, Cynthia	Oregon State University	USA
Vali, Gabor	University of Wyoming	USA
Yau, Peter	McGill University	Canada
Zipser, Ed	University of Utah	USA

Contents

Foreword

In 2003, the World Meteorological Organization (WMO) and the International Union for Geodesy and Geophysics (IUGG) agreed to conduct jointly a scientific review of the current state of knowledge on the impacts of aerosol pollution on rain, snow and hail. The review is concerned with a critical societal resource: water. Problems are frequently encountered in relation to too little water, such as those caused by droughts, or to too much water causing flooding and landslides. Under a changing climate and a growing global population, the availability of enough water to sustain life ranks amongst the highest priorities in many communities, especially in developing countries. Considerable evidence was presented in the recent Intergovernmental Panel for Climate Change (IPCC) Fourth Assessment Report in the sense that pollution aerosols can affect climate by altering the Earth's energy budget. These aerosols do so by absorbing or reflecting radiation in clear skies or by affecting clouds that in turn interact with solar and terrestrial radiation. However, less well understood are the effects of aerosols on precipitation in the form of rain, snow and hail.

Atmospheric suspended particulate matter, commonly known as aerosols, can originate from many types of human activity in the form of wind-blown sand and dust, biomass-burning particles and smog particles from fossil fuel combustion and industrial activities. The hydrological cycle is potentially very vulnerable to changes in climate and to associated changes in aerosol pollution, so a key question would be: *How far have we progressed in quantifying the impacts on the amount of water reaching the ground from the atmosphere and its temporal or spatial distributions*? It has been noted that even relatively small changes in the spatial distribution of rain might cause significant shortages in one country and abundance or even flooding in another.

WMO and the IUGG opportunely requested a review of the state of scientific knowledge on this subject as well as recommendations on the issues to be studied in order to advance the decision-making capabilities of those responsible for managing water supply problems. A group of experts under the leadership of Professors Zev Levin and William Cotton produced an assessment of the subject that was subsequently reviewed independently by a team of scientists led by Dr. George Isaac. The late Professor Peter Hobbs is also acknowledged for beginning this initiative. WMO and IUGG wish to thank all those who volunteered their time and efforts to contribute to this scientific review, which sets a benchmark in our current understanding of this complex phenomenon and provides a valuable introduction to the subject for the next generation of scientists. The review also makes specific recommendations for collective international actions that can be deemed essential in advancing our knowledge in this critical field.

A number of internationally coordinated projects have been proposed that, if carefully designed and implemented, could considerably contribute to unraveling the complex interactions occurring among aerosols, clouds and precipitation. It has been suggested that WMO and IUGG should take the lead in such projects, together with other organizations of the UN System. This is indeed a challenge that our respective organizations are particularly suited to address and one that they would be prepared to consider.

In this respect, WMO has a long history in organizing research assessments like the IPCC which it co-sponsors with UNEP and others that have led to highly successful environmental conventions, such as the Vienna Convention on the Protection of the Ozone Layer, and has also contributed to the global coordination of atmospheric and marine observations. Moreover, WMO has coordinated the international provision of weather and climate warnings for over a century. IUGG complements these capabilities with a broad programme of research in Earth systems undertaken by a global community organized under its scientific associations, two of which, the International Association of Meteorology and Atmospheric Science (IAMAS) and the International Association of Hydrological Sciences (IAHS), are particularly involved in these issues. It is therefore our sincere wish that the full strength of early 21st Century science and technology can be focused on international projects to enhance our understanding of aerosol impacts on precipitation and thereby contribute to reduce the societal impacts of hydrometeorological disasters.

(M. Jarraud)
Secretary-General
World Meteorological Organization

(T. Beer)
President
International Union of
Geodesy and Geophysics

Preface

Clouds are known to play a major role in climate through their direct interactions with solar radiation. In addition, precipitation from clouds is the overwhelmingly most important mechanism that replenishes ground water and completes the hydrological cycle. Changes in either the amounts and/or the spatial and temporal distribution of precipitation will have dramatic impacts on climate and on society. Increases or decreases in rainfall in one region could affect rainfall downwind. Similarly, changes in rainfall distribution will strongly affect semi-arid regions that are of dire need of water.

One of the factors that could contribute to cloud and precipitation modification is aerosol pollution from various sources such as urban air pollution and biomass burning. In 2003, the WMO and the IUGG recognized the potential danger from such effects and passed resolutions aimed at focusing attention to this issue. As a follow-up to this resolution, the WMO and IUGG formed an international forum composed of a number of experts to review the state of the science and to identify areas that need further study.

Prof. Peter Hobbs from the University of Washington in Seattle was appointed as the chairman of this forum. Unfortunately, Prof. Hobbs passed away in mid-2005 and the responsibility for the report was transferred to Prof. Zev Levin (Chairman) and William Cotton (Co-chairman). The organization of this report reflects in a large part, the layout that was set up by Prof. Hobbs.

Dr. George Isaac from Environment Canada, the lead reviewer of this document, was responsible for the review process and the selection of the review panel. We would like to acknowledge the very important contributions to this document made by George and his team.

In 2006, just after the first version of the report was completed, Dr. Yoram Kaufman, the lead author of Chapter 5, died in a tragic traffic accident. We are thankful to Dr. Didier Tanré for agreeing to step into Yoram's big shoes and help us complete this report. During the preparation of the final draft of this report in 2006, Dr. Brian Ryan, one of the reviewers, died suddenly from a stroke. We all appreciate Brian's contributions to the report and his death is a great loss.

Since the report is aimed to both the general public and to experts in the field, we took upon ourselves to include chapters that discuss in detail the scientific background and the most recent findings in this field. Each chapter was written by a number of people under the leadership of the lead author. Since the contributors to each chapter were numerous, we listed them at the beginning of the report. It should be noted that some of the contributors provided input to more than one chapter. We would like to thank all of them for their contribution.

Zev Levin
Tel Aviv, Israel

William R. Cotton
Fort Collins, Colorodo, USA

Executive Summary

The WMO and IUGG resolutions handed a mandate to IAPSAG to review the effects of aerosol pollution, including biomass burning, on precipitation. The initial setup and framework of IAPSAG were illustrated by List (2004). This document is a review of our knowledge of the relationship between aerosols and precipitation reaching the Earth's surface and it includes a list of recommendations that could help to advance our knowledge in this area.

Background

Life on Earth is critically dependent upon the continuous cycling of water between oceans, continents and the atmosphere. The atmospheric component of this cycle involves surface water moving into the atmosphere via wind-driven evaporation and biosphere-modulated evapotranspiration followed by transport and dispersion, multiple cycles of cloud formation/evaporation and ultimately removal as precipitation (rain or snow), fog water deposition or dew formation.

Precipitation (including rain, snow and hail) is the primary mechanism for transporting water from the atmosphere back to the Earth's surface. It is also the key physical process that links aspects of climate, weather, and the global hydrological cycle. Although precipitation is the parameter that has the most direct and significant influence on the quality of human lives in terms of the availability of water, the changes in temperature, past and future, have received the bulk of the public's attention regarding global climate change. Precipitation is a more difficult component of the climate to assess because it is much more variable (locally, seasonally, and possibly long-term) than temperature. It is affected primarily by atmospheric dynamics; however, it is also influenced by cloud microphysical processes associated with aerosol properties, which are primarily responsible for cloud drop and ice crystal formation. Changes in precipitation regimes and the frequency of extreme weather events, such as floods, droughts, severe ice/snow storms, monsoon fluctuations and hurricanes are of great potential importance to life on the planet. Thus, by influencing the hydrological cycle, natural and anthropogenic changes in atmospheric aerosols might have important implications for precipitation, which in turn could feed back to climate changes.

Aim

The main goal of this document is to review what is known about the relationship between aerosols and precipitation reaching the Earth's surface and to suggest ways to advance this knowledge.

An overview of the report

From an Earth Science perspective, a key question is how changes expected in climate will translate into changes in the hydrological cycle, and what trends may be expected in the future. With changes in climate, there will be changes in the atmospheric aerosol, whether from natural sources because of changes in winds or other atmospheric conditions, or from anthropogenic sources due to changes in activities as a response to climate change. The hydrological cycle is driven by a multiplicity of complex processes and interactions, many of which are inadequately understood and poorly represented in weather and climate models. We require a much better understanding and hence predictive capability of the moisture and energy storages and exchanges between the Earth's atmosphere, oceans, continents and biological systems, over a wide range of space scales from local to regional to hemispheric to global, and on time scales from hours to decades. These elements are discussed in the Introduction chapter.

To set the stage for the discussion on aerosols and clouds, we briefly summarize the complex interactions between aerosols and clouds and the mechanisms that lead to precipitation formation. This is outlined in Chapter 2.

Chapters 3 and 4 discuss the nature and the global, temporal and spatial distributions of aerosols, followed in Chapter 5 by a review of the measurement methods and instrumentation used to quantify precipitation.

In Chapters 6 and 7 we discuss the observations and model simulations of the effects of pollution aerosols on precipitation and identify some of the difficulties in estimating these effects. Finally Chapter 8 illuminates some of the parallels between the effects of pollution on precipitation and the attempts to artificially modify precipitation through cloud seeding.

Summary

From this report it is clear that aerosol-pollution affects the amount, and the spatial and temporal distribution of clouds, in general agreement with our physical understanding of cloud processes. However, much uncertainty remains regarding these processes. Moreover, because of the complex interactions between environmental meteorological parameters, aerosols, cloud microphysics and dynamics, clear causal relationships between aerosols and precipitation are difficult to identify and moreover, even the sign of the precipitation changes by aerosols are hard to determine in a climatological sense. Based on many observations and model simulations, the effects of aerosols on individual clouds (particularly in ice-free clouds) are more clearly understood than are the effects on precipitation.

Processes of Aerosol Effects on Precipitation

Atmospheric aerosols are highly variable in space and time. A subset of atmospheric aerosols is responsible for the formation of drops (cloud condensation nuclei) and ice crystals (ice nuclei). The initial sizes and concentrations of these hydrometeors determine the efficiency with which precipitation will initially form. A subset of the cloud condensation nuclei population, called giant cloud condensation nuclei and/or ultragiant aerosol particles, have been observed to produce large drops that serve as embryos for initiating coalescence growth and accelerate precipitation development. Modeling studies suggest their importance increases with increasing pollution. There is a point of diminishing returns, however, since modeled precipitation amounts tend to decrease steadily with increasing pollution (all else being equal) and the absolute potential increase in rainfall is small.

Ice nuclei are a smaller subset of atmospheric aerosols than cloud condensation nuclei, however, their role in precipitation formation in certain clouds is critical. This is because ice crystals grow fast to precipitation size embryos, due to their saturation vapour pressure being lower than water. The low concentrations of ice nuclei and the number of different physical mechanisms (see Chapters 2 and 6) that they can take to form ice in clouds makes it difficult for them to be measured. One would expect a high correlation between the concentrations of ice crystal and ice nuclei. However, measurements suggest that in many clouds the concentrations of ice crystals exceed that of the ice nuclei by several orders of magnitude. Many more studies are required in laboratories and in different field environments using recent improvements in measuring techniques .

Modifying the rate of growth of precipitating particles will likely affect the dynamics of the clouds, thus modifying the depth, size, lifetime and propagation of clouds or cloud systems. In fact modeling studies suggest that in deep convective clouds pollution not only affects cloud and precipitation development by ingesting aerosols into initial or primary cells, but also through the formation of secondary cells and cold-pool dynamics in a very nonlinear way, and often many hours after initial convective cell formation.

Larger concentrations of cloud condensation nuclei (CCN) produce larger concentrations of smaller cloud drops that are slower to grow into raindrops. There are reports that show that reduction of cloud drop size also delays the formation of ice, which then forms at higher altitudes and lower temperatures. This may lead to suppression of precipitation in shallow and short-lived clouds, such as those that form during winter over topographical barriers. On the other hand, modeling and observational studies indicate that when deep convective clouds are prevalent, the delay in precipitation can lead to greater amounts of supercooled water being transported aloft and that once it freezes it can invigorate the storm dynamics. In some environments the invigoration of the convective storm updrafts can lead to enhancement of surface precipitation, while in other environments reduced precipitation can result.

The Evidence for Aerosol Effects on Precipitation

Orographic clouds form when moist air is lifted over mountain barriers. This orographic lifting greatly amplifies precipitation over that which occurs for the prevailing weather systems over flat terrain. For many parts of the world orographic precipitation is the dominate water resource. Both observational and modeling studies suggest that orographic clouds are highly susceptible to modification of precipitation by pollution owing to the modest liquid water contents in them, the relatively short time the drops and ice crystals spend in the clouds and their large areal coverage. Some observational studies suggest that aerosol pollution may decrease orographic precipitation as much as 30% annually. In spite of many remaining uncertainties, this potential effect has major ramifications in watersheds, where orographic precipitation is the major water resource.

Global climate model (GCM) estimates of the change in global mean precipitation over the last 100 years due to the total aerosol effects (direct and indirect) vary between no change to –4.5%. A large contributor to this variability is the different methods of representing clouds and aerosol effects in the different models. The differences among models are larger over land, ranging from –1.5 to –8.5%. Recent coupled ocean mixed layer GCM simulations show that the cooling due to the direct and indirect effects by aerosols reduce surface latent and sensible heat transfer and, as a consequence, act to reduce surface evaporation and evapotranspiration. The concomitant reduction in atmospheric water vapour (a greenhouse gas) acts to counter greenhouse gas warming. On the other hand, a reduction in precipitation due to an increase in aerosol pollution leads to longer-lived clouds and more water vapour in the atmosphere. Thus, the potential influence of aerosols on climate could be far more significant than previously thought. Estimates of the consequences of coupled greenhouse gas warming and aerosol cooling in the future depend on uncertain estimates of future pollution emissions and greenhouse gas releases into the atmosphere, as well as the uncertainties associated with the parameterizations of cloud-aerosol interactions in GCMs.

The complexity of aerosol-precipitation interactions is highlighted in urban environments, where field measurements of the effects of urban pollution on precipitation from convective clouds downwind of cities such as St Louis, New York and Houston have failed to link the changes in precipitation to increases in aerosol pollution. Modeling studies suggest that urban land-use effects dominate over aerosol pollution on controlling the locations and amounts of precipitation. The importance of pollution aerosols to precipitation in urban regions depends on the unique properties of the meteorology of the urban environment, its interaction with rural physiography and urban land-use, and on the background aerosol concentrations and chemical properties.

It has also been shown that strongly absorbing pollution particles (e.g. black carbon) may reduce convection due to warming of the atmosphere, and reduction in surface fluxes of heat and moisture. This may result in shallower clouds, and in some cases, no clouds at all.

Recommendations

- That a series of international projects targeted toward unraveling the complex interactions among aerosols, clouds, and precipitation be implemented. A series of international workshops and field studies are needed to address the impacts on clouds and precipitation of aerosols from a range of sources including biomass burning, dust, and industrial pollution within different regional weather regimes in the tropics, middle latitudes, and the polar regions. All of these studies need to have an adequate scientific component, but should also deal with specific situations occurring in the developing world by addressing the impact of pollution/precipitation on economic, sociological, environmental and health issues.
- That the WMO/IUGG take the lead in such projects together with other UN and International Organizations. Some of those projects could be sponsored and financially supported by the countries involved. For example, the effects of an evolving industrial economy, such as China, on precipitation should be studied. Similarly, a study of the effects of biomass burning and dust in some of the African regions would be highly valuable for our understanding of these complex issues.

  It is also proposed to consider a project in a country where the relationship between pollution and precipitation can be explored with a large array of state-of-the-art scientific instrumentation.
- Since ice formation in clouds is not yet fully understood, it is recommended that further laboratory studies and in situ measurements be conducted to clarify the nucleation mechanisms.

  A workshop should be held on developing improved instrumentation for measuring ice nuclei (IN), small ice particles and precipitation.
- There is a special need for more in situ measurements to follow and link physical processes leading from aerosol effects on cloud growth and precipitation.

- Because both observational and modeling studies suggest that orographic clouds are highly susceptible to precipitation modification due to pollution aerosols, and because the ramifications on water resources are large, it is recommended that focused, coordinated observational and modeling campaigns are implemented to study aerosol precipitation interactions for selected watersheds where pollution effects are likely to be large. Examples include the planned Convective and Orographically induced Precipitation Study (COPS) experiment in Germany in 2007, SUPERCIP in the Sierra Nevada of California and the Colorado River Basin in the western U.S. It is recommended that similar studies be conducted in regions such as China and India, where the effects of mega-cities on precipitation could be evaluated. These studies should be basin-wide and include ground-based, airborne, and satellite measurements of aerosol properties, in situ microphysics measurements, surface and radar (particularly cloud radar) measurements of cloud structures and precipitation, basin-wide hydrological measurements of liquid precipitation, snowpack and runoff etc.
- A workshop should be held to determine a strategy for a focused, coordinated observational and modeling campaign to address the effects that aerosols have on orographic clouds.
- Because cloud-resolving models suggest a significant dynamic response to pollution aerosols, which then modifies simulated precipitation, particularly through secondary dynamic responses of clouds via cold pools and gravity waves, it is recommended that a coordinated observational and modeling campaign be organized to investigate the response of cloud systems to varying amounts and characteristics of pollution. These studies should include ground-based, airborne, and satellite measurements of aerosol properties, surface meteorological measurements, ground-based multiple Doppler radars and airborne Doppler radars. These campaigns should be conducted in regions of significant biomass burning or urban pollution sources, and in other locations where the meteorology varies such that different cloud responses can be expected.
- Because of the indications of strong influences of urban land-use and aerosol pollution on precipitation and lightning, it is recommended that a coordinated modeling and observational campaign be established in a number of large metropolitan regions where convection is prevalent. Crucial to site selection is the importance of finding locations where the local physiography (i.e. topography, land-water interfaces) is relatively simple and where generally widespread aerosol pollution is minimal. Attention should in particular be paid to surface measurements of fluxes and land-use properties.
- It is recommended that cloud-resolving model intercomparison studies be implemented for models that explicitly represent aerosol-cloud-precipitation interactions. Such studies should serve as a stimulus for model refinement research and evaluations of model performance. These should be done for a variety of regimes including warm and cold-season orographic cloud models, shallow convection and deep convection, and stratiform cloud

systems. These studies could be linked to some of the campaigns suggested above or to other research campaigns where implementation of an aerosol component could be accomplished easily and economically.

- It is recommended that existing and new statistical methods be applied to current and future data sets to distinguish aerosol effects on precipitation from meteorological influences. Numerical models should play an important role in this process.
- It is urged that aerosol-cloud-precipitation specialists collaborate with global climate model (GCM), developers to refine the representation (parameterization) of aerosol-cloud-precipitation processes in GCMs.
- There is a strong need to assemble data sets or climatologies that can be used for the assessment of climate simulations with GCMs, including aerosol-cloud-precipitation parameterizations. Some of the parameters that need evaluation include seasonal and annual precipitation amounts, regional precipitation climatologies, global and regional aerosol distributions, and top of atmosphere (TOA) radiation budgets.

Chapter 1
Introduction

Leonard A. Barrie, Ulrike Lohmann and Sandra Yuter

1.1 The Hydrological Cycle

Life on Earth is critically dependent upon the continuous cycling of water between the oceans, the continents and the atmosphere. The hydrological cycle depicted as a cartoon in Fig. 1.1 is dynamic and complex. The atmospheric component of this cycle involves surface water moving into the atmosphere via wind-driven evaporation and biosphere-modulated evapo/transpiration followed by transport and dispersion, multiple cycles of cloud formation/ evaporation, and ultimately removal as precipitation (rain or snow) or by fog water deposition or dew formation.

Precipitation (including rain, snow, and hail) is the primary mechanism for transporting water from the atmosphere back to the Earth's surface. It is also the key physical process that links aspects of climate, weather, and the global hydrological cycle. Moreover, precipitation is the parameter that has the most direct and significant influence on the quality of human life through of the availability of fresh water.

Changes in precipitation regimes and the frequency of extreme hydrologic events, such as floods, droughts, severe ice/snow storms, monsoon fluctuations and hurricanes are of great potential impact. The hydrological cycle affects and interacts with other components of the climate system such as glaciers, seasonal snow cover, polar ice cover, vegetative cover and last but not least, atmospheric aerosols.

In the context of atmospheric science, aerosol is a collective name for suspended particulate matter that ranges in size from molecular clusters of 1 nm to giant particles of ~20 μm diameter. The variation of aerosol size distribution and composition in the atmosphere depends on a complex combination of processes involving primary (direct release to the atmosphere) and secondary (formed in the atmosphere from chemical gas to particle conversion) sources, transport, dispersion, cloud processing and removal by precipitation as described in Chapters 2, 3 and 4. It is widely

L.A. Barrie (✉)
WMO, Geneva, Switzerland

Z. Levin, W.R. Cotton (eds.), *Aerosol Pollution Impact on Precipitation*,

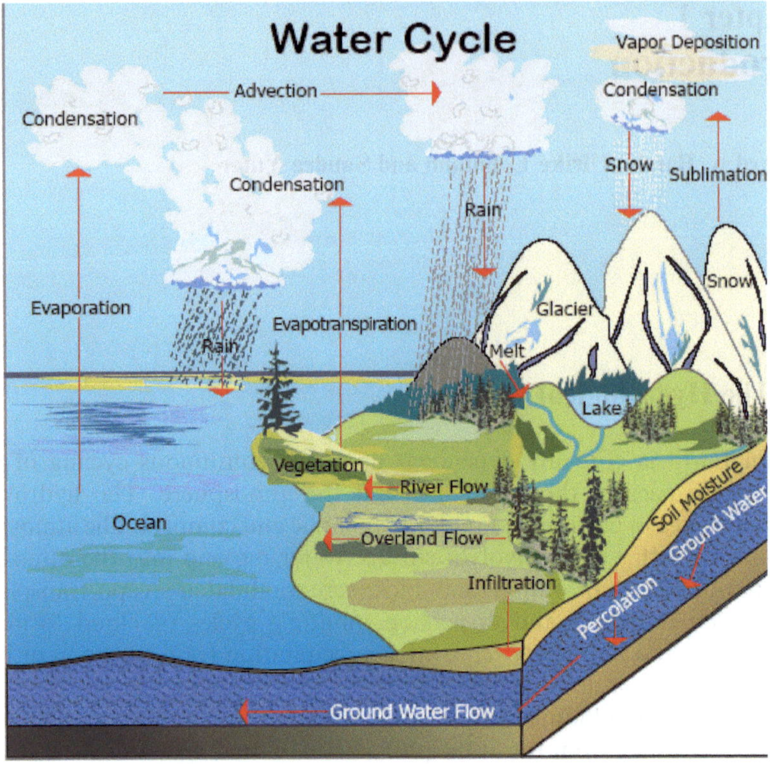

Fig. 1.1 Schematic of the hydrological cycling of *water* between atmosphere, land and oceans. Plots courtesy of S. Yuter, North Carolina State University

acknowledged that the presence of aerosols in the atmosphere is essential to the initial formation of clouds but that their influence on cloud growth and primarily on precipitation formation is less clear. The main goal of this document is to review what is known about the relationship between aerosols and precipitation reaching the Earth's surface and to suggest ways to advance this knowledge.

From an Earth Science perspective a key question is how changes expected in climate, such as warming or changes in atmospheric aerosols, will translate into changes in the hydrological cycle, and what trends may be expected in the future. The hydrological cycle is driven by a multiplicity of complex processes and interactions, many of which are inadequately understood and poorly represented in weather and climate models. We require a much better, understanding and hence predictive capability, of moisture and energy storages and exchanges among and within the Earth's atmosphere, oceans, continents and biological systems over a wide range of spatial scales from local to regional to hemispheric to global and of time scales from hours to decades.

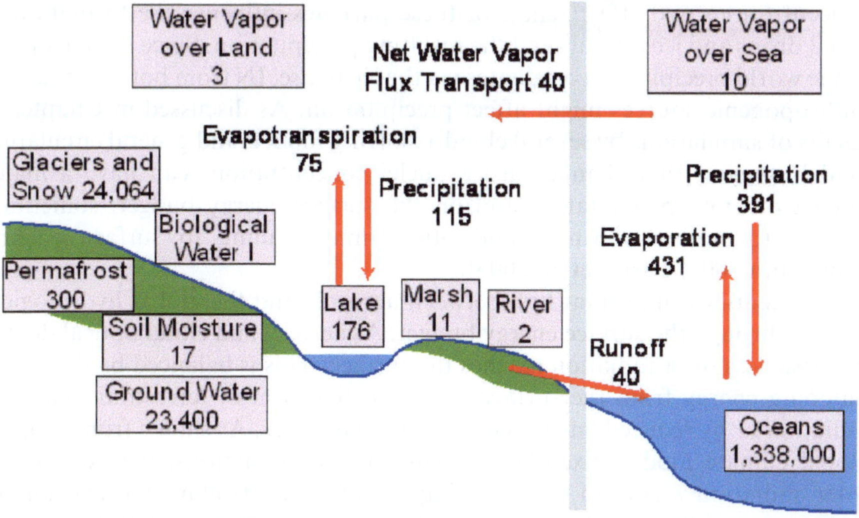

Fig. 1.2 A schematic diagram of the hydrological cycle. The numbers in the boxes are estimates of the total amounts of water in the various reservoirs in units of 10^{15} kg. The numbers alongside the arrows are estimates of average annual fluxes in units of 10^{15} kg/year (Oki 1999, with permission of Cambridge University Press)

Figure 1.2 is a more quantitative depiction of the hydrological cycle shown in Fig. 1.1. It shows the amounts of water in the various reservoirs and the rate of exchange between them. The hydrological cycle is closely linked to the atmospheric circulation and to temperature, which determines the maximum amount of water vapour in the atmosphere. However, the hydrological cycle in turn affects other components of the weather and climate system through complex feedback mechanisms. For example, latent heat released by the condensation of water to form clouds is an important component of the atmospheric heat balance.

The atmosphere is by far the smallest reservoir of water. Although accounting for only 0.3% of the total mass of the atmosphere, water vapour is an important greenhouse gas forming a naturally varying background upon which the effects of anthropogenically influenced long-lived gases such as carbon dioxide and methane are superimposed.

The residence time of water in the atmosphere given by dividing the amount of water vapour in the atmosphere (13×10^{15} kg) by the annual mean precipitation rate (506 kg/year) is 0.03 years (9 days). Nine days is by far the shortest residence time of water in any of its reservoirs. The residence time in lakes and rivers is on the order of a year and in the Antarctic ice sheet is ~200,000 years. Over a period of 9 days, water vapour can be transported over thousands of kilometers.

Another link between aerosols and the global hydrological cycle is by aerosols acting as cloud condensation nuclei (CCN) and ice nuclei (IN). The concentrations and effectiveness of these particles influence the formation of cloud drops and ice crystals and the resulting precipitation. Since in many parts of the world precipitation originates via the ice phase, IN from both natural and anthropogenic sources might affect precipitation. As discussed in Chapter 7, results of simulations by several cloud-resolving models and general circulation models suggest that changes in ice nuclei concentrations can have a major impact on both precipitation and on the surface energy budget, sometimes leading to surface warming and other times leading to surface cooling (Lohmann 2002; Carrió et al. 2005).

A potentially important link between aerosols and the global hydrological cycle is through the surface energy budget. As aerosol and cloud optical depths increase, less solar radiation reaches the surface. This is balanced by decreased outgoing energy from the surface, either in the form of terrestrial longwave radiation or by reduced latent and sensible heat fluxes. As shown from coupled global climate model/mixed-layer ocean model simulations, the decrease in solar radiation seems to be more important in controlling the amount of evaporation in the global mean than the increase in surface temperature resulting from the increase in greenhouse gases (Liepert et al. 2004). Aerosols could impact global atmospheric circulations through several mechanisms including changes in cloud cover, ocean and land-surface heating, rate of latent heat release, and precipitation efficiency.

1.2 Global Distribution of Clouds and Precipitation

Clouds cover about 60% of the Earth's surface (Rossow et al. 1993) and through scattering of solar radiation, play an important role in the radiative balance of the Earth. At any given location, daily cloud and precipitation variability can be large since clear sky conditions can transition to overcast skies and rain within a few hours. When temporal averages of 1 month or more are calculated, climatological patterns of cloud and precipitation start to emerge. Large-scale atmospheric circulation yields regional differences in cloud and precipitation distributions. The atmospheric circulation is driven by the combined effects of the latent heat released as water vapour condenses into cloud droplets, as well as the direct radiative forcing of water vapour, greenhouse gases and aerosols, and the indirect forcing of aerosols through their effects on clouds (IPCC 2001) and the varying surface reflectivity to solar radiation.

For a given region, the largest source of variability of clouds and precipitation is usually the seasonal cycle. Monsoons are seasonal circulations driven by temperature contrasts between ocean and land surfaces. Monsoon circulations reverse direction with the seasons. Interannual variations in large-scale atmospheric oscillations such as El-Niño-Southern Oscillation (ENSO) and the North Atlantic Oscillation (NAO) are associated with interannual variability of global precipitation distribution.

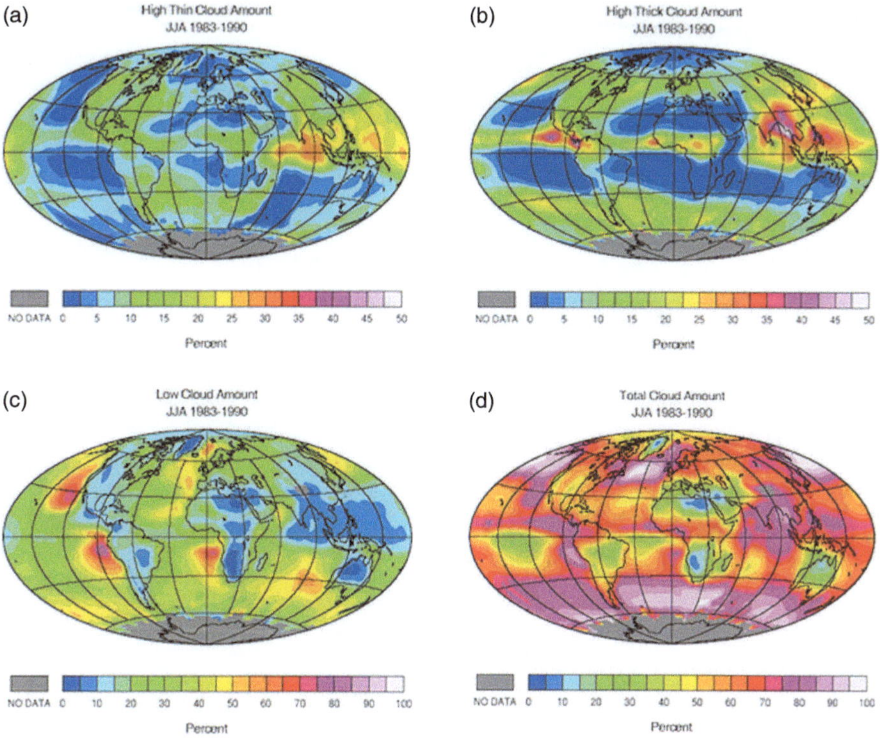

Fig. 1.3 Geographic distribution of cloud fractional coverage for June, July, August 1983–1990 for (**a**) high thin cloud (cloud top height >440 hPa and optical depth <9.38), (**b**) high thick cloud (cloud top height >440 hPa and optical depth >9.38), (**c**) Low cloud (cloud top height <440 hPa) and (**d**) total cloud. Plots courtesy of D.L. Hartmann, University of Washington

Hartmann et al. (1992) used the International Satellite Cloud Climatology Project (ISSCP) cloud cover, cloud-top height and cloud-type data sets (Rossow and Schiffer 1991) to yield the global patterns of clouds subdivided by cloud type (Figs. 1.3, 1.4 and 1.5).

High-cloud tops occur in regions with large-scale ascent in the tropics and in midlatitude storm tracks associated with baroclinic waves. Optically thick high clouds occur where convection is most active and cover a slightly smaller area of the globe than high, optically thin clouds. Low clouds are predominately oceanic and occur most frequently over the subtropical eastern ocean margins and over middle and high latitude oceans. The low clouds over the subtropical eastern oceans are associated with large-scale subsidence (descending branch of the Hadley cell) and lower than average sea surface temperature, which yield stratocumulus clouds trapped below an inversion. Low clouds over middle latitude oceans are usually stratus (Norris 1998).

Total cloud cover is greatest over middle latitude oceans in both the northern and southern hemispheres (Fig. 1.5). Minima in cloud cover occur over subtropical desert regions but smaller cloud amounts also occur over the Caribbean

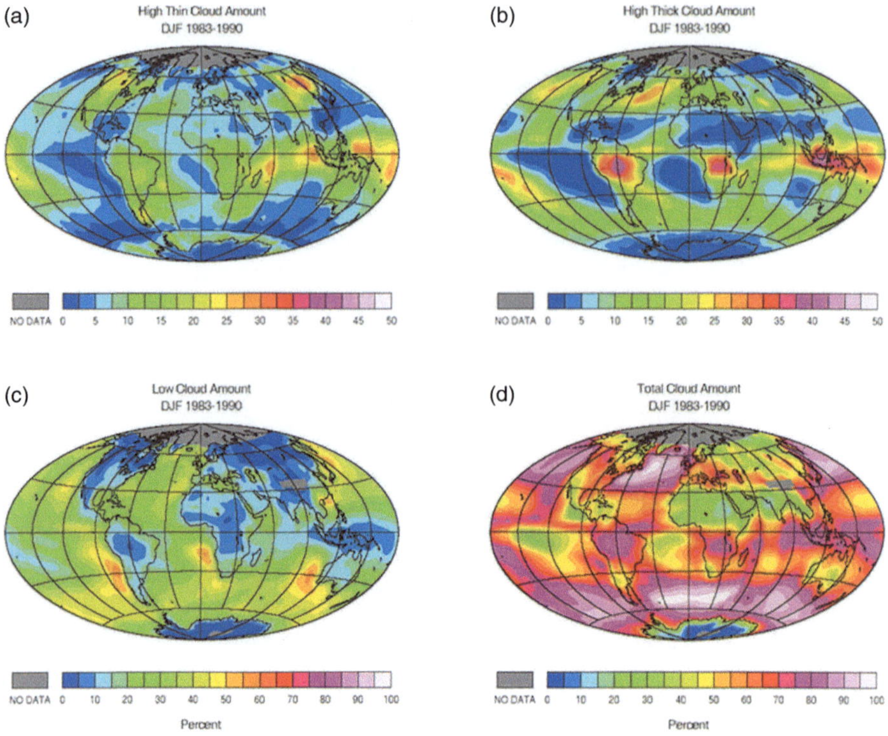

Fig. 1.4 As in Fig. 1.3 except for December, January, February 1983–1990. Plots courtesy of D.L. Hartmann, University of Washington

Sea and the southern subtropical zones of the Pacific, Atlantic and Indian oceans (Hartmann 1994). Cloud cover has seasonal maxima in regions of intense tropical convection such as the Bay of Bengal in June–August (Fig. 1.3) and over the Amazon and Indonesia in December–February (Fig. 1.4).

Precipitation forms within clouds, so it is not surprising that the global distribution of precipitation (Fig. 1.6) has many similarities to the global distribution of clouds with high tops (Fig. 1.5a). However, there are important differences between the cloud and precipitation maps related to the weak instantaneous correlations of cloud top properties and surface rainfall (see Chapters 5, 6 and 7).

Global average precipitation rate is estimated to be 2.6 mm day^{-1}, with higher average values over the ocean (2.8 mm day^{-1}) as compared to land (2.1 mm day^{-1}). On average, three quarters of the global precipitation falls over the world's oceans (Adler et al. 2003).

The zonally averaged precipitation (Fig. 1.7) is closely related to the global circulation pattern and cloud pattern with peak values at the latitude of the ITCZ for the ocean, peak values over land associated with Amazonia and Indonesia, and secondary maxima at middle latitudes associated with the storm tracks of baroclinic disturbances.

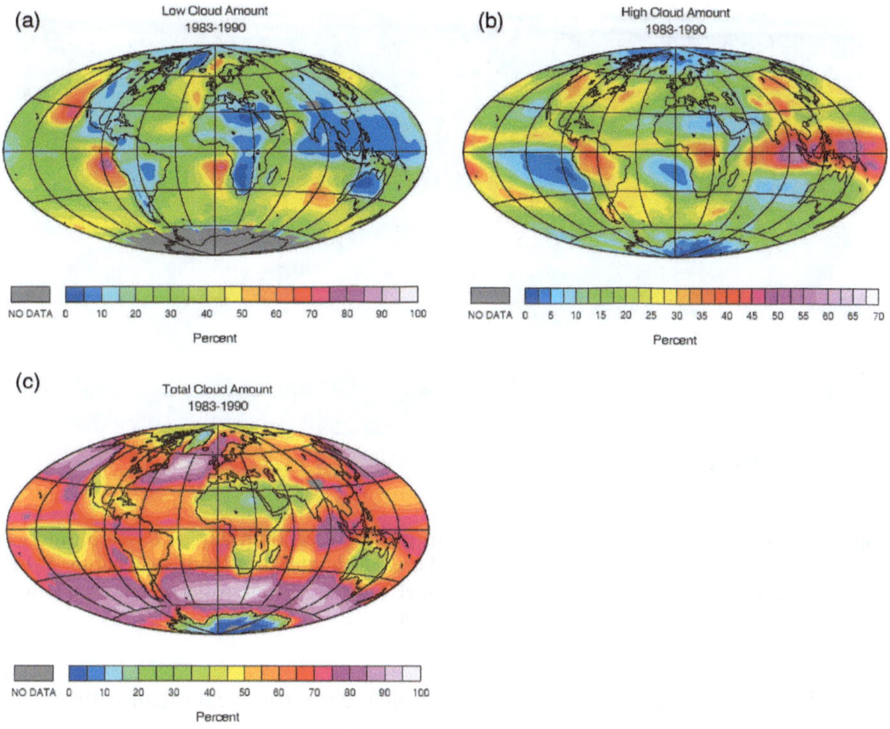

Fig. 1.5 Annual average cloud fractional coverage in percent estimated from ISSCP. (**a**) clouds with tops lower than 680 hPa, (**b**) clouds with tops lower than 440 hPa, (**c**) all clouds. Plots courtesy of D.L. Hartmann, University of Washington

The relative contributions of liquid-phase versus ice-phase precipitation mechanisms to the total precipitation at given locations need to be known in order to determine the potential impacts of aerosols on precipitation since the aerosols have different impacts on the two precipitation mechanisms. These contributions are simple to estimate for shallow clouds that contain only liquid-phase precipitation but are very difficult to determine for mixed-phase clouds, which yield the vast majority of global precipitation

In addition to the limitations arising from incomplete understanding of precipitation processes, direct determination of inadvertent effects of aerosols is also problematic due to large uncertainties in precipitation measurements (Chapter 5). The current accuracy and precision of precipitation measurements (rain and snowfall) in many parts of the globe makes the detection of effects of both deliberate and inadvertent aerosol seeding of clouds on surface precipitation impossible to determine.

Global estimates of tropical precipitation between 30°N and 30°S have the lowest uncertainties as compared to middle latitude and polar precipitation estimates (Sect. 5.5), so we will focus our discussion of global precipitation variability to tropical areas. Fig. 1.8 shows monthly global

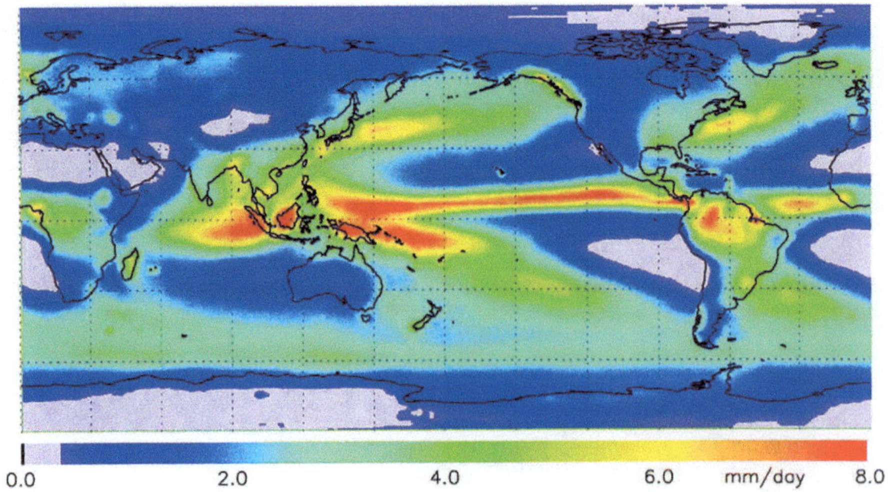

Fig. 1.6 The 23-year (1979–2001) annual mean precipitation (mm day^{-1}) based on merged global satellite observations, rain gauge derived areal estimates over land (see Chapter 5, Fig. 5.18 for network density), and numerical model outputs. From Adler et al. (2003) with permission of the American Meteorological Society

precipitation anomalies from a 23-year tropical climatological mean value based on GPCP (Global Precipitation Climatology Project) superimposed on a 12-month running mean. Also overlaid are the El Niño 3.4 SST index and months with significant volcanic eruptions. Monthly anomalies are typically smaller over the ocean, usually <0.4 mm day^{-1}, compared to larger variations that occur over land (Adler et al. 2003, Haddad et al. 2004).

Diverse studies have shown that ENSO is a major factor in the interannual variability of sea surface temperatures, large-scale atmospheric circulation, and the global precipitation distribution (U.S. CLIVAR Pan American

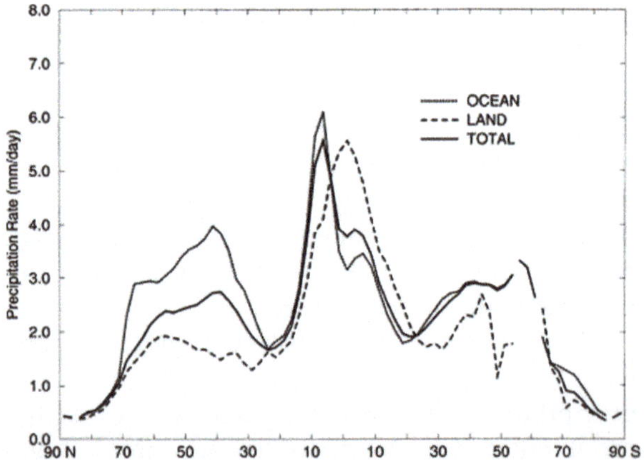

Fig. 1.7 Zonally averaged annual mean GPCP version 2 precipitation (mm day^{-1}) calculated for $2.5 \times 2.5°$ grid boxes. Total (*solid line*), ocean only grid boxes (*light line*) and land-only grid boxes (*dashed line*). Gaps in high latitudes occur where there are no land or ocean grid boxes. From Adler et al. (2003) with permission of the American Meteorological Society

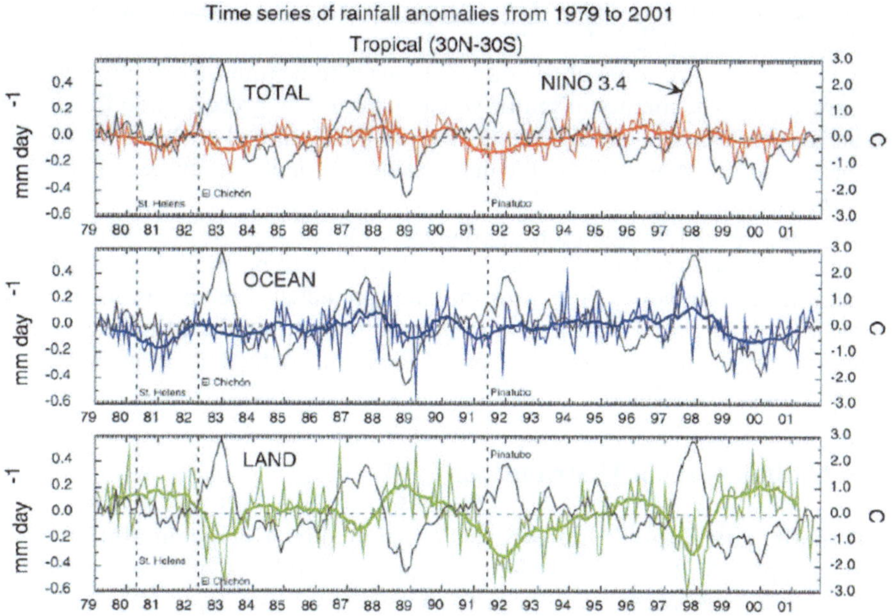

Fig. 1.8 Tropical (30°N to 30°S) averages of monthly precipitation anomalies (mm day^{-1}) for (*top*) total, (*middle*) ocean, and (*bottom*) land. Vertical dashed lines indicate months of significant volcanic eruptions. Black curves in all three panels indicate the El Niño 3.4 SST index in °C. From Adler et al. (2003) with permission of the American Meteorological Society

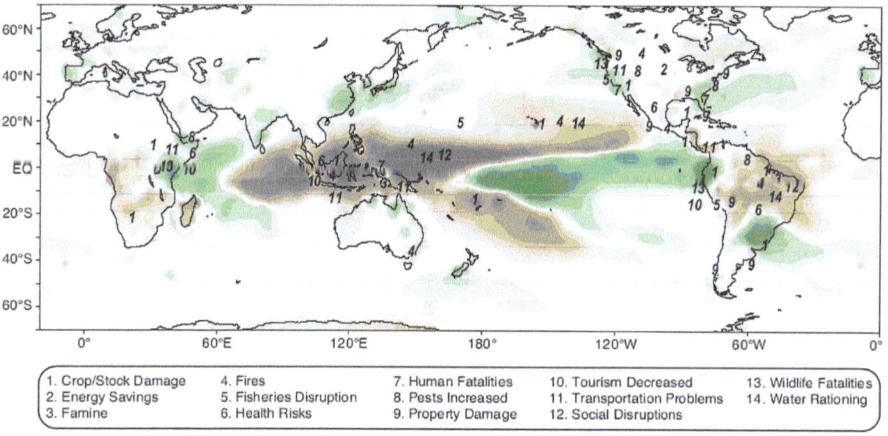

Fig. 1.9 Nature of global El Niño impacts during 1997–1998. Brown indicates dry regions while green indicates wet. From U.S. CLIVAR Pan American Implementation Panel (2002) with permission of W. Higgins

Implementation Panel 2002). Dry periods over the Indonesia and Malaysia region, South America and Africa are associated with the warm phase of ENSO (Fig. 1.9). These are some of the same regions where biomass burning aerosols are prevalent. The impact of major volcanic events on precipitation is difficult to isolate from that of ENSO (Adler et al. 2003). Removal of the ENSO signal on precipitation is a necessary step in assessing the long-term impact of aerosols on precipitation.

In contrast to many climate predictions, the 23-year record examined by Adler et al. (2003) in Fig. 1.8 shows no noticeable trend in global or tropical precipitation. However, the size of the predicted precipitation increase associated with global warming is smaller in magnitude than the interannual variations associated with ENSO and is likely not detectable in such a short record (Adler et al. 2003).

1.3 Global Aerosol Distributions

The spatial distribution of aerosols as represented by average monthly aerosol optical depth (AOD) in Fig. 1.10 illustrates the complexity of aerosols and the fact that they are transported long distances before being deposited. Details of

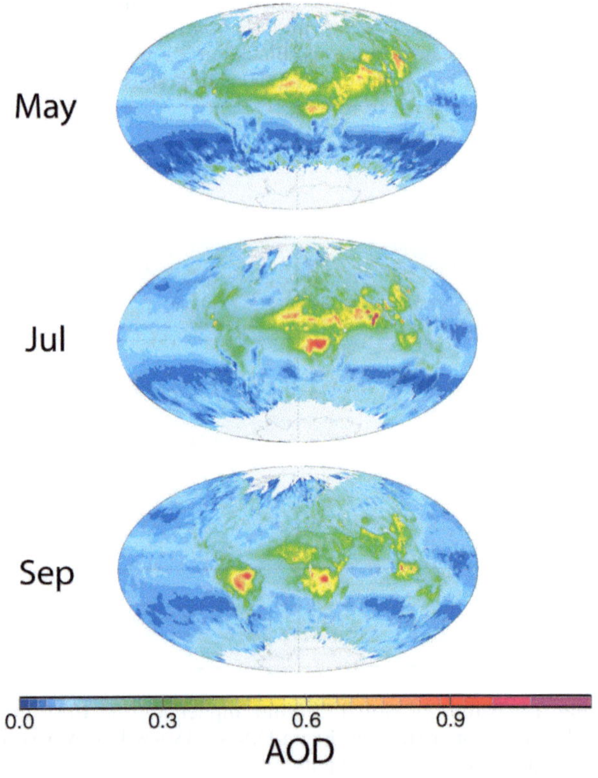

Fig. 1.10 Long term mean aerosol optical depth (AOD) in the northern hemisphere months of (**a**) May, (**b**) July and (**c**) September, compiled by combining data from satellites with surface-based aerosol sunphotometer data. Satellite observations were validated using ground-based AOD observations. Plots Courtesy of S. Kinne, MPI, Hamburg, Germany

these aerosols are given in Chapters 3 and 4. The main natural sources are divisible into soil dust, sea salt, marine biogenic sulphur, terrestrial biogenic and boreal biomass burning, while anthropogenic sources are those from industry, fossil fuel combustion and human-activity related biomass burning. Each aerosol type has its own characteristic sources, size distribution and effectiveness as cloud condensation nuclei (CCN) and ice nuclei (IN).

1.4 Aerosol-Precipitation Interactions: An Inherent Part of Climate Change

Aerosols, cloud properties and precipitation are recognized by the United Nations Framework Convention on Climate Change (UNFCCC) and the Global Climate Observing System (GCOS) as essential climate variables (GCOS-107 2006). The importance of anthropogenic aerosol impacts on clouds through effects on cloud albedo (indirect effect #1) has been acknowledged (IPCC 2001). The best estimate of this indirect cloud albedo effect due to anthropogenic aerosols from pre-industrial times to the present-day from different climate models varies between –0.5 to –1.9 W m^{-2}(Lohmann and Feichter 2005). Secondly, a reduction of drizzle production due to more and smaller cloud droplets may prolong cloud lifetimes and hence cloud coverage (cloud lifetime or indirect effect #2). The cloud lifetime effect, the semi-direct effect (heating of the air due to absorption of solar radiation by absorbing aerosol), and aerosol-ice cloud interactions in response to anthropogenic aerosols are considered to be part of the climate response rather than radiative forcing (IPCC 2001). Nevertheless, the total effect resulting from these aerosol-cloud interactions can be estimated from climate model simulations. These simulations are conducted such that one multi-year simulation uses pre-industrial aerosol and their precursor emissions (i.e. the anthropogenic sources are switched off) and another multi-year simulation that uses present-day aerosol emissions. Aerosols also tend to reduce the net radiation reaching the surface and thereby lower surface temperatures, indirectly affecting the hydrological cycle and atmospheric circulation. The impact of aerosols on the radiation balance and on the global mean precipitation is discussed in chapter 7.

1.5 Unraveling the Aerosol-Precipitation Factor in Long Term Observations: Ongoing Experiments of Opportunity

Changes taking place in certain regions of the globe, historically and in the future, may provide an opportunity to separate the aerosol effect on precipitation from other factors, and thereby give the possibility of good experimental design and adequate systematic observations for analysis. These studies may also be able address the distributions of air pollution and fresh water, which have economic, sociological, environmental and health impacts. It is recommended that WMO/IUGG, together with other UN, international and national organizations, participate in designing field studies such as the following:

- Indian subcontinent: document changes in precipitation in an evolving industrial economy dominated by the Indian Monsoon
- China: document changes in precipitation in an evolving industrial economy
- Amazonia: a history of biomass burning in Amazonia with simultaneous research studies
- Influence of Saharan Dust and biomass burning on the African monsoon
- Atlantic tropical storms development: testing the effects on storm development and the accompanying effects on precipitation
- Volcanic eruptions: investigate the massive perturbations of climate and general circulation through aerosols.

1.6 The Structure of this Review

This review addresses current knowledge of how natural and anthropogenic aerosols might affect clouds, precipitation and hence the hydrological cycle and climate. Diagnosis of the link between aerosol pollution and precipitation has not been possible due to the lack of studies. The report therefore focuses on available published evidence indicating possible links and on identifying effective ways to proceed in filling gaps in our understanding. In subsequent chapters we discuss in some detail the formation mechanisms of clouds and precipitation (Chapter 2), the sources and cloud relevant physical/chemical characteristics of atmospheric aerosols (Chapter 3), their global distribution and presence in various regions (Chapter 4), ground-based, airborne and satellite techniques for observing the effects of aerosols on clouds and precipitation (Chapter 5), the observed and model simulations of the effects of aerosols on clouds and precipitation (Chapters 6 and 7) and lessons learned about inadvertent aerosol effects on clouds and precipitation from many decades of attempts to modify clouds and precipitation by cloud seeding (Chapter 8). Finally, in Chapters 9 and 10 respectively, we summarize the main points of the report and add a list of recommendations to increase knowledge of the effects of aerosols on clouds and precipitation.

Chapter 2
Principles of Cloud and Precipitation Formation

William R. Cotton and Sandra Yuter

2.1 Introduction

In this chapter we provide an overview of the basic physical processes responsible for the formation of clouds and precipitation. A number of important concepts are discussed, and terms defined, which will be used in later chapters. For more detail on these topics the reader is referred to textbooks by Pruppacher and Klett (1997), Rogers and Yau (1989), relevant chapters in Wallace and Hobbs (2006), Cotton and Anthes (1989), and Houze (1993) and review articles by Stewart (1985) and Cantrell and Heymsfield (2005).

2.2 Formation and Structure of Clouds

Different cloud types are defined according to the phases of water present and the temperature of cloud top (AMS Glossary). If all portions of a cloud have temperatures warmer than 0°C it is referred to as a *warm cloud, or a liquid phase cloud.* In clouds extending above the 0°C level, precipitation may form either by ice phase or droplet coalescence processes. *Ice-crystal clouds* consist entirely of ice crystals. Analogously, a *water cloud* is composed entirely of liquid water drops. A *mixed-phase* cloud contains both water drops (supercooled at temperatures colder than 0°C) and ice crystals, without regard to their actual spatial distributions (coexisting or not) within the cloud. Most convective clouds extending into air colder than about −10°C are mixed clouds, though the proportion of ice crystals to water drops may be small until the cloud builds to levels of still lower temperature. Provided the temperature is not below about −40°C, supercooled droplets may coexist with ice particles. However, the liquid phase within unactivated aerosols (haze particles) may coexist with ice particles to very low temperatures.

W.R. Cotton (✉)
Colorado State University, Fort Collins, CO, USA

Z. Levin, W.R. Cotton (eds.), *Aerosol Pollution Impact on Precipitation*,

2.2.1 Dynamical Aspects of Cloud Formation

Clouds form when the saturation vapour pressure becomes lower than the actual partial pressure of the water vapour in the air. The difference condenses in the form of liquid water or ice, depending on the temperature. The saturation vapour pressure decreases when the temperature decreases (Clausius-Clapeyron). The most common way clouds form is therefore when a buoyant parcel of air is lifted (convective ascent) and cooled by adiabatic expansion. For an ascending parcel inside a cloud, the temperature decreases following a moist adiabatic lapse rate which is slightly less (0.65°C per 100 m) than in clear air adiabatic ascent (1°C per 100 m), because of the latent heat released by condensation. The corresponding rate of condensation depends on the temperature and pressure of the cloud cell. For example at 900 m and 20°C, it is in the order of 2 g m^{-3} km^{-1} of ascent. When the cloud base temperature and pressure are determined, the mixing ratio of condensed water at any level above cloud base can be derived as the difference between the water vapour mixing ratio at cloud base and the saturation water vapour mixing ratio at that level. This is referred to as the adiabatic water-mixing ratio. The actual condensed water-mixing ratio is generally lower than the adiabatic value. At cloud base the condensation of the available water vapour is not instantaneous, and the actual water vapour partial pressure can be momentarily higher than the saturation vapour pressure leading to supersaturation. Supersaturation plays a critical role near cloud base for the activation of CCN and IN that initiate cloud droplets (Sect. 2.2.2.1) or ice crystals (Sect. 2.2.2.2). Further up in the cloud, when cloud particles are numerous and big enough, the rate of condensation keeps pace with the production of supersaturation, so that supersaturations remain steady or declines. The actual condensed water content in a convective cloud is generally lower than one without mixing or what is called adiabatic water content. This is because the ascending air from the cloud base is continuously mixed with drier air entrained from outside the cloud. The condensed water content, either liquid or ice, is a key parameter for precipitation formation. Precipitation is most likely to form in the regions of largest condensed mixing ratio, i.e. in the least diluted cloud cells.

A second critical parameter in cloud formation is the Lagrangian time scale of the cloud particles, which is the time it takes a parcel of air containing an ensemble of cloud particles to form when entering a cloud, and evaporate or sublimate when the parcel reaches cloud top. Liquid phase precipitation is produced when the numerous but very small droplets, which have grown by condensation, collide and coalesce into a few much larger raindrops with a significant fall speed. Small ice phase particles can grow to precipitation size by any combination of vapour deposition, collection of small water droplets (riming) or collection of other ice particles (aggregation). These processes require significant time, sometimes comparable to or exceeding the lifetime of a cloud. If the Lagrangian time scale of the cloud particles is shorter than the time needed to produce precipitation, all the condensed water vapour may evaporate in the atmosphere without any precipitation reaching the surface.

The principal types of ascent, each of which produces distinctive cloud forms, are:

a) *Local ascent of warm, buoyant air parcels in a conditionally unstable environment*, which produces convective clouds. Because of the release of latent heat, a cloud convective parcel becomes warmer, hence positively buoyant, with respect to its environment. Air parcels within convective cells are thus able to rise into the atmosphere, until they reach a stable boundary such as a temperature inversion or lose their buoyancy by mixing. Convective clouds have diameters from about 0.1 to 10 km and air ascends in them with velocities up to a few meters per second, although updraft speeds of several tens of meters per second can occur within small volumes of large convective clouds. Within stronger updrafts, ascents of a few kilometers typically produce condensed water mixing ratios of a few grams per kilogram. Mixing ratios of more than 10 g kg^{-1} are possible in very strong updrafts within deep cumulonimbus clouds.

   The lifetimes of convective clouds range from minutes to several hours. The particles Lagrangian time scale may however be shorter. For example, in a shallow cumulus cloud with a depth of ~1.5 km and characteristic updraft speeds of 3 m s^{-1}, the Lagrangian time scale is $t_p = 1500$ m/3 m $s^{-1} = 500$ s $\simeq 8$ min. This represents the time available for initiation of precipitable particles. Once initiated, precipitation may continue over the remaining lifetime of the cloud. In a towering cumulus cloud with depth of ~10 km and updraft speeds of ~15 m s^{-1}, the Lagrangian time scale is $t_p = 10{,}000$ m/15 m $s^{-1} \simeq 660$ s $\simeq 11$ min, only slightly longer than shallow cumulus clouds. The main advantage that a towering cumulus cloud experiences over that of a shallow cloud in forming precipitation particles is associated with the greater amounts of condensate that is produced in deeper clouds. Because precipitation growth by collection is a non-linear function of the amount of condensate in a cloud (Kessler 1969; Manton and Cotton 1977), precipitation growth proceeds quite rapidly in cumulonimbus clouds relative to low liquid water content cumulus clouds. Supercell storms (~12 km depth) have lifetimes of several hours but their strong updrafts (~40 m s^{-1}), yield a Lagrangian time scale of only $t_p = 12{,}000$ m/40 m $s^{-1} = 300$ s = 5 min, which is shorter than that for shallow cumulus clouds. A characteristic feature of supercell storms is the bounded weak echo region where updrafts are so strong that there is not sufficient time to produce radar-detectable precipitation elements at mid levels of the storm (Browning and Ludlam 1962; Marwitz 1972)

b) *Forced lifting of stable air* to produce layer or stratiform clouds. Such clouds can occur at altitudes from ground level to the tropopause and extend over hundreds of thousands of square kilometers. Lifting rates range from a few centimeters per second to ~10 cm s^{-1}. Layer clouds generally exist over periods of tens of hours and their Lagrangian time scales are $\sim 10^4$ s ($\sim$3 h). Due to their long Lagrangian time scales, precipitation is likely in these clouds in spite of their small water contents ($\sim$0.5 g kg^{-1}).

c) *Forced lifting of air as it passes over hills or mountains* to produce orographic clouds. Updraft velocities depend upon the speed and direction of the wind and the height of the barrier, they can be several meters per second. Water contents are typically a few tenths of a gram per cubic meter of air, depending upon the altitude at which the air enters the cloud upwind and its maximum altitude above the mountain top. Orographic clouds may be quite transitory, but if winds are steady they may last for many hours. However, the relevant time scale that determines the time available for precipitation formation is not the time that it takes a parcel to ascend from cloud base to cloud top. Instead, it is the time that it takes a parcel to transect from the upwind lateral boundary to its downwind boundary, as shown in Fig. 2.1 For example, for a 15 m s^{-1} wind speed and a cloud of 18,000 m lateral extent, $t_p = 18{,}000$ m/15 m $s^{-1} = 1200$ s = min. This time scale is longer than that for cumulus clouds but considerably shorter than that for layer clouds. Since the liquid water contents of stable wintertime orographic clouds are low, production of precipitation requires efficient conversion of cloud droplets to precipitation. These conditions can occur in shallow maritime clouds passing over mountains near shore. Lagrangian time scales are longer for deeper mixed and ice-phase orographic clouds as compared to shallower liquid phase clouds. Note that the adiabatic liquid water content for orographic clouds is not determined

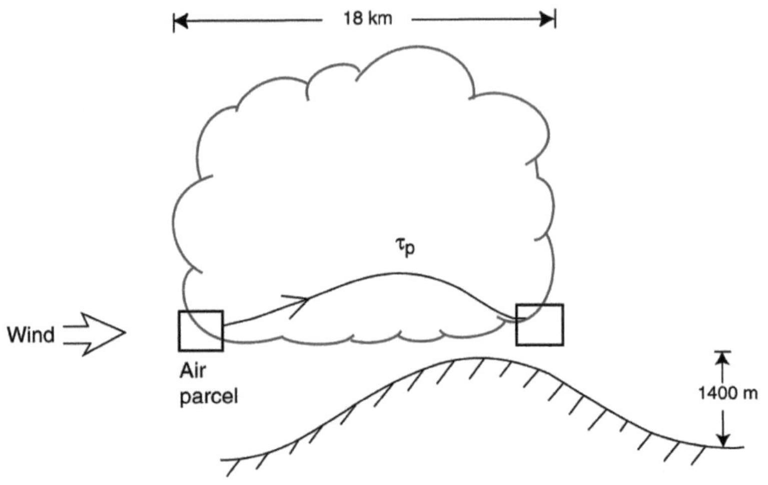

Fig. 2.1 Schematic of a stable orographic cloud indicating the trajectory of an air parcel through the cloud, which determines the Lagrangian time scale (t_p) for the development of precipitable particles. Adapted from Cotton and Anthes (1989) with permission of W.R. Cotton

by the ascent of a cloud parcel from cloud base to cloud top. Instead, it is determined by the vertical displacement of a parcel of air as it transects the cloud over the mountain barrier.

Vertical ascent is not the only way of lowering the temperature and the water vapour saturation pressure. *Cloud top radiative cooling* can lead to destabilization of cloudy layers. This is the main driving force for marine stratocumulus clouds. It is also important in fogs, stratus clouds and cirrus clouds. Longwave radiative flux divergence at cloud top creates cooling of the air, which, in turn, produces higher density air parcels that descend through the cloud layer causing vertical mixing. In the case of a marine stratocumulus layer in which the sea surface temperatures are slightly warmer than the overlying air, the descending cool air parcels can descend through most of the unstable sub-cloud layer, which enhances vertical mixing through the depth of the cloudy boundary layer. Typical lifetimes of stratus and stratocumulus clouds are ~6 to 12 h. The Lagrangian time scale for 1000 m deep clouds having vertical velocities of ~0.1 m s^{-1} is $t_p = 1000$ m/0.1 m $s^{-1} = 10^4$ s $\simeq$ 3 h. Thus, in spite of the fact that they have liquid water contents of only ~0.1 g kg^{-1}, the long Lagrangian time scales permit the formation of precipitation, in the form of drizzle, in stratus and stratocumulus clouds. Because of cloud top radiative cooling, the temperature at the top of a convective cloud may decrease more than expected from the adiabatic lapse rate, hence leading to a superadiabatic water mixing ratio.

As noted earlier, the above concepts of a cloud are based on simple "back-of-the-envelope" calculations. In general, clouds are very turbulent and thus the time-scales for precipitation formation can be much longer than simple Lagrangian parcel estimates. Likewise, the liquid water contents of clouds can be quite variable. In cumulus clouds it is not uncommon to find regions of high liquid water content, say 0.5 to 1.0 g m^{-3}, next to regions with hardly any condensate only tens of meters away. Often cumulus clouds exhibit considerable asymmetry in structure, with upshear parts of the cloud experiencing little mixing and turbulence, and downshear portions experiencing very large mixing and turbulence. In such a cloud hydrometeors undergo growth on the upshear side and experience evaporation and turbulence on the downshear.

Thus far we have only talked about individual cumulus clouds and the lifetimes of individual cells. But in fact precipitating clouds can affect neighboring clouds by exciting gravity waves, by latent heating induced buoyancy bores and by producing air chilled by evaporation of precipitation particles in the subcloud layer. As illustrated in Fig. 2.2 even small cumuli that produce precipitation that evaporates in the sub-cloud layer can form a cold pool. The chilled air is denser than surrounding air and spreads out beneath the cloud where it can lift the surrounding air sometimes enough to generate new cloud cells. In some cases neighboring precipitating clouds can produce cool outflows that run into each other and cause lifting of the air leading to the merger of the neighboring clouds (Fig. 2.3). The merged cloud cell is often wider and deeper than the parent clouds and is more likely to produce rain. As illustrated in

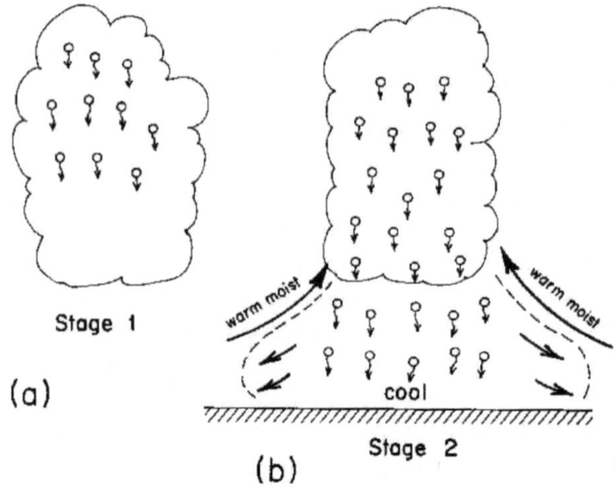

Fig. 2.2 (**a**) Illustration of droplets settling from the upper levels of a cloud, thus reducing the amount of liquid water content or water-loading burden on the cloud. (**b**) Illustration of the formation of an evaporatively-chilled layer near the surface, which can lift surrounding moist air sometimes to the LCL (Lifting Condensation Level) and LFC (Level of Free Convection). From Cotton (1990) with permission of W.R. Cotton

Fig. 2.4 the cold pool and leading edge gust front are important to the maintenance of an ordinary thunderstorm as an efficient engine. The fact that precipitating clouds alter their local environment is important for understanding how pollutants can influence clouds (Chapter 7) and how seeding clouds can perhaps alter precipitation on time scales greatly exceeding the lifetime of individual clouds (Chapter 8).

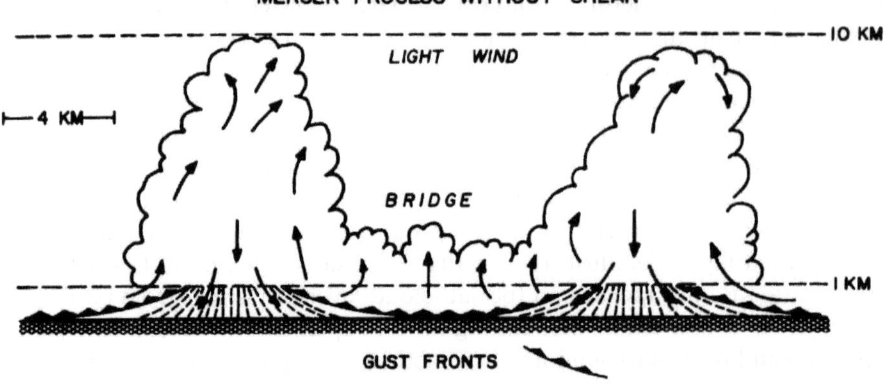

Fig. 2.3 Schematic illustration relating downdraft interaction to bridging and merger in case of light wind and weak shear. From Simpson et al. (1980) with permission from Springer Press

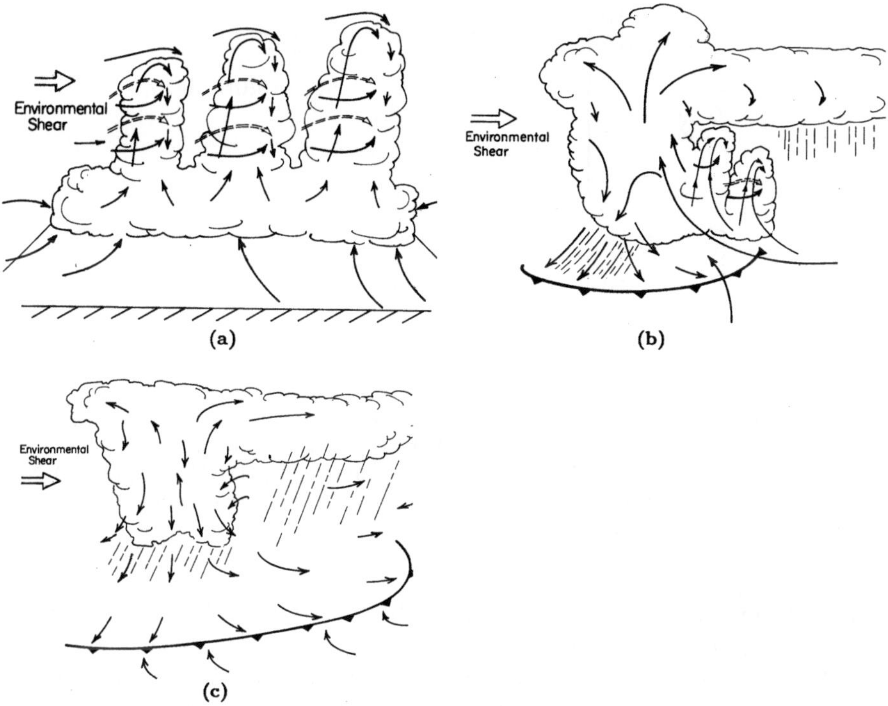

Fig. 2.4 Schematic model of the lifecycle of an ordinary thunderstorm. (**a**) The cumulus stage is characterized by one or more towers fed by low-level convergence of moist air. Air motions are primarily upward with some lateral and cloud top entrainment. (**b**) The mature stage is characterized by updrafts, downdrafts and rainfall. Evaporative cooling at low-levels forms a cold pool and gust front, which advances, lifting warm-moist, unstable air. An anvil at upper levels begins to form. (**c**) The dissipating stage is characterized by downdrafts and diminishing convective rainfall. Stratiform rainfall from the anvil cloud is also common. The gust front advances ahead of the storm preventing air from being lifted at the gust front into the convective storm. From Cotton (1990) with permission of W.R. Cotton

2.2.2 Liquid Phase Clouds

2.2.2.1 Cloud Droplet Formation

The supersaturations required to nucleate drops by the chance collisions of water vapour molecules (i.e. homogeneous nucleation) greatly exceed the observed supersaturations in the atmosphere. Consequently, droplets do not form in natural clouds by homogeneous nucleation. Instead, they form by *heterogeneous nucleation* onto atmospheric aerosol.

Köhler (1926) first determined the equilibrium vapour pressure above small solution droplets (Fig. 2.5). It can be seen from the curves shown in Fig. 2.5 that below a certain droplet size, the relative humidity of the air adjacent to a solution droplet is less than that which is in equilibrium with a plane surface

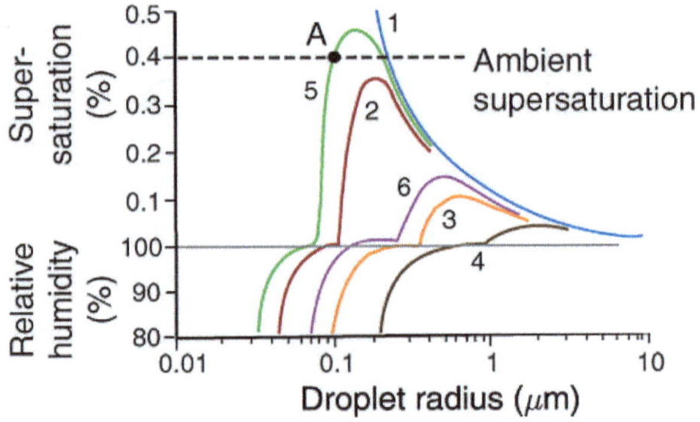

Fig. 2.5 Variations of the relative humidity and supersaturation adjacent to droplets of **(1)** pure water (*blue*), and adjacent to solution droplets containing the following fixed masses of salt: **(2)** 10^{-19} kg of NaCl (*red*), **(3)** 10^{-18} kg of NaCl (*orange*), **(4)** 10^{-17} kg of NaCl (*brown*), **(5)** 10^{-19} kg of $(NH_4)_2SO_4$ (*green*), and **(6)** 10^{-18} kg of $(NH_4)_2SO_4$ (*violet*). Note the discontinuity in the ordinate at 100% relative humidity (adapted with modification from Rasool (1973))

of pure water at the same temperature (i.e. 100%). As a droplet increases in size, the lowering of the equilibrium vapour pressure above its surface due to the dissolved material becomes increasingly less and the equilibrium vapour pressure over a small curved droplet (Kelvin curvature effect) becomes the dominant influence.

Suppose a particle of NaCl with mass 10^{-19} kg is placed in air with a water supersaturation of 0.4% (indicated by the dashed line in Fig. 2.5). As can be seen from the Fig. 2.5, the solution droplet will experience a supersaturation, and the droplet will grow by condensation. As it does so, the supersaturation adjacent to the surface of this solution droplet will initially increase, but even at the peak in its Köhler curve the supersaturation adjacent to the droplet is less than the ambient supersaturation. Consequently the droplet will grow over the peak in its Köhler curve and down the right-hand side of this curve to form a fog or cloud droplet. A droplet that has passed over the peak in its Köhler curve and continues to grow is said to be *activated.*

Now consider a particle of $(NH_4)_2SO_4$ with mass 10^{-19} kg that is placed in the same ambient supersaturation of 0.4%. In this case condensation will occur on the particle and it will grow as a solution droplet. At point A the supersaturation adjacent to the droplet is equal to the ambient supersaturation. If the droplet at A should grow slightly, the supersaturation adjacent to it would increase above the ambient supersaturation, and therefore the droplet would evaporate back to point A. If the droplet at A should evaporate slightly, the supersaturation adjacent to it would decrease below the ambient supersaturation, and the droplet would grow by condensation back to A in Fig. 2.5. Hence the solution droplet at A is in stable equilibrium with the ambient supersaturation. If the ambient supersaturation were to change a little, the location of A in

Fig. 2.5 would shift, and the equilibrium size of the droplet would change accordingly. Droplets in this state are said to be *unactivated* or *haze droplets.* Haze droplets in the atmosphere can considerably reduce visibility by scattering light.

A subset of the atmospheric aerosol discussed in Chapter 3 serves as particles upon which water vapour condenses to form droplets that are activated and grow by condensation to form cloud droplets at the supersaturations achieved in clouds (~0.1–1%). These particles are called cloud condensation nuclei (CCN). It follows from the above discussion that the larger the size of a particle with a given chemical composition, the more readily it is wetted by water, and the greater its solubility, the lower will be the supersaturation at which the particle can serve as a CCN. For example, to serve as a CCN at 1% supersaturation, completely wettable but water insoluble particles need to be at least ~0.1 μm in radius, whereas soluble particles can serve as CCN at 1% supersaturation even if they are as small as ~0.01 μm in radius. Most CCN consist of a mixture of soluble and insoluble inorganic and organic components (called *internally-mixed nuclei*). The solubility of a particle has an important effect on its effectiveness as a CCN. For example, the initial minimum dry radius of a particle that is activated by a supersaturation of 0.1% is 0.075 μm if the particle is completely soluble. However, if the ratio of the soluble mass to the total mass of the particle is only 0.2, the particle would need to have a dry radius of 0.13 μm to be activated by a supersaturation of 0.1%.

Another aspect of CCN activity that is often overlooked is the wettability of the aerosol particle (i.e. the ability of water to spread out over the surface of the particle) as measured by the contact angle of water on the particle. It is generally assumed that particles are completely wettable. However, this is by no means always the case (e.g. Knight 1971). For example, in the atmosphere there are a number of organic materials that are not wettable. To the extent that a substance has a non-zero contact angle, its ability to serve as a CCN will be hindered.

The surface tension of the solution formed by condensation onto a soluble particle will also affect the subsequent growth of the solution droplet. This is because for the Kelvin effect the energy barrier that has to be overcome for a droplet to be activated varies as the third power of the surface tension (Wallace and Hobbs 2006). The chemical components, concentrations, solubilities, and surface tensions of solution droplets that form in various environments are not well documented, and the effects of these parameters on droplet activation, individually and in combination are not well understood, but it is generally believed that surface tension effects are small.

Another factor of importance during droplet nucleation and vapour deposition growth is the diffusivity of vapour molecules near the surface of small droplets. Ordinarily diffusivities of water vapour that can be found in chemical handbooks do not take into account the kinetic effects of vapour diffusion when vapour molecules are within the mean free path distance of the droplet surface. To account for kinetic effects, the vapour diffusivities are modified with a so-called accommodation coefficient. There is strong disagreement in the literature regarding the value of the accommodation coefficient for atmospherically relevant conditions (Shaw and Lamb 1999; Laaksonen et al. 2005). For example, the existence of film forming compounds in the aerosol may significantly reduce the mass accommodation. This parameter has a very strong effect on the number of droplets activated, with the latter tending to decrease with increasing values of this coefficient. Its effect is much stronger than many effects associated with changes in aerosol composition mentioned above (e.g. Kreidenweis et al. 2003).

2.2.2.2 Cloud Condensation Nuclei (CNN)

Because of the uncertainties in predicting the CCN nucleating ability of atmospheric particles, an empirical approach is generally taken by measuring the concentration of particles that serve as CCN at various prescribed supersaturations; this is called the *CCN supersaturation spectrum*. The concentrations of CCN active at various supersaturations can be measured with a thermal gradient diffusion chamber, or other devices based on similar principles. A diffusion chamber may take a variety of geometric configurations, with the essential feature being that particles are statically or dynamically (with continuous flow) exposed to a steady supersaturation field created either by wetted plates held at different temperatures or by a streamwise gradient of temperature. By varying the temperature difference between the plates it is possible to produce maximum supersaturations in the chamber that range from a few tenths of 1% to a few percent, which are similar to the inferred supersaturations that activate droplets in clouds (Wieland 1956). Worldwide measurements of CCN concentrations have not revealed any systematic latitudinal or seasonal variations. At a given location, CCN vary by several orders of magnitude with time, depending on the proximity of sources, wind direction, air mass type, precipitation and cloudiness (Twomey 1960; Jiusto 1966; Radke and Hobbs 1969). Near the Earth's surface continental air masses generally contain larger concentrations of CCN than clean marine air masses (Fig. 2.6). For example, the concentration of CCN in the continental air mass over the Azores, depicted in Fig. 2.6, is about $\sim$300 cm^{-3} at 1% supersaturation, while in the marine air mass over Florida it is $\sim$100 cm^{-3}, and in clean Arctic air it is only $\sim$30 cm^{-3}. The ratio of CCN (at 1% supersaturation) to the total number of particles in the air (CN) is $\sim$0.2–0.6 in marine air; in continental air this ratio is generally less than $\sim$0.01 but can rise to $\sim$0.1. The very low ratio of CCN to CN in continental air is attributable to the large number of very small particles, which are not activated at low supersaturations. Concentrations of CCN over land decline by about a factor of five between the planetary boundary layer and the free troposphere (Squires and Twomey 1966; Hoppel et al. 1973; Hobbs et al. 1985a,b). Over the same height interval concentrations of CCN over the ocean remain fairly constant, or may even increase with height, reaching a maximum concentration just above the mean cloud height (Hoppel et al. 1973;

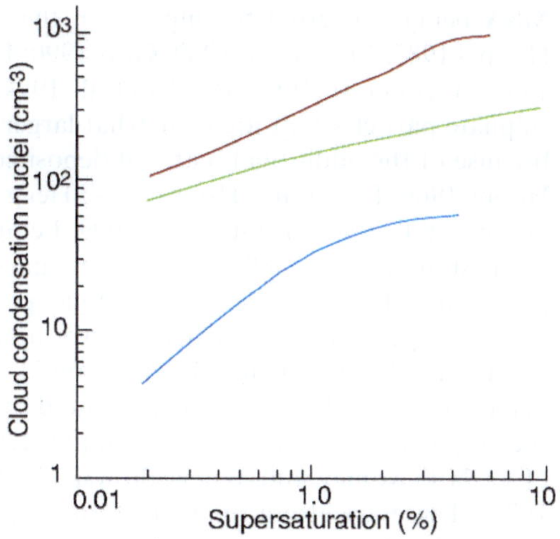

Fig. 2.6 Cloud condensation nucleus spectra in the boundary layer from measurements near the Azores in a polluted continental air mass (*brown line*), in Florida in a marine air mass (*green line*), and in clean air in the Arctic (*blue line*). Adapted with modification from Wallace and Hobbs (2006) and data from Hudson and Yum (1997)

Hudson 1983; Hegg et al. 1990). Ground-based measurements indicate that there is a diurnal variation in CCN concentrations at some locations, with a minimum at about 6 a.m. and a maximum at about 6 p.m. (Twomey and Davidson 1970, 1972).

The observations described above provide clues as to the origins of natural CCN. First of all it appears that the land acts as a major source of CCN, because the concentrations of CCN are generally high over land and decrease with altitude. Some of the soil particles and dusts that enter the atmosphere probably serve as CCN, but they do not appear to be a dominant source. The rate of production of CCN (active at a supersaturation of 0.5%) from burning vegetable matter is on the order of 10^{12} to 10^{15} per kg of material consumed (Eagan et al. 1974b). Thus, forest fires are a prolific source of CCN (Twomey 1960; Twomey and Warner 1967; Warner and Twomey 1967; Warner 1968; Hobbs and Radke 1969; Woodcock and Jones 1970; Stith et al. 1981). Although sea salt particles enter the air over the oceans, they do not appear to be a dominant source of CCN, even over the oceans (Twomey 1968, 1971; Radke and Hobbs 1969; Dinger et al. 1970; Hobbs 1971), although, because of their solubilities and large sizes, they may enhance precipitation by serving as giant CCN.

There appears to be a widespread and probably a fairly uniform source of CCN over both oceans and land, the nature of which has not been definitely established. A likely candidate is gas-to-particle conversion, which can produce particles up to a few tenths of a micrometer in diameter that can act as CCN if they are soluble and wettable. Gas-to-particle conversion mechanisms that require solar radiation might be responsible for the observed peak in CCN concentrations at ~6 p.m. Most CCN consist predominately of sulphates, although some organic material is usually present. Over the oceans, organic sulphur from the ocean, in the form of the gases dimethyl sulphide (DMS) and methane sulphonic acid (MSA), provides a source of CCN, with the DMS and

MSA being converted to sulphate in the atmosphere (Charlson et al. 1987; Hoppel 1987; Luria et al. 1989; Gras 1990; Hegg et al. 1991a,b; Ayers and Gras 1991; Ayers et al. 1991; Pandis et al. 1994). Evaporating clouds also release sulphate particles that are somewhat larger than non-cloud processed aerosol because of the additional material deposited onto them in clouds (Radke and Hobbs 1969; Easter and Hobbs 1974; Hegg et al. 1980; Hegg and Hobbs 1981; Hoppel et al. 1986; Birmili et al. 1999; Feingold and Kriedenweis 2002). It has been estimated that ~80% of sulphate mass globally is formed in the aqueous phase, and the remainder from the gas phase. As expected from the Köhler curves, larger sulphate particles serve more efficiently as CCN (Hegg 1985; Leaitch et al. 1986; Lelieveld et al. 1997). The production of sulphates (and maybe other soluble materials), followed by the release of these particles when the droplets evaporate, is an important mechanism for increasing the efficiency of CCN (Twomey and Wocjiechowski 1969; Hobbs 1971; Easter and Hobbs 1974). There are also a number of anthropogenic sources of CCN, which are discussed in Chapters 3, 4 and 6.

2.2.2.3 Effects of CCN on the Microphysical Structures of Clouds

In a cloud we are concerned with the growth of a large number of droplets in a rising parcel of air. As the parcel rises it expands, cools, and eventually reaches saturation with respect to liquid water. Further uplift and adiabatic cooling produces supersaturations that initially increase in proportion to the updraft velocity. CCN are activated as the supersaturation rises, starting with the most efficient CCN. When the rate at which water vapour in excess of saturation is equal to the rate at which water vapour condenses onto the CCN and droplets, the supersaturation in the cloud reaches a maximum value. The concentration of cloud droplets is determined at this stage (which generally occurs <100 m above cloud base) and is equal to the concentration of CCN activated by the peak supersaturation that has been attained. Subsequently, the growing droplets consume water vapour faster than it is made available by the cooling of the air, so the supersaturation begins to decrease. The haze droplets then begin to evaporate while the activated droplets continue to grow by condensation. The rate of growth of a droplet by condensation is inversely proportional to its radius, therefore the radius of the smaller activated droplets grow faster than that of the larger droplets. Consequently, in this simplified model, the size distributions of the droplets in the cloud become increasingly narrower with time (i.e. the droplets approach a *monodispersed* distribution). This sequence of events is illustrated by the results of theoretical calculations shown in Fig. 2.7.

Comparisons of cloud droplet size distributions measured a few hundred meters above the bases of non-precipitating warm cumulus clouds with droplet size distributions computed assuming growth by condensation for about 5 min show good agreement. The droplets produced by condensation during this time period extend only up to about 10 μm in radius. Moreover, as mentioned above,

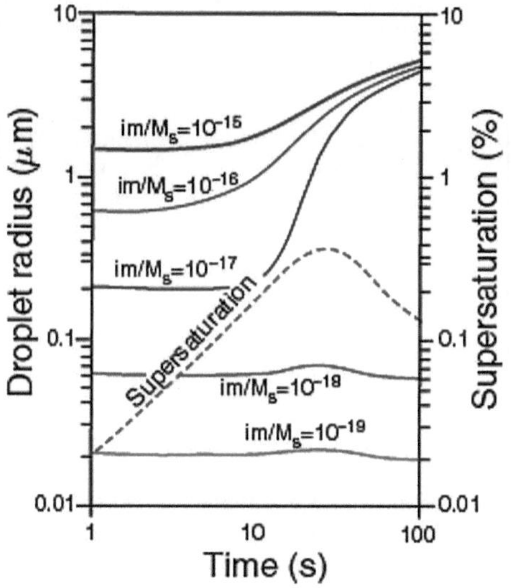

Fig. 2.7 Theoretical computations of the growth of cloud condensation nuclei by condensation in a parcel of air rising with a speed of 60 cm s^{-1}. A total of 500 CCN cm^{-3} was assumed with im/M_s values as indicated; m the mass of material dissolved in the droplet, M_s the molecular weight of the material, and i its van't Hoff factor (i.e. the number of ions produced by each molecule of the material when it dissolves). Thus, im/M_s is the effective number of kilomoles of the material in the dissolved droplet. One can think of the terms im as representing the particular chemical properties of the salt. Note how the droplets that have been activated (im/$M_5 = 10^{-15}$, 10^{-16}, 10^{-17}) approach a monodispersed size distribution after just 100 s. (i.e. ~60 m above cloud base). The variation with time of the supersaturation of the air parcel is also shown (*dashed line*). Adapted with modification from Howell (1949) with permission of the American Meteorological Society

the rate of increase in the radius of a droplet growing by condensation is inversely proportional to the drop radius, therefore, the rate of growth decreases with time. It is clear, therefore, as first noted by Reynolds (1877), that growth by condensation alone in warm clouds is much too slow to produce raindrops with radii of several millimeters. Yet rain does form in clouds that contain only water drops. The enormous increases in size required to transform cloud droplets into raindrops is illustrated by the scaled diagram shown in Fig. 2.8. For a cloud droplet 10 μm in radius to grow to a raindrop 1 mm in radius requires an increases in volume of one millionfold! However, only about one droplet in a million (about 1 L^{-1}) in a cloud has to grow by this amount for the cloud to rain. The mechanism responsible for the selective growth of a few droplets into raindrops in warm clouds is discussed in Sect. 2.3.1.

Since the concentration of CCN generally increases in passing from oceanic air sheds to continental air sheds to urban environments, the concentrations of cloud droplets likewise increase, and the droplet growth rate with altitude above

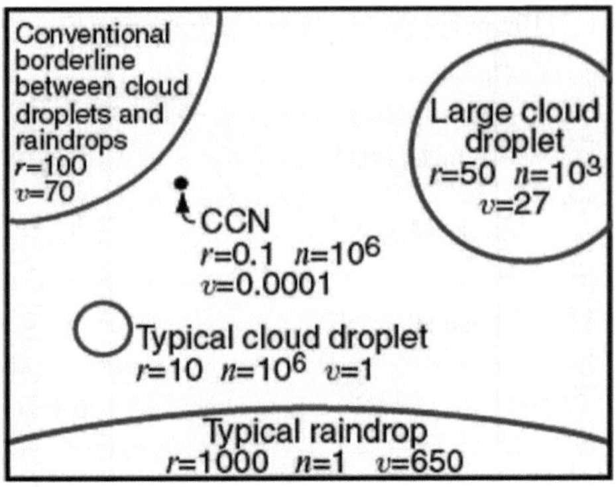

Fig. 2.8 Relative sizes of cloud droplets and raindrops; *r* is the radius in micrometers, *n* the number per liter of air, and *v* the terminal fall speed in centimeters per second. The circumference of the circles are drawn approximately to scale, but the black dot representing a typical CCN is twenty-five times larger than it should be relative to the other circles (adapted with modifications from Wallace and Hobbs 2006 with permission from the authors)

cloud base decreases, at least for non-precipitating clouds (e.g. Gerber 1996). For example, droplet concentrations in non-polluted, non-precipitating marine cumulus clouds are generally $<100\ \mathrm{cm}^{-3}$ while they can reach values over $1000\ \mathrm{cm}^{-3}$ in a polluted environment (Squires 1958) (Fig. 2.9). As we will see in Sect. 2.3.1, clouds with large concentrations of droplets are more colloidally stable and less likely to precipitate than clouds with small concentrations of droplets.

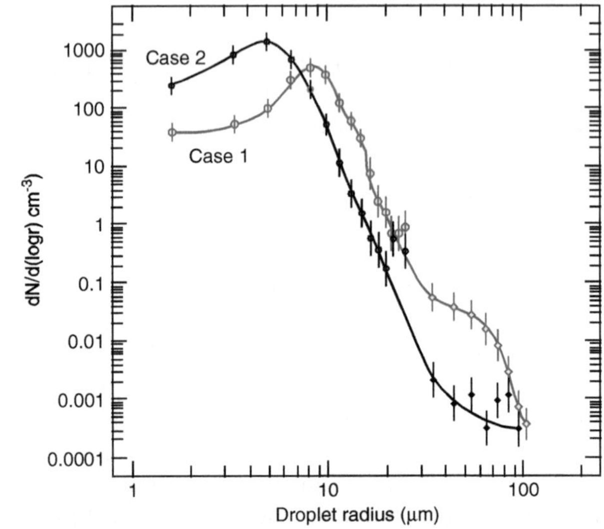

Fig. 2.9 Cloud droplet number distributions measured in stratocumulus clouds in the vicinity of the Azores by the FSSP-100 (*circles*) and PMS 1D (*diamonds*) cloud probes, averaged over 15 km of flight path for *case 1*—clean marine air (*light symbols* and *curve*), and averaged over 4 km of flight path for *case 2*—continentally influenced air (*dark symbols* and *curve*). The *vertical bars* are the geometric standard deviations of the droplet concentrations. From Garrett and Hobbs (1995) with permission of the American Meteorological Society

The effects of CCN from anthropogenic sources on cloud structures and precipitation are discussed in Chapters 6 and 7.

2.2.3 *Ice Phase Processes*

2.2.3.1 Nucleation of Ice

Ice particles can form either homogeneously or heterogeneously on some form of ice nuclei (IN). Homogeneous nucleation can take place either directly from the vapour or by freezing of cloud droplets. However, homogeneous nucleation of ice crystals from the vapour, or the chance formation of an embryo of ice-like structure of critical size, requires very high supersaturations with respect to ice and such low temperatures that it does not take place in the troposphere. On the other hand, homogeneous freezing of supercooled droplets by the chance formation of a cluster of ice-like embryos can occur in the atmosphere.

For homogeneous freezing to occur, enough ice-like water molecules must come together within the droplet to form an embryo of ice large enough to survive and grow. If an ice embryo within a droplet exceeds a certain critical size, its growth will produce a decrease in the energy of the system. On the other hand, any increase in the size of an ice embryo smaller than the critical size causes an increase in total energy. In the latter case, from an energetic point of view, the embryo is likely to breakup.

Since the numbers and sizes of the ice embryos that form by chance aggregations increase with decreasing temperature, below a certain temperature (which depends on the volume of water considered) freezing by homogeneous nucleation becomes a virtual certainty. Homogeneous nucleation occurs in about 1 s at about −41°C for droplets about 1 μm in diameter, and at about −35°C for drops 100 μm in diameter. An analogous freezing process occurs for unactivated droplets or haze particles at temperatures below −40°C, a process for ice formation in cirrus clouds (DeMott 2002). Hence, in the atmosphere, homogeneous nucleation by freezing generally occurs only in high clouds or high latitudes.

If a droplet contains a rather special type of particle, called a *freezing nucleus*, it may freeze by a process known as *heterogeneous nucleation* in which water molecules in the droplet collect onto the surface of the particle to form an ice-like structure that may increase in size and cause the droplet to freeze. Since the formation of the ice structure is aided by the freezing nucleus, and the ice embryo also starts off with the dimensions of the freezing nucleus, heterogeneous nucleation can occur at much higher temperatures than homogeneous nucleation.

If the particle that initiates freezing is contained within the droplet, it is called an immersion freezing nucleus. However, cloud droplets may also be frozen if a suitable particle in the air comes into contact with the droplet, in which case freezing is said to occur by *contact freezing*, and the particle is referred to as a *contact nucleus*. Laboratory experiments suggest that some particles can cause a drop to freeze by contact freezing at temperatures several degrees higher than if they were embedded in the drop (Fletcher 1962; Levkov 1971; Gokhale and Spengler 1972; Pitter and Pruppacher 1973).

Recent laboratory experiments by Durant and Shaw (2005) showed that as droplets evaporate, embedded aerosol particles become increasingly likely to penetrate the air-water interface layer and promote freezing. They interpreted this as evidence that contact freezing may be just as effective from the inside-out as from the outside-in.

Certain *particles* in the air also serve as centers upon which ice can form directly from the vapour phase. These particles are referred to as *deposition nuclei*. Ice can form by deposition provided that the air is supersaturated with respect to ice and the temperature is low enough. If the air is supersaturated with respect to water, a suitable particle may serve either as a *condensation-freezing* nucleus (in which case liquid water first condenses onto the particle and subsequently freezes) or as a *deposition nucleus* (in which case there is no intermediate liquid phase, at least on the macroscopic scale). It is generally thought that condensation-freezing is preferred at smaller supercoolings and large supersaturations, while deposition is preferred at large supercoolings and small supersaturation. In practice, it is not easy to distinguish between deposition and condensation-freezing modes.

2.2.3.2 Properties of Ice Nuclei

If we refer to an ice-nucleating particle in general, without specifying its mode of action, we call it an ice nucleus (IN). However, it should be kept in mind that the temperature at which a particle can cause ice to form depends, in general, upon the mechanism by which the particle nucleates the ice as well as upon the previous history of the particle. The basic distinction that has to be made is whether nucleation is from the vapour or from the liquid phase (Vali 1985).

Particles with molecular spacings and crystallographic arrangements similar to those of ice (which has a hexagonal structure) tend to be effective as ice nuclei, although this is neither a necessary nor a sufficient condition for a good ice nucleus. Most effective ice nuclei are virtually insoluble in water. Some inorganic soil particles (mainly clays) can nucleate ice at fairly high temperatures (i.e. above −15°C), and they probably play an important role in nucleating ice in clouds. Desert dust is known to be very good ice nuclei (DeMott et al. 2003a). For example, in one study, 87% of the snow crystals collected on the ground had clay mineral particles at their centers and more than half of these were kaolinite (Kumai 1951). Of course, it is only circumstantial evidence that they served as ice nuclei. Many organic materials are effective ice nucleators (Schnell and Vali 1976). Decayed plant leaves contain copious ice nuclei, some active as high as −4°C. Ice nuclei active at −4°C have also been found in seawater rich in plankton. In addition, some plant pathogenic bacteria have also been found to be effective ice nuclei at temperatures as high as −2°C (Vali et al. 1976; Yankofsky et al. 1981; Levin and Yankofsky 1983; Levin et al. 1987). Recently Von Blohn et al. (2005) have identified pollen as good ice nuclei at warm temperatures.

In some cases, after a particle has served as an ice nucleus and all of the visible ice is then evaporated from it, but the particle is not warmed above −5°C or exposed to a relative humidity with respect to ice of <35%, it may subsequently serve as an ice nucleus at a temperature a few degrees higher than it did initially (Roberts and Hallett 1968). This is referred to as *preactivation*. Thus, for example, preactivated nuclei may be transferred between sequential wave clouds and act at a higher temperature for the second and later clouds. However, preactivation is lost if the initially activated ice nucleus is either heated above at least −5°C and/or is dehydrated to <~35% relative humidity with respect to ice. Because preactivation can be lost if the particles are warmed or dehydrated prior to testing in an IN counter, this behavior creates a significant dilemma for IN measurements in almost any sampling scenario.

2.2.3.3 Variations of Ice Nuclei and Ice Particle Concentrations

Ice nucleus concentrations can sometimes vary by several orders of magnitude over several hours (see also Chapters 5 and 6). Historically ice nuclei concentrations have been assumed as predictors of ice particle concentrations. Laboratory studies in the 1940's and 1950's using systems for low temperature processing of ice nuclei collected on filters indicated that the variability in ice nuclei concentration was strongly a function of temperature. Fletcher (1962) derived an empirical relationship relating the increase of concentration of ice nuclei with decreasing temperature:

$$\ln N = a(T_1 - T) \tag{2.1}$$

where N is the concentration of active ice nuclei per liter of air, T is the air temperature, and T_1 is the temperature at which one ice nucleus per liter is active (typically about −20°C) and a varies from about 0.3 to 0.8. For $a=0.6$, equation (2.1) predicts that the concentration of ice nuclei increases by about a factor of 10 for every 4°C decrease in temperature. In urban air the total concentration of aerosol is on the order of 10^8 L^{-1}, and only about one particle in 10^8 acts as an ice nucleus at −20°C.

Later laboratory studies (e.g. Huffman and Vali 1973; Huffman 1973a,b) could not reproduce Fletcher's N-T empirical relationship. Measurements in a variety of field campaigns using a continuous flow diffusion chamber exhibited considerable variability in IN concentration (Rogers et al. 1998; Rogers et al. 2001; DeMott et al. 2003a, Prenni et al. 2007). The variability in ice nuclei concentration is likely not a simple function of temperature but also depends on thermodynamical, dynamical and aerosol characteristics (DeMott et al. 1994).

The activity of a particle as a condensation-freezing or a deposition nucleus depends not only on the temperature but also on the supersaturation of the ambient air. The effect of supersaturation on measurements of ice nucleus

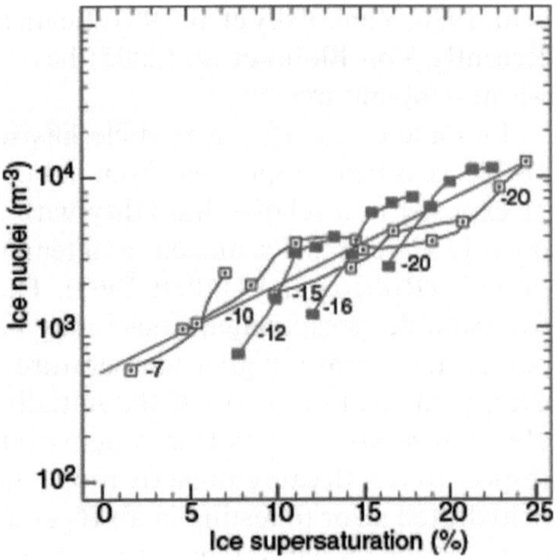

Fig. 2.10 Ice nucleus concentration measurements versus ice supersaturation; temperatures are noted alongside each line. All data were obtained at ground level. The *line* is equation (4.2). *Empty squares* are data from Rogers (1993); *dark squares* are from Al-Naimi and Saunders (1985). Adapted from Meyers et al. (1992) with permission of the American Meteorological Society

concentrations is shown in Fig. 2.10, where it can be seen that at a constant temperature the greater the supersaturation with respect to ice the more particles serve as ice nuclei. The empirical equation to the best-fit line to these measurements (the straight line in Fig. 2.10) is:

$$N = \exp\{a + b[100(S_i - 1)]\} \tag{2.2}$$

where N is the concentration of ice nuclei per liter, and S_i is the supersaturation with respect to ice, $a = -0.639$ and $b = 0.1296$ (Meyers et al. 1992). These measurements were obtained using a continuous flow diffusion chamber (CFDC), which with the limited data exhibits roughly a factor of ten higher concentrations of IN at warmer temperatures than older devices such as the filter-processing systems. Recognizing the need to allow vertical and horizontal variations in IN concentrations in mesoscale model simulations, Cotton et al. (2003) modified equation (2.2) to include the prognostic variable N_{IN}:

$$\mathrm{Ni} = N_{IN} \exp(12.96(\mathrm{Si} - 1)) \tag{2.3}$$

where $T < -5°C$. The variable N_{IN} can be deduced from continuous flow diffusion chamber data and used as a forecast variable in regional simulations (van den Heever et al. 2006; van den Heever and Cotton 2007). Equations (2.1), (2.2) and (2.3) only represent initial ice particle formation on IN and do not necessarily represent actual ice particle concentrations because other processes such as ice multiplication (see Sect. 2.2.3.4), sedimentation, breakup, and advection can greatly influence the concentrations of ice particles. Note that equation (2.3) allows for both horizontal and vertical variations in IN. Because the aerosol contributing to IN are large and large aerosol generally decrease with height (Georgi and Kleinjung 1968; DeMott et al. 2003b), we expect that IN concentrations generally decrease with height as well.

One should not therefore be surprised that many observations of ice particle concentrations do not show a good correlation with temperature (Gultepe et al. 2001; Field et al. 2005). Gultepe et al. (2001), for example, compiled data from the glaciated regions of stratus clouds for a number of field campaigns and found that ice crystals smaller than 1000 μm diameter do not show a good correlation with temperature and that the concentrations of these smaller ice particles varied up to three orders of magnitude for a given temperature (Fig. 2.11). On the other hand, measurements of ice particle concentrations in wave clouds where only "initial" ice particles are likely (Cooper and Vali 1981), showed ice particle concentrations increasing with decreasing temperature.

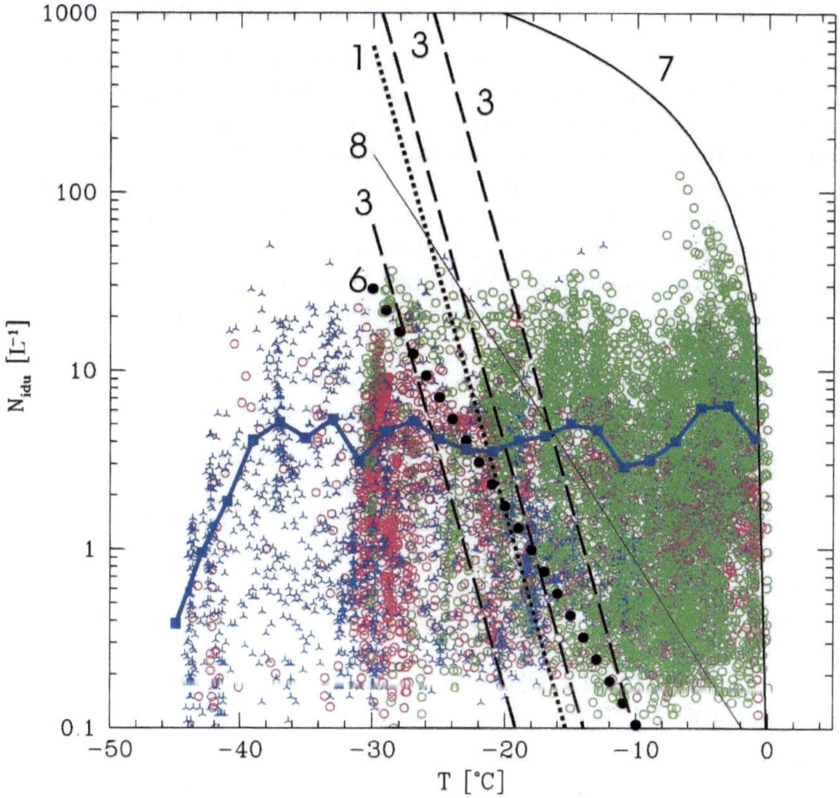

Fig. 2.11 Concentrations of ice particles, > 125 μm, from 2D-C probe measurements versus *T* at standard conditions during BASE (*open green circles*), FIRE.ACE (*blue triples*) and CFDE (*open red circles*) and from the earlier studies (black lines). Ice particles concentrations averaged over 2°C intervals for all projects is shown with a *blue line*. The *thick solid line* (**7**) is for Young (1974), dots (**1**) for Fletcher (1962), *dashed lines* (**3**) for Huffman and Vali (1973) *filled circles* (**6**) for a fit applied to Rogers et al.(1996) observations and the *thin solid line* (**8**) for Meyers et al. (1992). From Gultepe et al. (2001) with permission of the Royal Meteorological Society

In summary, we see that there remain many unanswered questions regarding the concentrations of ice nuclei, their composition, their activation relative to different environmental factors, and their relationship to ice crystal concentrations. Work has been hampered by severe difficulties in precise measurement of IN (see Chapters 5 and 6). Moreover, current field deployable devices for measuring IN do not take into account activation of IN by contact freezing. Fletcher's or Meyers' empirical curves are still used, and often misused, in numerical models, ranging from cloud-resolving models to GCMs. These curves at best only represent the concentrations of ice particles initially formed in clouds and probably rarely represent ice particle concentrations in most cloud systems.

2.2.3.4 Concentrations of Ice Particles in Clouds: Ice Multiplication

As we have seen, it has become increasingly evident that concentrations of ice crystals in "real" clouds are not always represented by the concentrations of IN measured or expected to be activated in such environments. In particular, it has been found that at temperatures warmer than −10C, the concentration of ice crystals can exceed the concentration of IN activated at cloud top temperature by as much as three or four orders of magnitude (Koenig 1963; Braham 1964; Mossop and Ono 1969; Auer et al. 1969; Mossop 1970; Mossop et al. 1967, 1968, 1972; Magono and Lee 1973; Hobbs 1969, 1974). The effect is greatest in clouds with broad drop-size distributions (Koenig 1963; Mossop et al. 1968, 1972; Hobbs 1974).

Some explanations or hypotheses that have been proposed to account for the high ice particle concentrations observed in some clouds are as follows:

- Ice multiplication by fracturing of fragile ice crystals, which may breakup during collision with each other. (Vardiman 1978).
- Fragmentation of large drops during freezing. (Mason and Mayban 1960).
- Secondary ice particle formation during ice particle riming. (Hallett and Mossop 1974; Mossop and Hallett 1974).
- Enhanced ice nucleation in the presence of spuriously high supersaturations. (Hobbs and Rangno 1985).
- Secondary ice particle generation during evaporation of ice particles (Oraltay and Hallett 1989; Dong et al. 1994).

The process that has been given the most attention and quantified in models is secondary ice particle formation by the rime-splinter process. Laboratory studies by Hallett and Mossop (1974) and Mossop and Hallett (1974), confirmed by Goldsmith et al. (1976), have indicated that copious quantities of splinters are produced during ice particle riming under very selective conditions. These conditions are:

- Temperature in the range of −3 to −8°C.
- A substantial concentration of large cloud droplets ($D > 24\ \mu m$).
- Large droplets coexisting with small cloud droplets ($D < 12.3\ \mu m$).

An optimum average splinter production rate of 1 secondary ice particle for 250 large droplet collisions occurred at a temperature of −5°C.

This process is consistent with field observations, where the greatest departure from IN estimates of ice crystals occurs when clouds contain graupel particles and frozen raindrops (Hobbs and Cooper (1987).

There is much indirect or inferential evidence that evaporation enhances ice crystal concentrations. This evidence is perhaps more intriguing than it is compelling. Some field studies have related unusually high ice nuclei numbers or unusual increases in ice crystal numbers to circumstances in which clouds were evaporating. Cooper (1995), for example, found a 100-fold increase in ice crystal concentrations in the evaporation region of orographic layer clouds. The largest ice enhancements in the Cooper study were observed in clouds with temperatures approaching the onset temperature for homogeneous freezing. Smaller enhancements were found in warmer clouds, and no enhancements were found at temperatures warmer than about −20°C. Further evidence of the possible role of evaporation nucleation has been presented by Field et al. (2001) and Cotton and Field (2002). They show observational evidence, and supporting parcel modeling calculations from wave cloud studies, that suggest ice had to form close to the downstream edge of wave cloud. Ice production coincident with the start of the liquid cloud, or earlier, would have suppressed the observed liquid cloud. Rapid ice crystal concentration enhancement versus expected IN concentrations in cumulus cloud were also observed to originate in close proximity to regions of cloud evaporation (Hobbs and Rangno 1985, 1990; Rangno and Hobbs 1994). The development of ice in a cumulus turret near its top at −18°C was followed by Stith et al. (1994). During the updraft stages low ice concentrations were observed in the turret (similar to what would be expected from primary ice nucleation), but during the downdraft stages the ice concentrations increased by an order of magnitude. This observation cannot be explained by rime splintering.

In summary, it is unlikely that all primary and secondary ice-forming processes have been quantitatively identified. Other mechanisms may sometimes operate, but their exact nature remains a mystery. In particular, our ability to measure small ice crystals has significant errors and needs improvement. Consequently, there are large uncertainties associated with our ability to simulate the affect of aerosols on the initiation of ice, and subsequent impacts on precipitation. This remains one of the critical problems in cloud physics.

2.3 Formation of Precipitation

Precipitation-sized particles can form by two mechanisms, known as the *collision-coalescence* and the *ice particle* mechanisms. The collision-coalescence mechanism can operate in clouds containing droplets, whether they are situated above or below the 0°C level. The ice particle mechanism can operate only in clouds with temperatures below 0°C that contain some ice particles.

2.3.1 The Collision-Coalescence Mechanism

This mechanism for rainfall production involves the initial formation of a few cloud drops with radii of about 20 μm by condensation onto CCN, followed by their rapid and substantial growth as they fall through the air and collide with smaller cloud droplets. If a drop increases in radius from 20 μm to a radius typical of raindrops, say 2 mm, its volume increases by a factor of one million. In other words, about one million cloud droplets have to combine to form one raindrop!

Due to the flow of air around a falling drop, a drop will not collect all of the droplets that lie in its path. Calculated values of the *collision efficiency*, defined as the ratio of the cross-sectional area over which droplets are collected to the geometric cross-sectional area of the collector drop, are shown in Fig. 2.12. This figure shows that the collision efficiency is negligible (<0.1) until a collector drop attains a radius of about 20 μm. Hence a few droplets need to grow to 20 μm in radius by condensation if a cloud is to form raindrops by the collision-coalescence mechanism. As the droplet grows by condensation alone, the rate of increase in the radius slows (Fig. 2.13). However, after a droplet has reached a radius of 20 μm its collision efficiency increases rapidly with increasing size (Fig. 2.13), so the droplet grows increasingly fast by collision. Moreover, as droplets get larger their cross sectional areas and fall velocities increase thereby increasing the collection kernel between large and small drops. Were it not for the fact that growth by collisions takes over as growth by condensation becomes negligible, many clouds would not rain.

Another factor effecting droplet growth by collection is the efficiency with which colliding drops coalesce. There is a good database of the efficiencies of drop collision efficiencies (Beard and Ochs 1993) but somewhat poorer knowledge of the coalescence efficiencies (Whelpdale and List 1971; Levin et al. 1973; Beard and Ochs 1993; Ochs et al. 1995), between drops over the range 1–300 microns. It is generally assumed that coalescence efficiencies are close to unity for small droplet collisions.

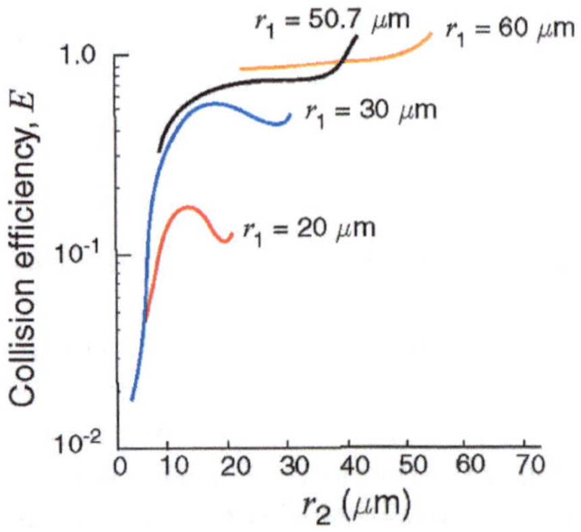

Fig. 2.12 Calculated values of the collision efficiency, E, for collector drops of radius r_1 with droplets of radius r_2 (adapted with modifications from Pruppacher and Klett 1997)

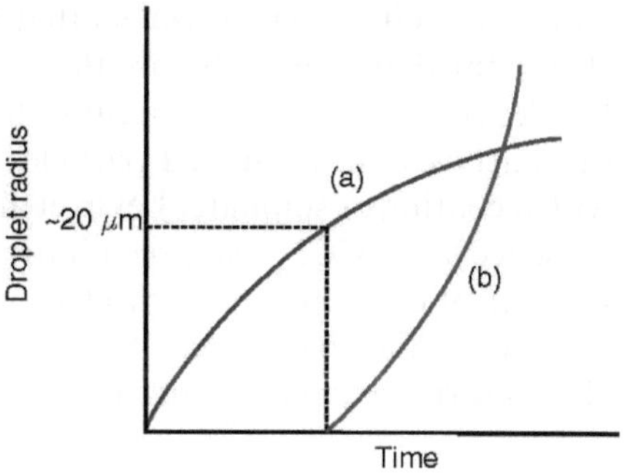

Fig. 2.13 Schematic curves of the growth of a drop (**a**) by condensation and (**b**) by collection of droplets

The basic problem is how do drops grow to a radius of 20 μm or greater fast enough to allow precipitation growth during the lifetime of clouds?

For an idealized cloud, the mean volume droplet radius (R_v) at any height above cloud base, can be derived from the adiabatic water content at that height, $q_c(h)$, and the number concentration of activated nuclei N as:

$$R_v = (q_c(h)/(4/3\pi\rho_w N))^{1/3} \tag{2.4}$$

For example, at a droplet concentration of 100 cm^{-3}, $q_c(h)$ must reach a value of 3.2 g m^{-3} (slightly less than 2 km of vertical ascent), for the mean volume radius to reach 20 μm. Since after CCN activation the droplet spectrum approaches a monodispersed distribution, there are no droplets significantly larger than R_v. The adiabatic assumption provides an estimate of the maximum droplet size, so that clouds growing in a polluted environment ($N > 500\ cm^{-3}$) should not be able to generate 20 μm radius droplets during their lifetime, based on the simplistic adiabatic model alone.

Four mechanisms have been hypothesized to explain how larger droplets or precipitation embryos can form:

- Role of giant cloud condensation nuclei.
- Turbulence influences on condensation growth.
- Turbulence influence on droplet collision and coalescence.
- Radiative cooling of drops to form precipitation embryos.

2.3.1.1 Role of Giant Cloud Condensation Nuclei (GCCN)

Observations reported by Woodcock (1953), Nelson and Gokhale (1968), Hindman (1975), Johnson (1976, 1982), and Hobbs et al. (1980) have shown the presence of potentially significant concentrations of aerosol particles of sizes as large as 100 μm. Their concentrations are $\sim 10^{-3}$ cm^{-3} (Woodcock 1953), that is, about one in 10^5 or 10^6 CCN are giant particles. Nevertheless, these particles can have a significant effect on the development of precipitation by serving as coalescence embryos (Johnson 1982; Feingold et al. 1999;

Yin et al. 2000a). The droplets that form are large enough for coalescence to start immediately even before the droplet reaches its critical size, based on the Köhler equation. This can occur if the nuclei are completely soluble, such as sea salt particles, or are mixed particles with a soluble coating (e.g. mineral dust with a coating of sulphate, Levin et al. 1996) or are very large and wettable. The presence of GCCN on precipitation formation has been investigated in a number of cloud resolving models, which show that their contribution to rain formation may be appreciable in polluted clouds but has little influence in clouds forming in clean airmasses (Feingold et al. 1999; Khain et al. 2000).

2.3.1.2 Turbulence Influences on Condensation Growth

While the adiabatic model assumes a uniform updraft and a constant droplet concentration, convective clouds are in fact made of a series of updrafts of diverse intensities. The least vigorous updraft at cloud base also produces the lowest droplet concentration during activation of the available CCN. Turbulence further contributes to entrainment of dry environmental air in clouds, hence reducing the liquid water content (LWC) below the adiabatic value and diluting the droplet concentration below its initial value after CCN activation. Convective cells with a significantly reduced concentration might generate bigger droplets than adiabatic cores if they experience further convective ascent (Baker et al. 1980; Telford and Chai 1980, and Telford et al. 1984). Turbulence also contributes to the continuous mixing of the convective cells, hence smoothing out their differences in terms of concentration and droplet growth.

Airborne measurements performed in cumulus clouds with the high resolution droplet spectrometer Fast-FSSP (see Chapter 6) reveal the occurrence of very narrow droplet spectra in cloud cores (Brenguier and Chaumat 2001). However, droplet spectra are often much broader than the narrow adiabatic reference, with droplet sizes extending from zero to the maximum predicted by the adiabatic model. This reflects the impact of mixing between convective cells that have experienced various levels of dilution with the entrained air. There are also airborne observations in stratocumulus clouds showing negative correlations between droplet concentration and droplet sizes (i.e. Fig. 7 in Pawlowska and Brenguier 2000). This corroborates the hypothesis that fluctuations of the updraft intensity at cloud base, or ascent of diluted cloud cells, might contribute to the formation of droplets bigger than predicted assuming adiabatic ascent. In deeper clouds, airborne cloud traverses at a given level all look rather similar in term of droplet concentration and sizes, suggesting that concentration/size correlations become progressively smaller due to continuous mixing.

Numerical simulations (Vaillancourt et al. 2002; Shaw et al. 1998) suggest that turbulence may also generate concentration fluctuations at the microscale level by inertia in regions of high vorticity. This can lead to droplet growth in microcells exceeding estimates in adiabatic cells and will exhibit the lowest concentrations. In situ measurements of the droplet spatial distribution at the microscale level (Chaumat and Brenguier 2001) did not support this hypothesis, however.

In summary, there is no observational evidence that turbulence contributes to the formation of precipitation embryos by enhanced condensation growth. In fact, stochastic processes induced by turbulence are not likely to enhance droplet growth because condensation is a cumulative process. To experience superadiabatic growth, droplets need to remain isolated in regions of higher supersaturation for a significant part of their ascent. The probability of this occurring are low, as mixing continuously redistributes droplets in the cloud. Turbulence is more likely to affect a discontinuous process like collision, since once droplets coalesce, they cannot be separated.

2.3.1.3 Turbulence Influence on Droplet Collision and Coalescence

Turbulence can influence the collision and coalescence process in 3 ways:

- By enhancing collision efficiencies.
- By enhancing the collection kernels.
- By producing inhomogeneities in droplet concentration.

The collision efficiencies we discussed earlier were calculated in laminar or stagnant flow. In turbulent flow droplets will be accelerating, and thereby be able to cross streamlines more efficiently than in laminar flow, resulting in enhanced collision efficiencies. Large droplets, having more inertia, will be affected more by turbulence than smaller drops. Calculations by Koziol and Leighton (1996) suggest that this effect is small for droplets smaller than 20μm diameter.

However, turbulence can also cause fluctuations on vertical fall speeds and horizontal motions, such that the collection kernel is enhanced relative to that defined in laminar flow (Pinsky and Khain 1997a; Khain and Pinsky 1997),

Because the collection rate is proportional to the square of droplet concentrations, inhomogeneities in droplet concentrations due to turbulence can produce enhanced droplet collection rates in regions where the droplet concentrations are locally enhanced, such as in regions of low vorticity (Pinsky and Khain 1997b).

2.3.1.4 Radiative Cooling of Drops to Form Precipitation Embryos

Consider a population of droplets that resides near cloud top for a sufficiently long time. The droplets will emit radiation to space quite effectively if the atmosphere above is relatively dry and cloud free. The droplets will thus be cooler than if radiative effects were not considered. The saturation vapour pressure at the surface of the droplets will therefore be lowered and the droplets will grow faster.

However as radiation cooling is proportional to the cross sectional area of a droplet, its effect is much greater on larger droplets than small ones (Roach 1976; Barkstrom 1978; Guzzi and Rizzi 1980; Austin et al. 1995). In a marine stratocumulus environment, where droplets are competing for a limited supply of water vapour, the larger droplets grow so rapidly by radiative enhancement that droplets smaller than 10 μm in radius evaporate, producing a bimodal size spectrum. (Harrington et al. 2000). This process is only effective in clouds where droplets reside near cloud top for time scales of 12 min or longer, such as fogs, stratus, and stratocumulus. Cumulus clouds with vigorous overturning, however, "expose" droplets that can radiate to space for too short a time for this process to operate.

2.3.1.5 Mature Phase of Liquid Precipitation Formation

Once precipitation embryos form they rapidly collect smaller droplets. They then transform into drizzle and raindrops in a matter of minutes if the liquid water content in clouds in great enough. The final size spectrum of the raindrops is determined by the liquid water content in the clouds, as well as their trajectories through updrafts and downdrafts in the cloud. Moreover, raindrops may breakup either spontaneously or by colliding with other raindrops and breaking up. Thus, the final size-spectrum of raindrops is influenced by the breakup process.

In summary, the ultimate control on the initiation, evolution, and intensity of rainfall from warm clouds is the temporal-spatial structure of a cloud's updrafts/downdrafts and the liquid water content (LWC). These properties, along with the initial activated droplet population, provide the boundary conditions (i.e. time scales) for particle growth that determines the intensity and duration of precipitation. Each droplet must first undergo nucleation, then condensation growth, followed by some mechanism of forming collection embryos, then stochastic collection, followed by continued raindrop collection of smaller drops and then breakup. All these steps take *time*. Thus critical to the formation of mature rain is whether or not the cloud lifetime and parcel lifetimes are long enough to allow all these steps to operate.

In simplest terms the clouds most likely to produce the most precipitation by exclusively liquid phase processes are maritime and warm-based. By maritime are meant clouds forming in a clean atmosphere with low CCN concentrations. If a cloud is warm-based then its saturation mixing ratio at cloud base is high. For example, a cloud with a base temperature of 24°C has a saturation mixing ratio of roughly 20 g kg^{-1}, whereas one with a base temperature of 5°C has a saturation mixing ratio of <6 g kg^{-1}. A cloud with a warm base has a depth of, say, 3 km it has a much greater potential of condensing a large amount of condensate than a cloud of similar depth having a higher, cold cloud base. We sometimes refer to warm-based, maritime clouds as being colloidally unstable (see Cotton and Anthes 1989).

By contrast, a cloud that is cold-based and forms in a continental or an aerosol-polluted environment, has a much smaller chance of producing precipitation and is said to be colloidally stable. The potential for liquid phase precipitation formation in most clouds will be between these two extremes.

2.3.1.6 Ice Particle Mechanisms

In regions of clouds lower than 0°C, supercooled water droplets can be present with or without coexisting ice particles. In Sect. 2.2.3, we discussed the nucleation of ice in clouds and ice multiplication mechanisms. The growth of ice particles to precipitation sizes in mixed clouds is now considered.

As early as 1789 Benjamin Franklin (1789) wrote "It is possible that, in summer, much of what is rain, when it arrives at the surface of the Earth, might have been snow, when it began its descent, but being thawed, in passing through the warm air near the surface, it is changed from snow to rain." This idea was not developed until the early twentieth century, when Wegener (1911) noted that ice particles would grow preferentially by vapour deposition in a mixed cloud. Subsequently Bergeron (1933) and Findeisen (1938) developed this idea in a more quantitative manner.

In a mixed cloud dominated by supercooled droplets the air is close to saturation with respect to liquid water and is therefore, supersaturated with respect to ice. For example, air saturated with respect to liquid water at −10°C is supersaturated with respect to ice by 10%, and at −20°C, the air is supersaturated with respect to ice by 21%. These supersaturations with respect to ice are much greater than the supersaturations of cloudy air with respect to liquid water, which rarely exceed 1%. Consequently, in mixed clouds dominated by supercooled water droplets, ice particles will grow from the vapour phase much more rapidly than droplets. This can lead to a variety of ice crystal types, some of which may be large enough to precipitate. This process is generally referred to the "vapour growth stage" of ice particle growth.

In a mixed-phase cloud, ice particles can also grow by colliding with supercooled droplets, which then freeze onto them. This growth process, which is referred to as *riming*, can produce heavily rimed ice particles, graupel particles and, if the cloud is deep enough, contains sufficient supercooled water, and updrafts are strong, hailstones can form. Finally, ice particles in a cloud may collide and aggregate with each other, leading to larger particles, which are called *aggregates*. Unfortunately the collision and coalescence efficiencies of ice particles riming cloud droplets are not well documented. Even more poorly quantified are the collision and coalescence efficiencies, and collection kernels among ice particles undergoing aggregation.

The growth of ice particles, first by deposition of water vapour followed by riming and/or aggregation, can produce precipitation-sized particles. If temperatures at ground level are below 0°C, these particles will reach the ground as snow; if the surface temperature is above 0°C, the ice particles will partially melt, or melt completely, and reach the ground as wet snow or rain. An increase in the concentration of cloud droplets, accompanied by decreases in the average size of the droplets (which can be caused by an increase in the concentration of CCN), may decrease growth by riming due to the lower collection efficiency of smaller droplets (Borys et al. 2000, 2003).

In deep convective clouds the rapidity of glaciation is dependent upon the presence of drizzle drops and large supercooled raindrops (Cotton 1972a,b; Koenig and Murray 1976; Scott and Hobbs 1977). When CCN concentrations are high, liquid phase precipitation collection processes are suppressed and supercooled raindrops are few in number. Small ice crystals must then grow to several hundred micrometers in diameter before they begin collecting cloud droplets. The riming process then proceeds slowly until the ice particles have grown to millimeter-size. Furthermore, since riming is suppressed in clouds forming in air masses with high CCN concentrations, secondary ice particle production by the rime-splintering process (Hallett and Mossop 1974; Mossop and Hallett 1974) is suppressed as well.

If the updrafts and liquid water contents in deep cumulonimbi are large enough, graupel particles, frozen raindrops and large aggregates can serve as embryos for hailstone formation (Heymsfield et al. 1980). At first the density of graupel particles is low, as the collected frozen droplets are loosely compacted on the surface of the graupel particle. As the ice particle becomes larger, it falls faster, sweeps out a larger cross-sectional area, and its growth by collection of supercooled droplets increases proportionally. As the growth rate increases, the collected droplets may not freeze instantaneously upon impact. The unfrozen water can thus flow over the surface of the hailstone filling in the gaps between collected droplets. The density of the ice particle, therefore, increases close to that of pure ice as the dense hailstone falls still faster, growing by collecting supercooled droplets as long as the cloud liquid water content is large. The ultimate size of the hailstone is determined by the amount of supercooled liquid water in the cloud and the time that the growing hailstone can remain in the high rainwater region. The time that a hailstone can remain in the high liquid water content region, in turn, is dependent on the updraft speed and the fall speed of the ice particle. If the updraft is strong, say 35–40 m s^{-1}, and the particle fall speed through the air is only on the order of 1–2 m s^{-1}, then the ice particle will be rapidly transported into the anvil of the cloud before it can take full advantage of the high liquid water content region. The ideal circumstance for hailstone growth is that the ice particle reaches a large enough size as it enters the high liquid water content region of the storm so that the ice particle fall speed nearly matches the updraft speed (Foote 1984). In such a case, the hailstone will only slowly ascend or descend while it collects cloud droplets at a very high rate. Eventually the hailstone fall speed will exceed the updraft speed, or it will move into a region of weak updraft or downdraft. The size of the hailstone reaching the surface will be greatest if the large airborne hailstone settles into a vigorous downdraft, as the time spent below the 0°C level will be lessened and the hailstone will not melt very much. Thus, a particular combination of airflow and particle growth history is needed to produce large hailstones.

In summary, hail growth is most likely if the convective storm updrafts are strong, if the liquid water contents are large, the storm is deep and long-lived, and the melting level is low relative to the ground (Foote 1984). There is some limited modeling work that suggests the concentrations of CCN and IN are factors in hail formation (Danielsen 1977).

2.3.2 *Summary of Cloud Microphysical Processes*

In preceding sections we have seen that the evolution of precipitation in clouds can take on a variety of forms and involve numerous physical processes. The evolution of ice-phase precipitation processes is greatly dependent upon the prior or concurrent evolution of the liquid-phase. These processes, in turn, are dependent upon the characteristics of the air mass (i.e. aerosols), the liquid-water production of the cloud, the vertical motion of air within the cloud, the turbulent structure, and the time scales of the cloud. Illustrated in Fig. 2.14 are the different precipitation paths that may occur, depending upon whether the cloud is a cold-based continental cloud or a warm-based maritime cloud. Remember from earlier discussions that the term maritime cloud is used to represent a very clean airmass and continental cloud to represent an airmass with much higher CCN concentrations. A polluted cloud would have still higher CCN concentrations. The figure does not note the speed by which these regimes can produce precipitation. We have pointed out earlier that a cloud with a vigorous warm rain process, referred to in the figure as a warm-based maritime cloud, will produce precipitation much faster than a cold-based continental cloud. The rapidity of glaciation of a warm-based maritime cloud is much faster than a cold-based maritime cloud, since the presence of drizzle drops and supercooled raindrops, once frozen, can rapidly transform a cloud from an all water cloud to an ice-dominated cloud. This should not be interpreted to mean that the largest precipitation elements, such as hail would, occur in a warm-based maritime cloud. In fact just the opposite can take place as a vigorous precipitation process lower in the cloud can deplete supercooled water amounts higher up and lower the trajectory of precipitation elements, leading to smaller sized hailstones (see Chapter 8)

Figure 2.14 illustrates the two distinctly different regimes. In fact there is a continuum of cloud types between these two extreme states. The heaviest precipitating clouds, and generally most efficient, are those that are warm-based and form in maritime airmasses. By contrast, clouds developing in a polluted airmass should be less efficient in producing precipitation than similar clouds with the same cloud base temperatures. This simple reasoning is valid only for single clouds and storms. As will be seen in Chapter 7, in some cases a suppression or retardation of precipitation in a primary convective cell could lead to a transformation of a cloud into a long-lived squall line through the interaction of cold-pools with wind shear, leading to self-propagating storms.

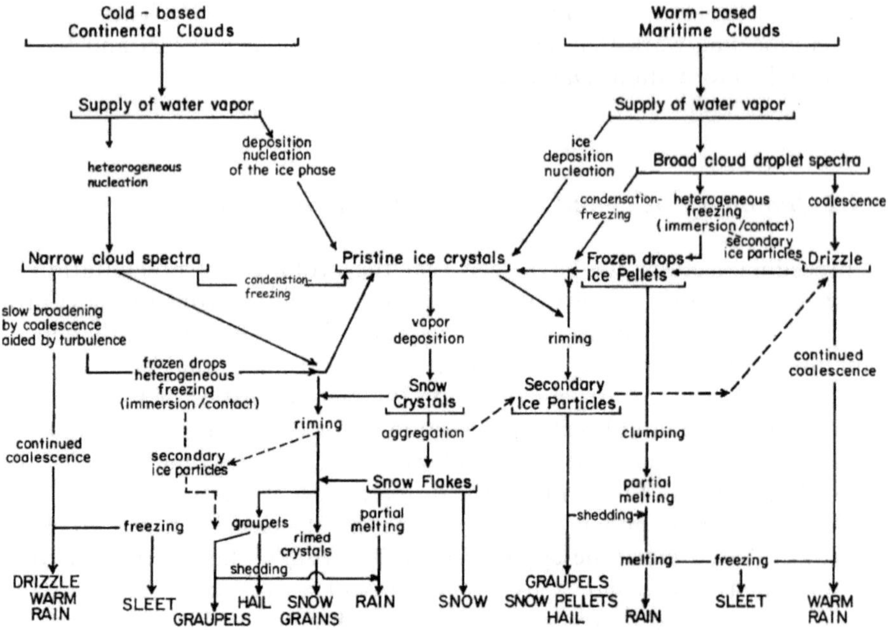

Fig. 2.14 Flow diagram describing microphysical processes, including different paths for precipitation formation. From Cotton and Anthes (1989) with permission of W.R. Cotton

2.4 Precipitation Efficiency

The term precipitation efficiency (PE) has many definitions in the literature and generally refers to the fraction of either the water vapour input at cloud base or condensed water in cloud that falls out as precipitation. Instantaneous values of PE vary from near zero early in the lifetime of a cloud before precipitation has commenced, to values exceeding 100% during the dissipation stages of a storm, when cloud-base moisture fluxes are near zero (Market et al. 2003). Doswell et al. (1996) suggest that PE is most meaningful when averaged over the storm lifetime. Market et al. (2003) propose that it is best to define a volume around a moving system and employ storm-relative winds in evaluating PE. In this way storm-averaged PE can be obtained for a moving system. PE estimates from different studies are problematic to compare because of differing definitions of PE, different data sources such as aircraft in situ versus radar measurements, and different time and spatial averaging (Hobbs et al. 1980).

Fankhauser (1988) presented one of the most detailed studies of thunderstorm precipitation efficiency. Data were taken from seven storms during the Cooperative Convective Precipitation Experiment (CCOPE) using aircraft, rawinsondes, a surface mesonetwork and radar histories. Calculated PE values ranged from 19 to 47%. Various environmental quantities, such as kinetic energy, cloud base area and height, and cloud base mixing ratio were found to be factors that had a strong positive correlation to PE levels. It was also found that variables such as the bulk Richardson number, the ratio of buoyant energy to the amount of wind sheer, and convective available potential energy (CAPE), a measure of potential updraft strength, did not correlate well with PE.

Another strategy for estimating precipitation efficiencies is to use drying ratio (DR), as proposed by Smith et al. (2003).

$$DR = \text{totalprecipitation}/\text{vapourflux} \tag{2.5}$$

DR encapsulates the moisture budget of the large-scale air mass transformation. Because PE requires an estimate of vertical air velocity at cloud base, it is subject to rather large errors in those measurements. Since DR uses the ratio of precipitation to water vapour flux, it is easier to quantify than PE. Moreover, DR can be estimated using hydrogen and oxygen isotope analysis, which permits evaluation using streamflow or sapwood collected near a stream. Isotope analysis indicated a DR value of 43% for the combination of the coastal and Cascade mountain ranges in western Oregon and 48% for the southern Andes (Smith et al. 2005; Smith and Evans 2007).

To date no one has made estimates of storm-averaged PE or DR for storms developing in environments containing different cloud-nucleating aerosol concentrations. Such work should be encouraged as a means of determining the impact of aerosols on precipitation while accounting for different water vapour and vertical air motion conditions.

Chapter 3
Sources and Nature of Atmospheric Aerosols

Meinrat O. Andreae, Dean A. Hegg and Urs Baltensperger

3.1 Introduction

In view of the nonlinear and spatiotemporally complex interactions between aerosols, clouds and precipitation, it is quite inadequate to represent aerosols by the conventional approach of atmospheric pollutant cycles, i.e. by specifying sources, burdens and sinks in terms of masses of particular aerosol components or "species". The basic information required for modeling the role of aerosols in cloud and precipitation processes is the 4-D distribution of the number concentration of particles that can be activated at the supersaturations occurring in clouds (cloud condensation nuclei, CCN) or that can initiate the formation of ice particles, especially at temperatures above the level of homogeneous freezing of cloud droplets (ice nuclei, IN). This information is difficult or impossible to derive from conventional studies of aerosol cycles, such as those that formed the basis of the section on aerosol sources and burdens in the IPCC-TAR report (Penner et al. 2001). Several problems present themselves:

Firstly, the very concept of "source strength" is difficult to define for some aerosol types, since the secondary aerosol species (e.g. sulphates, secondary organics) are not directly emitted, but are formed in the atmosphere from gaseous precursors. For these species, the "source" is a set of atmospheric reactions, not a surface emission process, and therefore the production rate depends sensitively on the efficiency of these reactions. These "sources" cannot be derived from measurements, but have to be inferred from measurements of the source strength of precursors, the (always incomplete) knowledge of atmospheric transformations, and a formal or informal inversion of measurements of atmospheric burdens. The source estimates for these aerosols therefore depend strongly on how the source processes are represented in the models.

Secondly, the conventional mass-based treatment of aerosol fluxes does not take into account the fact that the properties of aerosol particles can vary strongly with particle size, and that the cloud-physical effects of aerosols depend on number concentrations, not on mass concentrations. For example,

M.O. Andreae (✉)
Max Planck Institute for Chemistry, Mainz, Germany

Z. Levin, W.R. Cotton (eds.), *Aerosol Pollution Impact on Precipitation*,

the largest number concentration of cloud-active sea salt particles resides in the submicron size fraction, which represents only a very minor mass fraction of the sea salt aerosol. On the other hand, this fraction of the sea salt aerosol has quite different atmospheric sink mechanisms and lifetime than the coarse sea salt mode, which dominates the mass source strength and burden of the aerosol component. Similar considerations apply to mineral dust and primary biogenic aerosols (see Chapter 4).

A third problem arises from the fact that aerosol species often combine to form mixed particles, with properties and atmospheric lifetimes different from those of their components. Even though the direct radiative effects of these "internal mixtures" may be substantially different from those of "external" mixtures (Jacobson 2001; Chung and Seinfeld 2002), this distinction has been ignored in many analyses of the global aerosol system. However, for the effects of aerosols on clouds, even relatively small amounts of soluble material in or on otherwise insoluble particles can cause large differences, and the mixing state of the aerosol cannot be ignored.

It follows from these considerations that a meaningful analysis of the sources and burdens of cloud-active particles cannot be achieved through emission inventories, as it has been done typically for atmospheric trace gases. Rather, it must take the form of an atmospheric model that contains emissions of precursor gases and primary aerosol species, and accounts for transformations and sink terms of precursor species, as well as the various aerosol components. The model must carry either explicitly or in parameterized form information about the number/size distribution of soluble and wettable materials, ice-nucleating ability, and the interactions between gas phase and various condensed phases. Fundamentally, therefore, the issue of aerosol sources cannot be separated from an analysis of transport and atmospheric processes and must be treated interactively within the framework of a numerical model that represents these processes. Unfortunately, this type of approach is still in its infancy. One of the most complete analyses (suitable for long-term climate simulations) to date is represented in the aerosol-climate model ECHAM5-HAM (Stier et al. 2005). In this model, the major aerosol components sulphate, particulate organic matter (POM), black carbon (BC), sea salt and mineral dust are determined through an aerosol module that is interactively connected to a climate model. In a somewhat complementary approach, the distributions of ammonium sulphate and nitrate aerosols have been modeled using prescribed meteorology and a detailed description of gas/aerosol partitioning (Metzger et al. 2002a).

In the following sections we will present an updated inventory of emission fluxes, production rates, and burdens for the major aerosol species, providing information about particle numbers when possible. For selected aerosol types, we will provide spatial information in the form of 2D distribution maps. Ultimately, however, it must be acknowledged that this information just represents a convenient visualization of the spatiotemporally resolved distributions, which can be fully represented only in the form of aerosol modules in coupled aerosol-climate models. In addition, these data are results from one particular global aerosol model (ECHAM5-HAM) and do thus include the uncertainties connected to such simulations (Textor et al. 2004).

3.2 Natural and Anthropogenic Sources of Atmospheric Aerosols

Aerosol sources are classified into *primary* and *secondary* types. Primary aerosols are those that are emitted into the atmosphere directly as condensed solids or liquids, whereas secondary aerosols are formed within the atmosphere from gaseous precursors. Sea salt, mineral dust and soot particles are clearly primary particles, whereas organic particles from the oxidation of volatile organic compounds (VOC) and sulphates from the oxidation of SO_2 or other sulphur-containing gases are clearly secondary. There is a "grey zone" between primary and secondary in the case of some low-volatility organic compounds that condense onto aerosol near, but not directly within, emitting sources, such as some hydrocarbons in vehicular exhaust or condensable from biomass burning. We will discuss these materials together with the strictly primary emissions, taking into account that the definition of secondary aerosol formation involves chemical transformation from more volatile precursors (Fuzzi et al. 2006).

3.2.1 Soil Dust

Soil or mineral dust is the most conspicuous aerosol component in satellite imagery of global aerosol distributions (see Fig. 3.1). Many estimates of source strength and burden have been published; much of the older work has been summarized by Duce (1995), while the more recent literature has been reviewed by Tegen (2003) and Zender and Tegen (2004). The estimates of global annual source fluxes are in the range of 1000–2150 Tg a^{-1}, and have not changed much since the reviews of Andreae (1995) and IPCC-TAR (2001). There has been tremendous improvement, however, in the sophistication of the dust deflation and transport models used and in the techniques used for validation of the models, especially the use of remote sensing (Israelevich et al. 2002; Tegen et al. 2002; Werner et al. 2002; Luo et al. 2003; Zender et al. 2003; Miller et al. 2004; Tegen et al. 2004). The range of burdens predicted by these models, 8–36 Tg, remains still quite large, mostly due to differences in the way deposition processes are treated in the models. These differences result to a substantial extent from different assumptions regarding the size distribution of the dust aerosol.

There are major discrepancies in the size distributions of the mineral dust aerosol predicted by the models and observed by various techniques. Near the sources, the mass distribution peaks at large sizes, 10 μm or greater, in observation and models. At the source, about 10% of the dust is in the submicron fraction (Tegen et al. 2002; Luo et al. 2003). More relevant at the regional and

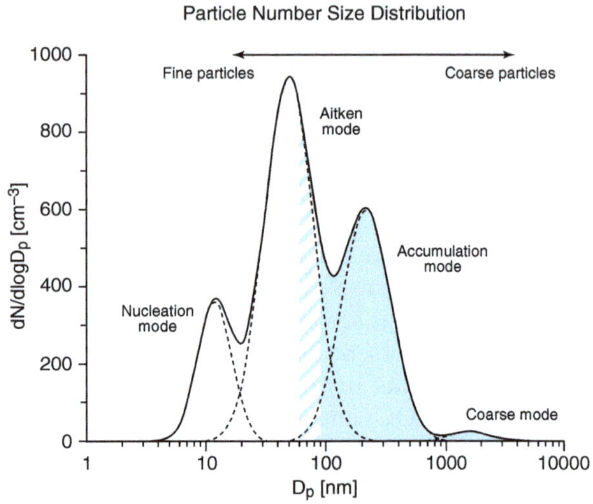

Fig. 3.1 Typical particle number size distribution. The shaded band represents the range of sizes activated as CCN at 0.3% SS, a typical median SS in clouds

global scales is the size distribution in the atmospheric burden after some atmospheric transport and processing has taken place. In the models, mass median diameters (MMD) are typically 2.5 μm or less. In a study using the NASA GISS atmospheric general circulation model (AGCM) with interactive treatment of dust flux and climate, Miller et al. (2004) obtain a substantially lower dust MMD (~1.6 μm) than in an off-line tracer model (~3 μm). Observations show at least twofold higher sizes than the model predictions, with some internal disagreement between different measurement approaches. Measurements of dust MMD by aerodynamic methods give a mean of 4.5±1.3 μm, in good agreement with the results from optical inversion methods (5±1.5 μm), but considerably below values obtained by optical counter methods (Reid et al. 2003). Since the current dust models are being constrained by data obtained by various combinations of these observational approaches, and in turn used to predict radiative forcing, substantial problems related to non-physical parameterizations may result.

Detailed analysis of dust source areas in recent years has shown that there are areas with particularly strong dust emission potential ("preferential source areas"), in particular dry lake beds in arid areas (Israelevich et al. 2002; Prospero et al. 2002; Tegen et al. 2002; Zender et al. 2003). These regions are also of special relevance to the issue of aerosol/cloud effects, as they contain a higher fraction of clay-size (1 μm or less) particles and therefore have the potential of releasing an over-proportionally higher number fraction of particles compared to the coarser desert dust particles. They also include soluble residues from the evaporation of the lake water, which renders them highly suitable as CCN (Formenti et al. 2003). Clay minerals are also known to be effective ice nuclei, although, presumably, these minerals from dry lakes may behave quite differently than ones from higher deserts due to the additional soluble content.

Dust emission shows substantial variability at all time scales from the last glacial maximum to the anthropogenically influenced present and in simulations of the future. This variability is related to changes in vegetation cover, soil moisture, and anthropogenic disturbance of the soil and vegetation surface, as well as to changes in climatic parameters such as wind speed and precipitation (Mahowald et al. 2002; Tegen et al. 2002; Werner et al. 2002; Luo et al. 2003). Variability at interannual to decadal time scales makes longer-term trends hard to discern. Earlier views that anthropogenic soil disturbance was responsible for as much as half of the present-day dust flux have been revised downward in the last few years, and the contribution of dust from agricultural activities is now estimated to be <10% (Tegen et al. 2004). Since dust affects precipitation through both radiative transfer and cloud microphysics, there is a feedback through soil moisture on dust emission. Dust exerts a negative forcing at the surface and a positive one in the atmosphere. The magnitudes of these forcings are dependent on the optical properties (mineralogy) of dust, which are still poorly known. Model calculations by Miller et al. (2004) suggest that increased dust loadings results in more rain in the Sahara, thus producing a negative feedback. In their model, which does not include the effects of dust on cloud physics, dust reduces *global* rainfall, thus lengthening the lifetime of aerosols, and increasing the burden of dust and other aerosols.

Dust fluxes and burdens in future decades will depend strongly on changes in land cover and climate, but reliable estimates cannot be made at present. One estimate suggests that dust fluxes may decrease with future land use and climate change by 10–60% by 2100 (Mahowald and Luo 2003), while another suggests smaller changes (10–25%), with increases or decreases possible depending on which climate/land-cover model is used (Tegen et al. 2004). Because of the strong dependence of dust mobilization on windspeed (increasing with a power of approximately 3–4), dust fluxes in future climates are highly sensitive to changes in the wind regimes in dust source areas.

For an analysis of the impact of dust aerosols on clouds and precipitation, the most important information would be the distribution of the number concentration of dust particles that can act as CCN, giant CCN (GCCN), and IN. Unfortunately, the published studies on dust budgets do not provide this information. The issue is further complicated by the fact that dust particles may or may not contain soluble materials already at the time of emission, e.g. in the form of evaporation residues, or that they may become coated with sulphate, nitrate, or other soluble salts during atmospheric processing, especially in clouds (Levin et al. 1996; Yin et al. 2002; Trochkine et al. 2003). Most of the dust models carry information about number size distributions, explicitly or implicitly through specification of the modal or bin characteristics of the mass/size distributions. The number maximum of the dust aerosol mode is generally well below 1 μm diameter. For example, Zender et al. (2003) show the number maximum at 0.6 μm diameter. Given the relatively high contact angles between water and silicates (Pruppacher and Klett 1997), we would expect that most of these particles would not be able to act as CCN at commonly found supersaturations, unless they were coated with soluble material. If they are not activated as cloud droplets, a substantial fraction of dust particles can remain as interstitial aerosol in convective clouds and reach the middle and upper troposphere, where they can act as IN. If, on the other hand, large dust particles contain appreciable amounts of soluble material or are coated with soluble substances, they can play an important role as GCCN (see Chapter 4). In summary, we find that the source fluxes and burdens of dust aerosols, and the CCN- and IN-active fractions of this aerosol, depend strongly and interactively on atmospheric processes, especially on their behavior in clouds.

3.2.2 Sea Salt

Sea salt aerosols are generated by various physical processes, especially the bursting of air bubbles entrained during whitecap formation and the tearing of droplets from wave tops (Schulz et al. 2004). The rates of particle production by both mechanisms are strongly dependent on wind speed, with production of spume droplets from wave tops occurring only at wind speeds >9 m s^{-1}. Under moderate wind regimes, sea salt particle number concentrations are typically about 10 cm^{-3} or less, while under high wind conditions (>10 m s^{-1}), their concentration can rise to 50 cm^{-3} or more (O'Dowd et al. 1997; Shinozuka et al. 2004).

In ocean regions where wind speeds are high and/or other aerosol sources are weak (such as the winter-time remote oceans of the Southern Hemisphere) sea salt may be the dominant contributor to CCN (O'Dowd and Smith 1993; Murphy et al. 1998a; Quinn et al. 1998; Gong et al. 2002; Shinozuka et al. 2004). By presenting a primary aerosol surface area, sea salt can also inhibit new particle formation.

Sea salt particles cover a wide size range (~0.05–10 μm diameter), with most of their mass in the supermicron range. The highest number concentration of sea salt particles, however, is in the submicron range, with a pronounced number mode at ~500 nm, and significant number concentrations still present at 100 nm (Campuzano-Jost et al. 2003). Unfortunately, field observations of sea salt particles presently only reach down to ~100 nm (O'Dowd et al. 1997; Murphy et al. 1998a; Campuzano-Jost et al. 2003), but laboratory studies of sea salt aerosol production by Mårtensson et al. (2003) indicate that the maximum of the number distribution is actually somewhat below 100 nm (~50–80 nm). Very small sea salt particles, with a number maximum around 20–30 nm, were found to be produced by coastal breaking waves (Clarke et al. 2003). This is supported somewhat indirectly by measurements of particle flux from the sea surface (Nilsson et al. 2001) that also yield a mode at or below 100 nm, but in this study the composition of these particles could not be identified as sea salt.

A substantial source of very small sea salt particles could be of importance to the marine CCN budget even if they are below the critical diameter for droplet activation at the low supersaturations in marine stratus clouds, because they can serve as nuclei for the growth of sulphate particles from the oxidation of dimethyl sulphide (DMS), which is produced from ocean biota (Charlson et al. 1987). Direct production of sulphate particles by homogeneous nucleation is thought to be only rarely possible in the clean marine atmosphere (Pirjola et al. 2000). On the other hand, larger sea salt particles also serve as a sink for gaseous H_2SO_4 molecules, thereby reducing the amount of sulphate available for the formation of accumulation mode particles (Gong and Barrie 2003). In summary, sea salt particles are only a minor fraction of the CCN population under most situations, except for the winter Southern Ocean, North Atlantic and North Pacific. There is evidence for the existence of fine and ultrafine sea salt particles, but their abundance and their direct or indirect role as CCN source remains unclear. More detailed studies on the size distribution, composition, and production rates of the submicron marine aerosol will be required to resolve these issues.

Several studies in the last few years have shown that “sea salt” aerosol actually contains a substantial amount of organic matter, consisting both of insoluble material (biological debris, microbes etc.) and water soluble constituents (Novakov et al. 1997; Middlebrook et al. 1998; Cavalli et al. 2004; O’Dowd et al. 2004). The fraction of organic components increases with decreasing particle size, and, in biologically active regions, may approach 90% in the size fraction around 100 nm (Cavalli et al. 2004). The chemical and isotopic composition of this organic matter suggests a marine biogenic source (Turekian et al. 2003). Organic materials and sea salt are present as internal mixtures, consistent with a production mechanism that involves fragmentation of organic-rich surface film layers during the bursting of air bubbles at the sea surface. Some evidence on dry particles suggests that the organics may be present as coatings on the surface of salt crystals (Tervahattu et al. 2002b). In this case, they could interfere with the uptake of water by the particles (Saxena and Hildemann 1996), and with the collision-coalescence process (Medina and Nenes 2004), but these findings may not be applicable to the more liquid-rich sea salt particles at high (especially super-saturated) humidities (Abbatt et al. 2005).

Since sea salt particles are very efficient CCN, characterization of their surface production is of major importance for aerosol impacts on clouds. Estimates of the total sea salt flux from ocean to atmosphere vary over a wide range. The most recent estimate (Schulz et al. 2004) is 2690 Tg a^{-1} in a mode centered at 2 μm diameter, plus 17,100 Tg a^{-1} in a mode at 11 μm diameter. At first glance, this appears considerably higher than most previous ones, e.g. 3300 Tg a^{-1} (Houghton et al. 2001), 6500 Tg a^{-1} (Grini et al. 2002), or 5900 Tg a^{-1} (Tegen et al. 1997). A large part of the differences, however, can be attributed to the choice of the upper size cut-off. With increasing cutoff size, a large mass flux of very coarse and therefore very short-lived salt particles is included in the estimate.

Because most of the sea salt mass is in the coarse mode, while most of the sea salt particle number is in one or more submicron modes, the conventional estimates for the mass source flux of sea salt are of little relevance for the discussion of sea salt particles as CCN, and it is necessary to analyze their number production rates and atmospheric distribution in a size-resolved model. This applies both to the submicron fraction, where most of the sea salt-CCN are found, and to the coarse fraction, which contains the giant CCN (see Chapter 4). There are several parameterizations available for the wind-speed dependent, size-resolved emission flux of particles from the sea surface, most of which have been summarized in Schulz et al. (2004). The best match with the observed size distributions (O'Dowd et al. 1997) is obtained using the Vignati and Gong-Monahan schemes (Vignati et al. 2001; Gong 2003), among which the Gong-Monahan scheme probably has the more faithful representation of the wind-speed dependence. Estimates of the number flux of sea salt particles have not been published so far, even though it must be assumed that they have been calculated in the various aerosol models. Below (Sect. 3.3) we are presenting an estimate that was retrieved from the model runs that formed the basis of the papers by Stier et al. (2005, 2006). Because of the strong dependence of sea salt particle production on wind speeds, the role of these particles in cloud and precipitation processes is expected to change with anthropogenically induced climate change.

3.2.3 Carbonaceous Aerosol

Carbonaceous material, consisting of organic compounds and near-elemental, sub-graphitic material ("soot") represents an important fraction of the atmospheric aerosol. Only a minor part of this complex mixture, in the range 10 to 40%, has been identified at the molecular level (Gelencsér 2004; Kanakidou et al. 2005; Decesari et al. 2006). The carbonaceous aerosol contains a large fraction of oligomers and highly polymeric matter from various origins. This unique chemical complexity is the cause of much of the prevailing uncertainties regarding the sources and properties of carbonaceous aerosol.

Carbonaceous materials represent a continuum from semi-volatile organic species (SVOC), chemically-identified particulate organic carbon (POC), polymeric POC, to black carbon (BC) or elemental carbon (EC). The BC or EC component is operationally defined as light-absorbing ("black") or thermally refractory ("elemental") carbon, and is generally thought to consist of aggregates of 20–50 nm granules with a graphite-like crystal structure and near-elemental composition with some surface functionalities (Smith et al. 1989). Such particles are also called "soot" particles (which term often includes the organic coating that is usually associated with these particles), and originate from various types of combustion (Andreae and Gelencsér 2006). Especially in biomass smoke, however, there are also other refractory organic compounds that operationally act like EC, and other light-absorbing organics that are determined as BC, so that at this time this aerosol component remains rather ill-defined. This is reflected in large discrepancies between the results of OC and BC/EC measurements from different laboratories (up to factors of 10 for BC/EC), even when identical samples and nominally identical analytical protocols are used (Ten Brink et al. 2004). The strong increase of the absorption with decreasing wavelength can sometimes be used to identify aerosols from biomass burning (Kirchstetter et al. 2004; Schmid et al. 2006).

As the most commonly used combustion-based chemical analyses provide carbon contents only, a conversion factor k is required to account for organic functional groups in particulate organic matter ($POM = k \times POC$). The k factor varies with the origin and age of the particles over a wide range: 1.3 (close to sources) to >2.0 (Putaud et al. 2000; Turpin and Lim 2001; Kiss et al. 2002; Russell 2003).

In addition to primary carbonaceous aerosols, there is a substantial amount of secondary organic aerosols (SOA) formed from a large variety of anthropogenic and biogenic gaseous precursors (Seinfeld and Pankow 2003). As a result, organic aerosols contain variable amounts of primary (PPOM: Primary particulate organic matter; PPOC: Primary particulate organic carbon) and secondary (SOA) material. It has been shown that in source regions the formation of combustion-derived SOA occurs readily and could represent the same amount as PPOM (Cooke et al. 1999). Once formed, carbonaceous aerosols are subject to continuous chemical modification in the atmosphere by photochemical and multiphase reactions (Maria et al. 2004). This includes reactions with OH and other radicals from the gas phase (Molina et al. 2004; Zuberi et al. 2005), oligomerization with and without acid catalysis (Claeys et al. 2004; Gao et al. 2004; Kalberer et al. 2004), and formation of humic-like substances (HULIS) by photochemical polymerization of biogenic precursors (Gelencsér et al. 2002).

3.2.3.1 Primary Biogenic Aerosols

Primary biogenic aerosol particles (PBAP) may enter the atmosphere from a number of sources. Plants shed various types of debris (cuticular waxes, leaf fragments etc.), and soil deflation releases a mixture of humic matter, plant decomposition products, fungi and microbes. Atmospheric aerosols over land and oceans contain a large variety of microbial particles (bacteria, fungi, viruses, algae, pollen, spores etc.). These particles cover a wide size range from tens of microns down to a few tenths of microns. Morphological studies using light and electron microscopy have shown that PBAP make up a large fraction of the aerosol down to ~0.2 μm at a variety of clean and polluted sites (e.g. the Amazon, Siberia, and even urban Central Europe (Matthias-Maser and Jaenicke 1995; Matthias-Maser et al. 2000; Graham et al. 2003a; Graham et al. 2003b; Posfai et al. 2003; Jaenicke 2005). Their number concentrations are in the range of tens to hundreds cm^{-3}. Unfortunately, it becomes quite difficult to unequivocally identify particles in the submicron range as primary biogenic, as they often have little or no diagnostic morphological or bulk chemical signatures. Chemical tracers studies suggest, however, that primary biogenic material contributes a large fraction of the pristine aerosol even in this size range (Simoneit et al. 1990; Graham et al. 2003b). Very small PBAP (20–50 nm) may also play a role as growth nuclei for CCN-sized particles over the remote oceans, through condensational growth by deposition of sulphate derived from the oxidation of DMS (Leck et al. 2002; Leck et al. 2004).

There is considerable evidence that primary biogenic particles may be able to act both as CCN and IN (Schnell and Vali 1976; Levin and Yankofsky 1983; Diehl et al. 2001; Bauer et al. 2003). This is consistent with their often relatively large size and soluble matter content, but surface properties are also thought to play a significant role (Bauer et al. 2003). It is therefore possible, that PBAP represent a significant fraction of CCN and IN in the clean atmosphere, and therefore are of importance for the regulation of cloud microphysical properties under pristine or remote conditions, but not enough is known at this time to assess their role with any confidence. Since their atmospheric abundance may undergo large changes as a result of land use change, they deserve more scientific study.

Since many of these PBAP are of large size, they are of special interest because they can already be activated at very low supersaturations (<0.02%). Therefore, they will be the first to activate at cloudbase, and they may therefore represent a major source of GCCN. This might enable them to play an important role in precipitation formation under circumstances when high concentrations of pollutant CCN otherwise would suppress warm rain production (Yin et al. 2000a; Rudich et al. 2002). To date, there are no reliable estimates on the rates of PBAP emissions. The estimates given in Table 3.1 have been calculated based on the commonly observed mass concentrations of 1–2 μg m^{-3} over remote, vegetated continental regions and an estimated atmospheric lifetime of 2 days for these particles.

Table 3.1 Global emission estimates for black carbon (BC) and primary particulate organic carbon (PPOC) from fossil fuel, biofuel, and open biomass burning

	Black carbon Tg C a^{-1}			Primary POC Tg C a^{-1}		
	Fossil fuel	Biofuel	Open burning	Fossil fuel	Biofuel	Open burning
Liousse et al. (2004) Juncker & Liousse (2005)	5.9	2.9	4.3[a]	6.4	8.7	8.2[b]
Bond et al. (2004)	3.0	1.6	3.3	2.4	6.5	25
Ito & Penner (2005)	2.8	2.0	3.5	2.4[c]	7.3[c]	22[c]

[a] Including savanna, open agricultural fires and tropical fires (Liousse et al. 1996b) and extratropical fires (Lavoué et al. 2000).
[b] Primary particulate organic carbon only, including savanna, open agricultural fires and tropical fires (Liousse et al. 1996b) and extratropical fires (Lavoué et al. 2000).
[c] Using a conversion factor of 1.3 for PPOC = >PPOM.

3.2.3.2 Primary Anthropogenic Carbonaceous Aerosol

Black carbon and a fraction of PPOM and SOA originate from combustion sources. Combustion sources using fossil fuel are found in industry, power generation, traffic and residential heating (Andres et al. 1996). It has, been shown by carbon-14 analysis that in cities the biogenic fraction may comprise about 60% of the carbonaceous aerosol. In summer this is mainly attributed to SOA from biogenic precursors, while in winter there is a substantial contribution from wood combustion (Szidat et al. 2006). In the intertropical zone, biomass burning (open savanna, forest and agricultural waste fires) and biofuel use form the main sources of anthropogenic carbonaceous aerosols. Biomass burning from natural fires provides a major source of carbonaceous aerosols, mainly in the mid- to high latitudes of the northern hemisphere (e.g. Lavoué et al. 2000). SOA is also formed by the conversion of natural organic gases such as terpenes and, probably to a lesser extent, isoprene (Claeys et al. 2004).

From both biomass burning and fossil fuel sources, POM (PPOM + SOA) is always the predominant fraction of the carbonaceous aerosol. However, the chemical composition of the aerosol as represented by the BC/POM ratio is highly dependent on the type of sources: on average the BC/POM ratio is higher for fossil fuel than for biomass burning sources. Also BC/POM ratio is linked to the type of burning: BC/POM is probably related to the temperature of the combustion process, and is thus higher for diesel emissions than for gasoline (Bond et al. 2004). Biomass combustion aerosols emitted under flaming conditions will have a higher BC/POM ratio than smoldering aerosols, which are almost entirely of organic origin (Reid et al. 2005).

At present the lack of adequate data for BC and PPOC concentrations precludes the use of inverse modeling. To estimate the emissions of combustion carbonaceous aerosols, the only suitable current method is the bottom-up approach. Emission estimates are based on two different data sets: one related to fuel consumption and the other to particulate emission factors.

Spatial distributions of fuel consumption (fossil fuel and biofuel) may be obtained from data provided by a variety of national and international organizations. They are complemented by regional inventories that make special efforts to account for regional source characteristics and may provide more detailed consumption maps (Streets et al. 2001; Reddy and Venkataraman 2002; Streets et al. 2003; Schaap et al. 2004).

Mapping the amounts of burnt biomass is a difficult task. In previous studies (Hao and Liu 1994; Hao et al. 1996) estimates of burnt areas for savanna and forest fires were based on statistical data. For emissions related to agricultural practices, the FAO (Food and Agriculture Organization of the United Nations) data are usually associated with assumptions of the burnt fraction (Yevich and Logan 2003). Qualitative improvements are now obtained from multi-year global distributions of fire pixel counts and burnt area given by satellite imagery, as these data allow taking into account the spatial and temporal variability of fires at different scales (Kajii et al. 2002; Generoso et al. 2003; van der Werf et al. 2003;

Hoelzemann et al. 2004; Ito and Penner 2004; Liousse et al. 2004; Michel et al. 2005). The need to couple the two existing satellite tools (fire counts and burnt area) is illustrated by the work of Michel et al. (2005) on biomass burning inventory for Asia using SPOT (Satellite Pour l'Observation de la Terre) imagery for burnt areas. The fire spatial and geographical distribution they obtained is totally different than that of Streets et al. (2003), derived from fire pixel count maps. Provided that current problems with detection algorithms can be solved, the uncertainty of such products might be reduced to <15% for burnt areas. Uncertainties now exist primarily on the estimates of fire properties (biomass density and combustion efficiency), and further improvement will necessitate accounting for regional differences and relating in a common effort local measurements, remote sensing techniques and fuel models.

Emission factors for BC and PPOC (or PPOM) particles depend on the nature of the fuel used and on the physical conditions of the combustion processes. For a given fuel, the type of use (industrial versus domestic) and combustion technology will be of high importance. The level of development of each country and the regulating standards they apply are also crucial. For biomass burning, BC and PPOC emission factors have been reviewed, based on a synthesis of the most recent experiments and existing papers (Andreae and Merlet 2001).

For fossil fuels and biofuels, problems originate from the complexity, heterogeneity, and uneven coverage of available information. To cope with this difficulty, global inventories have been then constructed following a standard procedure that may apply to all activities in all countries and is flexible enough to be easily improved (Cooke et al. 1999; Junker and Liousse 2006). The detailed sources were reduced to three main sectors of activities (traffic, industrial and domestic) and three levels of country developments (developed, semi-developed and developing) based on national gross income. This yields nine emission factors, each with a specific BC/PPOC ratio, for each fuel category of BC and primary POC. In an alternative approach the exhaustive description of fuel combustion activities, technology divisions and emission controls (mostly for developed regions) has been preserved (Bond et al. 2004). But numerous assumptions had to be made for regions where combustion technologies and emission controls were poorly known.

The few existing BC and PPOC emission inventories show significantly different budgets and different regional partitioning of the emissions. Most of them do not account for all sources (industrial versus biomass burning), or are restricted to one component of the aerosols (BC or primary OC only). Furthermore, the use of either PPOC or POM (including the SOA fraction) is often confusing.

These disagreements persist in the recent comprehensive fossil fuel and biofuel (BC and PPOC) global emission inventories by Bond et al. (2004) (B04), Liousse and coworkers (Liousse et al. 2004; Junker and Liousse 2006) (L05) and Ito and Penner (2005) (I05) (Table 3.1). Differences exceeding a factor of two sometimes exist between the various estimates. The B04 and I05

Table 3.2 Black carbon (BC) emission factors for the main fossil-fuel-related sources

	Black carbon emission factor (g C per kg fuel)	
	(Bond et al. 2004)	(Junker and Liousse 2006)
Developed countries industrial, coal	0.001–0.006	0.149
Industrial, lignite	0.015–0.03	0.3
Traffic, diesel	0.85	2
Traffic, gasoline	0.04	0.03
Developing countries		
Industrial coal	0.28–4.5 (1.12)	1.10
Industrial lignite	0.09	1.98
Traffic diesel	2–7	10
Traffic gasoline	0.14–0.6	0.15

emissions are significantly lower than the L05 values, although new sources were added that had not been taken into account before (such as off-road vehicle emissions and waste combustion). An important source of disagreement between these two studies probably arises from the choice of significantly different emission factors (EFs) (Table 3.2). In most cases I05 used the B04 EFs, which supports differences in EFs as the main cause of emission discrepancies between the studies.

To assess the role of the choice of EFs in the discrepancy, a calculation was conducted for the year 1996, with the B04 and L05 sets of EFs, but using the same UN fuel database and the same methodology. A comparison of the results shows about 10-fold larger emissions from lignite burning, 6-fold larger emissions from diesel use, and about 25% greater emissions from hard coal burning when the L05 EFs are used instead of the B04 values. Recent combustion chamber experiments (not yet published) have shown that for diesel emissions, the EF(BC) value of 2 for developed countries used by Junker and Liousse (2006) would be in the highest part of the acceptable range, whereas for coal industrial combustion, B04's EF(BC) value of 0.0024 would be too low. There is a critical need for further targeted experiments to resolve the discrepancies between these two studies.

A way to test the validity of these inventories is to apply them in a global model and to compare the modeled values with existing BC and POM measurements. Reasonable agreement can be obtained for surface BC and PPOM concentrations using the L05 emissions implemented in the TM-3 model (3-D Transport Model, Velders et al. 1994). As the BC global burden estimated by B04 is almost half that of L05 it might be expected that comparisons between model and measurements would be less satisfactory. Further improvements may come from simultaneous refinements in bottom-up estimates and combination with top-down approaches using distributions of emitted substances detected by remote sensing, such as in the recent study by Ito and Penner (2005).

Estimation of the source strength of the predominantly carbonaceous CCN from biomass and fossil fuel burning entails large uncertainties. In the case of biomass burning, the emission ratio of CN (particles larger than ~10 nm) to CO is a relatively robust parameter, with a characteristic value of about $25\pm10\ cm^{-3}\ ppb^{-1}$ (Guyon et al. 2005), corresponding to about 2×10^{13} particles emitted per gram CO. Of these particles, the fraction that can be activated at 1% SS (supersaturation) is very high, with most studies showing CCN/CN ratios of 60–100% for fresh or aged biomass smoke (Radke et al. 1988; Hudson et al. 1991; Rogers et al. 1991; Novakov and Corrigan 1996; Ross et al. 2003). Assuming an annual global emission of 700 Tg of CO from biomass burning (Andreae and Merlet 2001) and a $CCN_{1\%}$/CN ratio of 0.8 yields an estimate of 1.1×10^{28} $CCN_{1\%}$ (1% SS) for the annual emission from biomass burning. An alternative calculation, based on $CCN_{1\%}$ with an emission factor of 2×10^{15} per kg dry biomass, yields similar results. Estimates of CCN production from in situ measurements yield production rates of 6×10^{10} to 8×10^{12} CCN per gram of material burned (Warner and Twomey 1967; Eagan et al. 1974b).

Comparable emission factors are not available for fossil fuel combustion. Bond et al. (2002) have shown that residential coal burning produces 12×10^{15} particles kg^{-1} for lignite (mostly >100 nm) and 14×10^{15} for bituminous coal (mostly <100 nm). Given that residential burning tends to have considerably larger emission factors than industrial or power related combustion, the overall CCN emission factors for fossil fuel burning are likely to be considerably lower. Carbonaceous aerosol emissions are projected to decline over the 21st century to about two-thirds of their present values by 2050, but most likely with a significant rise in the BC/OC ratio in industrialized regions (Streets et al. 2004).

3.2.3.3 Secondary Organic Aerosols

As mentioned before, organic aerosols are comprised of primary and secondary particles. In the previous paragraphs, we have only considered emission of primary particles (PPOC). The treatment of SOA in emission inventories of carbonaceous particles is still an open question. There is no evidence for the direct formation of SOA particles by nucleation of organics from the gas phase in the present-day atmosphere. SOA-rich particles have been shown to form by condensation of organic gases with low vapour pressures on pre-existing particles, which may be as small as freshly nucleated sulphate particles in the ultrafine size class (Kerminen et al. 2000; Kulmala et al. 2000; Kerminen 2001; O'Dowd et al. 2002; Lihavainen et al. 2003; Maria et al. 2004; Zhang et al. 2004a). The resulting particle may consist almost completely of organic material, and the presence of the initial sulphate nucleus may not be readily detectable any more. This is most likely the case in environments such as the Amazon Basin, where nucleation events have never been observed, in spite of high concentrations of volatile organic compounds (VOC) and high rates of photochemical oxidation (Rissler et al. 2004). In the Amazon significant amounts of primary biogenic and other particles act as an efficient sink for condensable organics. On the other hand, in the presence of high concentrations of SO_2 and a high rate of H_2SO_4 production, new particle formation and growth by incorporation of organics occurs even at a high concentration of pre-existing aerosols, such as in the cities of Milan and Pittsburg (Baltensperger et al. 2002; Zhang et al. 2004a). Once incorporated into the condensed phase, organics can be made less volatile by oxidation and polymerization reactions (Gelencsér et al. 2002; Limbeck et al. 2003; Claeys et al. 2004; Gao et al. 2004; Kalberer et al. 2004).

A number of studies has been conducted in recent years to assess the magnitude of SOA formation from biogenic and anthropogenic precursors (Liousse et al. 1996a; Griffin et al. 1999; Kanakidou et al. 2000; Chung and Seinfeld 2002; Kanakidou et al. 2005; Tsigaridis et al. 2005). In spite of these efforts, the range of estimates remains huge: 2.5–79 Tg yr^{-1} for biogenic SOA and 0.05–2.6 Tg yr^{-1} for SOA from anthropogenic precursors. Ignoring the more extreme results of sensitivity studies, we can use the values from the review of Kanakidou et al. (2005) as a currently best guess: 19, 15, and 2 Tg yr^{-1} from terpenes, oxygenated VOC (OVOC) and anthropogenic VOC, respectively. There is no published estimate of the number of CCN produced in the atmosphere as a result of SOA formation from biogenic or anthropogenic precursors. Recently Heald et al. (2005) pointed out a large discrepancy between model-predicted and observed concentrations of organic aerosols in the upper troposphere. They suggested that a substantial amount of SOA production, presumably from anthropogenic precursors, is required to explain the observed levels of organic aerosols in the free troposphere. If it is mixed downward, this aerosol could make a significant contribution to the POM concentration in the lower troposphere.

3.2.4 Other Primary Anthropogenic Aerosols (Industrial Dust etc.)

This category includes aerosols from transportation (e.g. tire and brake detritus, road dust etc.), coal combustion (fly ash etc.), cement manufacturing, metallurgical industries, and waste incineration, but excludes carbonaceous aerosols to avoid double counting. Primary anthropogenic aerosols are prominent in uncontrolled emissions from old industrial and energy technology, but are now being controlled fairly tightly, especially in developed countries. Growing industrialization, especially in Asia, was expected to lead to significant increases of primary anthropogenic aerosols in some regions. On the other hand, increased awareness of the negative health impacts of aerosols in developing countries, coupled with readily available emission control technology, may lead to less significant growth in these emissions than had been previously anticipated (Wolf and Hidy 1997). To our knowledge, there are no recently updated global emission estimates for this aerosol type, so that the very rough

Table 3.3 Primary particle emissions and burdens for the year 2000[a]

	Mass emission	Low	High	Mass burden	Number prod.	Number burden
	$Tg\ a^{-1}$			Tg	a^{-1}	
Carbonaceous aerosols						
Organic matter (0–2 μm)	95	40	150	1.2	–	$310 \cdot 10^{24}$
Biomass burning	54	26	70	–	$7 \cdot 10^{27}$	–
Fossil fuel	4	3	9	–	–	–
Biogenic	35	15	70	0.2	–	–
Black carbon (0–2 μm)	10	8	14	0.1	–	$270 \cdot 10^{24}$
Open burning & biofuel	6	5	7	–	–	–
Fossil fuel	4.5	3	6	–	–	–
Industrial dust etc.	100	40	130	1.1	–	–
Sea salt						
d<1 μm	180	60	500	3.5	$7.4 \cdot 10^{26}$	–
d=1–16 μm	9940	3000	20,000	12	$4.6 \cdot 10^{26}$	–
Total	10,130	3000	20,000	15	$1.2 \cdot 10^{27}$	$27 \cdot 10^{24}$
Mineral (soil) dust						
<1 μm	165	–	–	4.7	$4.1 \cdot 10^{25}$	–
1–2.5 μm	496	–	–	12.5	$9.6 \cdot 10^{25}$	–
2.5–10 μm	992	–	–	6	–	–
Total	1600	1000	2150	18±5	$1.4 \cdot 10^{26}$	$11 \cdot 10^{24}$

[a] Range reflects estimates reported in the literature. The actual range of uncertainty may encompass values both larger and smaller than those reported here. Values are based on the following publications: (Andreae 1995; Andreae and Merlet 2001; Guelle et al. 2001; Penner et al. 2001; Gong et al. 2002; Luo et al. 2003; Bond et al. 2004; Liousse et al. 2004; Zender et al. 2004; Ito and Penner 2005; Junker and Liousse 2006; Stier et al. 2005; Stier et al. 2006)

estimates given in IPCC-TAR had to be reused in Table 3.3. Furthermore, no data are available on which a reliable estimate of the number source flux for this particle type could be based. The mass size distribution of these aerosols is dominated by the coarse mode, but a substantial submicron mode is also present, which contains significant amounts of soluble material (Kleeman and Cass 1998). Most of the primary particles from coal combustion are below 1 μm in diameter (Chen et al. 2004). (A fraction of these particles is carbonaceous, and therefore is represented in the category of primary carbonaceous aerosols below.) In Beijing, for example, primary aerosols have been estimated to account for more than half of PM2.5 (Particulate Matter in the size class with diameters <2.5 μm) (Zhang et al. 2004b). We conclude that primary anthropogenic particles could represent a significant source of CCN (and possibly IN), especially in polluted regions of the developing world.

3.2.5 Sulphates

Knowledge of the temporal and spatial details of anthropogenic and natural sulphur emissions is vital for estimating the atmospheric aerosol loadings of sulphate aerosols. The substantial growth of industrialization and energy production since the 1950s has resulted in increased burning of fossil fuels, causing extensive emissions of SO_2. Population growth also has contributed to growing emissions. On the other hand, local air quality concerns, structural change in energy supply and end-use, and the application of advanced technology and cleaning procedures have resulted in dramatically reduced SO_2 emissions in some regions. Regional changes in emission pattern are additionally driven by a shift of strong polluting industry from the developed to the developing countries.

3.2.5.1 Sulphate Precursor Emissions

The most abundant sulphur gases in the atmosphere are carbonyl sulphide (COS), dimethyl sulphide (DMS), sulphur dioxide (SO_2) and hydrogen sulphide (H_2S). They are all subject to oxidation by a variety of atmospheric species, including hydroxyl and nitrogen radicals, ozone, and excited oxygen atoms, and to photochemical decomposition (Seinfeld and Pandis 1998). The ultimate fate of these atmospheric sulphur compounds is the irreversible oxidation to sulphate (SO_4^{2-}), which forms particles mixed with water and ammonium (i.e. H_2SO_4, NH_4HSO_4, $(NH_4)_2SO_4$). A small fraction of DMS oxidizes to methane sulphonic acid (MSA), which condenses on existing particles. These atmospheric sulphur compounds have a variety of natural and anthropogenic sources, with anthropogenic fluxes dominating at present and into the foreseeable future (Dentener et al. 2006). The most important natural sources of DMS and SO_2 are the marine biosphere and exhalation from volcanoes, respectively. Intra-annual variability and climate-induced changes of the source strength and distribution of natural sulphur sources are poorly known. Gabric et al. (2004) used model simulations of the period of equivalent CO_2 tripling (2080) to derive the changes in oceanic DMS production and flux to the atmosphere. They predicted a globally integrated DMS flux increase of 14%. The greatest perturbation to DMS flux is simulated at high latitudes in both hemispheres, with little change predicted in the tropics and sub-tropics. The largest change in annual integrated flux is simulated in the Southern Hemisphere between 50 and 60°S. At this latitude, the DMS flux perturbation is most influenced by the GCM-simulated changes in the mixed layer depth. This indicates that future increases in stratification in the polar oceans will play a critical role in the DMS cycle and therefore the flux of sulphate aerosols in remote ocean regions.

Major uncertainties exist for the volcanic sulphur emissions, and current estimates are probably too low (Textor et al. 2004). Volcanic emissions are however released at mountain peaks mostly above the planetary boundary layer, where they have extended life times and disproportionate effects on cloud properties.

The major anthropogenic sulphur source is SO_2 emissions from the burning of sulphur-containing fuels, mainly coal. Metal smelting and other industrial processes also release significant quantities of SO_2. Global anthropogenic sulphur emissions are estimated to range between 65 and 90 Tg S in 1990 (Houghton et al. 2001). Reviews of most recent inventories indicate a most likely value of 75±10 Tg S in 1990 (Smith et al. 2001), and 71 Tg S in the AEROCOM inventory for 2000 (Dentener et al. 2006). Smith et al. (2001) estimate that 56% of 1990 world sulphur emissions are from coal, 24% from oil, 15% from industrial processes and 3% from biomass burning. Emission from bunker fuel burned by the international cargo and passenger ship fleet represents a contribution of 4–7% to the global anthropogenic SO_2 emissions (3.0 Tg S in 1996 (Endresen et al. 2003); 3.7 in 2001 (Endresen et al. 2005). The majority of these ship emissions occur in the Northern Hemisphere.

Globally, emissions have been roughly constant from 1980 to the present. However, a significant shift has occurred in the spatial distribution of emissions. Many climate studies to date have used the emissions pattern from the GEIA inventory for 1985 and simply scaled these backwards and forwards in time. This is clearly an oversimplification, as the regional pattern of emissions has changed dramatically over the last 30 years (Massie et al. 2004). Different inventories are similar at the global-mean level, but show marked differences at the regional level.

While 60% of global emissions in 1980 were from around the North Atlantic basin, this region contributed <40% of the global total by 1995 and will contribute even less in the future (Smith et al. 2001). As a result of the implementation of international emissions agreements, SO_2 emissions decreased 32% from 1980 to 1990 in Europe, and 3% from 1980 to 1995 in North America. In 1990, the spatial pattern of emission shows that the US, the USSR, and China were the main sulphur dioxide emitters (i.e. approximately 50% of the total). Historical data for Asia are more problematic, but the trends are clearly in the other direction. The data indicate a roughly 60% increase in sulphur dioxide emissions from 1980 to 1990. The two main reasons for this are the region's rapid economic growth and its high dependence on coal. It has been estimated that China's SO_2 emissions grew 3.6% per year from 1985 through 1996, and India's SO_2 emissions grew at a rate of 4.9% per year during the 1990s (Streets et al. 2000; Streets and Waldhoff 2000). This trend is confirmed by satellite data (TOMS) of aerosol optical depth (AOD). Large increases in AOD between 1979 and 2000 are evident over the China coastal plain and the Ganges River basin in India. For instance AOD increased by 17% per decade during winter over the China coastal plain (Massie et al. 2004).

In contrast, Carmichael et al. (2002) report that since 1995 Asian SO_2 emissions have declined from 38.5 Tg in 1995 to 34.4 Tg in 2000, a decrease of 2.3% per year. This remarkable change is almost entirely due to a reduction in emissions in China, which emits about two-thirds of SO_2 from Asia. This has been brought about by several factors: a marked reduction in industrial coal use from the closure of small and inefficient plants, a slow-down in the Chinese economy, the improved efficiency of energy use, the closure of some high-sulphur coal mines, a general reform in industry and power generation, and a rising awareness of the dangers of air pollution.

3.2.5.2 Transformation and Sulphate Particle Formation

When emitted into the atmosphere, sulphur species undergo transport and chemical reactions to form compounds that condense on aerosol surfaces or nucleate. Wet deposition removes gases and particles in the atmosphere and deposits them on the Earth's surface by means of precipitation. With dry deposition, particles and gases reach land and water surfaces without precipitation. All the processes controlling dispersion, residence time in the atmosphere, aerosol load and composition vary with weather and climate. Since sulphur emissions themselves influence climate, and so change the transport and sink processes, one cannot accurately model the influences of anthropogenic emissions unless the climate influences are also modeled in a consistent way. While this clearly requires an appropriate coupled climate-sulphur-chemistry model, the primary driving force must be reliable, spatially-resolved SO_2 emissions (Smith et al. 2001).

The percentage of SO_2 transformed into sulphate, according to simulations with global chemistry models, falls between 51 and 56% (four models) with one outlier calculating an efficiency of 83% (due to low dry deposition flux of sulphur dioxide) (Berglen et al. 2004). The COSAM (Comparison of Large Scale Sulphate Aerosol Models) model comparison (Barrie et al. 2001; Roelofs et al. 2001) reflecting the state of modeling in 1999 and the AEROCOM (Textor et al. 2004; Kinne et al. 2006) comparison performed in 2004 give a range of transformation rates between 45 and 67%, and 50–81%, respectively. This transformation rate depends on transport, sink processes and chemical transformation. The latter will change due to changes in the emission of oxidants and oxidant precursors. Only gas phase oxidation by OH will lead to formation of new particles, whereas aqueous phase oxidation and catalytic oxidation by metals take place on existing particles, thereby changing their CCN properties. Gas phase sulphate formation contributes between 10 and 22% to the total oxidation of SO_2 in the COSAM comparison and between 14 and 42% in the more recent AEROCOM comparison. In addition, a southward shift of the sulphur emissions toward regions with more incoming solar radiation and hence more OH would increase the fraction of SO_2 oxidized in the gaseous phase (Berglen et al. 2004).

Secondary particle formation from gas-phase sulphuric acid dominates the nucleation mode and contributes significantly to the particle number concentrations although contribution to total aerosol mass is small. Model calculations show that nucleation of sulphate particles is favored in regions with little available aerosol surface area, low temperatures, and high relative humidity. Thus, the maxima of the nucleation mode number concentration can be found in the upper tropical troposphere and in the remote regions of the Antarctic.

The soluble Aitken mode numbers are dominated by particles growing from the nucleation in the Aitken size range. Accumulation mode soluble numbers are highest in the lower troposphere, between 30°S and 60°N, close to the sources from biomass burning and fossil fuel use. Increased levels can also be found in the upper troposphere, attributable to convective detrainment of particles and their precursors, and evaporation of cloud droplets and ice crystals. Figure 3.1 shows the effect of the omission of SO_2 emissions from fossil fuel and biofuel use on the accumulation mode number concentration. Between 0 and 60°N sulphur dioxide from anthropogenic sources contributes about 25–30% of the total particle number burden.

Conventionally, additivity is assumed for aerosol emissions and for their climate impacts. However, due to the complex interactions between transport, aerosol and hydrological cycle, one cannot expect a linear relationship between emissions and aerosol load or aerosol effects. Graf et al. (1997) have shown that the efficiency by which sulphur precursors are transformed to sulphate depends on the source type. This efficiency depends additionally on the location and the season of the release. Barth and Church (1999) used the NCAR Community Climate Model (CCM3) to determine the contributions of southeast China to the global aerosol burden. Southeast China emitted 11.6% of the global anthropogenic sulphur emissions and contributed 9% to the global sulphate burden in their simulations, indicating a non-linear response between emission and burden. When anthropogenic sulphur emissions were doubled, the aerosol burden contributed by China increased by a factor of 2.2. Their simulations show a gas-phase oxidation rate from south-east-Asian emissions above average. Sulphate formed via this pathway is less susceptible to wet scavenging and thus contributes more efficiently to the atmospheric sulphate load than sulphate formed in clouds, which is already incorporated in cloud droplets and thus is more likely to undergo wet removal.

Stier et al. (2006) analyzed the response of the global aerosol system to changes in anthropogenic sulphur emissions. In a scenario without anthropogenic sulphur emissions (corresponding to a reduction by 55%), the sulphate burden decreases globally by almost 50% and the accumulation mode number concentration by 21%. However, not only is the sulphate budget affected, but the annual global-mean lifetimes of black carbon, particulate organic matter, and dust increases by 8.9%, 2.5%, and 1.5%, microphysical aging times increase by 164%, 84%, and 66%, and annual zonal-mean column burdens in high northern latitudes increase by up to 70%, 20%, and 20%, respectively. This emphasizes that changes in one particular anthropogenic emission have impacts on the cycles of other aerosol components, resulting in changes of their lifetimes, column burdens, and microphysical aging times. These results suggest that the separate treatment of the different aerosol components and their precursors can lead to non-negligible errors in estimates of the aerosol climate effects.

Concerning future emissions of sulphur, the SRES (IPCC Special Report on Emissions Scenarios) scenarios portray similar emission dynamics – at various future dates (between 2020–2030 and 2070, depending on the scenario and its underlying storyline), global SO_2 emissions reach a maximum level and decline thereafter. By 2030 sulphur emissions range between 40 and 160 Tg S, by 2070 between 20 and 165 Tg S, and by 2100 between 10 and 95 Tg S. Emission trajectories of the SRES scenarios reflect a combined impact of different scenario driving forces (local air quality concerns, structural change in energy supply and end-use etc.), which lead to a gradual decline in sulphur emissions in the second half of the 21st century.

3.2.6 Nitrates

The production of secondary aerosol nitrate is closely tied to the relative abundances of ammonium and sulphate. If ammonia is available in excess of the amount required to neutralize the stronger acid H_2SO_4, gaseous HNO_3 and NH_3 can enter the condensed phase, and their subsequent dissociation yields nitrate (NO_3^-) and ammonium (NH_4^+) ions. Submicron particles containing nitrate are potentially efficient CCN (Metzger et al. 2002a). In the presence of acidic accumulation-mode sulphuric-acid containing aerosols, however, HNO_3 deposits on larger, alkaline mineral or salt particles (Dentener et al. 1996; Murphy and Thomson 1997; Gard et al. 1998). This deposition of soluble material on otherwise insoluble dust particles increases their CCN activity.

Until recently nitrate has not been considered in assessments of the climatic effects of aerosols, and even current models often ignore the role of nitrate (e.g. Stier et al. 2005) or the sub-micron nitrate aerosol fraction (Derwent et al. 2003). Andreae (1995) estimated that the global burden of ammonium nitrate aerosol from natural and anthropogenic sources is 0.24 and 0.4 Tg (as NH_4NO_3), respectively, and that anthropogenic nitrates cause only 2% of the total direct forcing. Adams et al. (1999) obtained a value of only 0.17 Tg (as NO_3^-) for the global nitrate burden, which may be due to the fact that their model did not include nitrate deposition on sea salt aerosols. Their most recent estimate (Adams et al. 2001), predicts a nitrate burden of 0.38 Tg in the form of secondary ammonium nitrate/sulphate particles.

The importance of aerosol nitrate is likely to increase substantially over the next century. For example, the SRES A2 emissions scenario projects that NO_x emissions will more than triple in that time period while SO_2 emissions decline slightly. Using this scenario, Adams et al. (2001) predict a nitrate burden of 1.8 Tg for 2100, compared to a sulphate burden of 2.2 Tg for the same year. In their model run for 2100, the total anthropogenic forcing (direct and indirect) associated with nitrate actually exceeds that caused by sulphate.

3.2.7 Aerosol Number Fluxes

The calculation of the aerosol impacts on clouds, precipitation and hence on climate, and their interaction with the hydrological cycle requires size-resolved knowledge of the global distribution of aerosol number concentrations. Because primary aerosol emissions result in a direct increase of the atmospheric particle number burden, whereas secondary aerosol production often results only in the addition of mass to existing particles, the fluxes of primary aerosols play a key role in the context of cloud physical effects (Adams and Seinfeld 2003). Recently, the aerosol size distribution has been added to the prognostic variable space of global microphysical aerosol models to account for these effects. In addition to the emission mass fluxes, these models require as input the primary aerosol number fluxes and therefore the knowledge of the emission size distribution.

Natural sea salt and mineral dust aerosols are predominantly emitted in the accumulation and coarse mode size range, and their size distribution is mainly altered by size-dependent sink processes. A number of size-resolved emission parameterizations are available for sea salt and mineral dust (see above), and can be interactively calculated in numerical models. These parameterizations can also be used in future scenarios.

However, for primary emissions from combustion processes, the initial evolution of the size distribution is characterized by small-scale microphysical processes (Jacobson and Seinfeld 2004) that cannot be resolved in regional and global aerosol models. Therefore, the sub-grid scale evolution of the size-distribution and mixing state shortly after emission should be parameterized in a way that is appropriate to the model scale. Up to now, such parameterizations are not available and the emission size-distribution is generally estimated from measurements made on "slightly" aged plumes. For the AEROCOM aerosol model inter-comparison, an emission inventory has been provided (Dentener et al. 2006), based on mass flux estimates by Bond et al. (2004) for carbonaceous aerosols from fossil- and bio-fuel use, by Van der Werf et al. (2003) for vegetation fires, and by Cofala et al. (2005) for anthropogenic sulphur emissions (data and documentation available from http://ftp.ei.jrc.it/pub/Aerocom/). AEROCOM assumes that 2.5% of the total SO_2 emissions are emitted in the form of primary sulphate to account for sub-grid scale sulphate formation at the sources. In addition, source-specific size distributions are proposed (Table 3.4) for freshly emitted primary aerosols, which are then subject to further microphysical processing in global aerosol models. For comparison, the parameters of a log-normal fit (Stier et al. 2006) of natural sea salt (Schulz et al. 2004) and mineral dust (Tegen et al. 2002; Tegen et al. 2004) as used in the emission parameterizations, neglecting the super-coarse mode emissions are also given in Table 3.4.

The total emission number flux F_N of primary particles can be calculated from a given emission mass flux F_M and a log-normal emission size-distribution as follows:

$$F_N = F_M \frac{3}{4\pi\rho\left(r_{NMR}\exp(1.5\ln^2(\sigma))\right)^3} \tag{3.1}$$

Table 3.4 Global annual anthropogenic primary aerosol number fluxes as recommended by the AEROCOM emission inventory. For comparison, values for the natural emissions of sea salt (Schulz et al. 2004) and mineral dust (Tegen et al. 2002; Tegen et al. 2004) are also given

	Number median diameter (μm)	Standard deviation	Total mass flux (Tg a^{-1})	Total primary number Flux (a^{-1})
Fossil-fuel burning (BC + POM)	0.030	1.8	6.4[a]	4.8×10^{28}
Bio-fuel burning (BC + POM)	0.08	1.8	10.7[a]	4.2×10^{27}
Vegetation fires (BC + POM)	0.08	1.8	37.7[b]	1.5×10^{28}
Primary sulphate (H_2SO_4)	1.0	2.0	4.3[c]	5.1×10^{23}
Sea salt (accumulation mode)	≈0.4 varying with wind-speed	1.6	54.1	7.4×10^{26}
Sea salt (coarse mode)	≈4.4 varying with wind-speed	2.0	4955	4.6×10^{26}
Mineral dust (accumulation mode)	0.4	1.6	7.5	4.1×10^{25}
Mineral dust (coarse mode)	0.8	2.0	655	9.6×10^{25}

[a] Bond et al. 2004.
[b] van der Werf et al. 2003.
[c] Industry, fossil- and bio-fuels from Cofala et al. (2005) and vegetation fires from van der Werf et al. (2003) with 2.5% of SO_2 emitted as primary sulphate.

where ρ is the density of the compound and r_{NMR} the number median radius and σ the standard deviation of the log-normal emission size-distribution.

Based on the AEROCOM anthropogenic emission inventory and the proposed emission size distributions, the total anthropogenic primary aerosol number flux can be estimated as listed in Table 3.4. This flux is highly dependent on the choice of the emission size distribution. For example, the usage of the same emission mass flux for primary sulphate, but with the significantly smaller emission size-distributions used by Adams and Seinfeld (2003) results in a 6×10^4 larger number flux than using the large "fly ash" AEROCOM size distribution recommendation. This underlines the necessity of adequate emission size distributions and for careful adaptation to the respective model scales.

3.3 Physical and Chemical Nature of Atmospheric Aerosols

Knowledge of the various primary and secondary aerosols sources, combined with the size and composition dependent aerosol sinks, in principle allows prediction of the chemical and physical characteristics of the atmospheric aerosol. The properties most relevant to the impact of aerosols on precipitation are the size distribution and composition. They determine the interaction of the aerosol with water or, more specifically, their activity as either CCN or IN.

Before delving into these aerosols attributes, the concept of different aerosol regimes will be discussed. We shall use a series of aerosol "regimes", as may be found at various locales around the Earth: over the remote oceans and continents far from pollution sources ("remote marine" and "remote continental", respectively), over regions influenced by industrial pollution or biomass burning ("polluted continental" and "polluted marine"), and over areas dominated by mineral dust aerosols. Earlier classifications have attempted to separate the atmosphere into "marine" (equated with low pollutant levels and low CCN concentrations) and "continental" (high pollutant levels and high CCN concentrations) regimes. This view has been challenged, as it has become apparent that unpolluted continental regions may be just as clean, and have just as low CCN concentrations as remote marine areas (Squires and Twomey 1966; Hoppel et al. 1973; Twomey et al. 1978; Delene and Deshler 2001; Komppula et al. 2005; Roberts et al. 2001; Andreae et al. 2004). Conversely, polluted marine areas have aerosol concentrations just as high as polluted continental regions (Hudson and Xie 1999; Andreae et al. 2000; Raes et al. 2000; Hudson and Yum 2002). In this chapter, we will limit the discussion of the nature of aerosols to an analysis of generic properties within broad generic regimes; a further breakdown into regional characteristics is provided in Chapter 4.

3.3.1 Size Distributions

While many different types of distribution functions have been utilized over the years to represent the atmospheric aerosol, the discussion here is in terms of the ubiquitous lognormal function introduced and popularized by Junge and later Whitby and his colleagues (cf. Junge 1953; Junge 1963; Whitby et al. 1972; Whitby 1978). The distribution function itself can be expressed in terms of any of its moments, the most useful representation being dependent on the particular issue being addressed. For example, for exploring the dependence of cloud drop number concentration on aerosol concentration, the number size distribution (Fig. 3.1) has the most utility, whereas if one is considering the chemical nature of the aerosol, the mass distribution is commonly employed.

3.3.1.1 The Remote Marine Aerosol

This category includes both the sea salt aerosol and particulate oxidation products of biogenic DMS (Raemdonck et al. 1986; Charlson et al. 1987; Andreae and Crutzen 1997; Bates et al. 1998; Andreae et al. 1999; Bates et al. 2000; Bates et al. 2001). While there is substantial variability (cf. Heintzenberg and Covert 2000), the most salient features are ubiquitous. Perhaps most characteristic is the prominent mode in the number distribution in the size range from ~0.1–0.2 μm diameter, which contains most of the DMS-derived non-sea salt sulphate (the sulphate which is present in marine aerosols in excess of what originates from the sulphate in sea salt). This mode corresponds to the accumulation mode and is virtually global in extent (e.g. Kim et al. 1995; Quinn et al. 1995; Weber et al. 1998; Bates et al. 2002). Hoppel and his colleagues have argued persuasively that the characteristics of this mode are attributable to cloud processing of aerosol (Hoppel et al. 1986; Hoppel et al. 1990; Hoppel et al. 1994). A second smaller diameter mode commonly occurs ~30–50 nm, particularly in the most remote locales. It is most frequently referred to as the *aged-nucleation* or *Aitken mode*, and is thought to arise from in situ particle nucleation associated with gas to particle conversion under conditions of low pre-existing aerosol surface area (Wiedensohler and Covert 1996; Brechtel et al. 1998; Covert et al. 1998; Collins et al. 2000). The concentration of particles in these submicron modes ranges from a few tens per cm^3 to several hundred per cm^3, with concentrations in the Northern Hemispheric oceans generally (though not always) appreciably higher than those in the Southern Hemisphere oceans. Finally, a larger, or coarse, size mode, the so-called sea salt mode, is nearly always present (~0.5–3 μm diameter). While few in

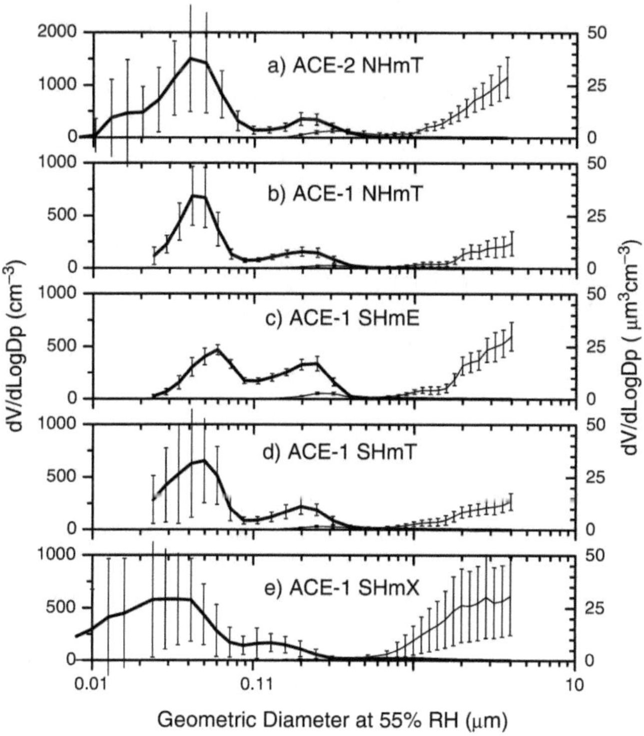

Fig. 3.2 Number (heavy line) and volume (light line) distributions at 55% RH for different air masses of origin during the first 2 ACE experiments. The vertical bars represent one standard deviation in mean number or volume in the size bin over the averaging period (23–114 hr). From Bates et al. (2002) with permission of the American Geophysical Union

number, the contribution of this mode to particle mass – and thus the volume distribution – is always very substantial and commonly dominant (Bates et al. 2002; Maring et al. 2003). These particles are not only sea salt, but also mineral dust that can be transported over very long distances and can have a marked presence even in remote marine air (e.g. Andreae et al. 1986; Arimoto et al. 1997). Non-sea salt sulphate is also always present in variable amounts in the coarse mode of the marine aerosol (Andreae et al. 1999; Andreae et al. 2000). Furthermore, the impact of even small numbers of large particles to marine precipitation can be quite substantial (e.g. Feingold et al. 1999). Some representative size distributions, taken from Bates et al. (2002), are shown in Fig. 3.2 and illustrate the features just discussed.

3.3.1.2 The Remote Continental Aerosol

Because of the vast amounts of anthropogenic emissions of aerosol particles and their gaseous precursors, combined with efficient long-range transport, it is nowadays very difficult to find areas of the Earth that are not measurably impacted by pollutant aerosols. This applies especially to the continental regions of the northern hemisphere, where most of the human activities are concentrated.

Aerosol particles have typical atmospheric lifetimes of 3–10 days, which implies that even after 10–30 days (3 lifetimes) about 5% of the initial burden is still left. Given that airmasses can easily travel several thousand km in 15 days, and that, at least in the northern hemisphere, few places are more than a few thousand km from major pollution sources, there are really no places where we can expect to find pristine conditions in the northern hemisphere (see Chapter 4). The remote continental aerosol nowadays consists of some natural material (dust, biogenic materials etc.), mixed with pollution aerosols at varying levels of dilution. For example, a study of the European aerosol climatology shows that the average PM2.5 concentration ranges from some 30 $\mu g\ m^{-3}$ at urban sites to about 4 $\mu g\ m^{-3}$ at the cleanest site, Sevettijärvi in northern Finland (Putaud et al. 2004; Van Dingenen et al. 2004). The composition of the aerosol at this remote site is, however, very similar to that at urban and regionally polluted sites, with a large fraction of non-sea salt sulphate, and about 4% of the combustion tracer black carbon. Pollution aerosols dominate even in most continental areas in the Southern Hemisphere, especially in the dry seasons, when biomass burning is widespread (Artaxo et al. 1988; Artaxo et al. 1994; Artaxo et al. 2002; Sinha et al. 2003).

Only under relatively rare circumstances do aerosol concentrations over the remote continents approach pristine conditions. This usually occurs when clean marine airmasses are advected over practically uninhabited continental regions. Examples are the rainy season in central Amazonia, some of the desert regions in Australia and parts of the boreal regions in Eurasia and North America. Unfortunately, there are very few studies that document these conditions carefully enough to be useful for a assessment of what aerosol concentrations (and composition) would be like in the absence of anthropogenic emissions. Size distributions from a variety of relatively unpolluted venues, such as rural Canada (Leaitch and Isaac 1991), the Southwestern U.S. and rural Midwestern U.S. (Kim et al. 1993a), and the Amazon Basin (Guyon et al. 2003a; Rissler et al. 2004; Krejci et al. 2005) and the high altitude site Jungfraujoch (3580 m asl) (Weingartner et al. 1999) are differentiable from more strongly polluted air and share some common characteristics. The mass accumulation mode of the remote continental aerosol tends to center in the size range from 0.1 to 0.4 μm diameter, thus showing more size variability than the remote marine aerosol, particularly on the upper bound, though tending to a slightly smaller modal size. In number concentration, the remote continental aerosol also shows more variability, but with concentrations usually appreciably higher than those of its marine counterpart, typically from a few hundred to a few thousand particles per cm^3 (Weingartner et al. 1999; Putaud et al. 2004; Van Dingenen et al. 2004). This difference is the result of two factors – the presence of pollution aerosols and the frequent occurrence of large particle number concentrations in the nucleation and Aitken modes. In the few instances when nearly pristine conditions could be sampled, number concentrations in the continental tropics (Roberts et al. 2001; Rissler et al. 2004; Krejci et al. 2005) as well as in the lower free troposphere during winter (Weingartner et al. 1999) were in the low hundreds, indistinguishable from marine values. From vegetated continental areas at northern mid-to high latitudes, no published estimates of pristine levels exist, but concentrations are estimated to be in the tens to low hundreds per cm^3 (M. Kulmala, personal communication 2005).

Besides the accumulation mode, separate Aitken or nucleation modes are frequently present, especially when the surface area of pre-existing aerosol is low and when small, but significant concentrations of gaseous H_2SO_4 are produced from SO_2 oxidation (Kulmala et al. 2000; Lihavainen et al. 2003). Despite their high number concentrations, these particles are of little direct consequence for precipitation processes, as they do not have enough soluble mass to act as CCN. Only when they have grown to sizes approaching 100 nm, e.g. by condensation of terpene oxidation products (Leaitch et al. 1999; Kerminen et al. 2000), are they able to nucleate cloud droplets.

The supermicron, or coarse mode, of the remote continental aerosol is commonly centered in the range from 2 to 5 μm diameter (well away from desert sources of dust). The number concentration in this mode is more variable than its marine counterpart, but generally significantly lower, with typical concentrations in the range from 0.1 to 1 cm^{-3}, as compared to 1 to 10 cm^{-3} in marine air. It consists of mineral and soil dust and a variety of primary biogenic particles, and dominates the aerosol mass size distribution under the cleanest conditions (Artaxo et al. 2002; Guyon et al. 2003b).

Because of the contrast between the size distributions of continental and marine remote aerosols, it must be noted that even the simple advection of continental air into the marine venue will alter the marine size distribution and chemical composition. This can be clearly seen in Fig. 3.3, taken from

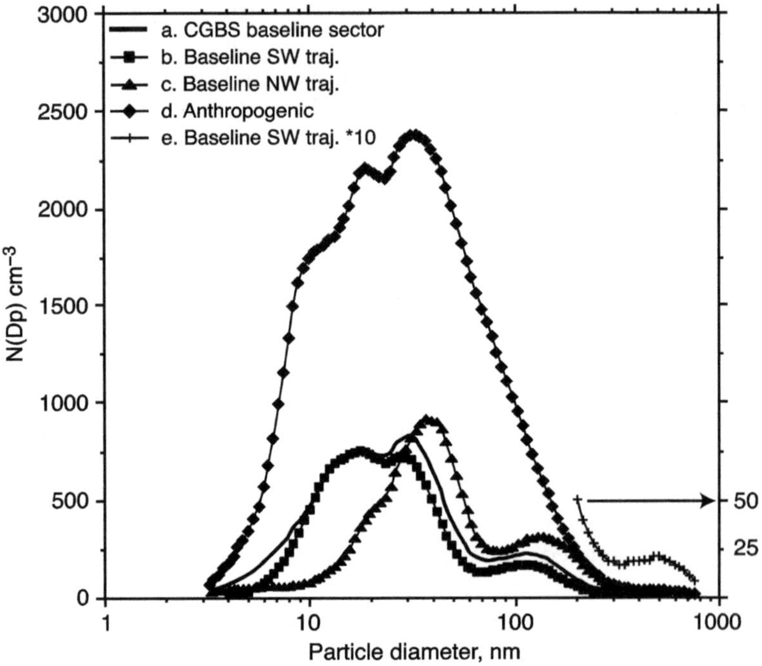

Fig. 3.3 Differential aerosol number size distributions from Cape Grim during ACE 1. CGBS in the legend refers to Cape Grim Baseline Station. Curve e is an expanded scale for curve b to show an additional mode above 200 nm. From Covert et al. (1998) with permission of the American Geophysical Union

Covert et al. (1998), which shows such impact for the Southern Ocean aerosol. For the submicron portion of the size distribution, the bi-modal structure characteristic of clean marine air is overwhelmed by a massive modal peak centered at 40–50 nm diameter. Similar changes have been reported for both the North and South Atlantic (Hoppel et al. 1990; Kim et al. 1995; Bates et al. 2000; Collins et al. 2000; Johnson et al. 2000), the Indian Ocean (Cantrell et al. 2000), the Pacific Ocean, (Moore et al. 2003), as well as by the review by Bates et al. (2002). For the coarse particle mode, advection of pollution generally has little effect unless soil dust is present (cf. Li-Jones and Prospero 1998; cf. Bates et al. 2002; Maring et al. 2003; Moore et al. 2003).

3.3.1.3 The Polluted Continental Aerosol

Much of the early work on industrial air pollution was done in the highly polluted Los Angeles Basin. Studies there show typically a tri-modal number size distribution with prominent nucleation, accumulation and coarse modes, and associated modal diameters of 0.01–0.02, 0.1 and 1.0 μm, respectively (e.g. Heisler et al. 1973; Sverdrup et al. 1975; Whitby et al. 1975). Later work confirmed this basic structure but showed that the nucleation and accumulation modes are not always fully resolved (e.g. John et al. 1990; Hering et al. 1997; Hughes et al. 1999). In Europe both a bi-modal submicron structure and a prominent, though variable, coarse mode have been commonly found (e.g. Birmili et al. 2001; Baltensperger et al. 2002; Petzold et al. 2002). Number concentrations vary in the range from thousands to tens of thousands per cm^3. The partition in number concentration between the nucleation and accumulation modes is highly variable, with the larger concentration switching back and forth between the two as a function of time of day, as well as season and various other variables.

3.3.1.4 Biomass Burning Aerosol

Besides industrial pollution, biomass burning emissions are the other main pollution source on a global impact scale, and occur mostly from the tropical regions (Crutzen and Andreae 1990), especially Brazil and Southern Africa. Andreae et al. (1996), summarizing a number of studies in Africa, report a large accumulation in the number size distribution at ~0.1–0.2 μm diameter for in-plume measurements. Regional haze associated with biomass burning was shifted to slightly larger sizes. There is also a coarse mode in both the plume and haze aerosol, which is composed to a large extent of resuspended soil dust, with some additional coarse ash particles (Reid et al. 2005). The number modal diameter is not clear, in part because volume rather than number distributions are reported. Nevertheless, it would appear to be at least 1–2 μm in size. Le Canut et al. (1996) also report an accumulation mode centered at ~0.15 μm and a coarse mode centered at ~3 μm. More recently, from the SAFARI-2000 campaign, Haywood et al. (2003a,b) reported an accumulation mode at a slightly larger volume modal diameter of 0.25 μm with a coarse mode at ~10 μm. In none of these studies was a distinguishable Aitken or nucleation mode reported.

Data on Brazilian biomass burning aerosol has recently increased sharply as a result of the SCAR-A and SCAR-B studies (Reid and Hobbs 1998; Reid et al. 1998) and the Large Scale Biosphere-Atmosphere Experiment in Amazonia – Smoke, Aerosols, Clouds, Rainfall and Climate (LBA-SMOCC) (Andreae et al. 2004; Guyon et al. 2005). Reid et al. (1998) have summarized work on both distinct fire plumes and the regional haze produced by such plumes. As in African emissions, there is no distinct nucleation mode in distinct plumes, but rather a single large accumulation number mode centered at 0.10–0.13 μm diameter, and a volume mode around 0.2–0.3 μm. In contrast, the regional haze associated with biomass burning showed an accumulation mode centered in the range 0.2–0.3 μm. In the fresh smoke, a coarse mode at ~3 μm was evident, while in the regional haze a significant coarse mode was not commonly present. Airborne measurements of fresh smoke from deforestation fires during the LBA-SMOCC campaign showed a single number mode near 0.10 μm, while in detrained smoke from pyrocumulus clouds the modal diameter had grown to

~0.13 μm (Guyon et al. 2005). At a ground site, where a mixture of fresh and aged smoke was present, the distribution was bimodal, with number modal diameters of 0.09±0.01 and 0.18±0.02 μm (Rissler et al. 2006). A nucleation mode with a diameter near 10 nm appeared sporadically.

Irrespective of locale, the submicron aerosol of tropical biomass burning is usually present as a large accumulation mode with number median diameters in the range of 0.1 to 0.2 μm and volume median diameters of 0.2–0.3 μm (Reid et al. 2005), and with a distinct coarse mass mode centered around 4 μm diameter (e.g. Radke et al. 1991). After some appreciable aging, the number mode shifts up to 0.2–0.3 μm. The coarse mode diminishes in aging smoke and is typically centered at a few micrometers in size. These characteristics contrast to some extent with the other major source of pollution, namely industrial emissions, where the accumulation mode and coarse mode are at somewhat smaller modal diameters and a distinct nucleation mode is frequently encountered.

3.3.2 Size-Dependent Chemical Composition and Solubility

3.3.2.1 The Marine Aerosol

The chemical composition of marine aerosols varies substantially with geographic locale. A review by Heintzenberg et al. (2000) found that anthropogenic pollution is ubiquitous over the oceans of the Northern Hemisphere due to long-range transport. The composition of the undisturbed remote marine aerosol is most easily assessed in data from the Southern Hemisphere. Sea salt is the single most common component of the aerosol, with latitudinal maxima in both number concentration and mass in the mid-latitudes, where strong winds associated with baroclinic systems predominate. The number size distribution of the sea salt aerosol extends well down into the sub-micron range and indeed commonly peaks there (McInnes et al. 1997; O'Dowd et al. 1997; Kreidenweis et al. 1998; Murphy et al. 1998a; Campuzano-Jost et al. 2003). It had long been thought that most remote marine sub-micron particles are primarily comprised of sulphate derived from the oxidation of dimethyl sulphide (DMS) emitted from the ocean surface and SO_2 from long-range transport of continental sources and ship emissions (Twomey 1971; Charlson et al. 1987; Andreae et al. 1988; Andreae 1990; McInnes et al. 1997; Andreae et al. 2000). However, numerous recent studies have revealed a substantial organic presence, even in remote marine air (Middlebrook et al. 1998; Murphy et al. 1998b), both in coarse mode and accumulation mode particles (McInnes et al. 1997; Matthias-Maser et al. 1999; Guazzotti et al. 2001; Mayol-Bracero et al. 2001; Cavalli et al. 2004; O'Dowd et al. 2004). The speciation of the organics is still not entirely resolved, but there are clearly substantial amounts of both soluble and insoluble carbon present.

While sea salt contributions to the marine aerosol mass are governed primarily by wind speed (Lewis and Schwartz 2004) (see Chapter 4), the contribution of organics and inorganics such as sulphate and nitrate are strongly a function of geography, induced largely by variability in the proportions of biogenic and anthropogenic emissions. The loadings of sulphate and carbon species increase dramatically in polluted marine air (Cachier et al. 1983; Rau and Khalil 1993; Andreae et al. 2000; Heintzenberg and Covert 2000; Putaud et al. 2000; Guazzotti et al. 2001). However, the dominant species depends on the source of the pollution. For example, pollution from either Europe or North America advected over the North Atlantic increases sulphate much more than the various carbon species, resulting in a marked increase in the inorganic mass fraction (Andreae et al. 2000; Putaud et al. 2000). In contrast, pollution from India transported over the Indian Ocean results in more organics and especially black carbon (Cantrell et al. 2000; Chowdhury et al. 2001; Mayol-Bracero et al. 2002a). While both the carbon and sulphate species are predominately sub-micron in size, the elemental and other insoluble carbon species tend especially to be disproportionately concentrated at the smallest sizes (Sicre et al. 1990; Rivera-Carpio et al. 1996; Neusüβ et al. 2000). This is of considerable interest since a number of these organic species act as surfactants, forming layers on the aerosol surface that might alter their CCN and IN properties (Facchini et al. 2000; Charlson et al. 2001; Feingold and Chuang 2002). Indeed, numerous surfactants, and even surfactant layers, have been found on marine aerosols (Barger and Garrett 1976; Gagosian et al. 1982; Mochida et al. 2002; Tervahattu et al. 2002a).

3.3.2.2 The Remote Continental Aerosol

Although remote continental aerosols often have a considerable contribution from soil dust (Jaenicke 1993; Eleftheriadis and Colbeck 2001), a substantial portion of the mass, particularly the sub-micron mass, is composed of sulphate and organic matter (Malm et al. 1994; Shrestha et al. 1997; Andreae et al. 2002; Henning et al. 2003; Putaud et al. 2004). While there are land sources of natural sulphur compounds (e.g. volcanoes and terrestrial biota), their magnitude is rather small compared to anthropogenic sulphur emissions (Penner et al. 2001). Small but significant amounts of nitrate are present even in remote locales, occasionally in excess of what would be expected from natural sources (Talbot et al. 1988; Talbot et al. 1990; Malm et al. 1994; Putaud et al. 2004). Much of the sulphate found in remote continental aerosols is likely of anthropogenic origin. The other major component of the remote continental aerosol are the carbonaceous compounds, mostly composed of organic matter (e.g. Artaxo et al. 1988; Talbot et al. 1988; Artaxo et al. 1990; Talbot et al. 1990; e.g. Malm et al. 1994; Zappoli et al. 1999; Krivacsy et al. 2001; Artaxo et al. 2002; Guyon et al. 2003b; Putaud et al. 2004; Kanakidou et al. 2005). This includes both primary biogenic aerosols and secondary organic aerosol from biogenic precursors. Indeed, in the most remote locales, the organic fraction can be as large or larger than that of sulphate (Artaxo et al. 2002; Graham et al. 2003a; Graham et al. 2003b;

Guyon et al. 2003b). There is a definite trend towards a larger soluble organic fraction as the aerosol approaches pristine conditions (Mazurek et al. 1997; Zappoli et al. 1999; Graham et al. 2003a). While this might well be in part due to evolution of industrial pollution, several studies have shown that the natural organics are qualitatively different from polluted air (Mazurek et al. 1997; Graham et al. 2003a). This aspect of the remote continental aerosol composition is of considerable importance when considering CCN activity. The few existing studies show,that the aerosol from the least polluted sites in the Amazon Basin, for example, is very efficient at nucleating cloud droplets (Roberts et al. 2001; Andreae et al. 2004; Rissler et al. 2004).

3.3.2.3 The Polluted Continental Aerosol

Polluted continental aerosols are by far the most widely studied. In industrial pollution, the most clear-cut trend with higher pollution levels is the absolute and relative increase in the abundance of black carbon (Malm et al. 1994; Zappoli et al. 1999; Neusüβ et al. 2002; Putaud et al. 2004), although the tendency is not universal (Molnar et al. 1999). Organic carbon also increases substantially in absolute and relative terms but the latter trend is much less clear than for the black carbon. In part, this variability may be due to the fact that organic aerosol has a distinct biogenic source (primary particles and SOA) which produces a significant natural background, which is not the case for BC. The inorganic portion of the aerosol mass is dominated by sulphate and nitrate, accompanied by an extensive suite of metal cations (Putaud et al. 2004).

The mass size distribution of the organics and inorganic in aerosols is also of interest. In accord with the changes in the number size distribution discussed in the previous section, most of the mass of industrial aerosol is confined to the sub-micron size range, for both organic and inorganic species. However, the carbon species, and particularly elemental carbon, tend to be disproportionately concentrated at the smaller sizes examined, especially below ~0.1 μm diameter (Hering et al. 1997; Maenhaut et al. 2002; Neusüβ et al. 2002; Jimenez et al. 2003). Additionally there is a tendency for the insoluble organics to be at smaller sizes than the soluble species (cf. Neusüβ et al. 2000). However, it should be noted that most of the particles, even in urban source areas, are largely internally mixed (e.g. Noble and Prather 1996).

The ratio of insoluble to soluble organics is generally higher in urban, industrial pollution than in background aerosols (e.g. Zappoli et al. 1999) (Turpin and Lim 2001). These characteristics of urban pollution may have implications for the formation of organic films on aerosols, with significant consequences for CCN properties. However, there has been relatively little recent work on this compared to numerous studies on marine aerosols. An illustrative example of some of the composition variability that can be seen in transitioning from background to polluted venues can be seen in Fig. 3.4.

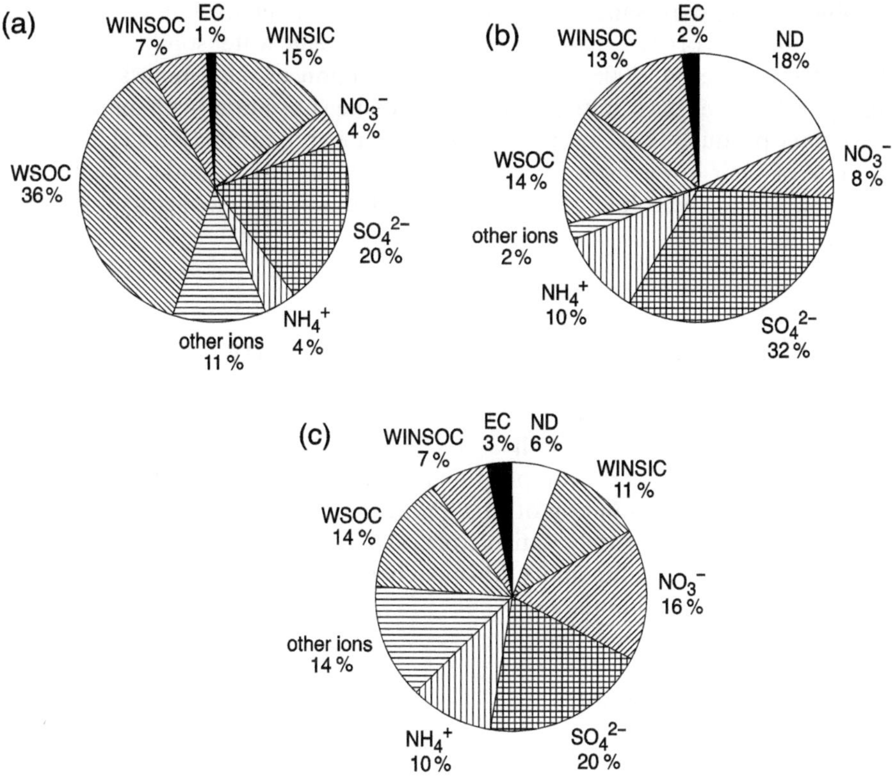

Fig. 3.4 Average mass balance at (**a**) remote, (**b**) rural and (**c**) polluted venues in Western Europe. EC = elemental carbon, WINSOC = water insoluble organic compounds, WINSIC = water insoluble inorganic carbon, WSOC = water soluble organic compounds, ND = not determined (Zappoli et al. 1999) with permission of Atmospheric Environment (@ Elsevier)

3.3.2.4 Biomass Burning Aerosol

For pollution arising from biomass burning, a somewhat different compositional picture arises. Much of the aerosol mass consists of carbonaceous species (Andreae et al. 1996; Cachier et al. 1996; Mayol-Bracero et al. 2002b; Sinha et al. 2003; Fuzzi et al. 2007; Reid et al. 2005; Decesari et al. 2006). A sizable fraction of this is black carbon, somewhat higher than that in industrial pollution. A more significant difference from industrial emissions is the relatively high organic fraction of the emissions (Andreae and Merlet 2001; Sinha et al. 2003). The organic composition of the biomass smoke is quite distinct from that of industrial emissions, with a substantial component attributable to dehydrated sugars – essentially cellulose breakdown products such as levoglucosan (Graham et al. 2002; Mayol-Bracero et al. 2002b; Gao et al. 2003; Decesari et al. 2006). However, from the standpoint of impact on precipitation, the key attribute of the pyrogenic organic emissions is their relatively high solubility, about half of the organic matter in these aerosols being water soluble (Reid et al. 2005; Decesari et al. 2006). The inorganic fraction of the biomass aerosol is minor in terms of mass initially, but can become substantial with aging due to secondary production of species such as sulphate and nitrate (Reid et al. 1998). In spite of their lower mass fraction, inorganic species can make a significant contribution to the solute content of pyrogenic aerosols (Roberts et al. 2002; Mircea et al. 2005).

3.3.3 Hygroscopic Growth, CCN and IN Activity

All the interactions of aerosol particles with water, including hygroscopic growth at RH $<100\%$, as well as CCN and IN activity, are in principle derivative from the aerosol composition and size (Dusek et al. 2006). However, our theoretical understanding of this linkage, even for the long-studied CCN activity, is still not entirely satisfactory, and our understanding of the IN linkage is at present rudimentary (Pruppacher and Klett 1997) (for more detail on ice processes see Chapter 2).

3.3.3.1 Hygroscopic Growth

At ambient relative humidity, water represents a significant component of most aerosol particles. The water content of a given particle is a function of its composition and size, and of the ambient relative humidity. Inorganic ionic components provide a large fraction of the soluble content of most aerosols, while the carbonaceous components include both water-soluble and insoluble species.

In most climate models, a fraction of emitted carbonaceous species, particularly the soot particles, is considered to be hydrophobic. These hydrophobic carbonaceous aerosols are then converted into hydrophilic aerosols by photochemical processes, condensation and coagulation, usually in a matter of hours to a day (Riemer et al. 2004). Once the carbonaceous aerosols are hydrophilic, they can act as CCN and thus are subject to in-cloud and below-cloud scavenging, whereas wet scavenging of hydrophobic carbonaceous aerosols is limited to impaction scavenging (see Sect. 4.7.2.1).

The equilibrium water content of inorganic species as functions of ambient relative humidity (equivalent to water activity for bulk solutions) is well known. Polynomial fits have been published for binary solutions (e.g. Tang and Munkelwitz 1994) and models for rigorous calculations for multicomponent mixtures are readily available (Kim et al. 1993b; Clegg et al. 1998; Nenes et al. 1998; Pilinis et al. 2000; Clegg et al. 2001; Metzger et al. 2002b). If particles composed of soluble inorganics are completely dried and then exposed to increasing RH, they take up no water until exposed to a relative humidity corresponding to the saturated-solution water activity, at which point a droplet of saturated solution forms. The water content progressively increases as RH is increased beyond this point. The wet-to-dry volume ratio at a particular RH can vary substantially with composition, with sulphuric acid (which is soluble in water in all proportions and thus has no deliquescence point) and sodium salts having relatively higher water content than neutralized sulphates. As RH is decreased, particles generally do not dry spontaneously at the deliquescence RH, but rather retain water to a lower RH. This hysteresis implies that most particles in the atmosphere, having been exposed to relatively high RH at some time, contain at least some water and rarely exist in a completely dry state.

Much less is known about the hygroscopic behavior of individual organic species and their mixtures found in atmospheric aerosols. Most of the constituents of organic aerosol matter remain unidentified, and any single compound rarely accounts for a substantial fraction of the mass. Laboratory studies have measured equilibrium water contents for dicarboxylic acids, sugars, humic and fulvic acids, model polymers, and several other species (Peng et al. 2001a,b; Prenni et al. 2001; Choi and Chan 2002; Chan and Chan 2003; Wise et al. 2003; Brooks et al. 2004; Hansen and Beyer 2004; Mikhailov et al. 2004; Chan et al. 2005; Svenningsson et al. 2006; Varutbangkul et al. 2006). The semi-volatile nature of many organic aerosol species makes them more difficult to study in a controlled laboratory setting, and techniques that are readily applied to non-volatile salts must be modified (e.g. Choi and Chan 2002). Furthermore, many relevant organic compounds are only sparingly soluble, and deliquesce at RH > 95%. It is difficult to control RH at these high values and thus laboratory water uptake experiments on such species are lacking, except for observations of their behavior at water supersaturations (e.g. Cruz and Pandis 1997; Corrigan and Novakov 1999; Giebl et al. 2002; Raymond and Pandis 2002; Hori et al. 2003; Kumar et al. 2003; Raymond and Pandis 2003; Shantz et al. 2003; Bilde and Svenningsson 2004; Broekhuizen et al. 2004).

It is generally adequate to model the density of multicomponent solutions assuming volume additivity (e.g. Brechtel and Kreidenweis 2000a). With respect to the water content of mixtures, the simplest treatment is the Zdanovskii-Stokes-Robinson (ZSR) assumption (Seinfeld and Pandis 1998), which states that the total equilibrium water content at a specified relative humidity (RH) is the sum of the equilibrium water contents for each individual species at the same RH. This approach can capture non-idealities that arise in individual binary solutions, but more sophisticated techniques are needed to represent solute-solute interaction effects. In other words, the ZSR approximation assumes that there are no interactions between species that either enhance or diminish hygroscopic growth beyond that expected for each component. Although some counterexamples have been documented in the literature (Tang 1997), the ZSR assumption appears to be adequate in many cases (Malm and Kreidenweis 1997; Prenni et al. 2003; Brooks et al. 2004). However, the deliquescence RH of a mixed salt is always lower than that of any of the components in the mixture (Wexler and Seinfeld 1991), an effect that is not captured with the ZSR treatment. Marcolli et al. (2004) demonstrate that mixtures of organic species deliquesce at very low RH and thus atmospheric organic aerosols, which are known to be mixtures of many soluble and insoluble species, are expected to retain water to very low RH.

Recently, several workers (Brechtel and Kreidenweis 2000a; Brechtel and Kreidenweis 2000b; Rissler et al. 2004; Kreidenweis et al. 2005; Koehler et al. 2006; Rissler et al. 2006; Svenningsson et al. 2006; Petters and Kreidenweis 2006) have suggested that information on particle hygroscopic growth below 100% RH can be extrapolated to predict the critical supersaturations required for activation to cloud drops. This method involves using growth factors obtained from Humidified Tandem Differential Mobility Analyzers (HTDMA) and deducing composition-dependent water activities of aqueous solutions of pure materials or mixtures, or directly extrapolating HTDMA wet-to-dry volume ratios into the supersaturated regime. Thus far the methods have produced reasonable agreement with direct measurements of critical supersaturations for particles composed of inorganic (Kreidenweis et al. 2005) and organic (Koehler et al. 2006; Svenningsson et al. 2006; Petters and Kreidenweis 2006) compounds and with field measurements on ambient aerosols (Rissler et al. 2004; Rissler et al. 2006), generally within the uncertainties of CCN measurements. When used to predict nucleated drop numbers in an adiabatic parcel model, even the least accurate extrapolations predict drop numbers within the range of uncertainties of those predicted using the most accurate thermodynamic models (Koehler et al. 2006). The validity of this approach has not yet been examined for some secondary organic aerosols.

3.3.3.2 Cloud Condensation Nuclei (CCN)

CCN are a subset of the atmospheric aerosol upon which water vapour condenses to form cloud droplets at the supersaturations achieved in clouds (<1%). It follows from the Köhler curves (Sect. 2.2.2.1) that the greater the particle mass and water solubility, the lower will be the supersaturation at which the particle can serve as a CCN, and the greater the chance that it will influence the droplet number concentration in a cloud. For example, to serve as a CCN at 1% supersaturation, completely wettable but water insoluble particles need to be at least 200 nm in diameter, whereas soluble particles can serve as CCN at 1% supersaturation even if they are as small as 30 nm. Most CCN consist of a mixture of inorganic and organic components. Most inorganic species, including those normally found in the greatest abundance in the atmosphere (e.g. ammonium sulphate), are highly water soluble. On the other hand, the solubilities of organic species are widely varying (e.g. Saxena and Hildemann 1996). Moreover, some organics, such as condensed hydrocarbons, may not even be wettable, i.e. unable to adsorb water on the surface of the particle. Water soluble organic compounds (WSOC) can influence aerosol activation by increasing the amount of soluble material (Shulman et al. 1996; Cruz and Pandis 1997).

Facchini et al. (1999) measured the surface tension of partially evaporated cloud water samples, and attributed the values lower than for pure water to dissolved organics. However, there is a limit to the extent to which surface tension can be reduced close to saturation because of dilution (Ervens et al. 2005). Lohmann and Leck (2005) have shown the presence of potentially surface-active particles from the ocean microlayer may account for an important fraction of CCN in a high Arctic aerosol. Another potential effect of organic material residing on the surface of the particle or solution droplet is to reduce the rate at which the water molecules condensing on the droplet stick to the surface (e.g. Cantrell et al. 2001; Feingold and Chuang 2002).

Particle size is an important consideration in determining its effectiveness as a CCN (Fitzgerald 1974; McFiggans et al. 2006; Dusek et al. 2006). However, it needs to be pointed out that an insoluble but wettable particle of 100 nm radius will require a supersaturation of at least 1% in order to activate (see Chapter 2, Fig. 2.5), whereas a 100 nm radius particle of ammonium sulphate will activate at a supersaturation more than an order of magnitude smaller (about 0.05%). A discussion of recent advances in knowledge of organics as CCN is given in the following section.

3.3.3.3 Carbonaceous Aerosols as CCN

The organic fraction of the fine aerosol mass is 20–50% at continental mid-latitudes (Saxena and Hildemann 1996; Putaud et al. 2004), up to 90% in tropical forested areas (Talbot et al. 1990; Andreae and Crutzen 1997; Roberts et al. 2001), and a significant fraction is water-soluble (Saxena and Hildemann 1996; Kavouras et al. 1998; Facchini et al. 1999). Thus, there is some expectation for the organic to influence CCN activity. In this section, measurements pertaining to the CCN activity of organic particles are summarized. In the subsequent section, results of "closure" experiments of ambient aerosols are used to examine the relative importance organic relative to inorganic components of aerosol particles.

Primary carbonaceous particles originating from fossil-fuel pollution may initially consist to a large extent of EC and water-insoluble substances (e.g. engine oil etc.). At that stage, most are probably not efficient CCN (Hudson et al. 1991; Weingartner et al. 1997; Zuberi et al. 2005), but some combustion particles do exhibit CCN activity (Petzold et al. 2005). Regardless of their initial composition, as they become partially oxidized and as soluble compounds (sulphates, nitrates, WSOC) deposit on them, they become CCN at realistic supersaturations (Tsigaridis and Kanakidou 2003). This process can occur on timescales of minutes to hours. Primary particles from biomass burning (pyrogenic particles) are already fairly efficient CCN near the source and become even more easily activated as they age (Warner and Twomey 1967; Hobbs and Radke 1969; Eagan et al. 1974b; Hallett et al. 1989; Radke et al. 1991; Rogers et al. 1991; Ross et al. 2003; Andreae et al. 2004).

Laboratory measurements of the CCN properties of SOA particles from monoterpene oxidation indicate a large variation in the CCN activity of the resulting compounds (VanReken et al. 2003). Field observations suggest that the condensation onto pre-existing particles of SOA from the oxidation of biogenic terpenoid compounds significantly enhances the CCN activity of those particles (Leaitch et al. 1999; Sun and Ariya 2006). Whether these SOA compounds form new particles from the gas phase or simply condense on existing particles remains unknown (e.g. Hoppel et al. 2001; Gao et al. 2001).

Other laboratory studies of activation of single-component aerosol particles at atmospherically relevant supersaturations have clearly shown that the solubility of the compounds is a key parameter for organic aerosol activation (Novakov and Corrigan 1996; Giebl et al. 2002; Raymond and Pandis 2002; Hori et al. 2003; Shantz et al. 2003; Broekhuizen et al. 2004), and that the lowering of the surface tension can dramatically reduce the activation diameter of a moderately soluble organic particle (Broekhuizen et al. 2004).

It has been shown that the ability of organic particles to activate can be substantially enhanced by adding small amounts of soluble compounds as inorganic salts or surface-active species (Hegg et al. 2001; Bilde and Svenningsson 2004; Broekhuizen et al. 2004; Abbatt et al. 2005). Broekhuizen et al. (2004), Bilde and Svenningsson (2004) and Henning et al. (2005) also found that the initial particle phase played a significant role in the critical supersaturation of particles containing only slightly soluble substances. If the particle was initially totally liquid the measured critical supersaturations were lower. Marcolli et al. (2004) recently demonstrated that multicomponent organic solutions favored the liquid phase. For complex mixtures found in the atmosphere, it may be expected that aerosol comprising slightly soluble organics would be unlikely to contain undissolved material. It is also important to note that even relatively small amounts of inorganic material will allow the activation of organic particles (Lohman et al. 2004; Ervens et al. 2005), though the abundance and nature of organics will affect both equilibrium water content and droplet growth kinetics. The recent work of Petters and Kreidenweis (2006) provides a method for connecting these results for model applications. Reviews of the carbonaceous aerosol and their roles in cloud droplet nucleation are found in Kanakidou et al. (2005) and McFiggans et al. (2006) respectively.

3.3.3.4 CCN Closure Studies

The classic description of the relationship between aerosol size, composition and CCN activity is the well-known Köhler theory (For a more detailed discussion see Chapter 2). If this theory is valid, then the CCN activity is specified by aerosol size and composition measurements, and direct measurement is unnecessary. This theory has been subject to laboratory trials in which the activation size of the particles of various sizes and compositions have been measured at supersaturated water environments in CCN chambers. Much of the early such work utilized inorganic salts and acids, since these were thought to dominate ambient water soluble aerosol composition. While experiments with these particles lent support to the applicability of the theory (cf. Fitzgerald 1973), more recent work on aerosol composition, which (as discussed above) attributes a very substantial role to carbonaceous species, led to questioning of the prognostic power of the Köhler theory (e.g. Shulman et al. 1996; Facchini et al. 1999; Charlson et al. 2001; Anttila and Kerminen 2002). Consequently, modified forms of the theory have been developed to take into account the lower water solubilities of many organics, their possible impact on haze particle surface tension, and even the influence of soluble gases that may co-condense with water during particle growth (Laaksonen et al. 1998).

A further complication of the organics is their surfactant properties and the consequent possibility that they could form films that would hinder water condensation/evaporation and thereby impose a kinetic limit to the thermodynamic Köhler theory (Gill et al. 1983; Cruz and Pandis 1998). However, laboratory experiments to quantitatively assess this phenomenon – and even some detailed theoretical assessments – have yielded mixed results. Most investigators have concluded that such a limitation is at best small (Cruz and Pandis 1998; Hegg et al. 2001; Anttila and Kerminen 2002; Raymond and Pandis 2003; Shantz et al. 2003; Broekhuizen et al. 2004). Hygroscopic growth factors were well reproduced at the high altitude site at Jungfraujoch using the chemical composition of the mixed particles and the Zdanovskii-Stokes-Robinson (ZSR) relation (Gysel et al. 2002; Weingartner et al. 2002). The temporal variability of the hygroscopic growth factors was mainly a result of varying fractions of organic/inorganic mass, and the monomodal distribution of the hygroscopic growth factor confirmed the presence of an internally mixed aerosol. Nevertheless, since there is some laboratory support for such limitation, and evidence for organic films has indeed been found on ambient aerosols (Mochida et al. 2002; Tervahattu et al. 2002a), the possibility of a decrease in droplet growth due to organic surface films must be considered when analyzing atmospheric data.

Evidence for a role of the organic aerosol as CCN has been found through the investigation of apparent changes in CCN efficiency and statistical correlations with organics (e.g. Bigg 1986; Hudson and Da 1996; Matsumoto et al. 1997). In a landmark study, Novakov and Penner (1993) and later Rivera-Carpio et al. (1996) combined CCN spectral measurements with size resolved composition measurements to show that both the sulphate and the carbonaceous compounds contributed to the CCN activity.

An effective method for investigating the CCN activity of ambient aerosols is the CCN closure experiment (e.g. Bigg 1986; Liu et al. 1996; Covert et al. 1998; Chuang et al. 2000; Snider and Brenguier 2000; Cantrell et al. 2001; VanReken et al. 2003; Mircea et al. 2005; Rissler et al. 2006; Broekhuizen et al. 2006; Ervens et al. 2007). In such an experiment, the aerosol size distribution and chemical composition as a function of size are measured simultaneously with the CCN activity. The linkage between the variables is the Köhler theory, and the closure constitutes a test of the theory and provides some level of confidence in our knowledge of the system. Ideal closure experiments would include the measurement of the individual composition of particles. Such experiments may not yet be possible, and instead there is a spectrum of closure experiments with varying degrees of measurement completeness. However, even the most rudimentary of these types of experiments have shed light on the CCN activity of atmospheric aerosols.

Closure studies have taken three different approaches. In the first, the aerosol size distribution and CCN activation spectrum have been measured and a certain soluble inorganic composition (commonly ammonium sulphate) is assumed. Comparison of predicted (based on Köhler theory) and measured CCN concentrations then test this assumption (Bigg 1986; Cantrell et al. 2000; Snider and Brenguier 2000; VanReken et al. 2003). In some instances (e.g. VanReken et al. 2003), reasonable closure has been achieved but in most instances, particularly in more polluted air, the predicted concentrations have substantially exceeded the measured CCN.

In the second, more constrained approach, the CCN activation spectrum, the aerosol size distribution and the size dependent aerosol hygroscopicity have been simultaneously measured (cf. Brechtel and Kreidenweis 2000a,b). In a practical sense, it is a more complete approach than utilizing size-dependent composition data since the latter must then be coupled with characteristic thermodynamic data to derive the activity of the aerosol chemical species. Recent studies that have utilized this approach (Covert et al. 1998; Zhou et al. 2001; Dusek et al. 2003; Rissler et al. 2004; Ervens et al. 2007) found modest discrepancies between predicted and measured CCN concentrations. While the differences were generally not large, they were systematic with the predicted concentrations nearly always in excess of the measurements. The Covert et al. (1998) study showed that the percentage magnitude of the overprediction varies directly with the extent of anthropogenic impact, consistent with the earlier work of Hudson and Da (1996).

The third type of closure study involves particle size, size-resolved composition and CCN activity measurements. Examples of this are Chuang et al. (2000), Cantrell et al. (2001) Mircea et al. (2005) and Broekhuizen et al. (2006). Chuang et al. (2000) overpredicted the measured CCN concentrations by as much as an order of magnitude, in part because the CCN activity and compositional measurements were not co-located and in part due to problems with the measurements. The other studies were able to achieve closure within 10–20% with a tendency to overpredict the CCN concentrations.

Cantrell et al. (2001), Broekhuizen et al. (2006) and Ervens et al. (2007) all obtained the best closure assuming no role of the organic components in the particles, despite the presence of significant organic mass fractions (20–90%). Cantrell et al. (2001) attributed their overprediction to the possibility that some of the organic components in the particles inhibited water uptake, while Broekhuizen et al. (2006) suggested that the limited resolution of the model and the lack of detail in the description of the mixing states could explain the slight bias in their closure. In a comprehensive analysis of size-dependent chemical composition, hygroscopic growth, and CCN activity, Mircea et al. (2005) showed that knowledge of aerosol WCOC composition in terms of classes of compounds and of their molecular weights and acidic properties were necessary to predict CCN activation.

A key question of the closure studies is how the various aerosol species are mixed? While some studies have not found this to be a critical assumption (e.g. Liu et al. 1996), later studies have demonstrated a strong sensitivity of the activity of either organic aerosols or inorganic aerosols to trace quantities of each other (e.g. Roberts et al. 2002; Bilde and Svenningsson 2004; Broekhuizen et al. 2004; Abbatt et al. 2005; Mircea et al. 2005; Broekhuizen et al. 2006).

The results of these closure experiments suggest a widely varying role for the organic compounds in the water uptake by the atmospheric aerosol. Studies in or near urban regions are able to explain most, but not all, of the CCN activity using the inorganic chemistry. In the case of a strong biomass burning influence (Mircea et al. 2005), there appears to be a strong role for the organic in determining CCN activity. And SOA particles from biogenic sources have at least some importance for water uptake. Recently, Petters and Kreidenweis (2006) have introduced a pseudo-empirical approach for modeling the CCN activation of multi-component aerosols that eliminates many of the thermodynamic details necessary for the description of water activity, and may provide a more streamlined approach to the representation of CCN activation in models. Still, the chemical components, concentrations, solubilities, and surface tensions of solution droplets that form in various environments are not well documented, and knowledge of the effects of these parameters on droplet activation individually, and in combination, is very limited (Kanakidou et al. 2005; McFiggans et al. 2006).

3.3.3.5 Source-related Global Distribution of CCN

Global microphysical aerosol models are required to globally predict the evolution of the size-distribution and the contribution of specific aerosol sources to the number of radiatively important aerosols and cloud condensation nuclei. In a study with the global microphysical aerosol-climate model ECHAM5-HAM, Stier et al. (2005) demonstrated that the global CCN number concentration is non-linearly related to the emission mass flux. While primary carbonaceous aerosol emissions disproportionately contribute to the CCN number, anthropogenic SO_2 emissions less than proportionally contribute to the number of CCN, because a significant fraction of the produced sulphuric acid condenses on pre-existing aerosols. The co-emission of anthropogenic carbonaceous and sulphur emissions generally produce less CCN in the vicinity of the source regions than would be obtained by summing the CCN produced by the individual emissions. Therefore, in contrast to results from traditional bulk mass aerosol models, the sum of the change of CCN due to individual emission changes is different from the total CCN change due to several combined emission changes.

As a measure of the contribution of specific aerosol sources to the potentially CCN-relevant size range, one can define the difference in CCN due to omission of the specific aerosol source with respect to a simulation with the complete set of emission. In Fig. 3.5a the annual zonal mean distribution of hygroscopic accumulation and coarse mode particles (an approximation of the CCN concentration) for the year 2000 is depicted, computed by the ECHAM5-HAM aerosol model (Stier et al. 2005). Also shown in the figure are the differences with respect to this reference scenario due to the omission of b) carbonaceous emissions from fossil-fuel use and industry, of c) sulphur emission from fossil-fuel use, industry, and bio-fuels, and of d) total anthropogenic emissions.

The zonal mean distribution of CCN (Fig. 3.5a) shows a pronounced maximum in the lower to mid-troposphere (1000–500 hPa) in the region of strongest anthropogenic emissions between 30°S and 50°N. The omission of carbonaceous emissions from fossil fuel use and industry (Fig. 3.5b) reduces the zonal mean number of CCN in the lower troposphere between 20 and 60°N by up to 30%. Fossil-fuel use and industry contribute (with 6.4 Tg) only 10% of the total carbonaceous mass emissions and a significantly smaller fraction of the total anthropogenic aerosol mass source. In the lower to mid-troposphere, sulphur

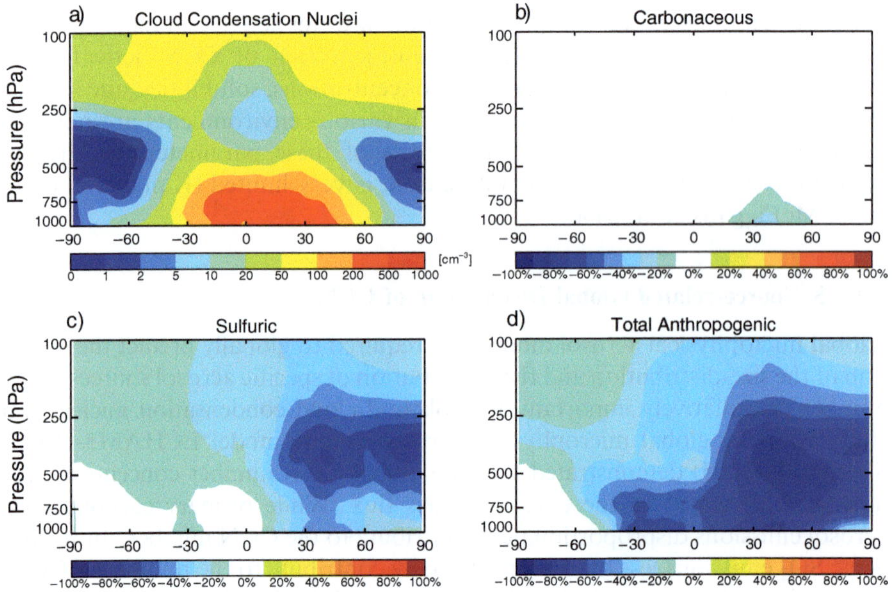

Fig. 3.5 (**a**) Total zonal annual-mean cloud condensation nuclei (cm^{-3} STP{1013 hPa, 273.15 K}) as simulated by the ECHAM5-HAM aerosol model. CCN are defined here as hygroscopic accumulation and coarse mode aerosol particles. Change in CCN due to the omission of (**b**) carbonaceous emissions from fossil-fuels and industry, (**c**) sulphur emission from fossil-fuels, industry, and bio-fuels, (**d**) all anthropogenic emissions including fossil-fuels, industry, bio-fuels, and vegetation fires. From Stier et al. (2005, 2006) with permission of the American Meteorological Society

emissions from fossil fuels, industry, and biofuels (Fig. 3.5c) contribute 30–50% of the total number of CCN between 20 and 60°N. Higher up in the NH mid-to-upper troposphere (750–250 hPa), they dominate the CCN numbers. The impact of anthropogenic sulphur emissions on southern hemispheric CCN in the lower troposphere is negligible. Total anthropogenic emissions (Fig. 3.5d), including fossil-fuel use, industry, vegetation fires and bio-fuel use, largely control the low to mid-troposphere CCN numbers between 30°S and 70°N contributing to 50–80% of total CCN numbers. Only the CCN at high southern latitudes are predominantly controlled by natural emissions.

3.3.3.6 IN Activity

Very little is still known about IN activity. The situation circa 2000 has been well summarized in the 3rd IPCC report (Houghton et al. 2001), but, unfortunately this field has seen relatively little activity in the last 30 years. Issues such as the precise mode of action of aerosol acting as IN (e.g. contact, immersion, deposition etc. nuclei), a lack of consistency in measurement techniques, in situ modification of particle IN efficiency, and secondary ice production, all contribute to the uncertainty in IN activity of atmospheric aerosols.

Nevertheless, a few facts have emerged. Firstly, there is no question that particles acting as IN are relatively rare compared to CCN. IN concentrations typically are measured at a few per liter or less, thus constituting no more than a few tenths of a percent of the normal total aerosol population. In part this is no doubt due to the limited chemical components that are known to act as IN. Best known are crustally-derived minerals, for which there is a large and still growing body of corroborative data (Kumai 1976; Hagen et al. 1995; Heintzenberg et al. 1996; Castro et al. 1998; Kreidenweis et al. 1998; DeMott et al. 2003a; Cziczo et al. 2004a; Sassen 2005).

Recently, experiments have been performed in a large-volume (84 m^3) chamber called AIDA (Aerosol Interaction and Dynamics in the Atmosphere), which can be operated at variable pressures and temperatures, between 30 and 90°C, and at high relative humidities. A unique feature of AIDA is the ability to supersaturate the gas phase with respect to ice and liquid water by quasi-adiabatic cooling, thereby realistically simulating cooling rates of air parcels in the atmosphere with updraft velocities between 0.5 and 6 m s^{-1}. Mohler et al. (2006) investigated the deposition mode ice nucleation efficiency of various dust aerosols (Arizona test dust (ATD) and two dust samples from the Takla Makan desert in Asia (AD1) and Sahara (SD2)) at cirrus cloud temperatures between 196 and 223 K. Deposition ice nucleation was most efficient on ATD particles with ice-active particle fractions of about 0.6 and 0.8 at an ice saturation ratio $S_i<1.15$ and temperatures of 223 K and 209 K, respectively. The desert dust samples SD2 and AD1 showed a significantly lower fraction of active deposition nuclei, about 0.25 at 223 K and $S_i<1.35$. Field et al. (2006), using similar samples, reported that at temperatures warmer than –40°C, droplets were formed before ice crystals formed, and there was generally no deposition nucleation observed. At temperatures colder than –40° C, dust samples exhibited dual nucleation events. The primary nucleation event occurred at ice saturation ratios of 1.1 to 1.3, and is likely to be a deposition nucleation mode. The secondary nucleation event occurred at ice saturation ratios between 1.35 and 1.5, and probably was a condensation mode, as there was some soluble material in the dust samples. For all samples, the ice activated aerosol fraction could be approximated by an exponential equation as function of S_i. This formulation of ice activation spectra may be used to calculate the formation rate of ice crystals in models, if the number concentration of dust particles is known. More experimental work is needed to quantify the variability of the ice activation spectra as function of the temperature and dust particle properties.

There is also evidence suggesting that carbonaceous aerosols (e.g. Karcher et al. 1996; Petzold et al. 1998; Strom and Ohlsson 1998) and various biogenic particles (see the review by Szyrmer and Zawadzki 1997) can act as IN. Nevertheless, the observed concentrations of IN are usually far below the concentration of particles containing any or all of these various favorable components. Differing degrees of IN deactivation due to pollutant deposition on IN surfaces or cloud processing have been invoked as an explanation for this phenomenon but clear evidence of this is wanting. Furthermore, it is also likely that favorable ice-forming substrates could be deposited as organic films on previously inactive particles. A combination of these processes could well explain the long-puzzling variability in the IN fluxes from pollution sources (e.g. Hobbs and Locatelli 1969; Braham and Spyers-Duran 1974). Once again, however, evidence of the impact of such secondary processes is sparse at best. For further discussion of IN activity, see Chapters 2 and 6.

3.4 Recommendations for Future Observations and Studies

- Emission inventories have been conventionally limited to estimates of mass emission fluxes for bulk aerosol components. To be useful in cloud and precipitation studies, size-resolved information about the numbers of particles emitted is required for the various primary aerosol sources, especially sea salt, mineral dust, pyrogenic aerosol and primary biogenic aerosol.
- Emission information must be provided in forms that are suitable for dynamic emission inventories. It is much more useful to have emission algorithms that permit calculation of an emission flux from model variables than a static array of emissions valid for one particular time.
- The abundance and rate of production of submicron sea salt particles is still under dispute. Since these particles can play a significant role in cloud microphysics and precipitation in remote marine regions, observational constraints to their source parameterizations are urgently needed.
- Primary biogenic aerosols probably play a significant role as cloud and ice nuclei at present in very remote regions, and must have been a key control on

cloud processes in the pre-anthropogenic atmosphere. Their characteristics and rates of emission need to be investigated.

- Due to the complexity of carbonaceous aerosol composition and emission processes, a critical concern is to define an adequate classification scheme for these aerosols based on observable characteristics.
- Building a reliable inventory for carbonaceous aerosols (BC, PPOC and SOA) remains a challenge because of uncertainties in emission factors and activity estimates. Emission parameterizations must be provided that include size distributions and number fluxes.
- The injection height of emissions also differs strongly from on aerosol source type to another, and even shows pronounced variability within some types of aerosol source. This applies especially to biomass burning and vegetation fires. Any changes in injection height due to shifts in agricultural or technical practices will have important consequences on the lifetime and fate of particles. Therefore, appropriate injection heights for the various aerosol types should be included in emission models.
- The rates of sulphate aerosol production are at least as dependent on the accurate representation of the processes leading from SO_2 to sulphate aerosol as they are on emission estimates of SO_2. Accurate knowledge of these processes and their correct parameterization in models is essential.
- Aerosol nitrate is expected to increase in climatic importance in the 21st century. Accurate measurement techniques and reliable modeling approaches for this component must be established.
- In order to assess the human impact on cloud physics, we need to know the aerosol, CCN, GCCN and IN distributions in the pre-anthropogenic atmosphere. The sources, characteristics, and fluxes of natural aerosol types must be investigated by carefully conducted field and laboratory studies.
- The ability of aerosol particles to act as CCN, GCCN and IN should be determined as a function of particle size, composition and supersaturation. Uncertainties are greatest for IN because of the different processes by which ice can be nucleated. Measurements on cloud-active particles should be conducted in the context of closure studies.

Chapter 4
The Distribution of Atmospheric Aerosols: Transport, Transformation and Removal

Sunling Gong and Leonard A. Barrie

4.1 Introduction

Because the time for air parcels to circle the Earth on winds in the troposphere is of the same order of magnitude as the residence time of atmospheric aerosols, there is no location on the globe that is not influenced by aerosol sources. Once released into the atmosphere from primary production or produced via gas-to-particle conversion (for source details, see Chapter 3), aerosols are subject to many processes that affect their global distribution, chemical and physical properties, and hence their influence on climate, weather, human health and ecosystems. They are dispersed in the atmosphere through processes of advection, convection and turbulence. They are also transformed and removed by physical and chemical processes involving clouds, precipitation as well as processes occurring in cloud-free air. During the life cycle of an aerosol in the atmosphere, gas to particle conversion and mixing of aerosols from different sources changes the chemical, physical and optical properties of the original aerosols.

4.2 Geographically Distinct Aerosol Regimes

Hemispheric-scale differences in land area (39% of the northern hemisphere is covered by land versus. 19% in the southern hemisphere) and in human population have led to large inter-hemispheric differences in primary aerosol source strengths (Chapter 3). This involves both natural and anthropogenic aerosols, but the greatest differences are seen for fossil fuel and aircraft emissions. According to data summarized in Chapter 5 of the IPCC 2001 assessment report (IPCC 2001), more than 90% of the aircraft emissions and ~99% of the fossil fuel emissions are emitted into the atmosphere over the northern hemisphere. More than 80% of the mineral dust is produced in the northern hemisphere, according to Jickells et al. (2005), and <20% of the dust flux into

S. Gong (✉)
Environment Canada, Toronto, Canada

the oceans occurs in the southern hemisphere. In contrast, biomass burning emissions are more evenly distributed between hemispheres, with a split of nearly 50%:50%. Sea salt is the only primary aerosol produced in greater quantities in the southern hemisphere (Chapter 3).

Consider the global atmospheric aerosol distribution from the perspective of the global distribution of mean annual aerosol optical depth (AOD, Fig. 4.1) obtained from a combination of satellite observations and verified using a large surface-based network of aerosol optical depth remote sensing instruments (Chapter 5). There are some biases in this distribution due to the dependence of satellite observations on sunlight. For instance, satellites cannot observe the polar regions in the dark winter half of the year, thus missing Arctic haze pollution (Barrie 1986; Barrie 1990; Christensen 1997), and they are unable to measure beneath clouds, thus possibly under-sampling the atmosphere where clouds are most frequent (Fig. 1.5a), such as in the roaring forties belt between 40 and 60°S of the southern oceans. In addition, the distribution of AOD does not completely correlate to that of aerosol mass, because the fine aerosol particles have a stronger optical effect (mass extinction coefficient), than larger particles. Therefore coarse sea salt and dust aerosols cannot be readily seen on this figure (see Sect. 4.2.6). Nevertheless, the geographical distribution of AOD in Fig. 4.1 helps us to organize a description of the global aerosol and aids in understanding their influence on weather, climate, air

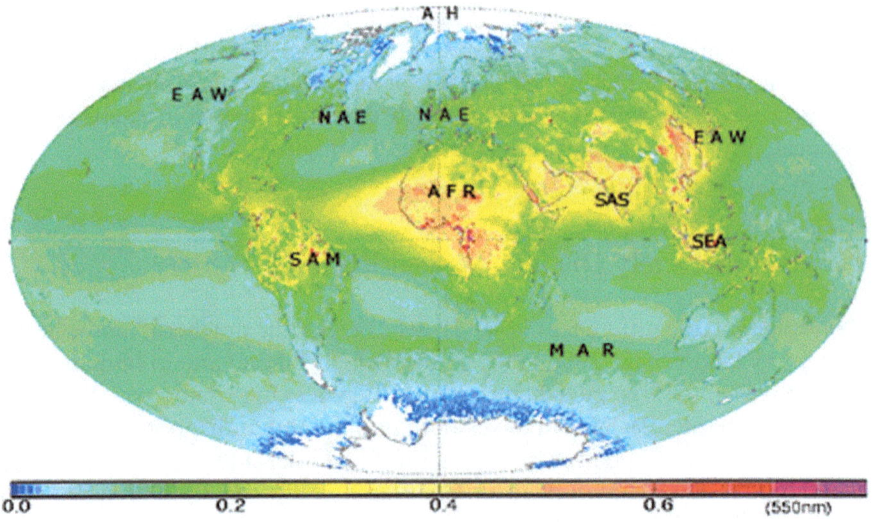

Fig. 4.1 A best estimate of the global distribution of annual average tropospheric aerosol optical depth (AOD) compiled by combining data from six satellites (operating for limited periods between 1979 and 2004). Observations for a region were selected using ground-based AOD observations as guidance. **AFR**: Africa, **SAS**: South Asia, **SEA**: Southeast Asia, **EAW**: East Asia and Western North America, **NAE**: Eastern North America, North Atlantic and Europe, **SAM**: South America, **MAR**: Marine and **AH**: Arctic Haze. Courtesy of S. Kinne MPI, Hamburg, Germany

quality and, in particular, precipitation reaching the Earth's surface. Seasonal counterparts to this figure can be found in Chapter 1 (Fig. 1.8).

For the purposes of this discussion, we divided the global aerosol into 8 geographically distinct regimes, each involving particles of one or more aerosol types (e.g. sea salt, marine biogenic, soil dust, biomass burning, anthropogenic sulphates and organics).

4.2.1 Africa (AFR)

The aerosol regime in Africa is a good example of a primary (soil dust) aerosol mixing with primary and secondary biomass burning aerosols. The soil dust component covers the Saharan desert source region and extends thousands of kilometers downwind. It affects not only the surrounding oceans but also other continents such as the Americas, Europe, the Middle East and south Asia. Transport across the Atlantic can affect visibility and deposition of nutrients in the Caribbean. There are outstanding questions regarding its role in influencing hurricane development over the tropical Atlantic (Evan, et al. 2006). For instance, does soil dust affect sea surface temperature? Soil dust is one of the variables that is widely recognized as important to hurricane development. The effects of soil dust on cloud and precipitation formation processes are only beginning to be understood. There is evidence that a proper treatment of Saharan dust can improve the forecast of winds, such as the easterly jet over West Africa, in numerical weather predictions (Tompkins et al. 2005).

Emissions from biomass burning sources in tropical and southern Africa drift over the tropical Atlantic into tropical South America as well as into the Indian Ocean off the southeast tip of Africa. A so-called "river of smoke" (Fig. 1.8), observed during the SAFARI 2000 field campaign flowed, from northwest to southeast over the subcontinent, and caused heavy haze and reduced visibility over Botswana and South Africa (Swap et al. 2003). The satellite image from the MODIS satellite in Fig. 4.2 from Kaufman et al. (2005a) separates coarse Saharan dust and fine bio-mass burning aerosols in central and west Africa. Superimposed on these major sources are lesser, but still quite significant, sources of fossil fuel aerosol resulting from human activities in the region.

The African monsoon occurs adjacent to the Saharan soil dust sources and is co-located with biomass and air pollution sources. The role of aerosol radiative forcing in the dynamics of this monsoon is thought to be significant, but is only beginning to be investigated (Caminade et al. 2006).

4.2.2 Eastern North America, North Atlantic, Europe (NAE) and Arctic Haze (AH)

Widespread fossil fuel combustion in eastern North America leads to a North American plume that moves eastward across the North Atlantic where it mixes

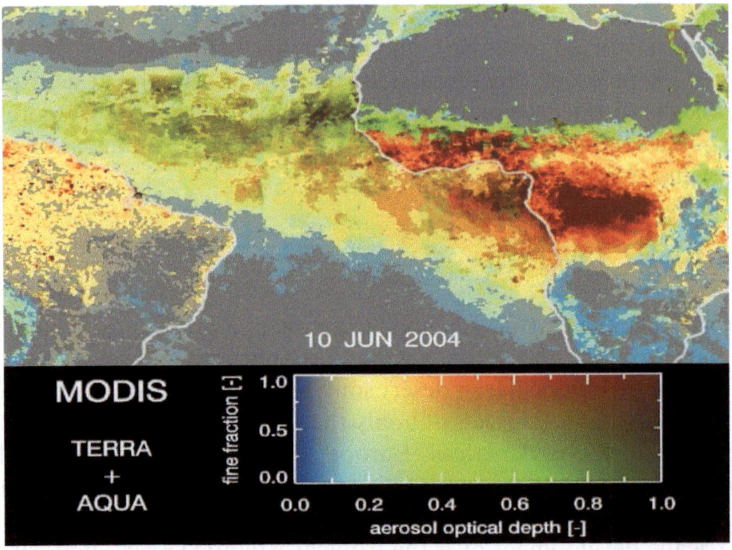

Fig. 4.2 MODIS aerosol composites for 10 June 2004. Blue represents clean conditions, aerosol optical thickness <0.1, and green and red show higher optical thickness corresponding to the coarse (*green*) and fine (*red*) modes. The fine fraction (*y axis*) varies from green for fine fraction of zero to red for fine fraction of one. Therefore pure dust is green, and pure smoke or pollution is red. From Kaufman et al. (2005a) with permission of the American Geophysical Union

with sea salt (see below) and eventually reaches Europe (Methven et al. 2006). Not evident in Fig. 4.1, because of a blind spot of aerosol satellite observations in the polar troposphere, is the reverse flow from Eurasia to North America over the Arctic from November to May that constitutes the Arctic haze (Barrie 1986; Barrie 1990; Christensen 1997). European pollution also drifts eastwards in mid latitudes joining pollution from regions of the former Soviet Union where soil dust and bio-mass burning occurs. The surface concentrations of aerosols in both North America and Europe from ground monitoring stations are discussed in Sect. 4.3.2.

4.2.3 South Asia (SAS) and Southeast Asia (SEA)

The aerosol regime in south Asia, including the Arabian Peninsula and Indian sub-continent, is a result of soil dust, pollution and biomass burning aerosols. Saharan soil dust advected from the west affects the Indian sub-continent, and this combines with local sources of dust. In addition, the Indian sub-continent has pollution from fossil fuel and biomass burning that leads to a highly absorbing aerosol (Ramanathan et al. 2005). It is advected out of the region, affecting clouds and precipitation in the Himalayas in summer and over the Indian Ocean in winter. There is documented evidence of transport as far as the Mediterranean basin (Lelieveld et al. 2002).

In the Atmospheric Brown Cloud (ABC) study, the presence of highly absorbing aerosol over the Indian Ocean was observed as far south as the Maldives. It had the effect of reducing solar radiation reaching the surface by as much as 20 W m^{-2} (Ramanathan et al. 2005) and increasing the stability of the atmosphere through redistribution of solar energy from the surface to soot-bearing aerosols aloft. This affects the evaporation of water vapour and hence the cloud regime. In addition, aerosol dust and pollution is thought to affect the Asian monsoon through direct radiative forcing of changes in regional circulation (Lau et al. 2005) and indirectly through aerosol impacts on clouds and precipitation formation (Chylek et al. 2006). The latter, which is the focus of this review, is more difficult to quantify due to the non-linearity of processes involved (Chapters 6 and 7). In any case, there are documented changes in precipitation in India over the last half century that raises the issue of direct aerosol pollution effects, as well as other climate change effects that modify the monsoon dynamics. For instance, river basins in central India (Sabarmati, Mahi, Narmada, Tapi, Godavari and Mahanadi) have been experiencing reduced rainfall from about the 1960s, while in others – Indus from 1954, Ganga from 1993, Brahmaputra from 1988, Krishna from 1953 and Cauvery from 1929 – there has been an upward trend in rainfall. The recent trend in rainfall across India is believed to be due to global warming (Singh, et al. 2005). Whether changing aerosol pollution in the region has played a role remains an open question.

Another aerosol source region that is prominent in Fig. 4.1 is south-east Asia centered on Indonesia and Borneo. Biomass burning aerosols coupled with air pollution from mega-cities are the main components of this aerosol regime. During the 1997 forest fires in Indonesia, approximately 42 Tg of aerosol mass were emitted into the atmosphere, and resulted in a –0.32 Wm^{-2} forcing averaged over the 4 months of the fire season (Davison et al. 2004). This is more distinctly seen in Fig. 1.10.

4.2.4 Eastern Asia and Western North America (EAW)

Another major aerosol regime is located in eastern Asia, mainly including China, Korea and Japan and the downwind region of the north Pacific as far as western North America. Soil dust from the Gobi and Taklimakan deserts and anthropogenic pollution from industry and fossil fuel burning are the main aerosol types. The soil dust source is seasonal, peaking in February to May and originating from the western and northern parts of the region. The mineral dust advects in the northerly and westerly winds over populated areas, often mixing with pollution. It also is lofted by convection into strong westerly winds of the middle troposphere (2–7 km) that periodically transport it in a matter of 5 to 10 days to North America (Jaffe et al. 1999) and even into the Arctic (Welch et al. 1991; Sirois and Barrie 1999; Sassen 2005). Air pollution in the region is persistent throughout the year and has increased greatly in the latter half of the 20th century, largely as a result of the economic growth in China. The southern part of this region is affected by monsoonal precipitation in summer where, as in Africa and south Asia, the effects of aerosols are poorly understood. Observations have shown that regional precipitation has changed substantially in China between 1951 and 2000 (Zhai, et al. 2005). Simultaneously, there is a decreasing trend in the total global radiation reaching the surface over most industrial regions (Che et al. 2005) reflecting an increase in aerosol loadings (details in Sect. 4.3.4). Whether these changes are linked remains an open question.

Natural biomass burning aerosols originating from fires ignited by lightning in the boreal region of Canada, Russia and Alaska are highly seasonal but climatologically persistent (see Fig. 1.8a). Occurring mostly in summer, the burning products sometimes loft to great heights and can even produce their own thunderstorm clouds into which pollution is mixed, scavenged and dispersed in the outflow regions. They can also drift long distances near the ground affecting air quality and human health. These aerosols appear most prominently in summer season in the northern hemisphere. A good example was an event during the summer of 2002, when over 360 fires occurred in the James Bay region of Canada. Sparked by a combination of lightning and dry conditions, these fires consumed hundreds of square kilometers of forest (8871 km^2) and created a huge plume of smoke (Fig. 4.3) that affected air quality in much of eastern Canada, and as far south as Washington, DC in eastern United States.

Fig. 4.3 In this satellite image taken on July 7, 2002, large fires in northern Quebec are identified as *red* spots, clouds are *white* and smoke has a *brownish tint*. MODIS image from NASA's Terra and Aqua satellites. Where the smoke plume enters the United States over the state lines of New York and Vermont, it measures almost 300 km wide. The smoke affected air quality and visibility in major cities including Montreal, New York, Philadelphia, Baltimore and Washington, D.C. (Canada Center for Remote Sensing)

Once airborne, boreal biomass burning aerosols can circle the hemisphere in 15–20 days in the strong westerly winds of the middle to upper troposphere. They are much longer lived than lower tropospheric aerosols. Their radiative and microphysical effects on clouds are poorly understood. Since northern boreal regions tend to receive most precipitation in the summer period these aerosols are potentially very important.

4.2.5 South America (SAM)

South American aerosols originate mainly from biomass burning in the Amazon and savannah regions, as well as from pollution emitted from mega-cities. Large urban areas such as Sao Paulo, Buenos Aires, Rio de Janeiro, Santiago and others are large sources of aerosol particles. These aerosols, when released in the tropical Amazon basin, drift westwards towards the Andes. When released further south, they drift over the south Atlantic on westerly winds. Recently, the influence of smoke aerosols on the microphysical properties of clouds and convection has been extensively reported (see Sect. 4.3.5).

The biomass burning occurs during the dry season that extends from June to November (Artaxo et al. 2002). Huge plumes of particles extending for 3 to 6 million square kilometers with AODs at 500 nm of 2 to 4 are constantly observed. This has strong effects on net energy input to the atmosphere and its distribution in the vertical over large areas of South America (Eck et al. 2003; Procopio et al. 2004). In addition, the aerosol pollution affects microphysics and occurrence of clouds (Andreae et al. 2004; Koren et al. 2004). The influence of smoke aerosols on the microphysical properties of clouds and convection has been extensively reported. One simulation result suggests that biomass burning aerosols, peaking shortly before the onset of the South American monsoon, may influence the development of the monsoon (Liu et al. 2005b). Monsoon precipitation is part of a complex weather and climate system that affects large portions of southeastern South America. The long-range transport is well documented (Andreae, et al. 2001). Deposition of particles occurs over most of South America, with important climatic and environmental effects.

4.2.6 Marine Aerosols (MAR)

Marine aerosols consisting of sea salt, and often mixed with marine biogenic sulphur, are another prominent aerosol optical depth maximum (Fig. 4.1) in the Earth's atmosphere particularly over the high wind "roaring 40's" belt between 40 and 60°S and in the north Pacific and north Atlantic in winter. The AOD-signal of marine aerosol is relatively weak in comparison to the marine aerosol mass present in this region since sea salt aerosols tend to have larger size than pollution aerosols and are thus less effective scatterers of light per mass unit.

A global model of the distribution of the coarse and fine fraction of sea salt (Fig. 4.4) shows clearly that they are prominent in, and downwind of, windy regions of the Earth's oceans (Gong et al. 2002). The fine particle distribution of sea salt (Fig. 4.4b) shows that advection deep into continental regions can occur in some places. For instance, the penetration of marine aerosols into Amazonia

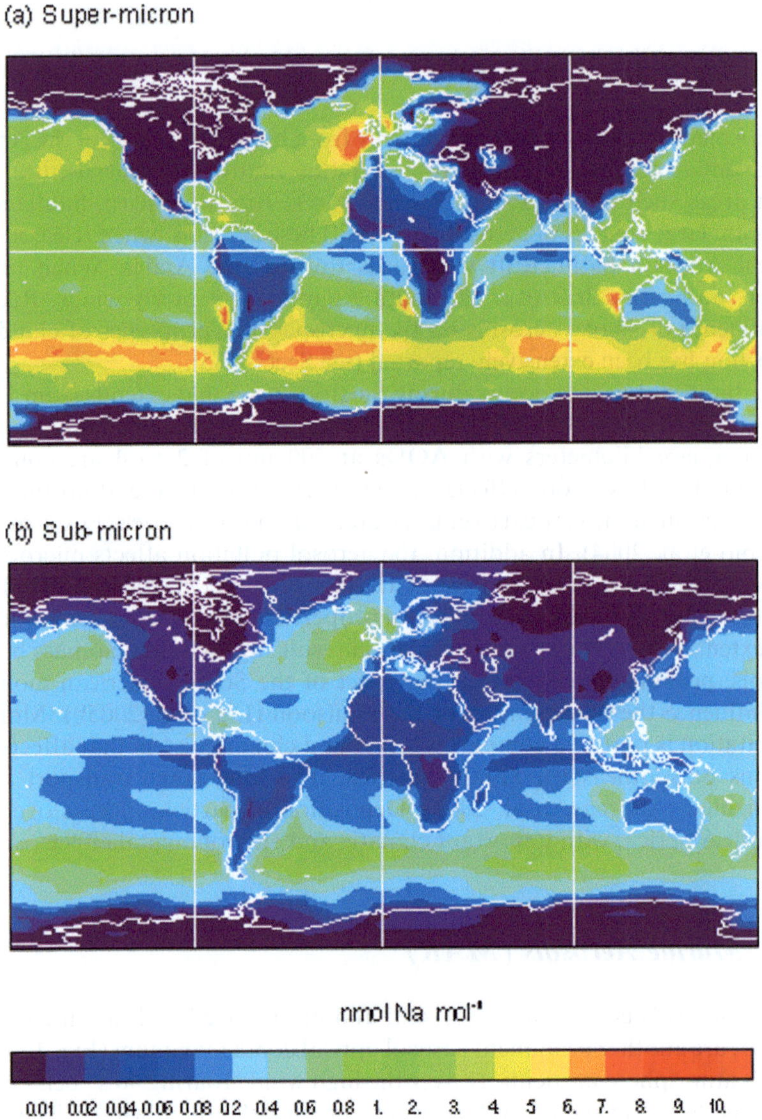

Fig. 4.4 Predicted global distributions of atmospheric sea salt Na^+ mixing ratios in January in the atmospheric surface layer (0–50 m) for two size ranges: (**a**) Super-micron and (**b**) Sub-micron. From Gong et al. (2002) with permission of the American Geophysical Union

is observed in precipitation 1000 km inland where it dominates the concentrations of major inorganic ions (Andreae et al. 1990). Coarse marine sea salt particles (Fig. 4.3a) provide giant cloud condensation nuclei that can dramatically alter the effect of pollution aerosols on clouds.

4.3 Aerosol Surface Observations and Regional Characteristics

This section reviews selected observational studies that characterize atmospheric aerosols and their chemical compositions in various regions. Chapter 5 covers additional aspects of in situ and satellite measurements of aerosols.

4.3.1 Global Monitoring Networks

The Global Atmospheric Watch (GAW) program of the World Meteorological Organization (WMO) consists of 24 global observatories and approximately 200 regional stations that measure the composition of the atmosphere including aerosols, their gaseous precursors and precipitation chemistry. Global observatories host facilities for a broad range of atmospheric observations and research and are often located in remote locations (Fig. 4.5) and Regional

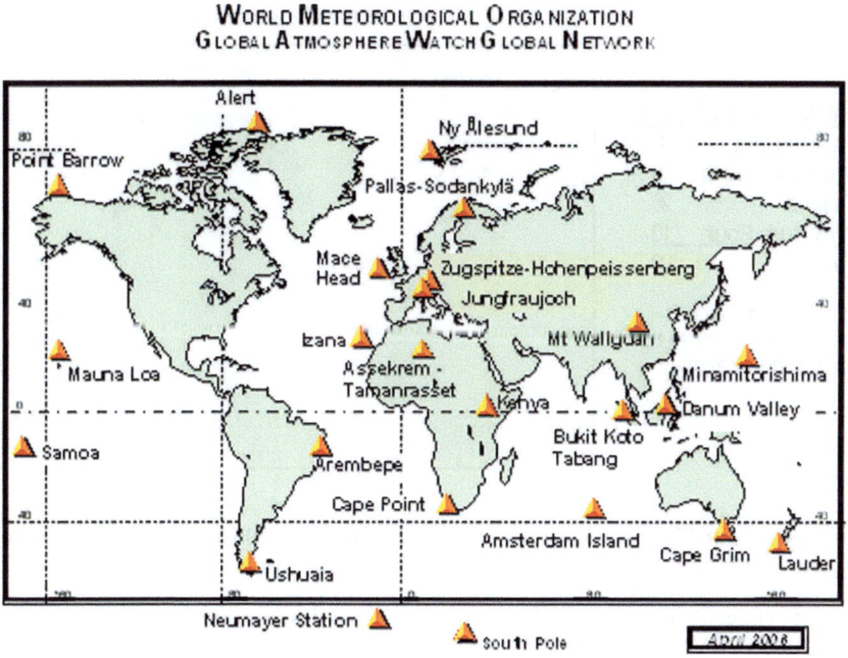

Fig. 4.5 The network of the WMO Global Atmosphere Watch (GAW) global observatories

stations measure at least one of the six GAW target variable groups and complement the global coverage. All observations are made according to GAW measurement guidelines (GAW 2003). Measurement of five core aerosol variables is recommended for all GAW stations: fine and coarse mass, fine and coarse aerosol composition, in situ aerosol scattering, in situ aerosol absorption and aerosol optical depth. In addition, a comprehensive list of optional observations is offered, including Lidar profiling, number size distributions, and cloud condensation nuclei. The GAW Global observatories Cape Grim, Australia and Mace Head, Ireland have unique long term records of cloud condensation nuclei which are directly related to cloud formation and initial cloud radiative properties.

Recently the GAW program initiated an international effort to identify contributors to long term AOD observations globally and steps needed in quality assurance and coordination to merge them into a global AOD network (GAW 2004). Such observations will form the surface-based component of an integrated global aerosol system. With satellite observations, they will yield global products such as the AOD map in Fig. 4.1, and be used in predictive weather and climate models. The surface AOD network in March 2004 (GAW 2004) is comprised of approximately 90 stations operated by about five different networks (Fig. 4.6).

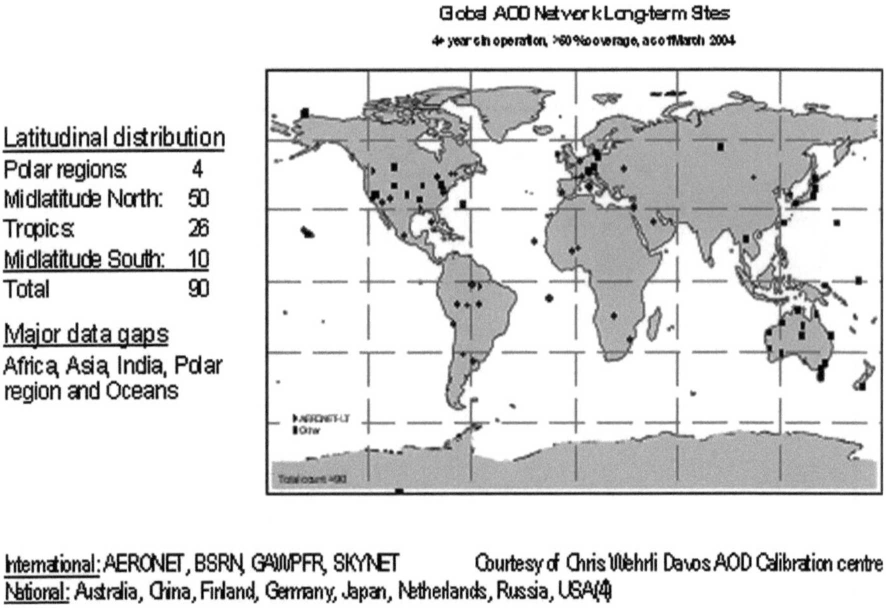

Fig. 4.6 The configuration of the long term surface-based aerosol optical depth network as of March 2004 (GAW 2004) that will be an important component of an integrated global aerosol observation system. It is comprised of a large number of international and national networks. Courtesy of C. Wehrli Davos AOD Calibration Center

In North America, several networks have been operational since the late seventies and early eighties. They were initiated mostly in response to acid deposition and visibility impairment problems. One of the goals of the Canadian Air and Precipitation Monitoring Network (CAPMoN) and the U.S.Clean Air Status and Trends Network (CASTnet) is a description of the aerosol chemical composition across North America. This is greatly augmented in the United States by the IMPROVE (Interagency Monitoring of Protected Visual Environments) network, which is operated by the U.S. Parks Service and the Environmental Protection Agency. It started with stations in most US National parks, but later expanded into national coverage, including some urban areas. IMPROVE focuses on the monitoring of aerosol chemical properties and their relationship to visibility impairment (Malm et al. 2002). It has a comprehensive aerosol chemical characterization network.

Since 1977, the European Monitoring and Evaluation Program (EMEP) has been routinely measuring the major ion composition of aerosols at stations mostly across western and northern Europe (Tørseth et al. 2002; Kahnert and Tarrasón 2003; Tørseth 2004). In addition, there is an extensive regulatory network for aerosol mass in Europe operated to monitor compliance with air quality standards. The stations tend to be in urban areas. Aerosol mass and chemical composition monitoring networks have more recently been developed in Asia, including projects underway for the Air Pollution in the Mega-cities of Asia Program and Asiairnet (http://www.asiairnet.org/).

4.3.2 Observations in North America

Figure 4.7 shows the isopleths of the integrated aerosol mass for d<10 μm (PM_{10}) and for d< 2.5 μm ($PM_{2.5}$) over the 3-year period from March 1996 through February 1999 from the IMPROVE network. Higher PM_{10} concentrations are found in the eastern United States. The highest concentration was observed in Edwin B. Forsythe National Wildlife Refuge in New Jersey at 23 μg/m^3, followed by the Southeast region, Sequoia National Park, and the Mid South, which experienced concentrations >15 μg/m^3. Outside of southern California and the Northern Rockies, the lowest PM_{10} concentrations occurred in the western United States where the concentration of PM_{10} was <8.0 μg m^{-3}. The spatial variations in $PM_{2.5}$ mimic those observed for PM_{10} (Fig. 4.7b), although the spatial gradients tend to be smaller compared with PM_{10}; this trend is consistent with the known longer residence time of fine aerosol particles in the atmosphere.

IMPROVE has provided data for aerosol sulphate and nitrate, organic carbon (OC), light-absorbing black carbon (BC), and mineral dust; that is, all of the major aerosol components except sea salt (Malm et al. 2004). Results on fine aerosol (<2.5 um) sulphate from the U.S. are summarized by the isopleths plot in Fig. 4.8. The average sulphate component of the fine aerosol measured

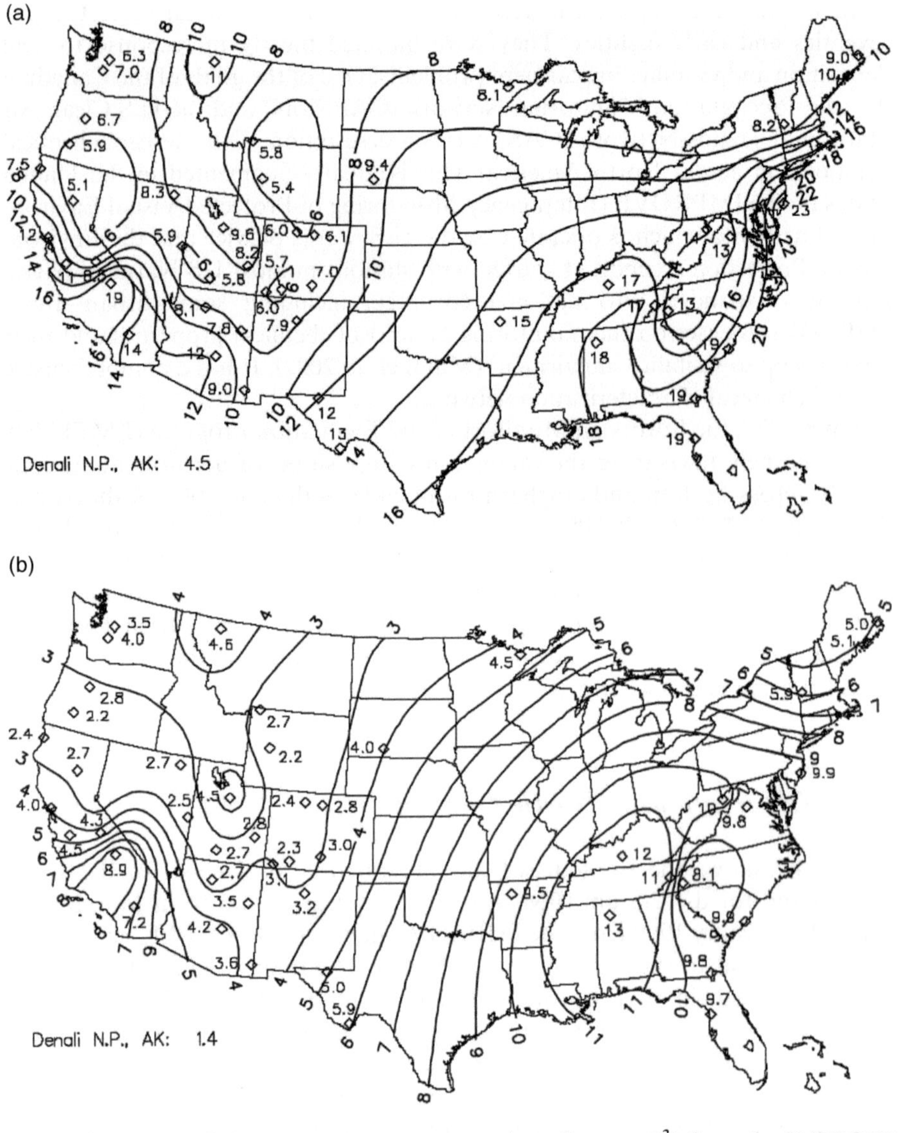

Fig. 4.7 Isopleths of the average mass concentrations (in $\mu g\ m^{-3}$) from the IMPROVE Network (**a**) PM_{10} and (**b**) $PM_{2.5}$. Since IMPROVE stations are all in the rural areas, hot spots in big cities are not specifically represented in these plots. From Malm et al. (2004) with permission of the American Geophysical Union

over the 3-year period March 1996 through February 1999 is shown. Since sulphate is a major component of fine aerosol mass, its spatial distribution and gradient across the United States is similar to that observed for total fine aerosol mass, as shown in Fig. 4.7b.

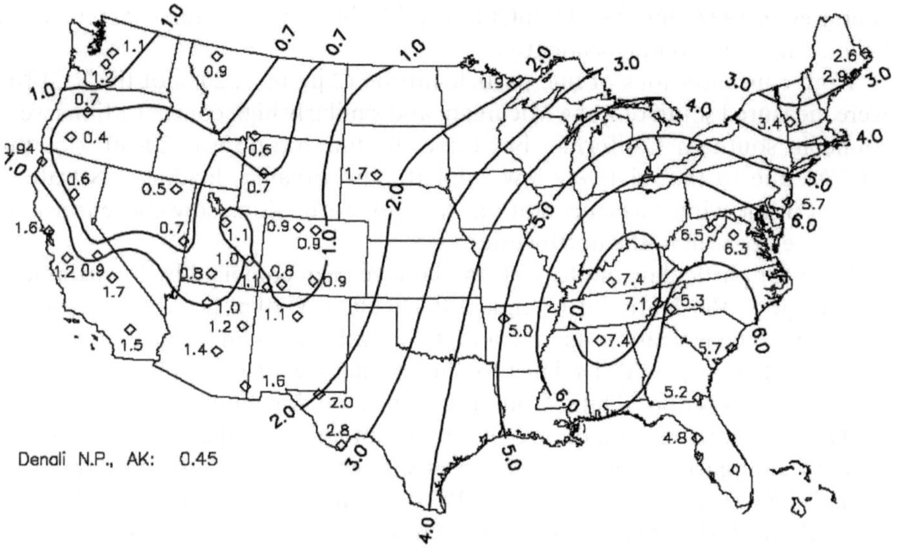

Fig. 4.8 Average fine sulphate aerosol concentrations (in μg/m^3) for each site in the IMPROVE Network, excluding Washington, D.C. From Malm et al. (2004) with permission of the American Geophysical Union

Sulphate concentrations (Fig. 4.8) typically average ~5 μg m^{-3} over a broad region of the central eastern United States, amounting to 50–60% of the aerosol mass. Sulphate tends to peak in the summer, presumably due to photochemical processing of SO_2 from coal-burning sources. Sulphate decreases towards the west, both in terms of absolute concentration and the percent contribution to reconstructed fine mass (RCFM). Organic matter also exhibits its highest concentrations in the eastern U.S., averaging 2–3 μg m^{-3}, with some enhancements in the Southeast due to emissions from fires and secondary particles formed from volatile biogenic organic compounds. Particulate organic matter (POM) accounts for the highest percentage (40–50%) of the RCFM in the Northwest and 20–30% of the mass over much of the country, and the concentrations there are highest during April to September when fires are most common. Fine soil particles are highest in April, especially in the southwestern/south-central states, roughly from south-central Texas to northern Utah, where the annual average fine soil concentrations are 0.5–2 μg m^{-3} and make up 30–50% of the RCFM. A secondary maximum in soil dust occurs in August over most of the western US, including the Northern Great Plains and Mid-South. The large-scale spatial and temporal patterns in dust loads observed at the IMPROVE surface sites led Malm et al. (2004) to suggest that long-range transport contributes significantly to the fine particle soil mass. While vertical variations in soil dust concentrations from long-range transport are known and may have influenced the IMPROVE results, the transport of dust from Africa

(Perry et al. 1997) and Asia (VanCuren and Cahill 2002) to North America has been demonstrated independently.

High concentrations of fine particle nitrate (2 μg m^{-3}, 20% of the RCFM) were measured south of Lake Michigan, and similarly high concentrations were found in southern California, but there the fine nitrate was 20–40% of the RCFM due to the relatively low sulphate and organics loadings. Nitrate is highest in the winter when conditions favor its formation. Low concentrations of nitrate are seen in the northwestern US.

Data for particulate matter in Canada have been compiled by Environment Canada into the National Chemical/Particulate Matter Database, with the purpose of investigating spatial and temporal trends in particulate matter (PM), particularly as PM relates to climate change, visibility, human health, and acid deposition. Data from the National Air Pollution Surveillance (NAPS) network in Canada are reported annually and they are periodically used to summarize trends, most recently from 1990 to 2001 (Environment Canada 2004). Analyses of the NAPS data showed 89% of the PM_{10} and 72% of the $PM_{2.5}$ mass was from open sources, including forest fires, and several sources, i.e. agricultural tilling, wind erosion, construction, and road dusts, that involved the generation of mineral particles.

The highest $PM_{2.5}$ concentrations occurred in southern Ontario where the peak levels occurred in June. Prairie sites showed variable levels of $PM_{2.5}$, with peaks tending to occur in August, while in Vancouver, the $PM_{2.5}$ loadings were of the same magnitude but fewer variables than those over the prairies. PM_{10} concentrations were highest at the prairie sites, most likely due to the presence of mineral aerosol (Brook et al. 1997). The chemical composition also varies geographically in Canada (Fig. 4.9). For example, the organic fraction in $PM_{2.5}$ decreases from 37.6% at Abbotsford on the west coast, to 25.8% at Egbert in Ontario, to 20.4% at St. Andrews in the east while the sulphate fraction increases from 16.9, 25.6, 27.0, to 31.5%, respectively (Vet et al. 2001). This fractional difference is primarily due to higher $PM_{2.5}$ concentrations from higher emissions of NOx and SOx in the east and transports of PM and precursors from up-wind regions.

4.3.3 Observations in Europe

PM_{10} and $PM_{2.5}$ measurements of the EMEP Monitoring Network shown in Fig. 4.10a indicate that the PM_{10}concentrations are the highest in Spain, where annual mean concentrations are typically above 20 μg m^{-3} but can be as high as >40 μg m^{-3}. Data from Switzerland and Germany indicate more moderate levels of PM_{10}, between 10 and 20 μg m^{-3}. Data are lacking in large parts of Europe, in particular the eastern and southeastern parts of the continent. The geographical distribution of annual mean concentrations of $PM_{2.5}$ in 2002, as measured at EMEP sites, is presented in Fig. 4.10b (Tørseth 2004). Due to the

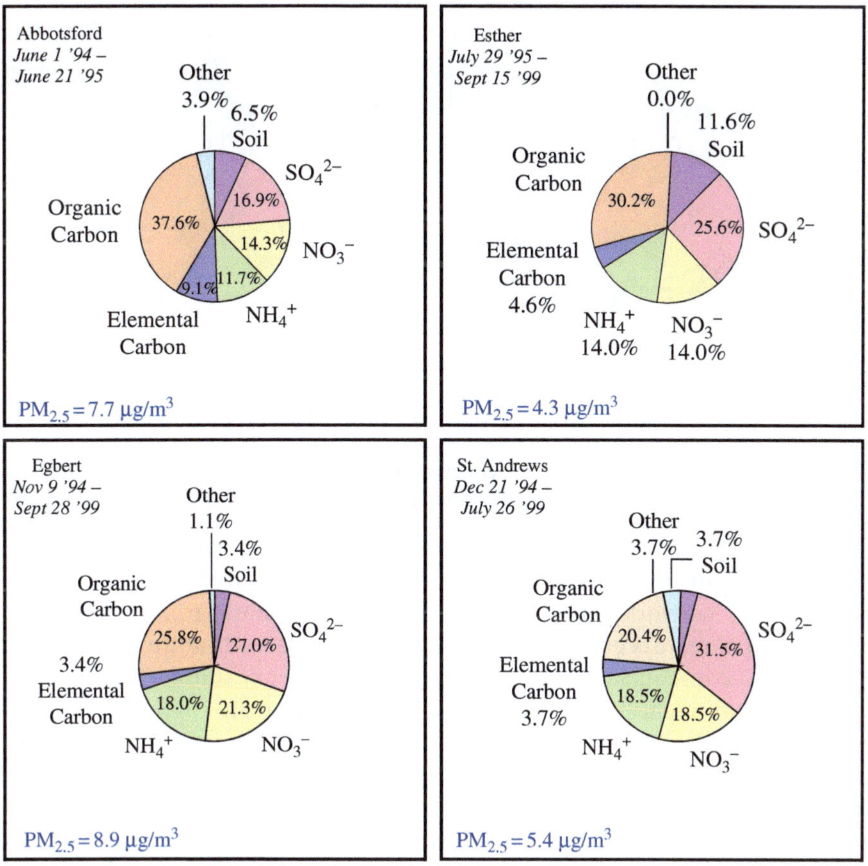

Fig. 4.9 Fractional composition of Rural $PM_{2.5}$ measured at Abbotsford (49.0°N, 122.3°W, BC), Ester (51.6°N, 110.2°W, AB), Egbert (44.2°N, 79.8°W, ON) and St. Andrews (45.9°N, 67.1°W, NB) of Canada. The size of the circle represents the magnitude of the total $PM_{2.5}$ concentration at that station (Brook et al. 1997) with permission from Air & Waste Management Association

limited number of monitoring stations, it is difficult to obtain a generalized spatial distribution pattern for the $PM_{2.5}$ mass. However, the available data indicate that $PM_{2.5}$ was generally between 5 and 15 $\mu g\ m^{-3}$ in most of Spain, Switzerland, and Germany, with higher values observed in northern Italy and eastern Austria. Additional modeling work showed the highest $PM_{2.5}$ would be expected in the Netherlands (Tørseth 2004).

In the future a growing network of continuous $PM_{2.5}$ observations, presently generated by national agencies, will provide European air quality observations for regulatory monitoring, thus increasing the capability of studying links between aerosols and precipitation.

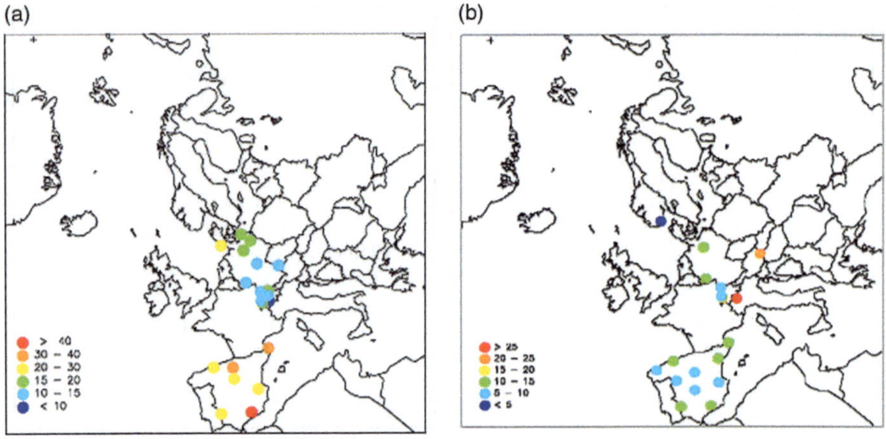

Fig. 4.10 (**a**) Annual averages of PM_{10}concentrations at EMEP sites in 2000 (Kahnert and Tarrasón 2003) and (**b**) $PM_{2.5}$ in 2002 at selected monitoring stations in Europe (Tørseth 2004)

Measurement results of aerosol sulphate in Europe, from the EMEP particulate matter network, can be seen in Fig. 4.11. Similarly to North America, aerosol sulphate concentrations in Europe coincide with anthropogenic SO_2 emissions, a consequence of coal as the primary energy source. Sulphate concentrations are highest in southern Europe, mostly at sites in Spain and Italy, and in central Europe, mostly in the eastern part of Germany, Poland, Austria and Hungary, where they can be higher than 1.35 μg S m^{-3} (or 4.05 μg $SO_4^{=}$ m^{-3}) on an annual average (Fig. 4.11). In Europe, sulphur dioxide emissions increased from the post war years and peaked mainly in the 1960s and

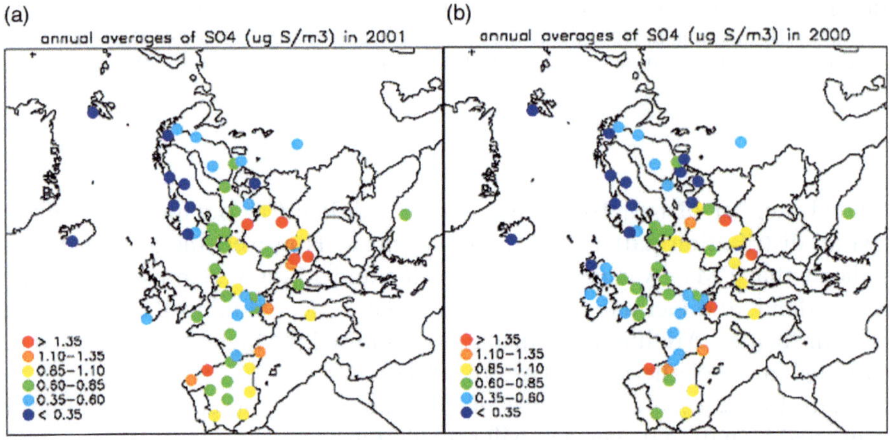

Fig. 4.11 The annual averages of sulphate mass (μg S m^{-3}) concentrations for 2000 (**a**) and 2001 (**b**) measured at EMEP sites (Kahnert and Tarrasón 2003)

1970s. Following emission control measures, gradual reductions took place in most countries and especially in Western Europe in the 1980s.

The Putaud et al. (2003) compilation summarized data from thirty-four sites which were classified as follows: natural background, rural background, near city background, urban background, free troposphere and kerbside based on distances from large pollution sources, vehicular traffic etc. The background PM_{10} and $PM_{2.5}$ mass loadings at those sites averaged $7.0 \pm 4.1\ \mu g\ m^{-3}$ and $4.8 \pm 2.4\ \mu g\ m^{-3}$, the latter being five to ten times lower than the maximum $PM_{2.5}$. The persistent background was said to be a result of anthropogenic substances, such as black carbon, transported long distances in addition to particles originating from natural sources. The regional background was concluded to be a major influence on PM concentrations in cities, and this background was at least partially responsible for the observed variability in the PM masses within the various categories of sites. The PM mass data, which showed PM_{10} concentrations occasionally exceeding $40\ \mu g\ m^{-3}$, were consistent with measurements reported in EMEP/CCC 5/2001 of PM_{10} reaching $50\ \mu g\ m^{-3}$ in Western Europe and up to $70\ \mu g\ m^{-3}$ in Eastern Europe (Fig. 4.12). Interestingly, during polluted periods, the increases in aerosol mass were mainly due to $PM_{2.5}$.

Indeed, the $PM_{2.5}$ masses at a subset of thirteen sites in the Putaud paper were found to be fairly similar, varying by roughly 4-fold, while in comparison, the particle number concentrations were more variable among this subset of data, differing by a factor of ten. The scales of these differences illustrate the limitations of mass-based data for problems more appropriately addressed by considering particle numbers, as is the case for most issues concerning aerosol/cloud interactions. Particle size distributions reviewed by Putaud et al. showed

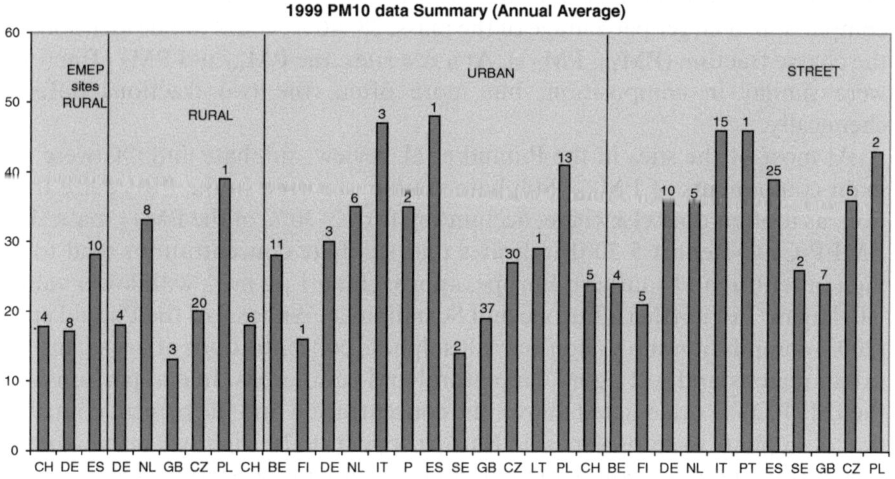

Fig. 4.12 Annual average PM_{10} concentrations in different European countries for various site types (units: $\mu g\ m^{-3}$) during 1999, from EMEP/CCC-Report 5/2001. The x-axis indicates the country and the number on the bar represents the number of sites under that category

diurnal influences from motor vehicle traffic and photochemistry, with influences from the regional background on particles about 100 nm in size. These particle size data showed that even the clean sites were affected by pollution aerosol and that during transport particle numbers are reduced by dilution and coagulation, the latter process removing smaller particles from the aerosol population as they collide with large ones.

Further analyses presented in the Putaud et al. compilation showed that the $PM_{2.5}$ masses were linearly related to particle number concentrations at the clean sites, suggesting that the small particles in clean air mostly originate from distant sources. This relationship did not hold at the more polluted sites, however, and consistent with what was noted above, the numbers of particles with diameters (d) >10 nm tended to be more variable than $PM_{2.5}$ mass. Furthermore, 70–80% of the particles had d <100 nm, and compositional analyses for size-separated samples indicated that these particles were mainly carbonaceous.

Seasonality was evident in the mass loadings at all but one of the polluted sites, with both PM fractions highest in winter. However, this trend was not seen at the natural or rural background sites. The elevated PM loadings in winter were thought to be driven primarily by meteorology, mainly large-scale stratification, but seasonality in emissions and long-range transport patterns, and the chemical processing of condensable species, such as ammonium nitrate and organics at the lower winter temperatures, were offered as additional explanations for the higher wintertime PM.

Chemical analyses showed that ~70% or more of the PM_{10} and $PM_{2.5}$ mass at the sites in the Putaud et al. compilation could be accounted for by the chemical measurements and some simple assumptions relative to elements not measured, such as non-carbon atoms in organic compounds (Table 4.1). In comparison, a larger percentage of the mass, 28–46%, could not be identified in the coarse fraction (PM_{10}–$PM_{2.5}$). At a few sites, the PM_{10} and $PM_{2.5}$ fractions were similar in composition, but more often, the two fractions differed chemically.

At most of the sites in the Putaud et al. review, sulphate and OC were the main components of $PM_{2.5}$. Sulphate mainly occurred in the $PM_{2.5}$ size fraction, as it often does elsewhere, accounting for 12–30% of the $PM_{2.5}$ mass. The EMEP/CCC-Report 5 2001 indicates that sulphate concentrations tend to be high in central and southern Europe, approaching 1 $\mu g\ m^{-3}$, with lower values ~0.4 $\mu g\ m^{-3}$ in northern Europe and Scandinavia. Annex 4 of the Putaud et al. (2003) compilation shows non-sea salt sulphate concentrations of 5–6 $\mu g\ m^{-3}$ in urban regions and 1–2 $\mu g\ m^{-3}$ in natural and rural areas. In comparison with the IMPROVE values cited above, the concentration and the percent contribution of sulphate to aerosol mass in Europe appears to be slightly less than for the eastern US and more comparable to mid-western and western states.

Carbonaceous particles were the most abundant component of $PM_{2.5}$ on a mass basis at the near-city and urban background and kerbside sites in the Putaud et al. paper, reaching 5 to 10 $\mu g\ m^{-3}$ at the most heavily impacted sites.

Table 4.1 "Grand average" composition (%) of the annual mean contributions to PM_{10}, PM_{10}–$PM_{2.5}$ and $PM_{2.5}$ adapted from EMEP/CCC-Report 5/2001

	PM10			PM2.5			PM10-PM2.5		
	Natural & rural background	Near-city & urban background	Karbside	Natural & rural background	Near-city & urban background	Karbside	Natural & rural background	Near-city & urban background	Karbside
Black carbon	6	5	13	8	8	17	4	3	7
Organic matter	16	20	22	23	22	29	4	8	10
Nitrate	7	15	10	5	17	11	8	7	8
Ammonium	6	7	4	8	10	7	1	1	0
Sulfate	19	13	10	29	17	43	3	5	5
Sea salt	8	4	3	3	3	1	20	9	6
Mineral dust	10	9	19	5	7	8	21	22	37
Unknown (*)	28	27	19	20	17	15	40	46	28

(*) calculated only when all the main aerosol components were measured.

With further reference to carbonaceous materials, Noone et al. (2003) drew attention to the uncertainties in the composition and sources of organic compounds, and they noted that studies current at that time were beginning to show that aerosol organics might play a more important role in the multi-phase atmospheric system than previously thought. These authors highlighted the potential importance of water soluble organic compounds (WSOC) for cloud formation and cloud properties, citing a study by Saxena and Hildeman (1996) showing that WSOC accounted for 20–70% of the total aerosol carbon.

Overall, mineral dust was the main contributor (20 to 40%) to the coarse particle fraction at all three general classes of the Putaud et al. sites, with sea salt accounting for 5–20% of the coarse particles (Table 4.1). Nitrate and black carbon contributed roughly equal proportions to the PM_{10} and $PM_{2.5}$ fractions, and when the PM_{10} concentrations exceeded 50 $\mu g\ m^{-3}$, nitrate and organic matter were the main contributors to the aerosol mass.

In addition to the recognition of WSOC as a source of CCN, studies conducted for EUROTRAC-2 showed that road traffic is an important source for nanometer-sized particles (Noone et al. 2003 and references therein). Even though these particles are rapidly lost during transport, the processes involved became better understood as a result of research conducted for the program. Another general conclusion drawn from the EUROTRAC studies was that the algorithms for describing particle emissions from surface processes are not universal but rather there are differences between the situations in Europe compared with North America.

4.3.4 Observations in Asia

Asia is an immense and diverse source of aerosol particles and trace gases. Mineral dust particles mix with pollutants from various industrial and energy-related sources, biomass burning and transport exhaust to produce an extraordinarily complex regional aerosol mix. In the late winter and early spring of 2001, two major field campaigns were mounted to study the atmosphere over Asia and the western Pacific: TRACE-P (TRAnsport and Chemical Evolution over the Pacific) and ACE-Asia, the third in a series of Aerosol Characterization Experiments (ACE).

Several findings from these programs concerning Asian-aerosol outflow are directly relevant to this aerosol/precipitation review. First, Jordan et al. (2003), Seinfeld et al. (2004a), and others have presented data showing that gas-phase species mix with Asian dust particles, making the dust particles more hygroscopic, and presumably affecting the formation of clouds over regional scales. In particular, the Seinfeld et al. (2004a) studies highlighted the fact that the structure of the atmosphere is highly complex and that layers of dust, pollution, and biomass burning products can be transported long distances either as distinct entities or mixed together. Data obtained with a single-particle mass spectrometer by K. Prather and colleagues aboard the research vessel Ronald H. Brown and reported in Arimoto et al. (2006) indicate that the processes involved in the mixing of sulphate, nitrate, and chloride with dust are competitive or selective in nature. These uptake and mixing reactions not only affect the physical properties of the mineral dust but also affect the number and size distributions of the ensemble aerosol population, especially nitrate-containing, but also sulphate-containing, particles. In TRACE-P Kittaka et al. (2004) showed that the mass of sulphate in ice clouds was much less than in liquid clouds and that sulphate particles could be released into the atmosphere following the oxidation of SO_2. Their study showed a close and direct linkage between meteorology and aerosols, that is, accurate predictions of clouds and precipitation were needed to successfully model the atmospheric sulphate budget.

ACE-Asia and TRACE-P also have provided information on new particle formation, particle distributions, Asia outflow, and how areas downstream could be affected by long-range transport, all topics relevant to this review. For example, a study by McNaughton et al. (2004) showed evidence of secondary aerosol particle formation over synoptic scales. These particles formed heterogeneously in a postfrontal situation when dry continental air mixed with maritime air; that is, nucleation occurred when mixing led to supersaturations of condensable species. Comparisons of model results with aircraft observations showed that in areas where dust was concentrated, it composed up to 90% of the super-micrometer particles, and even though dust impacts sometimes extended above 6 km, the influences were strongest in the boundary layer (Tsai et al. 2004). These authors also showed correlations between submicron aerosol mass loadings and ethyne, an indicator of combustion sources, and from this they concluded that anthropogenic emissions were the major source for the submicron particles. Similar conclusions have been reached by many others; indeed Clarke et al. (2004) suggest that even though interactions sometimes do occur, the coarse mode and fine mode particles can at times function as nearly independent phenomena, with coarse mode dust often nearly non-hygroscopic and the fine pollution aerosol much more hygroscopic.

Aircraft studies for TRACE-P and ACE-Asia have shown that aerosol properties vary strongly with altitude, and comparisons with ground-based measurements show that boundary layer measurements generally are not good indicators of the aerosol properties above the surface. More specifically, the aerosol population in the boundary layer (below $\sim$2000 m) is in many ways distinctly different from that in the free troposphere. Pollution aerosol is largely confined to the boundary layer, while dust is distributed much more evenly throughout the column (Clarke et al. 2004).

One drawback from the above analysis is that these campaigns were conducted over the marine atmosphere outside the mainland China downwind of the major pollution sources. Within the mainland continent, observational data are very limited. Recently, monitoring stations have been established in China to quantify air pollution levels and to better define the problems. Table 4.2 summaries the typical range of aerosol concentrations observed in China from

Table 4.2 Aerosol concentrations observed at Chinese stations (unit: μg m^{-3})

Station	Lat/Lon	TSP	Sulphate	Nitrate	BC	OC	Year
Lin An	30.3°N, 119.73°E East China	50–70	15–18	24	~2.0	10–12	2004
Zhenbaitai	38.3°N, 109.7°E Northwest China	45–50	4–9	~2.0	2–5	9–17	2003–2004
Lian Yun Gang	34.7°N,119.3°E East China	100			3–5	10–14	2003
Tong Liao	43.2°N, 122.2°E Northeast China	78			~2.0	~7.0	2003

several monitoring stations. Data from the newly developed stations (X.Y. Zhang, Center for Atmosphere Watch and Services, CMA, personal communication) as well as from field campaigns (Zhang et al. 2005a) show elevated aerosol concentrations. The total suspended particle (TSP) concentrations generally range from 50 to 100 μg m^{-3}.

High concentrations were also found for other aerosol species shown in Table 4.2. One notable example is the elevated organic carbon content at all sites. Increased aerosol concentrations over China are reflected in long term decreases in the total global radiation reaching the surface (Che et al. 2005). Over the latter half of the 20th century there have been significant decreases in global radiation (–4.5 W m^{-2} per decade), direct radiation (−6.6 W m^{-2} per decade), clearness index (−1.1% per decade), and the percentage of possible sunshine duration (−1.28% per decade). This is at least partially attributable to the increase in the aerosol loading in China. Changes in precipitation at the ground in China from 1951 to 2000 have been observed in the same areas where the total global radiation decreased (Zhai et al. 2005). Even though there is little trend in the country-wide precipitation for China, there are distinct regional and seasonal patterns. Annual total precipitation has significantly decreased over southern Northeast China, North China, and over the Sichuan Basin, but significantly increased in western China, the Yangtze River valley and the southeastern coast. There are many factors influencing the precipitation variations. Nevertheless, coherent changes in the radiative fluxes and precipitation patterns no doubt contain valuable information on the link between the concentrations of aerosols and patterns of precipitation in China. This needs to be investigated in more detail using extensive monitoring network observations and numerical models.

4.3.5 Southern Hemisphere Observations

PM_{10} and $PM_{2.5}$ have been determined gravimetrically, and more recently, with the use of TEOM monitors at a network of Australian monitoring stations that are mostly located near the major urban centers along the east and southeast coast. Annual median PM_{10} concentrations are typically 15–20 μg m^{-3}

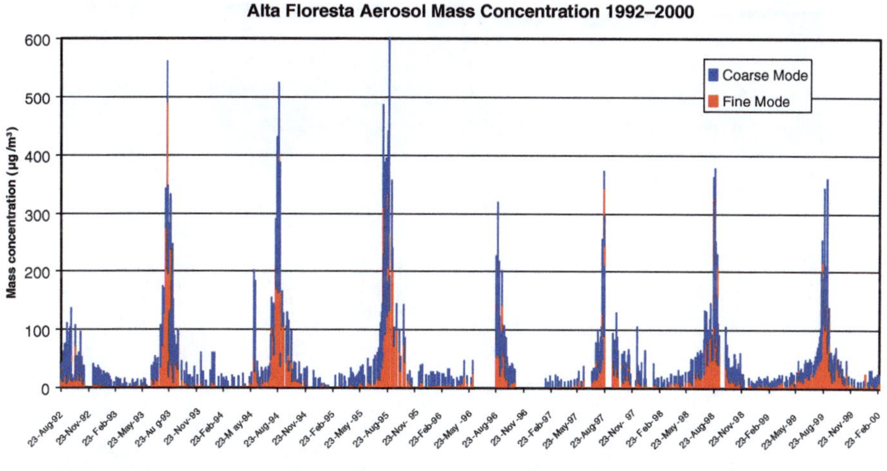

Fig. 4.13 Long term aerosol concentrations in the center of the biomass burning region in South America, showing that every year concentrations of PM_{10} particles exceeds 300 µg/m^3. From Artaxo et al. (2002) with permission of the American Geophysical Union

in coastal locations in New South Wales, Victoria, Queensland, and West Australia, but appear to be higher in Tasmania where the annual median of $PM_{2.5}$ concentrations range from 5 to 10 µg m^{-3}(DEH 2004).

Figure 4.13 presents long term aerosol concentrations in the center of the biomass burning region in South America, showing that concentrations of PM_{10} particles exceeds 300 µg/m^3 every year (Artaxo et al. 2002). These high aerosol loadings also shows in the AERONET sun-photometer network, presented in Fig. 4.14 with a time series of AOD in the central Amazonia. AOD at 500 nm peaks higher than 3, corresponding to a very high atmospheric aerosol loading. Areas covered by high aerosol concentrations occur not only in Amazonia but also in a large part of the rest of the continent. Figure 4.15 shows a MODIS image of smoke covering much of South America, with AOD ~3. The high concentrations of aerosols and the large geographical extent show clearly that smoke is a major driver of weather, climate and air quality in South America.

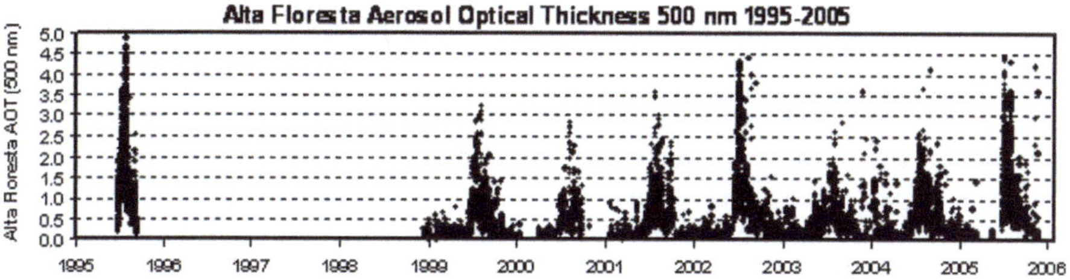

Fig. 4.14 Time series of aerosol optical depth obtained by an AERONET station in central Amazonia, showing AOD values higher than 3 at 500 nm

Fig. 4.15 MODIS image of smoke from biomass burning covering a large fraction of South America. Image taken in September 25, 2002. Curtesy of NASA

The forested areas of South America are also source of natural biogenic aerosol, both primary and secondary, the latter produced from precursors such as volatile organic compounds. Dry areas of Northern Chile and Southern Argentina are significant sources of dust. In addition, emissions from large urban areas such as Sao Paulo, Buenos Aires, Santiago, Quito, Lima, Rio de Janeiro and others strongly contribute to the overall regional aerosol burden.

4.4 Vertical Profiles

4.4.1 Vertical Profiles of Aerosols – Surface-based Network

The vertical distribution of atmospheric aerosols is an important property that determines their lifetime and long-range transport potentials as well as their climatic impact. However, due to limitations in measuring equipment and technology, information on the vertical distribution of aerosols over the globe is limited, not to mention the lack of data on size distribution and chemical compositions as a function of height. Ground-based Lidar networks for aerosols have been established in various locations (Chapter 5). Since these networks are in an early stage of operational monitoring, a global data set for the vertical profiles of aerosols is not readily available. However, some regional characteristics of aerosol vertical profiles can be deduced from the networks.

Figure 4.16 shows the mean aerosol Lidar extinction profiles for northern and southern European stations in 2000 (Matthias et al. 2004) and for a US station representative of the inner continental background in North America during 2000 (Ferrare et al. 2006). These profiles are compared with the global aerosol model outputs from AeroCom. The model results agree reasonably well with the observations. Both observations and models show a distinct aerosol layer at higher altitudes for the southern European stations, which is not present at the northern stations. The simulations attribute this layer to the presence of dust. Aerosol concentrations in the planetary boundary layer are higher at the northern stations; the simulations attribute this by increased

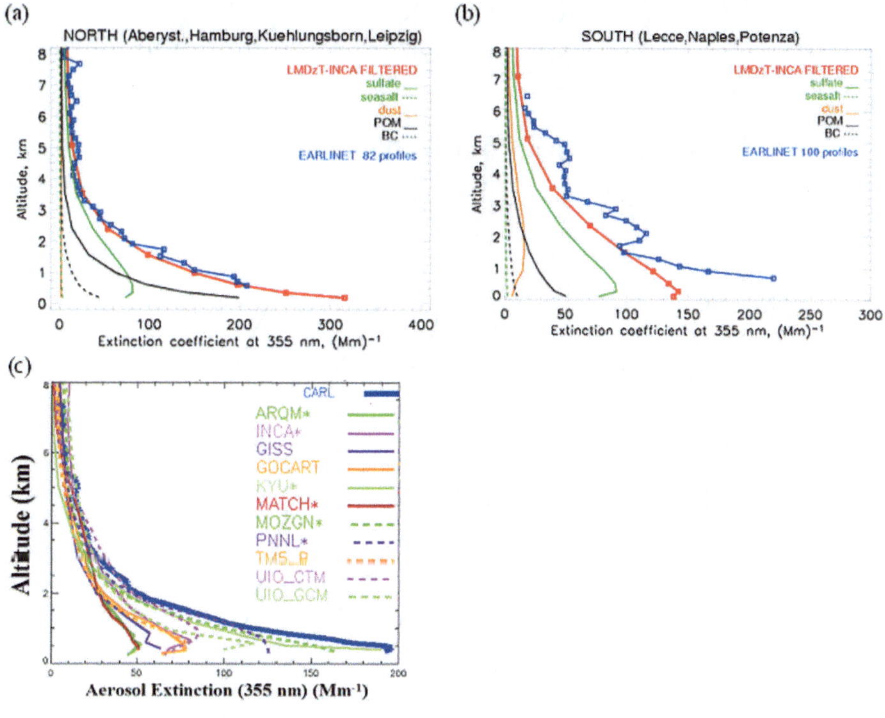

Fig. 4.16 Comparison of the mean aerosol extinction profiles from Lidar (at 355 nm) with the LMDzT INCA model for (**a**) northern and (**b**) southern EARLINET stations in 2000. The observed extinction profile is indicated by the solid blue line. The all-year-average modeled profiles are plotted with a solid red line. The modeled contributions from individual species are shown as well in this figure, as indicated in the legend (adapted from S. Guibert et al. presonal communication); (**c**) Average aerosol extinction profiles over the ARM Southern Great Plains (SGP) CRF (36.62 N, 97.5 W, 317 m height) site during 2000–2001 compared to AeroCom models (see Sect. 4.5.2). (Ferrare et al. 2006, with the author's permission)

sulphate concentrations. The aerosol profiles at the US station are similar to the northern European profiles. The average aerosol extinction profiles simulated by the AeroCom models (Sect. 4.5.2) typically show good agreement with the observations above about 2 km, but are not as good in the planetary boundary layer.

Figure 4.17 shows monthly mean extinction coefficient profiles observed at the AD-Net Lidar network site in Sri Samrong, Thailand (17.1° N, 100.0° E)

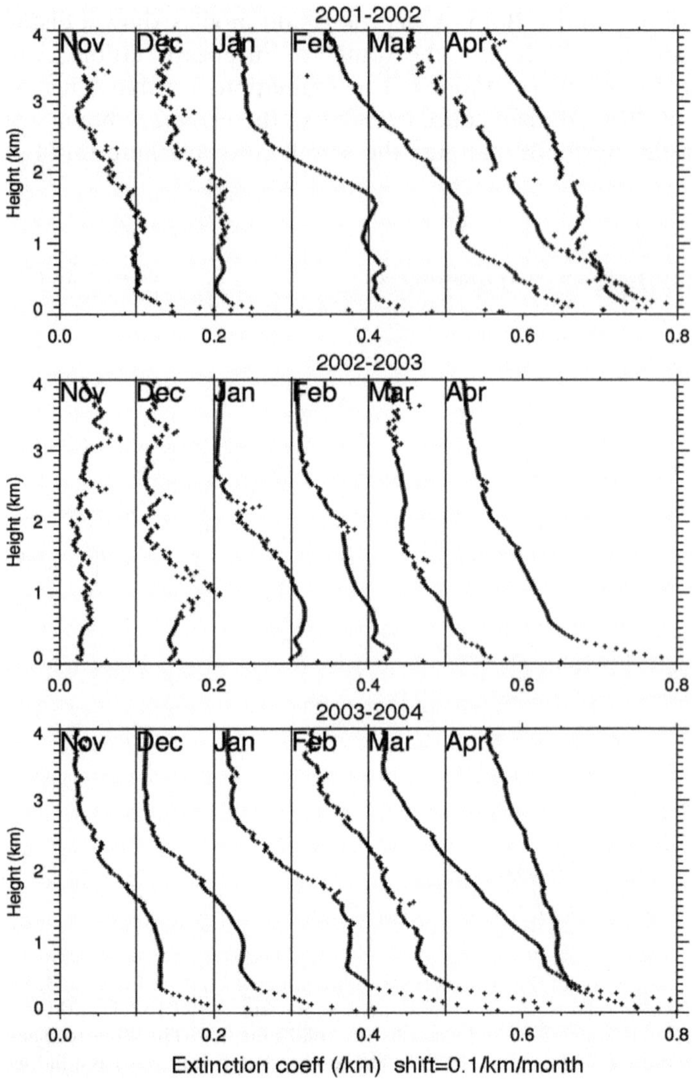

Fig. 4.17 Monthly mean extinction coefficient profiles in Sri Samrong, Thailand during dry seasons in 2001–2004 (Shimiz et al. 2005) with permission of by the National Institute for Environmental Studies, Tsukuba, Japan

during dry seasons in 2001–2004 (Shimizu et al. 2005). The boundary layer height increases from January to April, and the extinction due to aerosols also increases. A marked increase of aerosol extinction occurs from November to February during the burning season. Year-to-year differences are seen in the aerosol extinction and seasonal variation. The aerosol extinction in 2003 was lower than other years. This result was consistent with the co-located sky-radiometer data.

One of the most useful applications of Lidar data is to validate aerosol transport models. For this purpose, event-to-event comparisons using a network of continuously operated Lidars is effective. Continuity of the observations and suitable locations of the observation sites (depending on the scale of phenomena) are very important. Aerosol type classification methods using depolarization ratio, wavelength dependence, and the Lidar ratio (if available) are useful for the comparison. Figure 4.18 shows an example of Asian dust transport captured by the Lidars in Beijing, Suwon (37.1° N, 127.0° E), Nagasaki (32.8° N, 129.9° E), and Tsukuba (36.1° N, 140.1° E). Vertical structures and relative magnitudes of the dust storm were clearly observed at those stations. Downwind of the dust source regions, measuring the clouds and aerosol together using a polarization-Lidar, Sassen (2005) detected the

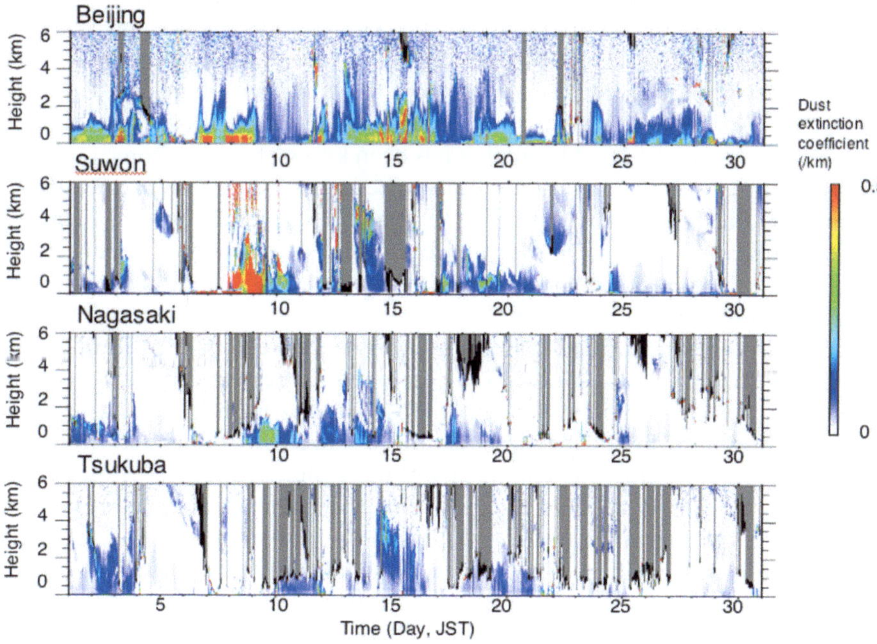

Fig. 4.18 Mie-Lidar dust extinction coefficient ($S_1 = 50$) at 4 locations in China and Japan, April 2002, with permission of by the National Institute for Environmental Studies, Tsukuba, Japan

long-range transported dust aerosols in 2004 spring in Alaska; these particles served as ice nuclei (IN) in the atmosphere and caused ice clouds to form at modest super saturations and temperatures, 4.5 and 6.5 km above sea level. The aerosol profiles peaked at a much higher height above the ground in Alaska than those observed at East Asian stations, reflecting the evolution of dust long-range transport.

4.4.2 Vertical Profiles of Aerosols – Space Borne

Unlike Lidars that profile the atmosphere above at a single geographic location, or airborne Lidars that measure above and below aircraft on regional scales along flight trajectories, space borne Lidar allows the study of aerosol transport in regions that are difficult or impossible to explore by other means. As such spaceborne Lidar may be the only means to capture the vertically profile of aerosols in those regions that is essential in understanding the global transport perspective. The era of space borne backscatter Lidars started with the successful launch of LITE (Lidar In-Space Technology Experiment) in September 1994 onboard the Space Shuttle Discovery. There were two purposes to the LITE mission: evaluate technological requirements of a space borne Lidar (Couch et al. 1991) and evaluate the science capability (McCormick et al. 1993). LITE convincingly demonstrated the value of space borne Lidar in retrieving the vertical structures of clouds (Winker and Trepte 1998; Omar and Gardner 2001) and aerosols on a global scale.

Several examples of the global transport of aerosols were evident from the LITE data. Aerosols from natural sources, such as Saharan dust, was measured on several orbits by LITE. Although it has been known for quite some time that large quantities of Saharan dust are transported across the Atlantic towards the Caribbean, the unique capabilities of a space borne Lidar proved ideal for tracking and quantifying the magnitude of these events (Powell et al. 1996). LITE showed these enormous plumes would stretch 100's of kilometers, reaching altitudes of >5 km (Fig. 4.19).

LITE also proved invaluable at pinpointing source regions of anthropogenic aerosols: from small local sources and combined net regional impacts, to large scale activity such as biomass burning. Observations from LITE showed biomass burning in South America extending hundreds of kilometers from the source region. The depth and complexity of these plumes are nicely captured by the Lidar, making it possible to apportion their impact and quantify their contribution to long-range transport. Using multiple orbits as LITE precesses around the globe made possible the tracking of regional transport. A good example would be the measurement of anthropogenic aerosols leaving the Eastern United States and riding the "gulfstream highway" towards Europe (Hoff et al. 2001). An example of a LITE validation track by a Canadian team is given in Fig. 4.20. This represents an orbital overpass along the Californian coastline, while aircraft measurements along the track were being made below.

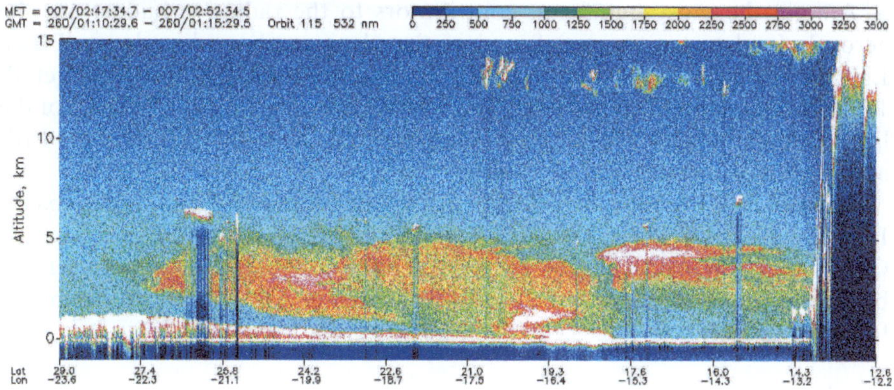

Fig. 4.19 LITE observation of a Saharan dust plume over the Atlantic. Significant structure (as the raw signal counts at 532 nm wavelength) is shown within the plume as it reaches altitudes of 5 km. Courtesy of NASA Langley Research Center

The most spectacular feature from the LITE data is the identification of the urban aerosol plumes from San Francisco and Los Angeles (the cities locations are obvious by the position of the plume origin) flowing out along the coastline for several hundred kilometers (Strawbridge and Hoff 1996). What is truly remarkable is the 60 s trip for a space borne Lidar to go from San Francisco to Los Angeles. Other Lidar data from various aircraft and ground sites provided validation of the measurements, thus increasing confidence in the LITE's performance.

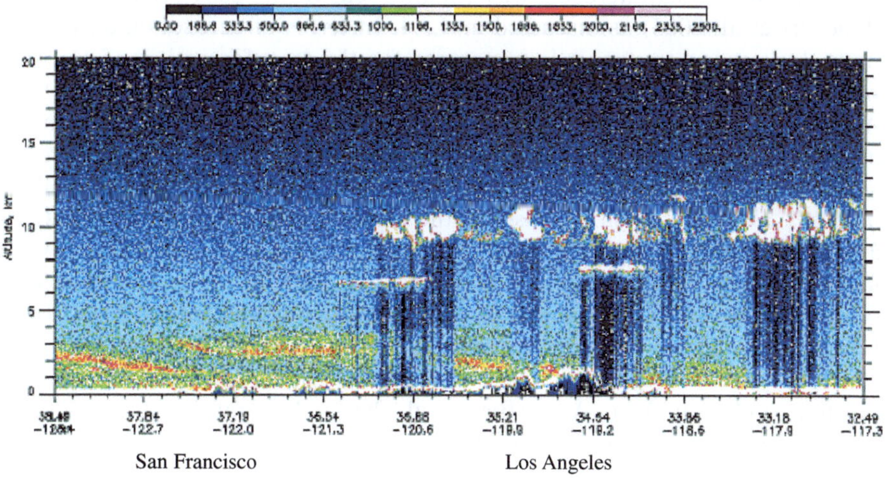

Fig. 4.20 A 2 min snapshot from LITE along California's coast showing the long-range transport of pollutants from the Los Angeles and San Francisco areas. Courtesy of NASA Langley Research Center

Among the most significant contributors to the radiative budget are the aerosols (and water vapour) found in the planetary boundary layer (PBL). Lidars have been used for several years to determine PBL height (Melfi et al. 1985; Piironen and Eloranta 1995; Strawbridge and Snyder 2004) because of the large gradient in aerosol concentration that occurs between the top of the PBL and the free troposphere.

The LITE technology provided the backbone for more recent satellite-based Lidars; e.g. CALIPSO (Cloud-Aerosol Lidar and Infrared Pathfinder Satellite Observation) (Winke et al. 2002) which will provide 4-D distributions of aerosol backscattering in the atmosphere after its launch in April 2006 (See Chapter 5 for more details).

4.5 Transport of Atmospheric Aerosols

Transport is responsible for the spread of aerosols from their source locations to downwind destinations caused by dispersion and convection related to turbulence and by advection on winds. Dispersion is controlled by small-scale turbulent processes that depend on atmospheric stability and advection by the large-scale general circulation patterns.

4.5.1 Characteristics of Global Aerosol Transport

Long-range horizontal transport is important, for example, for dust aerosols being transported from Africa and Asia to North America (Prospero 1999; Husar et al. 2001; Gong et al. 2003b; Gong et al. 2006). The spatial and vertical distribution of Asian aerosols was a main focus of the 2001 ACE-Asia study (Huebert et al. 2003) as discussed in 4.3.4. In terms of aerosol-cloud interactions, long-range transport is important for the transformation of aerosols, with consequences for their ability to act as CCN and IN (Levin et al. 1996). A study by DeMott et al. (2003), in which African dust aerosols were sampled in the east part of the US, concluded that these particles were very effective IN.

Pathways for trans-Pacific and trans-Atlantic transport differ in the following ways. For trans-Pacific transport, the highest concentrations in the outflow are in the boundary layer (0–2 km), but the strongest outflow fluxes are in the lower free troposphere (2–5 km) and reflect episodic lifting of pollution over central and eastern China ahead of eastward moving cold fronts. This frontal lifting, followed by westerly transport in the lower free troposphere, is the principal process responsible for export of both anthropogenic and biomass burning pollution from Asia (Bey et al. 2001). In contrast, trans-Atlantic transport of North American pollutants takes place in the lower troposphere year round, and transport in the middle and upper troposphere is also important in summer (Li et al. 2002). Compared to trans-Pacific transport of Asian pollution to North America, which takes place mainly in the free troposphere followed by subsidence (Jacobi et al. 1999; Yienger et al. 2000), the relatively short distance between North America and Europe, and the prevailing westerly flow extending down to the surface, favor transport in the boundary layer. Over eastern Asia, in contrast, westerly winds do not extend down to the surface, so that lifting to the free troposphere is a prerequisite for trans-Pacific transport (Bey et al. 2001).

Long-range transport to the Arctic is also of importance, because the background atmosphere, especially in summer, is very clean. Girard and Curry (2001) point out that the concentration of IN, aerosol number concentrations, the slope of the aerosol size distribution, and the aerosol solubility, may impact substantially the cloud phase and total water content. A mesoscale model simulation by Lohmann et al. (2003) showed that the addition of anthropogenic aerosols to the background aerosol in the Arctic not only impacts liquid water clouds and drizzle, but can also alter snowfall rates. In this study, depending on the assumed snow crystal shape, which determines the accretion rate, the total amount of precipitation reaching the surface after 7 h of simulation can be larger or smaller than the polluted case compared to the clean background.

The transport of atmospheric aerosols from their source regions is also complicated by the seasonal variation of transport patterns and removal processes (Barrie 1990). The strong seasonality of Arctic aerosols is a function of their atmospheric lifetime, which is controlled by the variations in transport and removal processes (Barrie 1986). Because of stable thermal stratification in the atmospheric surface boundary layer and <10 mm liquid H_2O/month of frozen precipitation in the cold half of the year from October to May, aerosol residence times are much longer in winter ($\sim$3 to 7 weeks) than in summer ($\sim$3 to 7 days). Therefore, the seasonality of any aerosol species in the Arctic is more likely dominated by factors such as meteorology, transport and removal processes specific to the major transport pathways, and less to seasonal variations of emissions in the mid-latitude source regions of these species.

In addition, inter-continental transport is also modulated by climate fluctuations. By studying the transport of Asian dust to North America for the last 44 years with a numerical model, Gong et al. (2006) identified shifts in the inter-continental pathways for dust aerosols related to the El Niño Southern Oscillation. In that study, an analysis of the variability of the 44-year zonal mass transport flux (i.e. concentrations multiplied by westerly/easterly wind speeds) was conducted for eight typical El Niño and eight La Niña years. The zonal mass transport flux indicates the major transport routes and direction of Asian dust across the Pacific. Most trans-Pacific transport of Asian dust aerosol occurs in the middle troposphere between 2 and 5 km (Zhao et al. 2006). Fig. 4.21 shows the averaged dust transport flux (filled contours) for El Niño (Fig. 4.21a) and La Niña (Fig. 4.21b) years integrated from 3 to 10 km. On the same plot, 8 year anomalies (dashed contour lines) from the 44 year averaged values are also superimposed.

A sharp difference between zonal transport during El Niño and La Niña years is most clearly seen in the center of the transport path. During El Niño years, trans-Pacific transport is centered at 45°N, while during La Niña years it

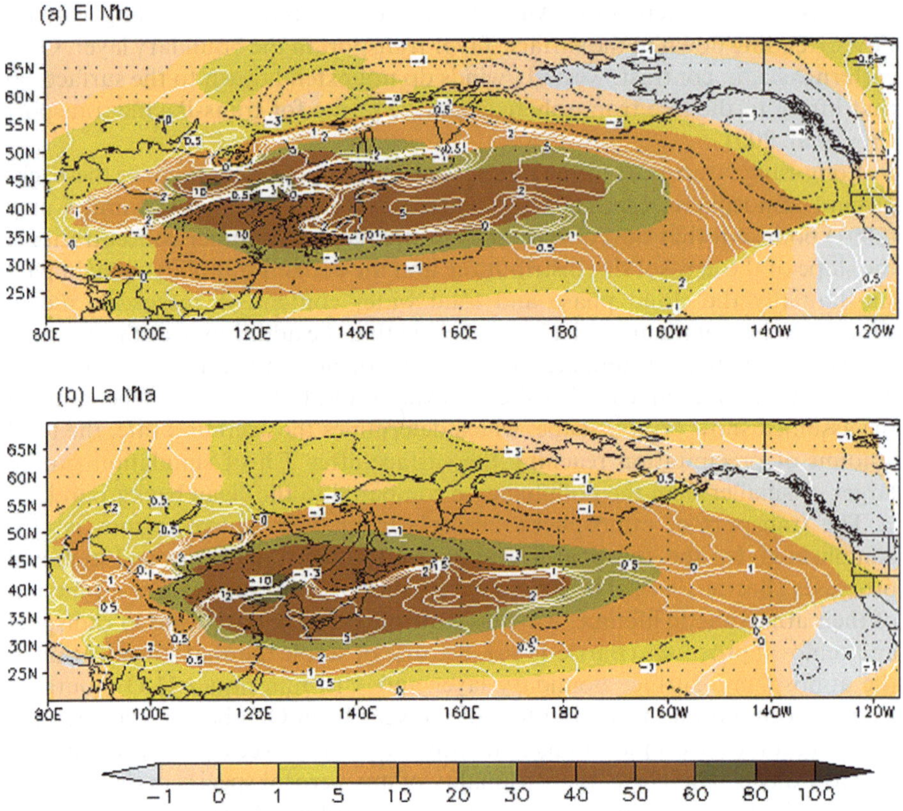

Fig. 4.21 The zonal transport fluxes (μg.m^{-2}.s^{-1}) of Asian dust in spring of (**a**) 8 El Niño years (1966, 1973, 1983 1987, 1987, 1992, 1998 and 2003) and (**b**) 8 La Niña years (1965, 1971, 1974, 1976, 1986, 1989, 1999 and 2000). The contour lines are for the anomalies relative to the 44-year mean of Asian dust aerosol in spring. From Zhao et al. (2006) with permission of the American Meteorological Society

is around 40°N. The zonal transport fluxes of Asian dust aerosol through 125°E in spring for the same El Niño and La Niña years indicate a shift of the main out-flow point in Asia from 45 to 40°N during La Niñas. In contrast, in the eastern Pacific (140°W) where the outflow reaches the coast of North America, peak transport moves southward from 45 to 40°N during El Niño years and to lower heights (5000 m during El Niño versus 6500 m during La Niña years).

Interannual variability of African dust transport over the north tropical Atlantic has been investigated by Chiapello et al. (2005) who analyzed in situ surface concentration data from Barbados beginning 1966, along with the Total Ozone Mapping Spectrometer (TOMS) and Meteosat dust optical thickness (DOT) records covering the last two decades. Their analysis shows a large regional impact of drought conditions in the Sahel on dust emissions and transport both in winter and in summer.

4.5.2 Numerical Modeling Studies - AeroCom Results

As many quantities of aerosols, including their transformation and spatial distribution on the global scale, cannot completely be covered by observations, numerical models are used to integrate the current knowledge and to better understand aerosol processes. AeroCom (http://nansen.ipsl.jussieu.fr/AEROCOM/), an initiative to compare aerosol modules in global models among each other and to observations, started in 2003 to systematically identify and quantify uncertain elements in global aerosol simulations. An analysis of the simulated aerosol life cycles for 16 global models shows large diversities, especially in spatial distributions, aerosol composition, and water uptake (Textor et al. 2006). Results from this study are summarized below.

The zonally averaged vertical concentrations of the total (component-combined) aerosol mass for the AeroCom models are shown in Fig. 4.22. The aerosol mass concentrations are dominated by the natural aerosols: dust contributes 50–80% of the mass in all models, except two models where sea salt is dominant. All models show two maxima, one in the northern hemisphere resulting from dust and the other in the southern hemisphere from sea salt in the " roaring forties" of the southern oceans.

Models agree least on the vertical transport of sea salt, followed by dust, particulate organic matter, black carbon and sulphate. Of the total aerosol mass, 14% is above 5 km altitude; this varies from 5 to 26%. The mass composition close to the surface, i.e. below 1 km, is dominated by sea salt in eight models and by dust in seven models. In this layer contributions to the total mass vary for sea salt between 20 and 80% and for dust between 15 and 70%. All models have a dust mass maximum in the upper PBL between 1 and 2.5 km. Vertical transport is relatively weak for the natural species. The sea salt mass decreases strongest with height. The all-models average mass fractions of sea salt and dust above 5 km are 9% (varying from 0.001 to 25%), and 14% (varying from 3 to 28%), respectively. The global annual average mass fractions above 5 km height are indicated in Fig. 4.23. This height has been chosen as representative for aerosols above clouds where they have increased life times due to the absence of wet removal processes. The ranges of vertical transport, i.e. the differences for each model between the species with the largest and the smallest mass fractions above 5 km, respectively, are indicated by the gray shadings in Fig. 4.23. These ranges, which indicate the degree of similarity of the vertical transport among the species within a given model, differ among models; the all-models-average range is 25%, varying from 10 to 45%.

The vertical transport of black carbon and particulate organic matter are more similar to each other than to the other aerosol components in most models as they have similar source regions. The two species are well dispersed in the vertical with the maximum situated in the upper PBL between 1 and 2.5 km in all models. The all-models average mass fractions above 5 km are 22% (varying from 6 to 38%) and 21% (varying from 6 to 40%) for black carbon and

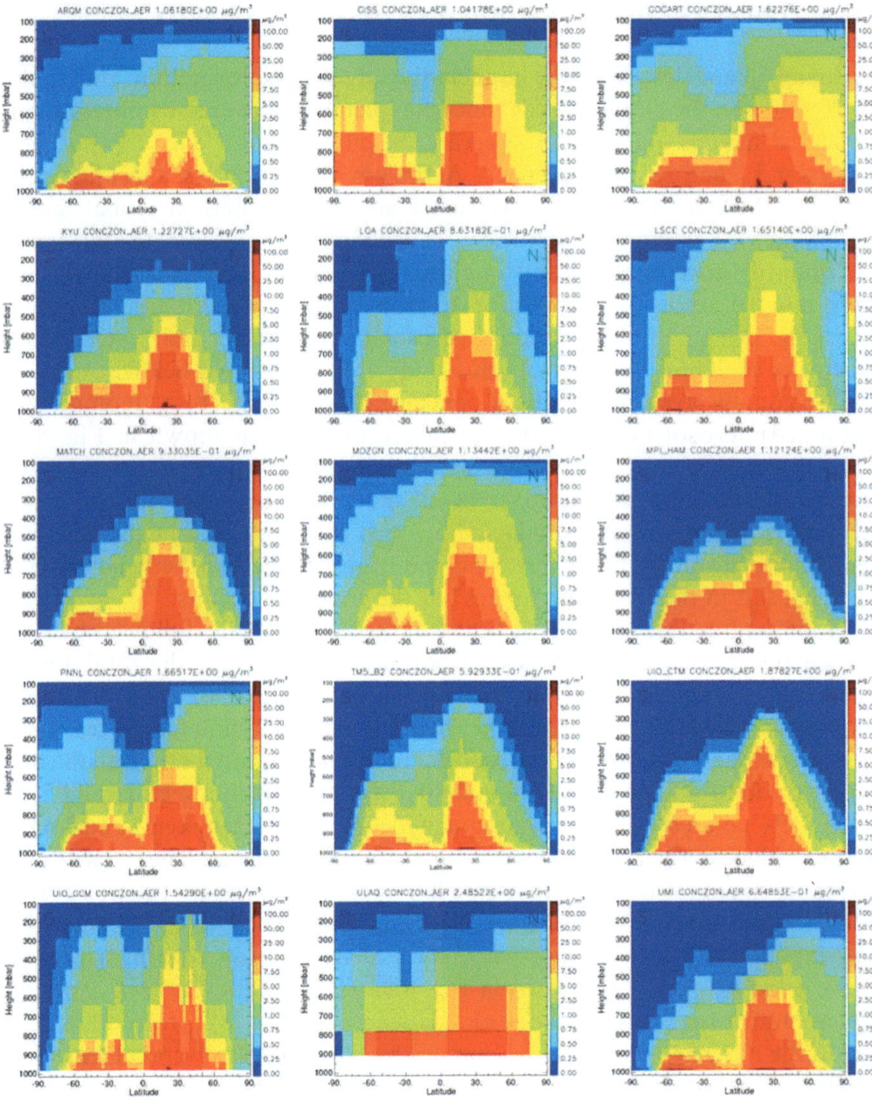

Fig. 4.22 Zonally and annually averaged concentration of total aerosol in (μg m^{-3}), the model name and the global average are given on top of the plots. Note the non-linear color scale. The white shading of lowest layer above ground in some models indicates that no data have been available in this layer (Textor et al. 2006)

particulate organic matter, respectively. In eight models, the vertical transport is stronger for black carbon than for particulate organic matter, in six models the situation is reversed, and in two models similar.

Sulphate is present at the highest altitudes of all species in most models. The all-models-average sulphate mass fraction above 5 km is 33% (varying from

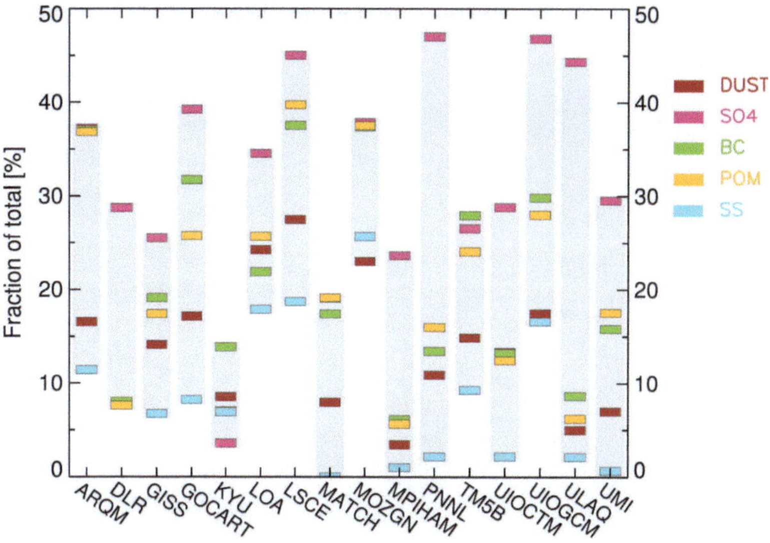

Fig. 4.23 Global annual average mass fractions in (%) of total mass above 5 km altitude for the AeroCom models. The gray shadings frame the range for each model (Textor et al. 2006)

4 to 47%). The main sulphate burden is situated in the upper PBL between 1 and 2.5 km in nine models, in the lower free troposphere between 2.5 and 5 km in three models, in the upper free troposphere between 5 and 10 km in one model, and above 10 km in the tropopause region in three models. In general, the sulphate-contribution to the total mass composition becomes increasingly important with height due to the removal of dust and sea salt, and due to chemical sulphate production at greater altitudes within the atmosphere. AeroCom predictions confirm the results of an extensive analysis of sulphur cycle simulations in large scale atmospheric models which was performed in the COSAM exercise (Barrie et al. 2001; Lohmann et al. 2001; Roelofs et al. 2001). The uncertainty in predicting the global sulphate distribution is related to vertical mixing of emitted sulphur species from the planetary boundary layer into the free troposphere. In addition, cloud physics and cloud distributions play a major role as they influence cloud-related processes, i.e. wet deposition and the aqueous oxidation of SO_2 to sulphate. Furthermore, model agreement is low for the dry deposition of SO_2. The chemical production of sulphate at high altitudes explains why its mass fractions at higher altitudes are larger than those of dust, although these components have similar atmospheric residence times.

A similar diagnostic as for the vertical transport can be performed for the horizontal distribution of aerosol. The mass fractions in polar regions can serve as an indicator for the horizontal transport, i.e. for meridional long-range transport, because rolar regions are far from the aerosol sources. We show the mass fractions of aerosols south of 80°S and north of 80°N in Fig. 4.24. The

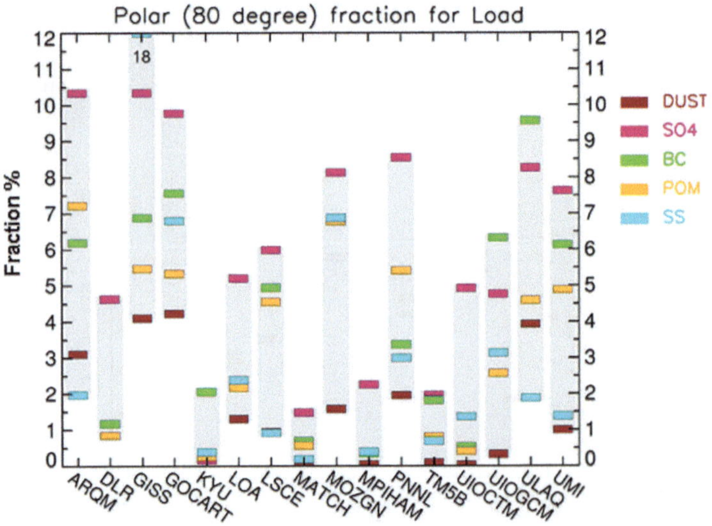

Fig. 4.24 Global annual average mass fractions in (%) of total mass in polar regions (south of 80°S and north of 80°N) for all AeroCom models. The gray shadings frame the range for each model (Textor et al. 2006)

all-models-averages of the mass fractions at the poles in relation to global aerosol are 2.4% (varying from 0.3 to 8%) for total mass of five aerosol types, 6% (varying from 0.2 to 10%) for sulphate, 4.2% (varying from 0.4 to 10%) for black carbon, 3.3% (varying from 0.2 to 18%) for sea salt, 3.3% (varying from 0.3 to 7%) for particulate organic matter, and 1.5% (varying from 0.03 to 4%) for dust, respectively. The AeroCom results show major differences in aerosol transport and more observations especially in remote regions and vertical profiles are needed to constrain aerosol distributions.

Within the AeroCom model intercomparison, simulated aerosol parameters (optical depth, concentrations, and extinction coefficients) have been compared to various observational data sets (see AeroCom web interface for individual comparisons). This analysis however did not allow for a simple ranking of models, because model performance depends strongly on the parameter, region and process under investigation. A model that performs well for one data set may show inconsistent results for another.

Models perform reasonably well when compared to the component-combined aerosol optical depth (AOD) (Kinne et al. 2006). The annual global mean of the simulated AODs are within 0.11 to 0.14, at the lower end of global averages suggested by AERONET (~0.135) and satellite composites (~0.15). More detailed comparisons, however, revealed that larger differences in regional distributions and in compositional mixture remain.

Aerosol size and its micro-physics are also simulated by some sophisticated models within the AeroCom. The results show a large diversity in the fine aerosol fraction (radius smaller than 0.5 microns), especially for dust and sea salt (Textor et al. 2006) , indicating a large variations in aerosol size predictions. Therefore, verification of the resulting size distribution is still in an early stage. A comparison of the simulated global annual mean composition of ambient aerosol has also shown major disagreement among models, especially in regard to the water content.

The global aerosol models compare less well to measured aerosol surface concentrations than to AOD, in terms of both spatial and temporal correlation and bias. Results indicate that these models overestimate the surface sulphate concentration by about 40% (correlation coefficient of 0.6) and underestimate POM concentrations by about 15% when compared to station data from the US (IMPROVE network) and Europe (EMEP network) (Guibert et al. 2006). Simulated black carbon concentrations match the observed level of concentration, but the comparison is so far only based on the IMPROVE network. Mean correlation between modeled and observed monthly mean concentrations is smaller than for sulphate, with 0.49 and 0.58 for BC and POM, respectively.

For dust the data from the 14 remote sites in the AEROCE network have been used as available from the University of Miami Aerosol Group for the years 1983 to 1998. Observed dust concentrations are slightly underestimated by the models (12%), but average correlation is quite good ($r = 0.73$). However, the limited number of dust measurements would not allow judgment on the performance of the dust modeling in AeroCom. This is also true for the other aerosol components, and more data and more detailed analysis are necessary for reliable conclusions about individual and ensemble model performance.

4.5.3 Numerical Modeling Studies - DMIP Results

An intercomparison study involving eight mineral dust/emission models over Asia (DMIP) has been evaluated (Uno et al. 2006) in order to quantify model differences resulting from the variety of implemented dust emission schemes, model resolutions, different deposition parameterizations and numerical methods. The models were run for two major dust storms developed in spring 2002 over East Asia. The model results were evaluated against in situ concentration measurements, satellite aerosol index data, Lidar observations and synoptic visibility data in the region.

Figure 4.25 shows that all models produce surface concentration patterns that generally coincide well with regions of observed maximum concentrations. However, the concentration levels from models are quite different. For example, differences of about one order of magnitude appear over the downwind Beijing region. Similar variations also occur in the source areas. A large variability between models exists in the amounts of dust emitted into the atmosphere. During one of simulated dust storm, models produced total dust emissions ranging from 27 to 336 Tg.

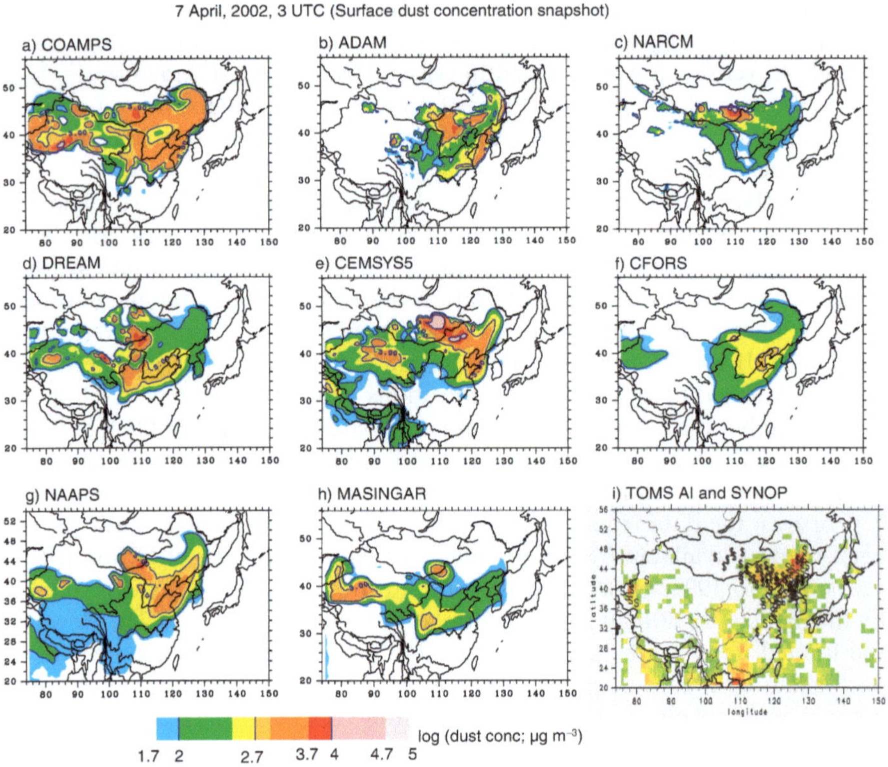

Fig. 4.25 Surface dust concentration (Log µg/m^3) valid for 7 April 2002 at 03 UTC, as predicted by eight dust models participated in DMIP. The lower-right corner map shows observations. From Uno et al. (2006) with permission of the American Geophysical Union

Within the DMIP exercise, simulated vertical concentrations were compared with the Lidar measurements available in the region. As Fig. 4.26 indicates, most of the models showed increased concentrations in the second half of the March 2002 period, when a major dust event approached Beijing. However, there are significant diversities in details in the vertical-time structure of the models. Most of models overestimated the vertical extent of concentration, with some of them retaining too much dust aloft.

This study demonstrated that although general agreement of models is achieved in reproducing major geographic extent and position of maxima of the surface concentration, they experience large differences in concentration values and in emission fluxes. Variations in vertical dust structure are also obvious. There are numerous possible reasons for the diversities, such as different parameterizations and numeric solutions incorporated in such dust models. The differences are also due to a serious lack of relevant dust measurements that may improve both model dust initialization and model verification.

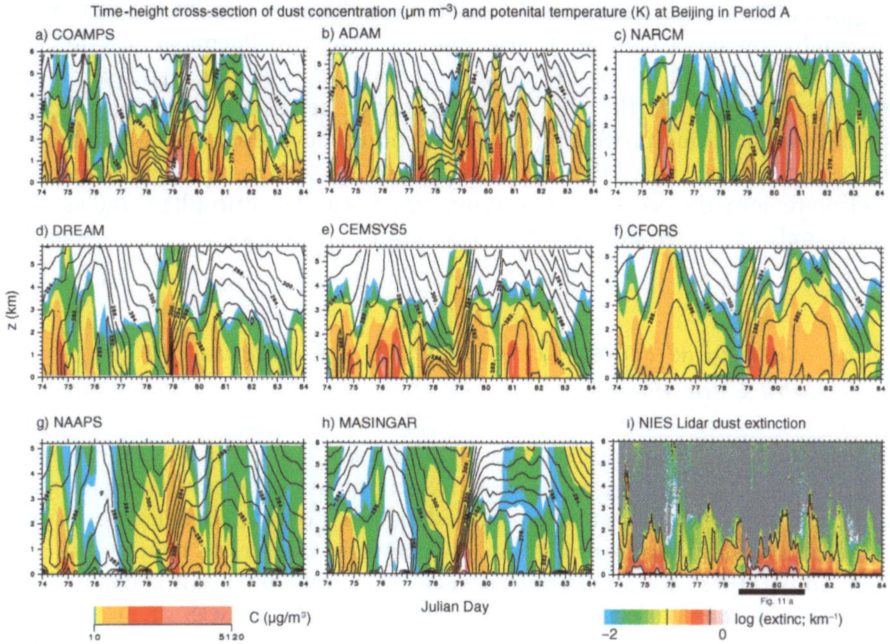

Fig. 4.26 Time-height cross sections of dust concentration in Beijing during the March 2002 dust episode, as simulated by eight models; Lidar data is also shown for comparison in the lower-right corner map. From Uno et al. (2006) with permission of the American Geophysical Union

4.6 Physical and Chemical Tranformations

There is increasing evidence that aerosol particles are predominantly a conglomerate of different internally mixed chemical substances (i.e. each particle is composed of more than one chemical species) (Murphy and Thomson 1997; Cziczo et al. 2004b; Kojima, et al. 2004). However, the knowledge about the composition of aerosols on the global scale is still incomplete and most GCMs (global climate models) still treat aerosols as external mixtures (i.e. each particle is composed of only one chemical species) in terms of their optical properties (e.g. Lohmann et al. 1999; Textor et al. 2006). Advanced aerosol modules in some GCMs have been expanded to include aerosol internal mixtures (e.g. Ghan et al. 2001; Gong et al. 2003a).

After aerosol particles are formed they undergo chemical and physical transformations as they move through the atmosphere. Collision and coagulation due to Brownian motions is most important for the nucleation and Aitken-mode aerosol particles. Larger particles can collide if they have a substantial difference in gravitational settling velocity and phoretic forces. Also vapours can condense on pre-existing aerosol particles, which is especially important for converting aerosol particles from being externally mixed to being internally mixed. Chemical transformation refers to chemical reactions on the surface or in the volume of aerosol particles that can increase aerosol mass, and/or change the ability of aerosols to act as CCN or IN. In-cloud oxidation and cloud processing can also produce new aerosol mass and modify the size distribution. All these processes make aerosols more hygroscopic. This section reviews the processes modifying aerosols in the atmosphere and assesses their implications to the cloud and precipitation formations.

4.6.1 Aerosol Microphysics - Nucleation, Condensation and Coagulation

When condensable species such as H_2SO_4(g) are formed in the gas phase by chemical reactions, they participate in two competing processes: nucleation and condensation on pre-existing particles depending on the particle size, vapour concentrations and source rates, and concentration, as well as relative humidity and temperature. It is apparent that nucleation and condensation can occur at the same time. Condensation is responsible for the growth of atmospheric particles and for changing the aerosol chemical composition. When an aerosol exists in a supersaturated vapour environment, the vapour condenses on the particle population, resulting in a change of the particle size distribution. The supersaturation is usually produced by cooling of saturated vapours via adiabatic expansion (atmospheric clouds) or by gas-phase chemical reactions yielding products of low vapour pressure (e.g. in photochemical smog or in oxidation of biogenic hydrocarbons). In the case of soluble particles, condensation can take place even in unsaturated conditions (Hinds 1982). On the other hand, higher saturation ratios may be needed for condensational growth of very small particles, because the attractive forces between surface molecules are smaller due to sharply curved surfaces, so that vapours more easily evaporate (Kelvin effect).

The condensational growth of aerosol particles is accompanied by simultaneous mass and energy transport. Analytical expressions for the condensational growth rate of a single particle have been presented by a number of researchers (Mason 1971; Kulmala, et al. 1989; Kulmala 1993). Once the growth rate of a single particle is known, the change in particle size distribution can be obtained by solving the condensation equation (Seinfeld and Pandis 1998; Park and Lee 2000). The condensational growth rate of a single particle is proportional to particle surface area (free-molecule regime) or to particle diameter (continuum regime). Its implication is that the particle size distribution becomes narrower as a result of condensational growth.

Nucleation events in highly polluted regions can produce CCN (Laaksonen et al. 2005). 2.5 years of continuous nucleation observations from San Pietro Capofiume, Italy (44°39'N, 11°37'W), showed that nucleation events occur frequently in the Po Valley region, even though the region is rather polluted with high pre-existing particle concentrations. The nucleation events are often very intensive, and the newly formed particles can grow to sizes as large as 100–200 nm in diameter within a few hours, and thus constitute an important source of CCN.

Coagulation is a process whereby aerosol particles come into contact because of their relative motion and then stick together to form larger particles. During the coagulation process, the number of particles decreases and the average particle size increases, while the total mass of particles is conserved. High coagulation rates among disparate particles have another important implication, which is the change of mixing state of particles. Particles originated from different sources often have different size and chemical compositions. Therefore, selected coagulation between unequal-sized particles will convert externally to internally mixed aerosols. This results in changes of optical properties and hygroscopic growth characteristics of the particles.

To simulate the change in the particle size distribution due to simultaneous nucleation, coagulation and condensation, it is necessary to resort to numerical methods. Among the most widely used numerical methods for aerosol microphysics is the modal dynamics method (Whitby and McMurry 1997), which expresses the particle size distribution as a sum of two or three log-normal distribution functions, and the sectional method (Gelbard et al. 1980), in which the entire particle size range is divided into a finite number of sections (Gong et al. 2003a).

4.6.2 Chemical Modification by Non-reactive Heterogeneous Uptake

For the adsorption of semi-volatile organics onto aerosols, particularly for persistent organic pollutants (POPs), the partitioning of the organics between the gas and aerosol phases is often modeled using the Junge-Pankow partitioning theory (Junge 1977; Pankow 1987; Pankow 1994; Pankow 1994) under non-saturated vapour pressure conditions. The incorporation of organic species into the particle phase increases with pre-existing aerosol mass and is generally favored at lower temperatures (Pankow and Bidleman 1991). This temperature dependency is likely a contributing factor to the observed increase in the percentage of aerosol mass attributed to organic compounds with height in the troposphere (Jaffe et al. 2005).

Some of the oxygenated compounds can be directly emitted, particularly from bio-mass burning sources. However, most hydrophilic organic compounds are often formed in the atmosphere through gas phase photochemical oxidation of less hydrophilic precursors, for example, as hydrocarbons are oxidized to alcohols, ketones, aldehydes, and finally to acids as the oxidation progresses toward a breakdown of the entire molecule. This results in the originally hydrophobic aerosols becoming more hygroscopic and more effective as CCNs.

4.6.3 Chemical Modification by Reactive Heterogeneous Uptake

There has been a great deal of attention given recently to heterogeneous and multiphase reactions occurring on the surfaces and within tropospheric particles (Ravishankara and Longfellow 1999). For example, increased relative humidity enhances the reaction of $CaCO_3$ in dust with HNO_3 (Goodman et al. 2000). The reacted particles have been shown to be more hygroscopic than the un-reacted dust and are thus better CCN (Krueger et al. 2003; Laskin et al. 2005). There is also evidence that SO_2 can be adsorbed onto mineral particles and oxidized to sulphate, creating mixed particles that may have complex hygroscopic behavior depending upon which cations are associated with the sulphate (Usher. et al. 2003). Dust is also frequently associated with organic matter, although the mechanisms creating such mixtures and their effects on dust hygroscopicity are not well understood.

Deviations from model predictions by the Junge-Pankow partitioning theory, which is based on equilibrium, are often observed for organic gases and organic particulate matter in the atmosphere (Jang and Kamens 2001; Jang and Kamens 2001). The main reason for the deviation is that the absorbed product can further react within the host aerosols, thereby significantly increasing the capacity of the aerosols for continued uptake of the organic gases. Many aldehydes form polymers in aqueous solution under acid catalysis. Such polymerization has been proposed to occur in atmospheric aerosols (Olson and Hoffmann 1989; Tobias and Ziemann 2000; Jang and Kamens 2001; Jang et al. 2003; Noziere and Riemer 2003; Gao 2004; Kalberer et al. 2004). Recent laboratory studies have shown that for certain aldehydes, polymerization can indeed occur at significant reaction rates (Liggio et al. 2005a,b). Liggio et al. (2005b) observed the process of polymerization of glyoxal, a dialdehyde commonly encountered in the atmosphere from both anthropogenic and biogenic precursors, on a time scale of 1 to 3 h in acidic aerosols, and estimated that in a typical urban-suburban atmospheric environment, the glyoxal uptake alone would be sufficient to account for up to 10% of aerosol organic contents. Oligomers up to trimers were observed under conditions that can be directly applied to the atmosphere in less than 1 h. This polymerization leads to a significant growth in the size and mass of the aerosols. It has been demonstrated that freshly formed secondary organic aerosols from anthropogenic and biogenic precursors in a smog chamber continued to undergo oligomerization to higher-molecular-weight compounds over 28 h, and the particles exhibited increasing thermal stability with aging (Baltensperger et al. 2005). Interestingly, although the hygroscopicity of the particles (at 85%) increased during the first few hours, it remained virtually unchanged during the remainder of the experiment, and was similar to that observed for humic-like substances (HULIs) that have been identified in atmospheric aerosols (Brooks et al. 2004; Baltensperger et al. 2005).

4.6.4 Black Carbon Aging Processes

Arguably the most complex constituent of the tropospheric aerosol is the carbonaceous fraction, which includes both soot and myriad organic species. Many studies have shown that soot, although initially quite hydrophobic unless formed during combustion of high-sulphur containing fuels (Popovitcheva et al. 2001), becomes more hydrophilic as it ages through interaction with atmospheric gases such as SO_2, sulphuric acid, and nitric acid, and through oxidation (Zuberi et al. 2005), in addition to non-reactive uptake and condensation/coagulation processes as discussed above. Chand et al. (2005) measured the hygroscopic growth of particles produced from burning of biomass in the laboratory, and noted that the growth was less than observed for aged smoke particles from similar origins. It has also been demonstrated that acidic particles catalyze the formation of secondary organic aerosol species from the gas phase on their surfaces (Jang et al. 2002; Gao 2004), a mechanism whereby an acidic inorganic particle could become internally-mixed with organic species.

The aging process is dominated by the condensation of sulphuric acid with the time scale of about 8 h near the source region in summer during the day, while in winter ammonium nitrate becomes more important at the same time scale (Riemer et al. 2004; Riemer et al. 2004). Equations of first order decay rate as a function of height and time of the day can be derived from these data. Such equations have been implemented into the Canadian climate model (Croft et al. 2005) where the BC aging is assumed to be a first order decay and the aging e-folding time is parameterized as the coagulation of insoluble BC particles with soluble particles, the condensation of sulphuric acid and organic compounds, and the oxidation of a BC coating by ozone. A modal aerosol module coupled with the GCM ECHAM5 model was developed by Stier et al. (2005), using a detailed microphysical approach for aerosol modeling through condensation and coagulation (Vignati et al. 2004). The impact of BC aging processes from different schemes differs substantially on the removal and activation processes of BC aerosols (Croft et al. 2005). Liu et al. (2005a) suggest, based on a modeling study that included a detailed representation of aerosol microphysics, that most carbonaceous aerosols in the troposphere are internally mixed with sulphate, nitrate, and secondary organics, and are thus more hygroscopic and have significantly shorter lifetimes than predicted using typical hydrophobic-to-hydrophilic conversion rates. More work is clearly needed before atmospheric aerosol aging processes and their consequences for particle hygroscopicity and CCN activity are adequately represented in chemical cycle and climate models.

4.6.5 Aerosol Thermodynamics

The gas-particle partitioning and solubility of organic and inorganic species is critically important to our ability to accurately predict the water uptake of aerosols and clouds/precipitation formation. Nenes et al. (1998) compared the equilibrium composition and partitioning of aerosols as calculated by different thermodynamic models – including their ISORROPIA, as well as the SCAPE model of Kim et al. (1993) and SEQUILIB of Pilinis and Seinfeld (1987) – for representative urban, marine, remote continental, and non-urban continental mass loadings taken from Heintzenberg (1989) and Fitzgerald (1991). These thermodynamic models consider the inorganic aerosol system of sulphate, nitrate, ammonium, chloride, sodium, potassium, calcium, magnesium, carbonate, and water. They generally agree within several percent, except at high relative humidity, where differences in concentrations of particular ionic species differed by as much as 30%. The discrepancies in ionic content relate to the varying treatments, especially of bisulphate, equilibrium constants, thermodynamic data, water content, and deliquescence. Using the coupled IMAGES (Intermediate Model of Global Evolution of Species)-SCAPE2 model, Rodriguez and Dabdub (2004) found that nitrate concentrations predicted compare better with observations than those calculated with IMAGES alone. They also find that over polluted continental regions the presence of sea salt and dust aerosol potentially increases the formation of aerosol sulphate by 20–80%. Aerosol nitrate formation is enhanced by 14–60%, whereas ammonium formation is decreased by 20–60%.

Organic aerosols and condensable organic species are observed in all environments, and they are the dominant aerosol-phase species in some environments. Organic species no doubt also contribute to the composition of many primarily inorganic particles (Abdul-Razzak and Ghan 2000). However, the state of the science is such that the great majority of the organics present in aerosol are unidentified. For those that have been identified, the lack of measurements of solubilities and their dependence on inorganic solute concentration is imposing, not to mention the lack of a suitable unified thermodynamic formulation for mixed inorganic/organic aerosol (Saxena et al. 1995; Cruz and Pandis 1997; Turpin et al. 2000; Clegg and Seinfeld 2004). The impact of the chemical compositions on aerosol activation is discussed in Chapter 7.

4.7 Removals of Atmospheric Aerosols

The average residence times of atmospheric aerosols are on the order of a few days to about two weeks, depending on their size and location. Aerosols are removed from the atmosphere by wet and dry deposition. Wet deposition removes aerosols from the atmosphere in precipitation with increasing efficiency for soluble aerosols. In the absence of precipitation, smaller aerosols are deposited to the Earth surface within turbulent eddies. Larger aerosols settle out of the atmosphere very quickly under gravity, and some surfaces are more efficient at capturing aerosol than others. Both wet and dry removal processes need to be parameterized in global climate models, as these result in rather large uncertainties in budgets. We will first examine some removal pathways before looking at the differences in atmospheric residence for different aerosol components.

4.7.1 Size-dependent Dry Deposition of Aerosols

The process of particle dry deposition includes several mechanisms. Transported from the free atmosphere down to the viscous sub-layer that envelops all surfaces is the first process. This happens due to passive transport within turbulent eddies, or for larger particles through sedimentation. They are then transported across the viscous sub-layer by Brownian diffusion, phoretic effects, inertial impaction, interception or sedimentation. Once transported across the viscous sub-layer, particles will then interact with the surface, either sticking to it or bouncing off (Ruijgrok et al. 1995). A small fraction of very fine particles may also diffuse into leaf stomata. After collected by leaf surface, water-soluble particles may be gradually removed by wetting, and insoluble particles may become embedded in epicuticular waxes. Thus, the characteristics of the atmosphere, the nature of the surface, and the physical/chemical properties of the atmospheric aerosols all affect dry deposition (Ruijgrok et al. 1995; Zufall and Davidson 1998; Wesely and Hicks 2000). Although the dry deposition process is a slower process compared to wet deposition, it is a continuous event happening anytime over any surfaces. The accumulated amount of dry deposition can be much more important for some aerosol species/sizes (Ginoux et al. 2001), especially for the coarse aerosol types, sea salt and dust. It has been estimated that dry deposition contributes about two-thirds of the total deposition flux, as shown by the AeroCom models (Textor et al. 2006).

Most measurements on aerosol flux made between 1970 and the 1990's are made with bulk sampling due to the limitations in aerosol measurement techniques (Sehmel 1980; Nicholson 1988; Sievering 1989). The development of more sophisticated measurement techniques in the 1990's made it possible to measure size-resolved aerosol physical and chemical properties. The number of studies making direct aerosol fluxes using eddy correlation methods also greatly increased (Gallagher et al. 2002). Based on the measurements, dry deposition models were developed to calculate the dry deposition velocity (V_d) as a function of particle size (e.g. Giorgi 1986; Zhang et al. 2001; Nho-Kim et al. 2004). Estimates from several earlier models (Ruijgrok et al. 1995) revealed that these models differ from each other greatly and the largest uncertainty is for the 0.1–1.0 μm particle size range, which is not important for mass fluxes but is important in terms of number concentrations.

Dry deposition velocity (V_d) generally increases with particle size for particles larger than 2 μm because the collection efficiency by interception and impaction increases with particle size. The collection efficiency is a parameter defined as the ratio of the total number of aerosols reaching the surface to the total number of aerosols actually sticking to it. Particles larger than 2 μm tend to have a deposition velocity as high as a few cm s^{-1} over rough forests (e.g. Hofschreuder et al. 1997). V_d also increases with the decrease of particle size for particles smaller than 0.1 μm since the collection efficiency by Brownian diffusion is larger for smaller particles. Available measurements show that V_d of ultra-fine particles (<0.01 μm) in field conditions can be higher than 5 cm s^{-1} (Schery and Whittlestone 1995). Deposition velocities for particles in the size range of 0.1–2 μm are lowest and can vary by several orders of magnitude. Many earlier laboratory experiments suggest that V_d for particles of this size range should be on the order of 0.01 cm s^{-1} or less, over both smooth and rough surfaces (Nicholson 1988). Higher values have been obtained in many recent field studies that have investigated some trace species considered to be representative of particles in this size range. Field observations for sulphate and other sub-micron particles showed that the dry deposition velocities are one to two orders of magnitude higher than in some earlier studies (Gallagher et al. 1997; Gallagher et al. 2002), and references therein), especially for sub-micron particles over rough vegetated surfaces.

The particular particle size for which the minimum V_d is predicted to occur is determined by the relative magnitude of the collection efficiencies by Brownian diffusion and interception. Since the collection efficiencies by Brownian diffusion and interception are different over different canopies, it follows that the minimum V_d will appear at different particle sizes over different canopies. Most measurements show that the minimum should be located at particle sizes around 0.1–0.3 μm. However, some individual field studies show a minimum V_d located at particle sizes close to 1 μm. In order to locate the exact size for minimum V_d, more measurement studies on size-resolved V_d over different canopies are needed. Meteorological conditions and surface characteristics need to be measured simultaneously in order to relate the dry deposition process to different controlling factors.

4.7.2 Wet Deposition

Wet deposition is usually split into two categories, in-cloud and below-cloud (precipitation) scavenging. In-cloud scavenging includes contributions from both nucleation and impaction scavenging while below-cloud scavenging only includes contributions from impaction scavenging.

4.7.2.1 Below-cloud Scavenging

For practical considerations, the below-cloud scavenging of aerosols is typically represented by a scavenging coefficient (S_c) in aerosol mass continuity equations in large scale models. Most S_c parameterizations currently used in large-scale atmospheric models are treated as a function of known precipitation properties (e.g. $S_c = aPR^b$, with PR being the precipitation rate). Such a simple representation causes large uncertainties as shown in a comparison of model results from many different large-scale models (Rasch et al. 2000). S_c values for particles ranging in size from 0.001 to 20 μm differ over three to five orders of magnitude. Since most aerosol number concentration is associated with small particles, while most mass is associated with large particles, S_c is expected to be different for the bulk number concentration and the bulk mass concentration. However, S_c will be the same for the number and mass concentrations of a specific particle size. With the development of size-resolved aerosol modules within large-scale models, new parameterizations of S_c as a function of aerosol size have been developed (e.g. Gong et al. 2003a; Jung et al. 2003). Particles ranging from 0.001 to 20 μm have to be included in climate and air-quality models and S_c values for this size range can differ by three to five orders of magnitude.

Jylhä (1999) derived the scavenging coefficients from radar reflectivity based on the fact that both are functions of the hydrometeor size distribution. Although many studies show an increase of the scavenging efficiency with precipitation intensity, a measurement study by (Chate and Pranesha 2004) actually obtained the largest S_c values for sub-micron particles when the precipitation intensity was weakest. This implies that precipitation intensity alone may not be able to characterize S_c for all particle sizes. Other precipitation properties, such as total surface area of hydrometeors, may be needed in parameterizing S_c. For example, (Zhang et al. 2004c) showed that S_c for any size particles has a strong dependence on total droplet surface area. This is especially important for in-cloud impaction scavenging since the cloud droplets are much smaller compared to the rain drops below the cloud, and thus the total droplet surface area inside the cloud-layer is much larger. For the same reason, including the droplet surface area in the parameterization of S_c is more important for weaker precipitation than for strong precipitation.

S_c is also a function of the collection efficiency (E_c), a parameter defined as the ratio of the total number of aerosols collected by falling drops to the total number of aerosols within the swept volume of falling drops. Usually E_c is much smaller than 1 because particles can follow the streamlines around the drop. It can however be much larger than 1 under certain conditions, e.g. for charged particles. For small aerosol particles with little inertia Brownian diffusion is important to bring them into contact with the drop, thus increasing their collection efficiency. The Brownian diffusion of particles decreases rapidly as particle size increases; this mechanism is most important for particles of diameters smaller than 0.1 μm. Large aerosols experience inertial impaction because their inertia prevents them from following the streamlines around falling droplets. Inertial impaction increases with the increase of particle size and is important for particles larger than 2 μm. Interception and inertial impaction are closely related, with interception occurring as a result of a particle's size irrespective of its mass, while inertial impaction occurs due to its mass and not its size. For particles within the size range of 0.1–2 μm, commonly referred to as "Greenfield gap" in the literature (Greenfield 1957), neither the Brownian diffusion nor the inertial impaction plays an effective role. Thus the collection efficiency is smallest in this size range (Slinn 1984; Pruppacher and Klett 1997). The exact depth, width and position of this minimum depend on the properties of the aerosols and hydrometeors, and on ambient conditions. Within the Greenfield gap, thermophoresis dominates over diffusiophoresis, so that in the case of single-phase hydrometeors, phoretic scavenging becomes more efficient with decreasing relative humidity of air (with respect to either water or ice) (Pruppacher and Klett 1997).

Many experimental and theoretical studies have been dedicated to estimating S_c and E_c values under a variety of conditions (e.g. Facchini et al. 1999; Jung et al. 2003; Laakso et al. 2003; Zhang et al. 2004c; Chate 2005). S_c and E_c for very small and very large particles estimated from measurement studies generally agree well with theoretical studies. However, significant discrepancies between observations and theoretical estimates have been found for sub-micron particles, with measured values being one to three orders of magnitude higher than predicted (Davenport and Peters 1978; Radke et al. 1980; Schumann 1989; Volken and Schumann 1993; Laakso et al. 2003; Chate 2005). This is because the magnitudes of S_c measured for this size range of particles are very low (e.g. 10^{-6}–10^{-3} s^{-1}), and any collection mechanisms that were not treated properly in the theoretical studies would cause large uncertainties such as those discussed above.

Collection of aerosols by snow crystals is more complicated than by liquid drops due to the irregular shape of ice elements. Large uncertainties exist in the values of collection efficiency between experimental results and theoretical predictions, especially for sub-micron particles. The differences between collection efficiencies by rimed and unrimed snow crystals seem to be small ((Murakami et al. 1985) and references therein), although earlier laboratory studies showed an enhancement of collection by rimed snow crystals. Collection efficiencies depend on the shape of the particles and generally increase with the temperature. Charged snow crystals and holes associated with them may also increase the collection efficiencies (see a review by Jennings 1998).

Considering the scarcity of field measurement data on below-cloud scavenging, detailed cloud microphysics models such as the one described in Zhang et al. (2004a) are needed to develop and verify new parameterizations. More field measurements on size-resolved $S_c$under different precipitation intensities are also needed to improve existing parameterizations and to develop new parameterizations.

4.7.2.2 In-cloud scavenging

In-cloud scavenging includes both nucleation scavenging and post-nucleation processes including Brownian diffusion, impaction scavenging and phoretic effects. Very few measurements are available, and most studies are based on numerical simulations. Although impaction scavenging inside a cloud layer removes little aerosol mass, it removes a substantial number of small particles if the cloud lasts for more than a few hours (Zhang et al. 2004c). The mechanisms of in-cloud impaction scavenging are the same as below-cloud scavenging. The differences are that cloud droplets are much smaller than raindrops and only very small aerosols (i.e. $<0.1\ \mu m$) exist as interstitial aerosols within a cloud-layer. Thus, carefully parameterizing the collection efficiency of small aerosol particles by small droplets is very important for in-cloud impaction scavenging if the interstitial aerosol number concentration inside the cloud layer is a concern. In this section nucleation scavenging is only briefly discussed since some discussions will be presented in Chapter 7.

Earlier studies used a simple power law to link the concentration of CCN to the aerosol number concentration C under a specific supersaturation S: $N_{CCN} = CS^k$ (where k is an empirical constant and varies with the origins of the aerosols, thus implicitly accounting for the influence of aerosol chemical composition). Although the power law approach was refined in several different ways, it still has large uncertainties due to the large variability of C and k with different aerosol source types (Hegg and Hobbs 1992). Other forms of parameterizations using pre-assumed aerosol size distribution for predicting droplet concentration through the nucleation process were made available more recently (e.g. Abdul-Razzak and Ghan 2002 and references therein).

Cloud droplet nucleation from aerosols cannot be properly represented in models with modal representations of the size distributions. Abdul-Razzak and Ghan (2002) extended their earlier parameterizations, which assume single or multiple log-normal aerosol distributions, to sectional models. The parameterization performs reasonably well when compared to detailed numerical simulations in a parcel-model framework, except for weak updraft conditions. Nenes and Seinfeld (2003) developed a parameterization which includes two steps to predict the number concentration of nucleated drops. The first step involves the representation of the aerosol number and the chemical composition distribution with respect to size and the calculation of the number concentration of droplets that can potentially form at a certain level of supersaturation. In the second step, the CCN spectrum is included within the framework of an adiabatic cloud parcel model, with a specified updraft velocity, to compute the maximum supersaturation of the cloud parcel. The parameterization seems to perform better than earlier parameterizations. Evaluations of the above parameterizations using field measurements have not yet been performed.

The influence of organic aerosols on CCN concentration has also gained some attention (see discussions in Chapter 3). With modifications, several parameterizations for sectional aerosol models described above can also handle the nucleation of organic aerosols. Aerosols can also be scavenged by serving as ice nuclei as discussed in Chapter 3.

Note that in-cloud scavenging may occur in non-precipitating clouds. If these clouds evaporate, the scavenged aerosols will be returned to the free atmosphere and ultimately no removal from the atmosphere occurs via wet deposition. However, the aerosol properties might be changed, and this process is referred to as "cloud processing".

4.7.3 Residence Times of Various Aerosols

The residence time τ cannot be measured globally, but has to be obtained from model simulations. It is independent of model differences in emissions strengths and reflects the integral of all simulated aerosol properties and processes. The residence time is controlled by the aerosol properties, as it depends on particle size and solubility, as well as on the model-specific transports and parameterizations of aerosol processes. In addition it reflects the spatial distributions of aerosols, particularly as they relate to relative humidity, precipitation, and surface properties.

Some common features in deposition patterns have been found using global simulation results (Ginoux et al. 2001; Zhao et al. 2003; Han et al. 2004; Satake et al. 2004). Large particles usually deposit rapidly close to their sources (e.g. within the continent). On the other hand, small particles, especially sub-micron particles, can be transported far away from their sources (e.g. Asia to North America, Africa to Atlantic Ocean).

Figure 4.27 shows the simulated residence times from the AeroCom models (Textor, et al. 2006). Sea salt has the shortest τ of about half a day, followed by sulphate and dust with about four days, and particulate organic matter and black carbon with about six and seven days, respectively. The largest model diversities in the residence times are found for the coarse aerosol types, in particular for sea salt.

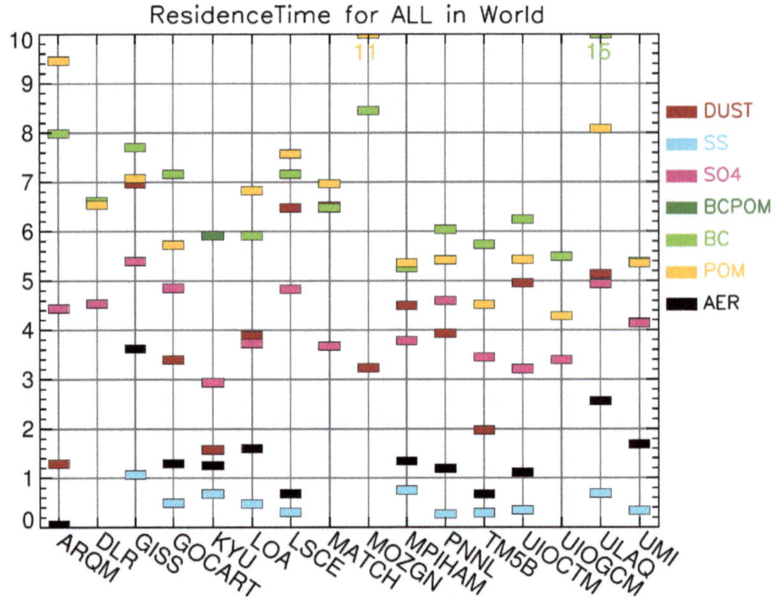

Fig. 4.27 Tropospheric residence times (days) in the AEROCOM models for different aerosol components (Textor et al. 2006)

The residence time also varies with geographic location. The size dependence of sea salt residence time was investigated by Gong et al. (2002) for diameter ranges from 0.01 to 40.96 μm. For two size ranges of sea salt aerosols, the residence time follows a similar pattern; that is, τ is larger on either side of the equator than at middle or high latitudes. There are two persistent subtropical anticyclonic regions in the Southern Hemisphere. One is in the Pacific Ocean and another in the Indian Ocean. Three climatologically persistent subtropical anticyclonic regions prevail in the Northern Hemisphere. Two are in the Pacific Ocean and one in the Atlantic Ocean. These anticyclones are regions of long aerosol residence time. They are located on either side of the ITCZ where tropical cyclones with strong precipitation scavenging of aerosols are frequent. The removal of sea salt is less in the anticyclonic regions, resulting in a residence time peak.

4.8 Recommendations

To better understand the transport, transformation and removal of atmospheric aerosols, the following activities are recommended:

- Development of high resolution atmospheric models with explicit treatment of aerosols, clouds and precipitation formation processes that effectively utilize observations from surface-based networks, aircraft and satellites through advanced data assimilation techniques
- Identification and analysis of systematic long term measurements of both aerosol distributions and precipitation rates in regions where great changes in both aerosols and precipitation have been observed in order to find statistically significant correlation between them.
- Field measurements of dry deposition fluxes for specific particle sizes over different natural surfaces (e.g. using eddy-correlation method above canopies rather than using surrogate surfaces) are needed in order to better understand this process and to improve existing models or develop new models. Meteorological conditions and surface characteristics need to be measured simultaneously in order to relate the dry deposition process to different controlling factors.
- Development of a suitable parameterization of aging and transformation processes and incorporation into weather and climate models to account for: (i) the secondary organic aerosol fraction, (ii) the hygroscopic nature of particles and (iii) modifications of aerosol optical properties by physical and chemical transformation processes.
- Studies of cloud droplet activation as a function of chemical compositions of aerosols and cloud dynamics such as vertical velocity and turbulent mixing.

Chapter 5
In Situ and Remote Sensing Techniques for Measuring Aerosols, Clouds and Precipitation

Didier Tanré, Paulo Artaxo, Sandra Yuter and Yoram Kaufman

5.1 Introduction

In this chapter we review the methods presently available and expected in the near future to measure the effect of aerosols on clouds and precipitation, and the limitations of current measurements that make it difficult to assess this connection.

The short lifetime of clouds and aerosols, the large variability of cloud properties, and the variability in the aerosol chemical and physical properties, makes it difficult to decipher the impact of aerosols on cloud properties and on the onset of precipitation. An array of satellite instruments, and surface networks that routinely measure aerosol, clouds and precipitation are designed to overcome some of these obstacles by generating large data sets of precipitation, cloud properties and the nearby aerosols. The aerosol properties include measures of the column concentrations (expressed by the aerosol optical depth) and size (given by distribution of the aerosol in 2–3 size modes or measurement of the Angstrom coefficient). A major remaining obstacle that is difficult to overcome concerns the precise measurement of precipitation and its co-location with cloud and aerosol measurements.

The measurement tools of aerosols and precipitation, therefore, include in situ measurements from the ground or from the air, remote sensing from the surface or from space and indirect assessments using other attributes. The use of measurements to deduce the effect of aerosols on precipitation includes the following groups: long term statistics of precipitation with assessment of aerosol based on location and time of the measurements (Warner 1968; Givati and Rosenfeld 2004); and individual case studies (Rosenfeld 2000).

D. Tanré (✉)
CNRS, University of Lille, Lille, France

Z. Levin, W.R. Cotton (eds.), *Aerosol Pollution Impact on Precipitation*,

5.2 Measuring Aerosol Properties from the Ground

It is not easy to collect and analyze atmospheric aerosols with high precision and with high reproducibility and specificity (McMurry 2000). A number of factors make this task difficult. The dynamic range of aerosol particles in the atmosphere is ~1 nm to 100 μm, and they occur in a large variety of shapes and different chemical compounds. In particular, our knowledge of the composition of organic compounds is rudimentary and a significant fraction of the aerosol mass is semi-volatile. In this section issues of aerosol sampling and physical and chemical characterization of aerosols are discussed. We will focus on the properties that are relevant to aerosol-cloud interactions. More information is found in the GAW Measurement Guide compiled by the WMO/GAW Scientific Advisory Group for Aerosols (WMO 2003). Comprehensive description of the methods of aerosol sampling analysis can be found in Hinds (1999), Baron and Willeke (2001) and Heard (2006).

5.2.1 Ground Based Aerosol Sampling and Mass Analysis

Particle mass is one of the most commonly measured aerosol properties. Sampling artifacts are of concern for some chemical species, particularly those exhibiting evaporative losses of semi-volatile organic compounds (Baltensperger et al. 2001). The very broad range of particle sizes makes unbiased sampling and analysis of aerosol properties difficult. Aerosol particles may be divided into three general size ranges: PM_{10} (particles smaller than 10 μm aerodynamic diameter), $PM_{2.5}$ (<2.5 μm aerodynamic diameter), and $PM_{1.0}$ (<1 μm). The ideal aerosol sampling inlet would draw in 100% of the particles in a specified size range and transport them all without modification to a detector or collector. A broad range of instruments has been developed over the past 10 years to measure aerosol mass concentration in certain size ranges (Artaxo et al. 2002). The *gravimetric method* is the most common, in which a filter (Teflon, Nuclepore, quartz fiber, nylon etc.) is exposed to a flow of air loaded with particles for ~1 to 24 h, after the aerosol sample is conditioned in a temperature and humidity controlled environment, and the mass of aerosol deposited onto the filter is weighed using a microbalance. This method provides high accuracy, but is very labor intensive and cannot be done in real time. Sampling artifacts can affect aerosol concentrations, as in situations where semi-volatile compounds are present in significant amounts. This can either increase the collected aerosol mass or decrease the mass of already collected aerosol particles. New real time PM_{10}, $PM_{2.5}$ or $PM_{1.0}$ instruments, such as the Tapered Element Oscillating Microbalance (TEOM), measure the gravimetric mass of the particles with high time resolution (<30 min) and are now widely used in air pollution monitoring networks. In its original configuration the air had to be heated to 50°C or more, which resulted in some loss of semi-volatile organic and inorganic compounds. New versions of this instrument correct for the particle losses in real time, and provide gravimetric mass with good accuracy.

Aerosol monitors using beta attenuation (also called beta gauges) are widely used for real time aerosol mass measurements in urban areas, although they can have large errors due to the different beta radiation attenuation coefficients needed for different aerosol elemental composition and size. New instruments derive PM_{10} or $PM_{2.5}$ using light scattering from small lasers, but they require careful calibration and are prone to errors when aerosol types or size distributions change. Some aerosol samplers, such as the Dichotomous Partisol-Plus (Model 2025 Sequential Air Sampler; Patashnick et al. 2001), collect aerosols in two size ranges: $PM_{2.5}$ and PM_{10} ($2.5<d_p<10$ μm) on two Teflon filters that are suitable for gravimetric, ionic and trace element analysis. Stacked Filter Units (SFU) collect aerosols in the same two size ranges on sequential Nuclepore filters for subsequent gravimetric and elemental analysis (Hopke et al. 1997). High volume samplers can also be used to collect large amounts of PM_{10} and $PM_{2.5}$ for analysis of various organic and inorganic compounds when a large amount of material is required (Artaxo 2002).

5.2.2 Aerosol Size Distributions and Particle Counting

Measurement of aerosol size distribution for the whole size range (from 1 to 2 nm to several hundred micrometers) requires different approaches and different instruments. Cascade impactors separate particles in several size ranges for subsequent analysis. For example, the MOUDI (Microorifice Uniform Deposit Impactor) cascade impactor acts as a 12-stage impactor when the inlet stage and final filter are included, with the stages having 50% cut-points ranging from 0.056 to 18 μm in aerodynamic diameter (Marple et al. 1991). New versions, such as nano-MOUDI, can collect particles down to 10 nm. Many other types of cascade impactors, such as the Berner impactor, are widely used to collect aerosols. Condensation particle counters (CPCs) grow particles in a highly supersaturated environment, allowing them to be detected optically. Particles down to 3 nm can be detected and counted, and very low particle number concentrations can be measured. There are CPCs that use butanol as the condensing vapour and some new instruments are water-based. However, size distribution measurements are not possible using a CPC alone.

Particle number size distributions can be measured with a scanning mobility particle spectrometer (SMPS), which is an adaptation of the differential mobility particle sizer (DMPS). Both methods can measure aerosol size distributions from 3 to 800 nm with a time resolution of a few minutes. They include a differential mobility analyzer (DMA) to separate particles according to their mobility, which is a function of size and a condensation particle counter (CPC) to count them. Coarse mode particles can be sized optically with the

Aerodynamic Particle Sizer (APS) spectrometer or with an optical particle counters (OPC). The latter instrument measures the intensity of light scattered by individual particles as they traverse a tightly focused beam of light.

At relative humidity >70–80%, water can comprise a large fraction of the fine particle mass. Water uptake capacity of aerosol particles is critically important in determining the CCN activities and optical properties. The water uptake capacity can be measured with instruments such as the Humidified Tandem Differential Mobility Analyzer (HTDMA), which measures the hygroscopic diameter growth of aerosol particles at varying relative humidity (RH) (McMurry and Stolzenburg 1989). The HTDMA works with two DMA working in tandem, where the first selects a particular narrow size range in a dry air flow that is then conditioned to a particular relative humidity, while the second scans the size range to analyze the ratio of the wet and dry diameter (growth factor) of the particles. In general, HTDMA can measure growth rates for relative humidity up to 80%, but new methods can measure the growth factor up to 97% relative humidity. These instruments can separate hygroscopic from hydrophobic particles based on their water affinity, a critical property for determining their ability to act as CCN.

5.2.3 Individual Particle Analysis

Particles collected on filters can be analyzed by scanning electron microscopy (SEM), to provide size, morphology and elemental composition. Basically, a finely focused high energy electron beam scans the particle, and secondary electrons from the sampled particle are collected to form an image of the particle (Van Grieken et al. 1991). X-rays emitted by the interaction between the high energy electron beam and the sample are analyzed by a Si(Li) X-ray detector to obtain elemental composition information (Artaxo et al. 1992). Scanning electron microscopy can be used to measure particles >0.1 μm in diameter, with a mass detection limit for elements of ~0.5%. The measurement and quantification of the elemental mass is complicated by particle matrix effects. Environmental SEMs allow analysis of particles under near atmospheric conditions and different humidity and temperature conditions, thus also allowing the measurements of volatile particles. Automated SEM can measure and quantify thousands of particles on a filter in a few hours. Transmission electron microscopy is used to observe particles <0.1 μm diameter. Optical microscopy can be used to observe and count large particles (larger than 0.5 μm). Laser Microprobe Mass Analysis (LAMMA) measures organic and inorganic ions by the use of a laser beam to vapourize particles that are then analyzed in a time of flight mass spectrometer. Measurement of the elemental composition of individual particles is a valuable addition to bulk measurements, especially when the effects of aerosols on clouds are of interest, because of the potential to measure the internal versus external mixture of the particles (Van Grieken et al. 1991). The presence of internally mixed particles, containing some soluble material in them, can change CCN efficiencies in the atmosphere. For example, dust particles that are normally not efficient CCN can become good CCN if some soluble material accumulates on their surfaces (Levin et al. 1996).

5.2.4 Measurements of Optical Properties of Aerosols – Light Absorption and Scattering

Quantification of light scattering and absorption by aerosols is needed to assess the climatic impacts of aerosols (e.g. Ogren 1995; Ramanathan et al. 2001). The aerosol single scattering albedo (ω_0) is a critically important aerosol parameter that expresses the ratio between scattering and extinction of radiation and can be derived from simultaneous measurements of scattering and absorption coefficients. To measure light scattering, integrating nephelometers measure the amount of light scattered by a sample volume of air loaded with particles. Nephelometers of various types are used, some with a single wavelength (generally 500–550 nm), while others use three wavelengths that can provide useful information on aerosol size distributions and Angstrom coefficients with high time resolution and sensitivity. Instruments are available that measure both forward and backward scattering coefficients.

Measurements of light absorption by aerosols are notoriously difficult. Several instruments are available to measure the light absorption coefficient, which is related to the so-called "black carbon" or "elemental carbon". The Aethalometer and the Particle Soot Photometer (PSAP) are two widely used instruments that measure aerosol absorption on a filter. New versions, operating at several wavelengths, allow measurement of the absorption coefficient as a function of wavelength. These instruments suffer from multiple scattering processes in the filter substrate, resulting in high uncertainty of the data. The Multi Angle Absorption Photometer (MAAP) uses transmitted as well as scattered light for the quantification of the absorption coefficient, resulting in a more accurate reading (Petzold and Schönlinner 2004). Absorption coefficients are most commonly determined from measurements of particles collected on filters using integrating spheres or photometers. Photoacoustic spectroscopy measures the absorption coefficients of suspended particles in real time, and can obtain a reliable measurement of aerosol absorption; however, their sensitivity still needs to be improved. Individual light-absorbing particles can also be measured by incandescence (Baumgardner et al. 2005). Aerosol absorption coefficients can be inferred from mass measurements, using real time elemental carbon measurements that can nominally separate the "organic" from the "elemental" carbon component using a temperature/vapourization profile. The method reveals large differences. It is difficult to distinguish unambiguously between elemental, black or organic carbon based only on the temperature profile because the elemental composition affects this separation. Commercially available quasi-real time organic and elemental carbon monitors are available, but the methods are far from being widely accepted, and intercomparison between measurements is problematic. This area requires important advances in terms of measurement methods and instrumentation.

5.2.5 Elemental Composition of Aerosol Particles

Aerosol source apportionment methods use elemental composition information to infer and quantify aerosol sources (Artaxo 2002). The use of factor analysis and chemical mass balance methods requires the knowledge of the elemental composition of aerosols with as much detail as possible (Artaxo et al. 1999). Varieties of techniques are used for the determination of chemical composition of bulk or size-segregated samples. X-Ray Fluorescence (XRF) uses an X-ray beam to provide information on aerosol elemental composition. The newest instruments use polarized X-rays with excellent detection limits. Detection of light elements ($Z<20$) requires analysis in vacuum. About 15–20 elements can be routinely measured by XRF in a non-destructive analysis lasting about 5 min. Synchrotron radiation sources provide the best detection limits, but are a rather expensive analysis (Bukowiecki et al. 2005). Particle Induced X-ray Emission (PIXE) uses a beam of protons with energy between 2 and 3 MeV and X-ray spectrometry to measure routinely the concentration of trace elements (Artaxo et al. 1993). The detection limits are ~0.1–10 ppm or a few ng/m^3. For optimum results using PIXE analysis, samples should be collected on the thinnest possible backing materials composed exclusively from low atomic number elements, such as carbon, hydrogen, oxygen or nitrogen. Polycarbonate Track Etched (PCTE) and Teflon filters are often used, as are impaction surfaces such as Kapton and Mylar (Johansson and Campbell 1988). Blank values should be carefully taken into account for reliable measurements of trace elements.

Multi-element analysis by means of inductively coupled plasma-mass spectrometry (ICP-MS) provides excellent detection limits, in the range of a few parts per trillion (ppt). The aerosol sample is extracted from filters and mixed with water and nitric acid and is injected into the plasma torch via a nebulizer. The dissociated elements are detected by mass spectrometry for a very fast and precise quantitative determination. In INAA (Instrumental Neutron Activation Analysis) spectrometry a flux of neutrons in a nuclear reactor is used to produce unstable nuclei and gamma-ray spectrometry to measure trace elements with high accuracy and very low detection limits. Ion chromatography (IC) is commonly used to measure nitrate, sulphate and a broad range of ionic compounds, including organic components. Ion chromatography has also recently been adapted for semi-continuous measurements of particulate and gaseous species in commercial instruments. Detection limits and precision are excellent, but calibration and blank control must be followed closely.

Real time sulphate and nitrate measurements are now possible with commercial instruments. Flame photometric detectors (FPD) measure sulphates by detecting 394 nm light produced by excited-state S_2 molecules formed when sulphur compounds are burned in a hydrogen-rich flame. A few minutes time resolution is possible, with 1 μg/m^3 detection limit for sulphate. Particulate nitrate concentrations can be measured with a single stage impactor that is 95% efficient for particles >0.1 μm aerodynamic diameter and the collected particles are analyzed by flash vapourization using a chemilluminescent NO_x analyzer.

5.2.6 Real-Time Aerosol Mass Spectrometry

New approaches have recently been developed to measured aerosol composition in real time, either in single particles or bulk composition (see review by Sullivan and Prather 2005). Particle analysis by laser mass spectrometry (PALMS and similar instruments) makes it possible to determine the ionic composition of single particles (e.g. Prather et al. 1994, Noble and Prather 2000). Aerosol particles are drawn by vacuum into the instrument. A fraction of these particles is detected by a visible light laser (532 nm), the scattered light triggers an excimer laser (193 nm) that ablates and ionizes individual particles. A complete positive or negative mass spectrum of particles is obtained by using a time-of-flight mass spectrometer.

A different approach to mass spectrometry was adopted in the Aerodyne Aerosol Mass Spectrometer (AMS) that measures properties of an ensemble of particles. The AMS electron impact ionization (EI) and quadrupole mass spectrometry (QMS) applies thermal desorption for aerosol characterization. The AMS consists of three main parts: an aerosol inlet, a particle sizing chamber, and a particle composition detection section. The different sections are separated by small apertures and are differentially pumped. The aerosols are sampled through an inlet that focuses them into a narrow beam. Size-dependent particle velocities created by expansion into vacuum are used to determine particle size through a particle time-of-flight measurement. The focused particle beam is modulated by a rotating wheel chopper. Time-resolved particle detection, after a known flight distance, gives the particle velocity from which the particle aerodynamic diameter is obtained. Detection is performed by directing the particle beam onto a resistively heated, roughened surface under high vacuum (~10^{-7} Torr). Upon impact, the volatile and semi-volatile components in/on the particles flash vapourize. The vapourization source is integrally coupled to an electron impact ionizer at the entrance of a quadrupole mass spectrometer. When the quadrupole is tuned to a representative mass, bursts of ions are produced that are averaged to give a size-resolved mass distribution (e.g. Jayne et al. 2000; Jimenez et al. 2003). Refractory component are not measured by this technique.

5.2.7 The Counterflow Virtual Impactor (CVI)

The counterflow virtual impactor (CVI) has been utilized both in aircraft experiments and on the ground to study aerosol/cloud interactions, cloud physics, and cloud impacts on climate (Ogren et al. 1985; Noone et al. 2003). At the CVI inlet tip, cloud droplets or ice crystals larger than a minimum aerodynamic diameter are separated from the smaller interstitial aerosol and impacted into a dry carrier gas. This separation is possible via a counterflow stream of gas out of the CVI tip, which assures that only larger particles (cloud droplets or ice crystals) are sampled. The non-volatile residual nuclei remaining after droplet evaporation and water vapour removal are sampled downstream of the inlet with selected instruments. These may include a condensation particle counter, an optical particle counter, and various chemical techniques. Inclusion of a water vapour sensor allows determination of bulk cloud condensed water content (Strom et al. 1997; Twohy et al. 1997). CVI's have also been applied in combination with cloud activation instruments and particle composition measuring instruments (e.g. Cziczo et al. 2004a).

5.2.8 Cloud Condensation Nucleus Measurements

A fraction of the particles in the atmosphere can serve as cloud condensation nuclei (CCN) for cloud droplet formation. The fraction depends on the size, composition and surface properties of the particles, and on the water supersaturation of the air (Roberts and Nenes 2005). CCN are linked to cloud microstructures and precipitation formation although this last link in the chain is still under discussion and is the main aim of this assessment. In addition, the role of the CCN in climate feedback remains largely unknown. Measurements of CCN are very challenging (e.g. Twomey 1960) because they depend on supersaturation, which is a function of both size and chemical composition. Recent measurements in Germany by Dusek et al. (2006) show that size is the most important parameter in determining CCN efficiency. Significant improvements in measurement techniques are needed in this area, and even basic concepts need improvements.

A few instruments are available for measuring the CCN number concentration (or *activation spectrum*), including those based on the static thermal gradient diffusion cloud chamber (SDCC), continuous flow spectrometers (CFC), and other principles (Nenes et al. 2001). The SDCC consists of two parallel plates, held at different temperatures with their facing surfaces wetted, the lower plate being several degrees colder than the upper plate. By varying the temperature difference between the plates, it is possible to produce maximum supersaturations in the chamber that range from a few tenths of 1% to a few percent, which are similar to the inferred supersaturations that activate droplets in clouds (Wieland 1956). Small water droplets form on those particles that act as CCN at the peak supersaturation in the chamber. The concentration of these droplets can be determined by photographing a known volume of the chamber and counting the number of droplets visible in the photograph, or by measuring the intensity of light scattered from the droplets (Radke and Hobbs 1969). By repeating the above procedure with different temperature gradients in the chamber, the CCN activation as a function of supersaturation can be determined. Figure 5.1 shows the CCN spectra for different types of aerosol particles in Amazonia, expressed as CCN efficiency defined as the ratio of CCN to CN.

Several CCN instruments were developed and provided CCN spectra and concentration through several experimental designs. The Desert Research Institute (DRI) CCN spectrometer (Hudson 1989) measures nuclei that activate at critical supersaturations down to 0.002%. This instrument continuously provides more than 100 channels of CCN and large nuclei resolution with less than a few seconds time resolution. The Wyoming CCN counter (Delene and Deschler 2000) is a static thermal-gradient diffusion chamber, taking a grab sample every 30s. Both the top plate and the bottom plate of the chamber are covered with water-saturated blotter paper.

Recently a commercially available cylindrical continuous flow thermal gradient diffusion chamber has been built (Roberts and Nenes 2005), employing a novel technique of generating a supersaturation by establishing a constant

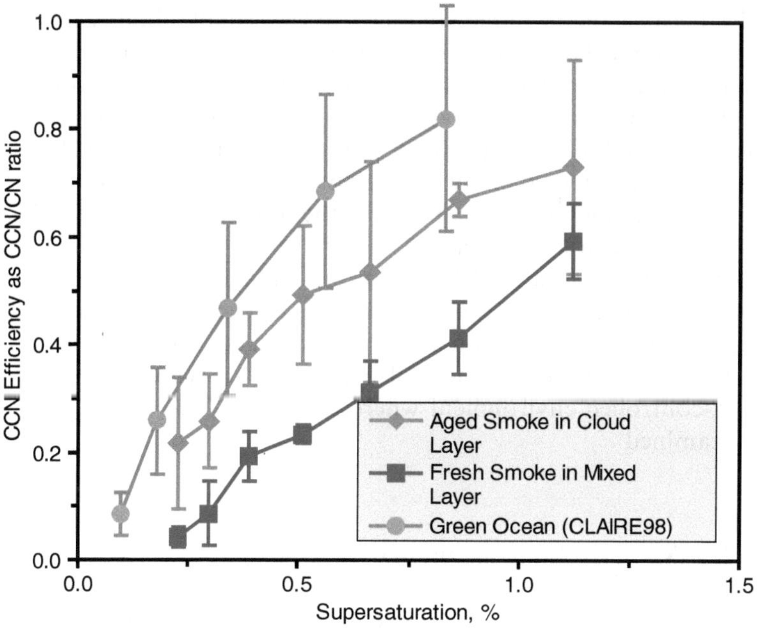

Fig. 5.1 CCN spectra for different types of aerosol particles in Amazonia. *Green ocean* measurements refer to natural biogenic conditions during the wet season. Most of the CCN are biogenic particles in the wet season in Amazonia. *Fresh* and *aged smoke* refers to conditions during heavy biomass burning emissions. (Andreae, personal communication)

streamwise temperature gradient so that the difference in water vapour and thermal diffusivity yields a quasi-uniform centerline supersaturation. The droplet spectrum at the outlet of the growth column is detected using a standard optical particle counter (OPC). This design allows for atmospheric CCN measurements with high time resolution for supersaturations in the range from ~0.07–3%, achieved by varying the flow rates and temperature gradient along the growth column.

In the dynamic CCN spectrometer the aerosol sample is exposed to a varying supersaturation field and the CCN spectrum is obtained by analyzing the outlet droplet size distribution. Time resolution for CCN spectrometers can vary from 1 to 30 s. Although these two types of counters provide some measurement of the CCN concentrations and their dependence on supersaturation, they both suffer from deficiencies, among which is the inability to correctly measure giant CCN that fall out in the chambers before they can be measured.

Giant CCN (GCCN) are wettable particles greater than ~5 μm in diameter that can initiate collision and coalescence in a very efficient way, and thus efficiently produce precipitation embryos. Their effect on cloud droplet spectra can be critical. Due to their large size and low number concentration, it is difficult to measure GCCN. Giant aerosol particles can be measured with large uncertainties from aircraft by optical scattering sensors (PMS FSSP, PCASP), because their sample volumes are very small (about 1 m^3 in 12 h of flight). Since typical concentrations of these particles are a few hundreds per cubic meter, the sampling statistics by these probes is poor. Another method of sampling GCCN is to sample air by an inlet tube into an aircraft using a low turbulence inlet (Huebert et al. 2003). This approach is only good for particles up to 7 μm in diameter, so many ultra-giant particles are not sampled. A third method of sampling GCCN is to sample them by impaction on a glass slide. With slide sample areas of 1 cm^2, 1 m^3 of air can be sampled in 1–2 min of aircraft time (Levin et al. 2005). The disadvantage of this approach is that tedious, labor-intensive microscopic analysis of each slide is required. However, new methods that use computer-controlled counting systems can automatically size the particles. It is also possible to set up the system in a humidity-controlled environment where the hygroscopic growth of particles can be examined.

5.2.9 Ice Nucleus Measurements

A significant fraction of supercooled clouds contains ice crystals, while some clouds (e.g. cirrus) are generally completely glaciated. The presence of ice in clouds is very important for the development of precipitation. It is also an important component of the aerosol indirect effect on climate (see Chapter 2). At temperatures warmer than about –40°C, ice can form by heterogeneous nucleation on particles called ice nuclei (IN) (see Chapters 2 and 6). For clouds with temperatures below about –40°C, ice can form both by heterogeneous nucleation on insoluble particles and by homogeneous nucleation. In spite of the importance of IN in clouds, there is a great deal of uncertainty in measurements of their concentrations and even in their characterization. Heterogeneous ice nucleation can take place via a number of processes (see Chapter 2): *immersion freezing*, *contact freezing*, *condensation-freezing* and *deposition*. In many of the instruments that measure IN activity or spectra, ice may be formed by one or more of these processes, some of which depend on the supersaturation of the air as well as on temperature, without being able to differentiate between them. For example, in expansion chambers cooling is produced by compressing the air and then suddenly expanding it. In case of mixing chambers, cooling is produced by refrigeration. Moist air is then introduced into a cloud but the amount of supersaturation is not well controlled.

A traditional way of measuring IN activity was to collect aerosols from air samples of about 300 liter on Nuclepore filters. Filters are then inserted into an IN chamber made up of two parallel plates held at different temperatures below the 0°C. Supersaturation with respect to ice is reached just above the sample. The concentration of ice crystals formed at each temperature is recorded through a downward looking microscope or a video camera. The number of crystals detected as a function of temperature is then plotted and a best-fit curve is obtained (e.g. Gagin 1975).

Another method for measuring the IN activity spectrum is via drop freezing measurements. This uses an isolated chamber, the lower horizontal plate of which is temperature controlled. Aerosol samples are collected on a filter and washed into a beaker with a known quantity of water. About 100 drops from this water are placed on the cold stage. The number of drops that freeze at each temperature as the plate is cooled down, is recorded. Knowing the volume of air sampled and the amount of water used in the washing process, a qualitative estimate of immersion freezing spectra can be obtained (e.g. Vali 1985). A more recent development is the continuous-flow diffusion chamber (CFDC), which measures ice nucleation by in situ aerosol particles under defined temperature (between –5°C and –80°C) and humidity conditions (Rogers et al. 2001). The instrument focuses on an aerosol particle stream in the central portion (approximately 10% of volume) of a vertically downward-oriented laminar flow between two ice-coated cylindrical walls held at different temperatures. The inter-wall temperature gradient and vapour field expose the aerosol stream to constant temperature and relative humidity conditions for a period (5–7 s), enough to initiate ice nucleation and for growth to occur. A hydrophobic material replaces ice on the warmer wall in the lower one third of the CFDC, which helps to evaporate liquid particles, including cloud droplets, leaving the ice crystals to pass through. The ice crystals that grow in the CFDC are detected and counted optically. By changing the supersaturation conditions, different modes of ice nucleation (e.g. deposition, condensation-freezing, immersion-freezing and homogeneous-freezing) can be detected

(DeMott et al. 2003a). It should be noted that none of these techniques simulate contact nucleation, since the droplets must remain in the chamber for a long enough time to scavenge the aerosol particles acting as IN. A few prototype devices have been built to simulate contact nucleation (Meyers et al. 1992) but none has been implemented for field measurements.

5.3 Airborne Instrumentation for Characterizing Aerosols and Clouds

5.3.1 Introduction

There are important weakness and technical difficulties in measuring aerosol and cloud properties in airborne platforms. In situ measurements of aerosol vertical profiles necessitate the use of airborne platforms with instrumentation suitable to fast and accurate analysis. Isokinetic inlets are required when using airplanes to correctly represent the natural conditions in still atmosphere. The adequacy of the inlet and the necessary high time resolution make measurements of aerosol from aircraft particularly challenging. Aircraft used for atmospheric measurements typically fly at speeds from ~40–>200 m s^{-1}, while the flow speed through filter samplers or aerosol counting/sizing instrumentation is typically <3 m s^{-1}. Variations in pitch, roll and yaw, together with turbulence generated at high aircraft speeds, make the conditions for aircraft sampling very difficult and with large associated uncertainties, especially for sampling of particles >0.5 μm (Huebert et al. 2003). In addition, many airplanes are pressurized and sampling of aerosols inside the cabin requires special care to prevent cabin air from contaminating the samples.

The aerosols samples collected on board the airplane are then analyzed by one of the techniques mentioned in Sect. 5.2. Instruments capable of conducting real time measurements with time resolution better than a few minutes are desirable. In spite of these constrains, aerosol mass spectrometers, nephelometers, PSAP (Particle Soot Photometer), CPCs (Condensation Particle Counters), SMPS (Scanning Mobility Particle Sizer), DMPS (Differential Mobility Particle Sizer), OPCs (Optical Particle Counter), and high time resolution instruments have provided critically important information in recent experiments. Bulk aerosol composition is still a challenge in aircraft measurements, since large samples require high volume pumping over long time periods, something that is not always feasible aboard aircraft. New real time instruments providing aerosol composition such as PILS (particle-into-liquid sampler) is helping to advance the knowledge on the chemical composition of aerosols and CCNs in airborne experiments.

Distortion of airflow around the aircraft fuselage can enhance or deplete certain sizes of particles, generating errors that may be larger than those inherent in the measurement sensor itself (King 1984; Twohy and Rogers 1993). However, instruments located on aircraft wings can measure particle size distributions in real time. The PMS FSSP-100 (for measuring cloud droplets) and the PMS-PCASP-100 (for measuring aerosol particles) are examples of such instruments. Cloud droplets and ice particles can also be collected, counted, sized and identified using airborne platforms.

5.3.2 Cloud Physics Aircraft Instrumentation

Particle Measuring Systems (PMS), have manufactured a set of aerosol and cloud droplet probes based on single droplet light scattering and photodetector imaging (Knollenberg 1981). These instruments came to dominate airborne cloud microphysics observations for more than two decades. However, these instruments are no longer manufactured or supported by PMS. The major instruments developed and still used today are the PMS PCASP-100X (Aerosol sizes 0.1–3.0 μm), PMS FSSP-100 (Cloud particle sizes 2–47 μm), PMS OAP-2D2-C (cloud and drizzle particle sizes 25–800 μm) and the PMS OAP-2D2-P (Precipitation particle sizes 200–6400 μm).

Particle and droplet instruments made by Droplet Measurement Technologies (DMT) also rely on light scattering and occulation of photodiode arrays. The Cloud Droplet Probe (CDP) provides a size distribution similar to the FSSP-100. CDP can measure cloud particles in the range of 1–50μm. The Cloud Aerosol Precipitation Spectrometer (CAPS) provides a large number of particle measurements, including cloud droplet and particle aerosol.

5.3.3 Liquid Water Meters

Cloud liquid water is a very fundamental property of clouds. It can be measured in-flight by the Johnson and Williams (JW), King liquid water probes (King et al. 1978) and the DMT Liquid Water Content (LWC-100) sensor. The temperature of the sensor is maintained constant. For a given air flow past the wire, the higher the liquid water content, the greater the current required to maintain a fixed wire temperature. Liquid water content at 100 Hz is determined as a function of current through the probe and the true air speed. The range of measurements is 0–3 g m^{-3} at 5–50μm droplet diameter. The very fast responding and versatile Gerber Scientific optical cloud water sensor (the Particulate Volume Monitor PVM-100) is an optical cloud microphysics probe designed to characterize the liquid water content (LWC), droplet surface area (PSA), and droplet effective radius (Re) of the smallest droplets in a cloud. The PVM makes these measurements optically on a cloud volume of $\sim$10 cm^{-3}, such that the measurements are independent of air speed. The accuracy of the PVM is estimated to be better than 10% for droplet spectra with volume medium diameter (VMD) smaller than 30 μm, with a precision on the order of 0.002 g m^{-3} (Gerber et al. 1994). Another instrument that is gaining wide acceptance on many research aircraft is the

Nevzorov total water content/liquid water content probe (Korolev et al. 1998). This is a constant-temperature, hot-wire probe designed for aircraft measurements of the ice and liquid water content of clouds. It consists of two separate sensors for measurements of cloud liquid and total (ice plus liquid) water content.

Recently there have been intercomparison studies using the major LWC probes in wind tunnels (e.g. Wendisch et al. 2002; Strapp et al. 2003; Twohy et al. 2003). Such studies are necessary to document the differences between different types of probes.

5.3.4 Cloud Particle Shapes

2D probes can measure cloud particle shapes in real time. The cloud particles enter a sensitive volume where they occult a laser beam casting a shadow on a photodiode array. Because the size range of cloud particles is several orders of magnitude, and the size of the arrays in the early 1980's was rather small, several 2D instruments had to be flown to sample the full range of cloud particles likely present in precipitating clouds. These digital images provided a substantial quantitative advance, especially as recognition analysis allowed separation of the images by crystal habit and particle dimensions. However, the relatively small number of pixels in the array detector limited the resolution of these devices. It was not always possible to decide whether the imaged particle was liquid or frozen. The cloud imaging probe (CIP from DMT) is a new design based on older optical imaging probes. Improvements include a solid-state diode laser, on-board Pitot tube, an on-board digital signal processor, and a synchronous RS-422 data channel providing statistics and compression of image data. The Stratton Park Engineering Company (SPEC) Cloud Particle Imager (CPI) (Lawson et al. 1995; Lawson et al. 2001) produces digital, near photographic quality, images of the cloud particle at aircraft speeds. The Cloud Integrating Nephelometer model CIN-100 is designed for aircraft mounting and measures the optical extinction coefficient and asymmetry parameter. It has a large range of detection, spanning from 5 to 3000 micrometer (Gerber et al. 2000). The Small Ice Detector (SID) developed for the Meteorological Research Flight at Farnborough employs a set of six high sensitivity detectors arranged about an illumination beam (from an Nd:YAG laser) to assess particle shape, and a further two detectors to provide trigger and particle size signals. The instrument is designed to classify cloud particles in the range from ~2 to ~50 μm diameter, and is intended to provide, for the first time, an ability to discriminate between micrometer sized ice crystals and super-cooled water droplets (Hirst et al. 2001).

5.3.5 Remaining Issues

Measurement of aerosols and cloud properties using airborne platforms is difficult. Maintaining isokinetic sampling conditions over a large range of aircraft speeds is not easy, and efficient collection of large particles remains a challenge. The time resolution of aerosol measurements must be improved in order to better describe the rapid changes in aerosol population over different heights. Better instruments are needed to measure CCN and ice, including the capability to distinguish between drops and very small ice crystals (<15 μm). These are vital in order to improved characterization of CCN and ice within vertical profiles under a range of environmental conditions. Despite attempts at instrument de-icing, icing can still be a problem in supercooled clouds, as even light icing will distort the sample flow, resulting in particle size sorting, shadowing and possible particle shattering. Additionally, several droplet instruments have significant "dead time" after a particle is observed, before they are ready to measure the next particles. At high droplet concentrations, this can result in significant errors. Light scattering instruments are calibrated with spherical non-absorbing particles. Hence, the calibration is not strictly valid for absorbing and non-spherical particles. The airflow surrounding any instrument attached to the exterior surfaces of an aircraft will change as the aircraft maneuvers. Therefore, mounting locations should be selected to minimize these effects.

Despite satisfaction with recent instrument developments, the path forward to the next generation of airborne instruments remains uncertain. We must now consider measuring clouds ranging from the local to the global scale. For example, on the smallest scales the interplay between aerosols, water vapour supersaturation, and condensation on CCN, turbulence, coalescence, electrical charge separation and the production of precipitation remains observationally shrouded. To address such complex systems it will be necessary to observe down to the inter-droplet spacing. While in-cloud measurements of temperature and humidity using radiation transfer and tunable diode laser technology show promise, much more work is needed in this area. Tomorrow's instrumented research aircraft will need, in addition to in situ instruments, a compliment of remote sensing tools that permit a 3-D perspective of cloud structures and motions. Microwave imagers, such as the NCAR ELDORA Doppler radar, proportioned to much shorter wavelengths, is a powerfully research tool. Such miniaturized 3-D imagers can then be used in real time to depict the optimal locations to employ in situ observing tools. Characterizing the 3-D nature of clouds is particularly critical in radiation transfer calculations (e.g. Kasyanov 1999).

5.4 Surface-Based Remote Sensing of Aerosols and Clouds

5.4.1 Measurement of Aerosol Optical Depth

Aerosol optical depth (AOD) is a quantitative measure of the extinction of radiation by aerosol scattering and absorption along a path, often between a point of observation and the top of the atmosphere. Column AOD is a measure of the integrated aerosol load through the atmosphere, and is an important parameter for evaluating aerosol amount, variability, and direct radiative forcing. AOD can be determined from the ground by measuring the spectral transmission of solar radiation through the atmosphere, using rather simple and relatively inexpensive instruments pointed directly at the sun called sun photometers or filter radiometers. Routine ground-based AOD observations are of utmost importance for the calibration and validation of AOD retrievals from satellites. In addition, they are needed to correct for aerosol effects in the retrieval of other satellite products. The Ångström exponent, that gives an indication of the column integrated aerosol size distribution, can be derived from simultaneous AOD measurements at several wavelengths.

AOD is not directly measured, but is retrieved from observations of atmospheric spectral transmission. The solar irradiance I at a given wavelength can be expressed as $I = I_0 \exp(-m\tau)$ with I_0 the extraterrestrial (top-of-the-atmosphere) solar irradiance of the sun, m the air mass and τ the total optical depth. The air mass equals 1 for a vertical path and is roughly proportional to $1/\cos z$ with z the zenith angle of the sun during the observation. The total optical depth τ at a given wavelength results from several components such as scattering by gas molecules, τ_R (Rayleigh scattering), extinction by aerosol particles, τ_A, absorption of trace gases, τ_G, such as ozone, and possible cloud contamination. Thus, the AOD can be obtained from the total optical depth by subtracting modeled estimates of the other components $\tau_A = \tau - \tau_R - \tau_G$. Because AOD is essentially a difference between two large numbers, it is very sensitive to calibration errors and to a minor degree also to the methods chosen to model the other components (WMO 2003).

The current state of long-term observational networks for AOD was reviewed by a group of international experts gathered at a meeting in Davos, Switzerland in March 2004 (WMO 2005). There are a number of international and national networks that have extensive records of AOD measurements including AERONET (which encompasses AEROCAN, PHOTONS, AERO-SIBNET, BSRN, GAW-PFR, and SKYNET). National networks are supported by Australia (BoM), China (CMA), Finland (FMI), Germany (DWD), Japan (JMA), Netherlands (KNMI, including Surinam), Russia (Sibrad), and USA (networks including ARM, SURFRAD, CMDL, USDA). Figure 5.2 presents the Global AOD network of long-term sites. These comprise 90 stations with a continuous record for the past 4 years and temporal data for at least 50% of the day. (In any given year many more stations may be operating if short term campaign observations are taken into account. For instance, AERONET reported 110 stations operating during 50% of 2003). Half of the stations fall within the AERONET project, while the other half is maintained mainly by WMO Members and SKYNET. Hemispheric coverage corresponds roughly to the landmass distribution, (1/3SH, 2/3NH), with Australia, Europe and North America accounting for more than 50% of stations. Major gaps exist in Africa, India, Latin America and the polar regions.

The AERONET and the GAW-PFR networks have particularly important roles in worldwide aerosol monitoring. AERONET is by far the largest network. During the workshop in Davos (WMO 2005), the GAW-PFR network was assigned a special role in the harmonization of worldwide AOD data in the process of forming a federation of diverse networks coordinated under the WMO/GAW umbrella. These two networks will therefore be described in more detail below.

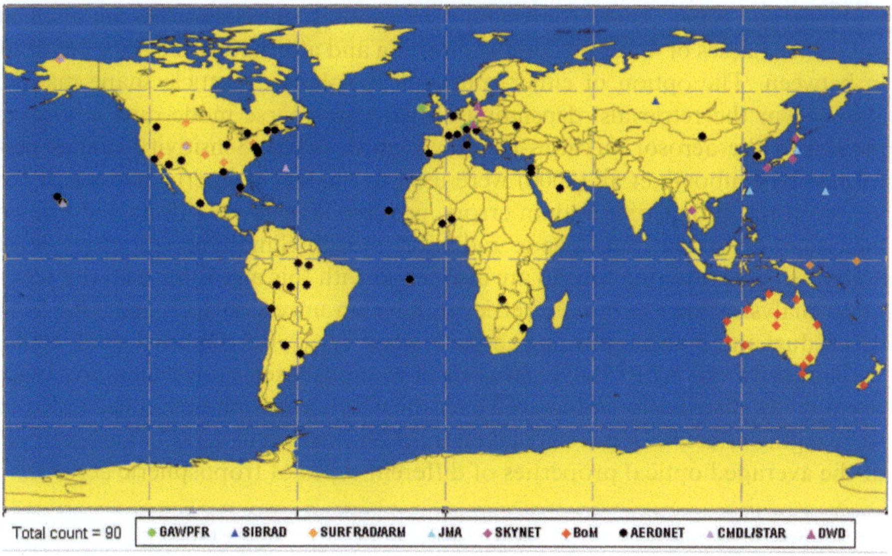

Fig. 5.2 Global AOD network of long-term sites, i.e. with a continuous record for the past 4 years and temporal data coverage of more than 50%

5.4.1.1 The AERONET Network for Monitoring Aerosol and Cloud Properties

The AERONET network of ~200 well calibrated sunphotometers and sky radiance radiometers distributed around the world (Holben et al. 1998, 2001), measures the aerosol optical properties in key locations that are representative of a wide diversity of aerosol conditions. Both direct sun and diffuse sky directional radiance measurements taken throughout the daylight hours are utilized to characterize total column integrated aerosol optical properties. The spectral aerosol optical depths are measured from 340 to 1020 nm in 7 wavelengths for the standard CIMEL sun/sky radiometer and also to 1640 nm in the extended wavelength CIMELs. The AERONET spectral aerosol optical depth (AOD) measurements are highly accurate (~0.015 uncertainty; Eck et al. 1999) and are used to validate satellite retrievals of AOD from all the current satellite sensor systems, including MODIS, TOMS, SeaWiFS, MISR, and AVHRR (Hsu et al. 1999; Remer et al. 2002a; Chu et al. 2002; Diner et al. 2002; Torres et al. 2002; Kahn et al. 2005; Abdou et al. 2005). In addition to validation, the aerosol size distributions and single scattering albedo retrievals from AERONET are utilized in algorithm development for some of these sensors, such as MODIS and TOMS (Remer et al. 2002b).

The CIMEL sky radiance almucantar measurements (measurements along a conical surface with an angle equal to the sun zenith angle) at 440, 675, 870, and 1020 nm in conjunction with the direct sun measured τ_a at these same wavelengths are used to retrieve aerosol size distributions following the methodology of

Dubovik and King (2000). Almucantar sky radiance measurements are made at optical airmasses of 4, 3, and 2 in the morning and afternoon, and once per hour in between. The option of either spherical or spheroid particle shape may be assumed in the retrievals, depending on whether coarse mode dust aerosol is present in the aerosol mixtures (Dubovik et al. 2002). Sensitivity studies performed by Dubovik et al. (2000) were used to analyze the perturbations of the inversion resulting from random errors, possible instrument offsets and known uncertainties in the atmospheric radiation model. Simultaneous retrievals of aerosol single scattering albedo are also made with this algorithm and the sensitivity analysis shows that these retrievals have an uncertainty of ~0.03 for both desert dust and biomass burning aerosols when $\tau_{a440} \geq 0.5$ (Dubovik et al. 2000).

The use of the AERONET network of ground-based radiometers to obtain average size distribution integrated over the whole atmosphere has been proven to be highly effective. Figure 5.3 taken from Dubovik et al. (2002) is an example of the averaged optical properties of different types of tropospheric aerosol.

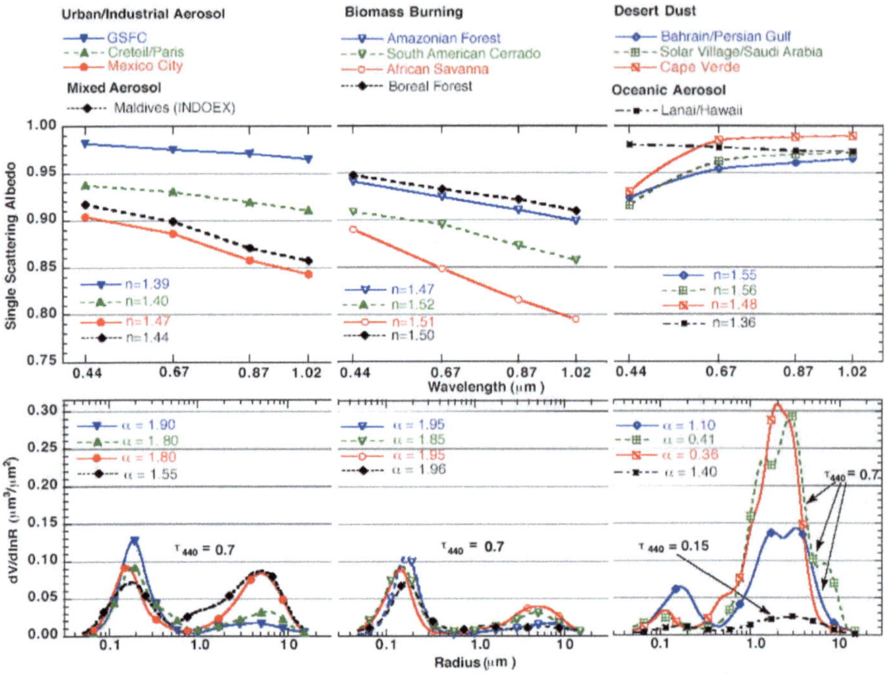

Fig. 5.3 The averaged optical properties of different types of tropospheric aerosol retrieved from the worldwide- multi-year AERONET network of ground-based radiometers. Urban/industrial, biomass burning and desert dust aerosols are shown for AOD(440 nm) = 0.7. Oceanic aerosol is shown for AOD(440 nm) = 0.15 since oceanic background aerosol loading does not often exceed 0.15. The figure shows the single scattering albedo (*top*), the size distribution (*bottom*) and values of the real refractive index (*n*) and the Angstrom exponent (α), estimated using optical depth at two wavelengths 440 and 870 nm. From Dubovik et al. (2002) with permission of the American Meteorological Society

The aerosol size distributions from all of the biomass burning sites are dominated by accumulation mode particles, however the sub-micron particle size is somewhat smaller in the savanna and cerrado sites, due perhaps to the phase (flaming or smoldering) of combustion, but also possibly due to differences in aerosol aging, fire intensity, and ambient relative humidity and temperature.

Based on AERONET data, recent analysis of the effects of aerosols on cloud cover, (Kaufman and Koren 2006) show a decrease in cloud cover with increasing column integrated aerosol absorption (Fig. 5.4) and an increase in cloud cover with increasing non-absorbing aerosols. This work shows that the effect of aerosol on clouds depends on the composition and radiative properties of the aerosols. This result is consistent with studies that showed that biomass burning aerosols strongly inhibit cloud formation (Andreae et al. 2004; Koren et al. 2004).

AERONET's distribution across the globe with standardized instrumentation and processing algorithms provides similar characterizations, aerosol climatologies and aerosol models for marine, urban/industrial, aeolian dust,

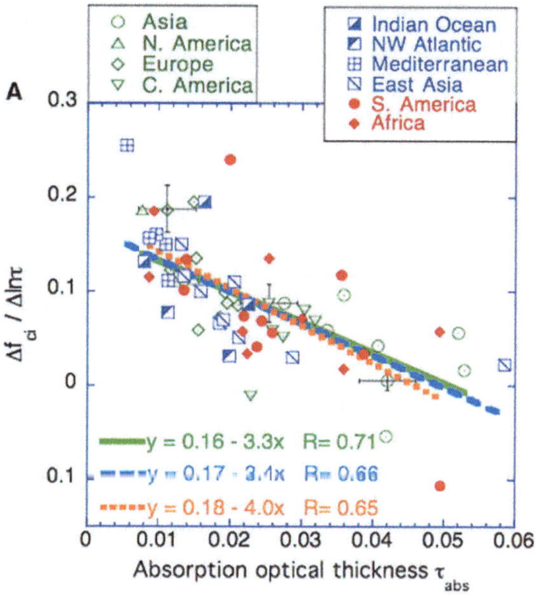

Fig. 5.4 Regional analyses of the AERONET data for the effect of aerosols on cloud cover. Fraction of cloud cover per unit of AOD ($\Delta f_{ci}/\Delta \ln t$) was plotted as a function of absorbing aerosol amount. Each point represents an analysis of more than 3000 measurements from a given location and 2 calendar months averaged over 3 to 5 years. *Green symbols* indicate continental sites, *blue symbols* indicate marine sites, and *red symbols* indicate biomass-burning sites. Error bars are printed for three representative points and indicate the average uncertainties in the least-squares fit used for individual points and the estimated error in tabs (From Kaufman and Koren 2006). Reprinted with permission of the AAAS

biogenic, arctic and mixtures of these aerosol types. Polarization observation capabilities, along with non-spherical modeling of scattering, show great potential for more universal characterization of aerosols. AERONET data in combination with other web-based data sets such as back trajectories, lidar profiles, satellite and chemical transport and global climate models can be used to address issues of radiative forcing, public health and long term environmental impacts.

5.4.1.2 The GAW Precision Filter Radiometer (PFR) Network

A new generation of sunphotometers called "Precision Filter Radiometers" (PFR) was designed at the World Optical Depth Research and Calibration Center (WORCC) in Davos with emphasis on instrumental stability and ruggedness in field use. A series of 34 instruments were built by PMOD/WRC for deployment in the trial network and for use as traveling standards (GAW 2003). The remaining instruments were sold to different research groups and national weather services; PFR instruments are now in operation at 19 stations ranging from the tropics to beyond the polar circle and from sea level to sites above 3,500 m. The PFR instrument is presently not available commercially. Measurements are taken continuously at 1-min intervals. The ASCII data sets are self-contained and suitable for further evaluation into AOD results or Langley calibration runs either at the stations or at a centralized data center. During the trial phase, data are transmitted monthly to WORCC for evaluation, quality control and archiving. Instantaneous values of multi-spectral AOD are calculated using exo-atmospheric calibration coefficients and individual air masses for Rayleigh scattering, ozone absorption and aerosol components. Ångström exponents α are calculated by log-log regression of AOD from 368 to 862 nm.

In the GAW-PFR network, a new calibration method was tested as a possible alternative to classic Langley extrapolations at high altitude stations. A reference instrument obtained its exo-atmospheric values in October 1998 during a stratospheric balloon flight at 40 km height, and its radiometric stability has been monitored since then by repeated spectral comparisons with a silicon trap detector that is traceable to a standard at the Physikalisch-Technische Bundesanstalt in Berlin (Wehrli 2000). Additional reference instruments were calibrated by Langley extrapolations at Jungfraujoch and Mauna Loa. All field PFR radiometers are calibrated at WORCC by comparison with 1 or 2 reference instruments before deployment in the trial network. In situ Langley calibrations are routinely made at the high altitude or arid sites; for other sites, parallel measurements with a traveling standard, exchange of instruments or re-calibration at Davos are used.

GAW-PFR stations are co-located with other networks at several sites. A first comparison during 2000 at Mauna Loa with AERONET individual AOD results showed excellent agreement within 0.005. In 2001 the PFR at Bratt's Lake participated in a 3-month, multi-network comparison (McArthur et al. 2003) where average differences smaller than 0.01 per observation were found for direct-pointing instruments at most wavelengths.

5.4.2 Aerosol Vertical Profiles Retrieved by Lidar

Lidars are useful tools for measuring vertical profiles of aerosols. There are various lidar techniques for measuring aerosols, air pollutants, water vapour, temperature, wind velocity etc. The lidar methods used for aerosol observations are Mie-scattering lidar, Raman lidar, and high-spectral resolution lidar (HSRL). Mie scattering lidar is the simplest lidar technique, which detects elastic backscattering from aerosols. Mie lidars, however, have limitation in quantitative measurement because the lidar equation cannot be solved without an assumption on aerosol optical characteristics or some additional constraint such as independent optical depth measurement, for example, with a sunphotometer. Usually, the lidar equation is solved with an assumption on the extinction-to-backscatter ratio, or lidar ratio (or S1). An additional feature of Mie scattering lidar is the depolarization ratio measurement, which detects polarization components parallel and perpendicular to the polarization direction of the transmitted laser. The depolarization ratio is a good index of nonsphericity of the scatterers. It is useful for detecting mineral dust particles and distinguishing water and ice clouds. Mie lidar measurements at multiple wavelengths also provide useful information for characterizing (or classifying types of) aerosols.

Raman lidars and HSRL can measure extinction and backscattering coefficients independently. Consequently, the profile of the lidar ratio is also obtained with these methods. Raman lidars for aerosol measurements utilize Raman scattering of atmospheric molecules such as nitrogen. Because the profile of atmospheric molecules is known from radiosonde observations and models, the extinction coefficient profile of aerosols is derived from the Raman scattering signals. Using the Mie-scattering signals simultaneously measured with the Raman signals, the backscattering coefficient is also determined (Whiteman 2003). Raman lidars can be extended for multi-wavelength measurements and for water vapour profile measurement. It is also possible to have rotational Raman channels for measuring temperature profiles. A weak point of Raman lidar method is the low sensitivity. Raman lidars, generally, require larger laser pulse energy and larger telescope size. HSRLs use a similar method but utilize Rayleigh scattering of atmospheric molecules (Liu et al. 1999). HSRLs have a higher sensitivity, because Rayleigh scattering cross sections of molecules are much larger than Raman scattering cross section. HSRLs are technically more complex, because they need very high-resolution filters for separating spectra of Mie scattering and Rayleigh scattering.

Currently, several Lidar networks are operated in the world (Fig. 5.5), which include the Micro-Pulse Lidar Network of NASA (MPLNET), the European Aerosol Research Lidar Network (EARLINET), the Commonwealth of Independent States Lidar Network (CIS-LiNet) and the Asian Dust Network (AD-Net).

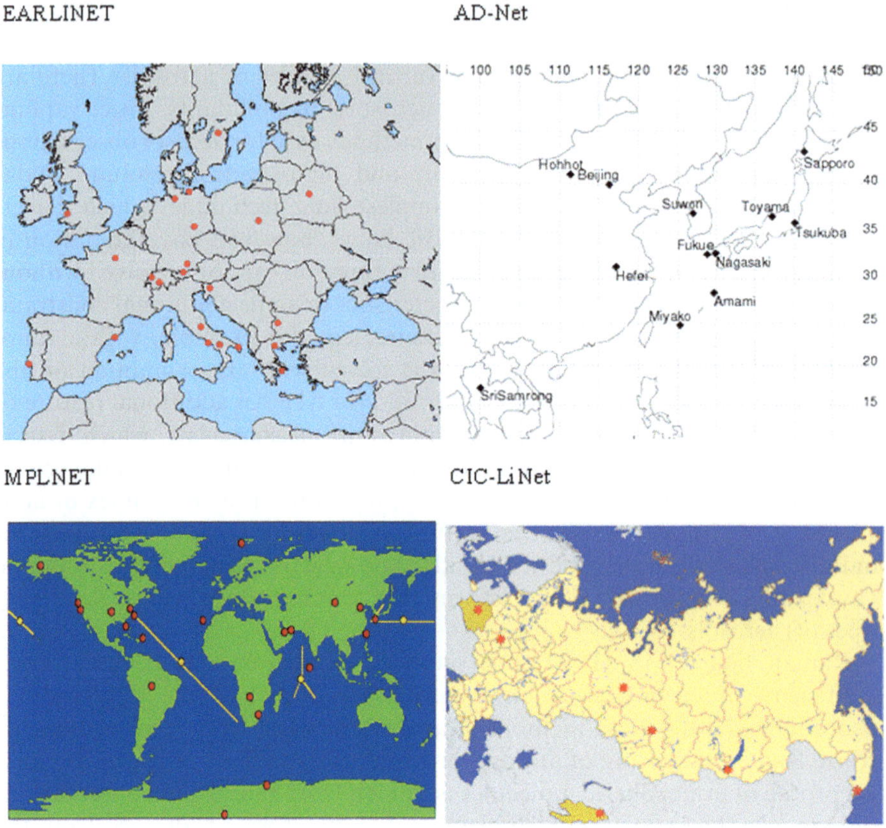

Fig. 5.5 Global lidar networks

The Micro-Pulse Lidar Network (MPLNET) operated by NASA is the network of the micro-pulse lidar (MPL) co-located with the sun/sky photometer sites in the NASA Aerosol Robotic Network (AERONET). MPL is a single-wavelength (523 nm) eye-safe lidar capable of unattended operation. At present, MPLs are operated at 22 sites around the world. The combined lidar and sunphotometer measurements are able to produce quantitative aerosol and cloud products, such as optical depth, sky radiance, vertical structure, and extinction profiles (http://mplnet.gsfc.nasa.gov/).

The Regional East Atmospheric Lidar Mesonet (REALM) is a network of lidar research groups on the east coast of the United States (http://alg.umbc .edu/REALM/).

The European Aerosol Research Lidar Network, EARLINET established in 2000 is a highly instrumented lidar network (http://www.earlinet.org/). The primary purpose of EARLINET is to build a quantitative comprehensive statistical database of the horizontal, vertical, and temporal distribution of aerosols on a continental scale. Measurements are also performed to address important processes, for example, dust events, forest fire events, and volcanic eruptions.

The Commonwealth of Independent States Lidar Network (CIS-LiNet), established in December 2004, is the lidar network in CIS regions (http://www.cis-linet.basnet.by). The basic concept of the network is similar to EARLINET.

The Asian Dust Network (AD-Net) (http://www-lidar.nies.go.jp/AD-Net/index.html) is a voluntary network of lidar researchers in the East Asia region for exchanging data and information. Various groups with various instrumentations are involved in AD-Net. There are multi-wavelength Raman lidars in Tokyo and Gwangju, Korea, and a high-spectral-resolution lidar in Tsukuba, Japan. The National Institute for Environmental Studies (NIES) Lidar Network (http://www-lidar.nies.go.jp/), which is a part of AD-net, is a network of automated two-wavelength polarization lidars (backscattering at 1064 nm and 532 nm, and depolarization at 532 nm). Currently, 13 lidars (Japan (8), Korea (1), China (1), and Thailand (1)) are operated continuously, year-round. Many of the lidars are co-located with the Skyradiometer Network (SKYNET) (http://atmos.cr.chiba-u.ac.jp/)

5.4.3 Surface Based Remote Sensing of Clouds

AERONET also measures the cloud optical depth, the most important of all cloud optical properties and vital for any cloud-radiation parameterization. Its impact on radiative fluxes and therefore climate is matched only by cloud fraction. Cloud optical depth is also a fundamental optical variable in any cloud-resolving model. Progress on cloud-radiation parameterization may halt because without good optical depths, the radiation calculations could be wrong.

There are ambiguities in definitions of cloud fraction. Visual cloud fraction is a nice thing to know but climate models need a *radiative* cloud fraction, and not a visual quantity. By radiatively effective cloud fraction we mean the one that forces plane-parallel radiative transfer calculations to give the same radiation as the measured ones. Stephens (1978) showed that radiatively effective cloud fraction is the cloud property actually required in GCM cloud parameterizations and it is always different from a simple visual one. The AERONET network has begun monitoring *cloud* optical properties, such as cloud optical depth and (effective) cloud fraction, using AERONET "idle time" inappropriate for aerosol study. When the sun is blocked by clouds, a radiometer instead of going "to sleep" is forced to look straight up and measure zenith radiance. Several radiometers have now been equipped with this "cloud mode".

There have been two major problems with inferring cloud optical depth from measurements of zenith radiance: (i) lack of a unique relationship to determine cloud optical depth; and (ii) a strong influence of 3D cloud structure on measured radiance. To solve these problems, a new method that exploits the huge jump in vegetated surface reflectance across 700 nm wavelength has been proposed (Marshak et al. 2004). The idea was to use zenith radiance measurements in two narrow spectral bands on each side of the jump. In the RED spectral region (~670 nm), the chlorophyll in green leaf absorbs 90–95% of solar radiation; thus, the vegetation albedo is low. In contrast, in the NIR spectral region (~870 nm), a green leaf reflects 90% of incident radiation; thus its albedo is high at this wavelength. If AERONET measurements are combined with surface reflectance properties, then zenith radiances at RED and NIR wavelengths can be used to determine the overlying cloud properties. This is illustrated in Fig. 5.6 using surface data. The left panel shows a calculated set of curves for one particular choice of RED and NIR surface reflectances, and for a range of possible cloud optical depths τ and cloud fractions N_c. Also plotted are three groups of 10 data-points measured by an AERONET radiometer on July 28, 2002 at the DOE/ARM site in Oklahoma at three different times. The plot shows that the 3 clusters of AERONET measurements correspond to three different pairs of cloud properties – ($N_c=0.9$; $\tau=28$), ($N_c=0.8$; $\tau=22$), and ($N_c=0.4$; $\tau=12$). The right panel shows an image from DOE/ARM's "all sky camera." Note that $N_c=0.8$ is not a horizontal cloud

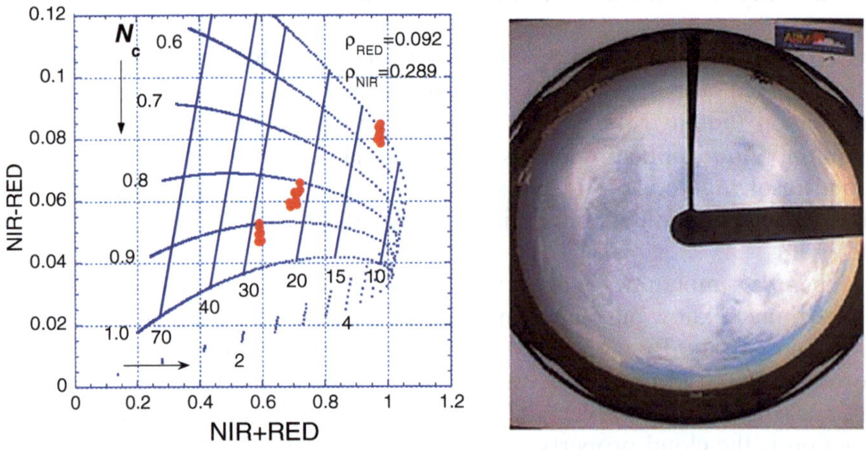

Fig. 5.6 *Left panel*: DISORT calculated values of RED and NIR radiances for a wide range of optical depth τ and cloud fraction N_c for SZA = 62°±3°, surface albedos $\rho_{RED}=0.092$ and $\rho_{NIR}=0.289$. When N_c is constant and τ is varying, the set of calculated values define the cloud fraction isolines. When τ is constant and A_c is varying, the set of values define the optical depth isolines. The red dots are data points from AERONET measurements at the ARM site on July 28, 2002 and show three different clusters of cloud properties. Measurements were taken around 13:45, 13:58, and 14:11 UT, respectively. *Right panel*: Total Sky Image taken at 14:00 UT with SZA = 62.750. From Marshak et al. (2004) with permission of the American Meteorological Society

fraction that would be seen in the surface camera, or the satellite image, but a "radiatively effective" fraction that accounts for the horizontal inhomogeneity of the clouds. For details, see Marshak et al. (2004).

Recently, surface-based remote sensors have been used to examine aerosol effects on clouds. Although these methods can only be applied at a limited number of locations, they yield high temporal resolution data. They offer some interesting insights, and are complementary to the global satellite view. Satellite borne probes measure cloud properties in cloudy pixels and aerosol properties in adjacent, cloud-free pixels, with the implicit assumption that these aerosols are representative of the aerosols entering cloud base.

5.5 Satellite Measurements of Aerosol and Clouds

5.5.1 The Large Scale Picture of Aerosol Distribution

Radiant energy reflected and emitted by the Earth's surface and atmosphere carries with it a signature of the atmospheric and surface properties. By measuring sunlight transmitted directly or diffusively to the surface through the aerosol and cloud layers, surface based networks of radiometers can derive aerosol and cloud properties. By measuring the reflected light's spectral, angular and/or polarization properties, satellite sensors can also quantify several atmospheric and surface properties.

To get a perspective on our remote sensing capability from satellites and surface radiometers, we start with our experience with the human eye. Our eye is sensitive to a narrow range of the solar spectrum, with receptors in the blue, green, and red. We get depth perception using the eyes' slightly different angles of observation. By analogy, aerosol remote sensing started using a single wavelength and single angle of observation (Fraser et al. 1984; Husar et al. 1997), corresponding to the viewing ability of a color blind person using a single eye (e.g. METEOSAT, and GOES). The TOMS instruments, flown since 1978 have two channels sensitive to ultraviolet (UV) light that were discovered to be excellent for observations (Herman et al. 1997a,b) of elevated smoke or dust layers above scattering atmosphere. A two-channel technique was used for the AVHRR with simultaneous analysis of aerosol and clouds (Nakajima et al. 2001; Sekiguchi et al. 2003; Mishchenko et al. 2003). These instruments provide a long-term (currently 20 years) series of aerosol and clouds properties (Fig. 5.7) (Torres et al. 2002) that can be used to assess regional changes in the aerosol concentration. However, calibration and characterization of the instrument performance is difficult (Vermote et al. 1995), and the TOMS analysis requires an assumption of the height of the aerosol layer (Torres et al. 2002). Note that the geostationary satellites, METEOSAT and GOES, have an advantage over the AVHRR and TOMS of being able to generate long term series with the full diurnal cycle, albeit at a single wavelength. This approach has not yet been completely explored.

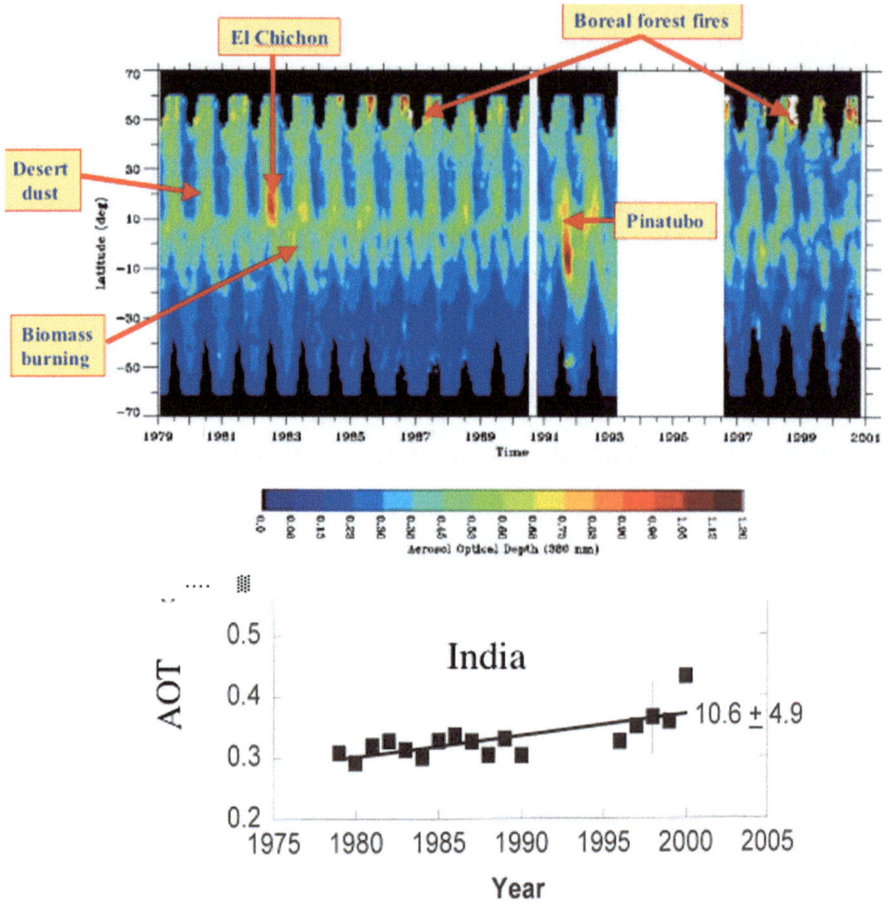

Fig. 5.7 Twenty year time series of aerosol optical depth from the TOMS satellite UV measurements. *Top*: for the globe as a function of latitude (Torres et al. 2002, with permission of the American Meteorological Society), *Bottom*: time series for India. From Massie et al. (2004) with permission of the American Geophysical Union

The first remote sensing instrument designed for aerosol measurements, POLDER (Polarization and Directionality of the Earth Reflectances, Deschamps et al. 1996), uses a combination of spectral channels in a range wider than the human vision (0.44–0.91 μm). The instrument is a wide-angle camera that observes the same target on the Earth at 14 different angles. POLDER also measures light polarization (that we cannot see) to detect fine aerosols over land, taking advantage of the difference between the small and spectrally neutral polarized light reflected from the Earth's surface and the spectrally decreasing polarized light reflected by fine aerosols. Two instruments, MODIS (MODerate resolution Imaging Spectroradiometer) and MISR (Multi-angle Imaging SpectroRadiometer), on the Terra satellite (King et al.

1999) have been measuring global aerosol distributions and properties since 2000. Over the ocean, MODIS uses the aerosol spectral signature in a wide range (0.47 to 2.1 μm) to distinguish small particles, which in high concentrations are typically associated with anthropogenic pollution or smoke, from coarse particles usually identified as natural sea salt or dust (Tanré et al. 1997; Kaufman et al. 2002a). This information is used to derive the fraction of aerosol optical depth that is due to anthropogenic sources (21±7% over the oceans) (Kaufman et al. 2005, 2005b). Over land MODIS uses the 2.1 μm channel to observe the surface cover properties, to estimate surface reflectance at visible wavelengths, and to derive the aerosol optical depth (Kaufman et al. 1997; Remer et al. 2005) from the residual reflectance at the top of the atmosphere. The brightness of the earth-atmosphere system at the top of the atmosphere is affected by the angle at which the light is reflected by the surface and atmosphere. MISR (Diner et al. 1998) takes advantage of this fact by detecting the reflected light at different viewing angles (nadir to 70° forward and backward) along the satellite's track in a narrower spectral range (0.44–0.87 μm), and is thus able to separate the aerosol signal from that of the surface, and to derive information about particle size and shape (Martonchik et al. 1998; Kahn et al. 2001). A mixed approach using two view directions but a wider spectral range (0.55–1.65 μm) is used by ATSR (Veefkind et al. 1998) to derive the aerosol concentration and type. These instruments, POLDER, MODIS, MISR and ATSR have controlled calibration and characterization that results in known and smaller errors than the previous generation of instruments (King et al. 1999).

The realization that aerosols affect surface temperature, Earth's radiation budget and precipitation patterns, creates a demand for more informative space borne observations. This is now achieved with the completion of the so-called A-Train in April 2006 (Fig. 5.8) that is designed to measure aerosol, clouds and precipitation, with PARASOL (Polarization and Anisotropy of Reflectances for Atmospheric Sciences coupled with Observations from a Lidar (POLDER-like instrument launched Dec. 2004) and MODIS for passive aerosol and cloud observations in the visible to IR, the OMI instrument for aerosol observations in the UV to visible, the CALIPSO lidar for aerosol profiling in two spectral channels and cloud top height detection, and CloudSat radar for profiling cloud liquid water and structure.

In parallel with the development of aerosol and cloud measurements, the Tropical Rainfall Measuring Mission (TRMM) (Kummerow et al. 2000) was launched in 1997 to measure tropical-subtropical rainfall, and in so doing, acquired the first accurate, representative, and consistent ocean climatology of precipitation. TRMM carries the first space borne precipitation radar (PR), producing new measurements of precipitation vertical structure (~250 m resolution). TRMM also carries several additional instruments to conduct new observations of convection, frontal zones, precipitating storms and tropical cyclones.

The set of measurements described above covers information from aerosol distribution and properties, to cloud properties and precipitation. The satellite data are evaluated and complemented by ground-based remote sensing of

Fig. 5.8 The A train: a constellation of satellites that jointly measure aerosol, clouds and precipitation properties. Photo from NASA collection

aerosols (e.g. AERONET and MPL networks – Holben et al. 1998), aerosol effects on clouds (at the ARM sites – Feingold et al. 2003; Kim et al. 2003; Garrett et al. 2004) and precipitation (Kummerow et al. 2000). The data are used to evaluate aerosol transport models (e.g. Chin et al. 2002) and as input to climate models that study the effects of aerosol on clouds and precipitation (e.g. Menon et al. 2002). Statistics have been acquired of the aerosol effect on cloud microphysics and albedo (Kaufman and Nakajima 1993; Kaufman and Fraser 1997; Rosenfeld and Lensky 1998; Nakajima et al. 2001; Bréon et al. 2002; Sekiguchi et al. 2003; Matsui et al. 2004). Some issues relating passive remote sensing aerosol retrievals in the vicinity of clouds have been raised (e.g. Coakley et al. 2005). It is recognized that the cloud properties (cloud droplet size distribution, extended cloud fields or broken clouds) being studied depend on the sensor resolution and on the method of investigation; which in turn affects the strength of the aerosol impact on clouds (Rosenfeld and Feingold 2003).

The accuracy of satellite retrievals is higher over the dark ocean surface than over the land with its varying albedo. Densely vegetated land areas, e.g. the Amazon Forest, are dark and relatively uniform and are therefore some of the best targets for measurements of the aerosol impacts over the land. Surface conditions affect both the aerosol and cloud retrievals and the measurements of liquid water content from TRMM. Therefore, the statistical analysis of aerosol-cloud interaction can be done globally over the oceans and over some continental regions, in particular the dark forests. The increase in cloud fraction and cloud development, with an increase in aerosol loading, is in agreement with the notion that an increase in CCN loading modifies cloud microphysics and inhibits precipitation (see Chapter 6 for more details), although the changes in cloud microphysics due to increase in aerosol on inhibition of precipitation has not yet been statistically established. New instruments are being designed or proposed for future launches with several critical innovations:

- Spaceborne instruments that combine a wide spectral range with a wide viewing angle range and with polarization (e.g. the Glory mission, Mishchenko et al. 1997; Advanced PARASOL (Tanré, private communication); Next-generation MISR (Diner, private communication)). Such instruments are expected to improve derivation of fine and coarse aerosol concentrations, their sizes, shapes and refractive index. Refractive index, an optical property, is sensitive to the aerosol composition and water uptake (Chowdhary et al. 2002).
- Derivation of aerosol absorption over the land (Kaufman et al. 2001) and over the oceans by measuring the aerosol spectral attenuation of the bright glint (Kaufman et al. 2002b).
- Measuring the vertical distribution of cloud microphysics using slant angle spectroradiometers (Martins et al. private communication).
- New strategies and instruments to measure precipitation from space (Smith et al. 2003).

It is recognized that there is a need for coordination of the aerosol, clouds and precipitation measurements and coordination with in situ and network measurements, as well as development of models that can integrate large data sets (Diner et al. 2004; Seinfeld et al. 2004b).

In the last decade, we have witnessed a major step in the amount and quality of satellite and surface-based remotely sensed data of aerosol, clouds, and precipitation. We are probably at the peak of research benefits from this new array of satellite and surface measurements. Specific plans for the next 2 decades are presently limited in scope. More attention is needed to ensure that the degree of sophistication in the aerosol, clouds, and precipitation measurements and the degree of coordination among them continues to increase.

In the following we describe in more detail some of the satellite instruments used to measure the effect of aerosol on clouds and precipitation. One of the advantages of the multi-instrument system to measure clouds and aerosol is the ability to intercompare the results. Figure 5.9 shows an example of such intercomparison for 4 satellite observations of the pollution over the Indian subcontinent.

5.5.2 Remote Sensing of Clouds and Aerosols by the POLDER Instrument

The POLDER instrument (Deschamps et al. 1994) measures the solar radiance reflected by the Earth's surface and atmosphere in the visible and near infrared (440 to 910 nm). The two main characteristics of POLDER are the ability to measure the linear polarization of the radiance in three spectral bands, and

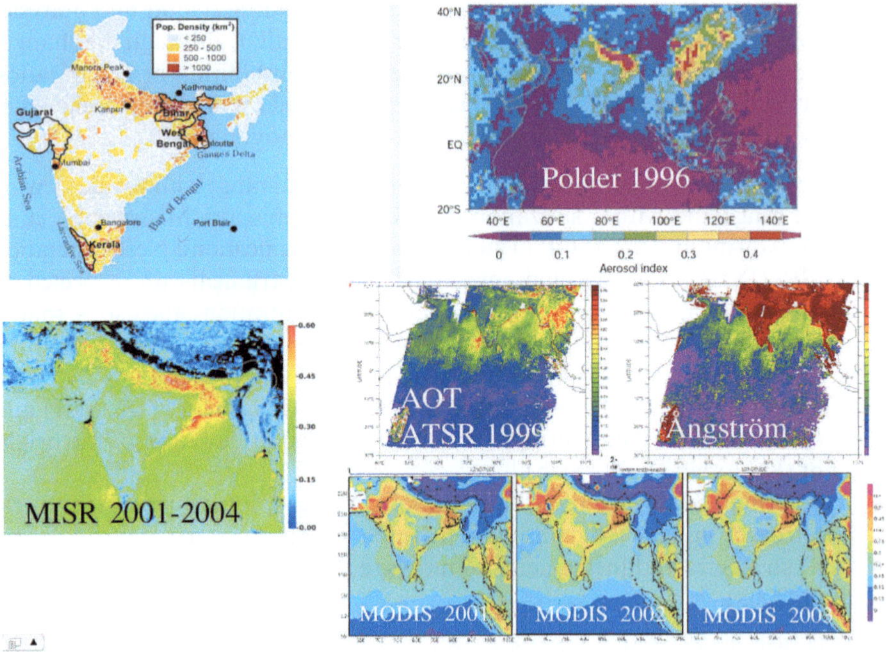

Fig. 5.9 Intercomparison for 4 satellite observation (MODIS, Polder, ATSR and MISR) of the pollution over the Indian subcontinent. *Top left* – population density per km^2, *top right* – POLDER measurements of the Aerosol index that is similar to the fine aerosol optical depth, *center right* – ATSR optical depth and Angstrom exponent for 1999, *bottom right* – MODIS aerosol optical depth for 2001 through 2003, *left bottom* – MISR optical depth averaged over 2001–2004. From DiGirolamo et al. (2004) with permission of the American Geophysical Union

to acquire the directional variation of the reflected radiance: the instrument concept is based on a wide field of view lens and a bi-dimensional Charge-Coupled Device (CCD) that provides an instantaneous field of view of $\pm 43^\circ$ along-track and $\pm 51^\circ$ cross-track. As the instrument flies over the target, up to fourteen views are acquired; these can be composited to infer the directional signature of the reflectance. This signature provides information on the surface, aerosol, and cloud characteristics. A limitation of POLDER is the rather coarse spatial resolution of $\sim$6 km, which affects the ability to account for scene heterogeneity.

The POLDER instrument flew onboard the ADEOS 1 and 2 platforms in 1996–1997 and 2003, respectively. Unfortunately, due to the failure of the satellite solar panels, the measurement time series are limited to respectively 8 and 7 months. Another very similar instrument was launched in December 2004 onboard the microsatellite PARASOL to be part of the A-train. Algorithms have been developed to process the measurements in terms of aerosols (Deuzé et al. 2001; Herman et al. 2005) and cloud characteristics (Buriez et al. 1997; Parol et al. 1999).

5.5.2.1 Remote sensing of Aerosol Properties

Over oceans the satellite algorithm (Herman et al. 2005) assumes spherical or non-spherical particles, non-absorbing particles and a size distribution that follows a combination of two log-normal aerosol size distributions in the accumulation and coarse modes respectively. When the geometrical conditions are optimal, the shape (spherical or not) of the particles is derived. The refractive index retrieval is next attempted from the polarization measurements. If the real part of the refractive index of the coarse mode is retrieved when spherical particles are present (close to 1.35, indicating hydrated particles), the derivation of the refractive index of the accumulation mode is very tentative. Comparisons of AOD with ground based AERONET measurements show excellent agreement, with typical RMS errors on the order of 0.05, including errors due to cloud cover, with no significant bias (Goloub et al. 1999). The fine-mode optical depth can also be compared to AERONET measurements, albeit with some uncertainty in the aerosol radius cutoff. Statistical results indicate a low bias of 0.02 with a standard deviation of 0.02.

The retrieval of aerosol properties over land surfaces is based on polarized reflectance measurements. Polarized reflectance of land surfaces is small and fairly constant, although it does have a very strong directional signature (Nadal and Bréon 1999). Scattering by sub-micron (accumulation mode) aerosol particles generates highly polarized light (Deuzé et al. 2001), which makes the polarized satellite radiances more sensitive to the presence of aerosols than to total radiances. On the other hand, larger aerosol particles, such as desert dust, do not polarize sunlight and are therefore hardly detected from polarization measurements. Results for September 2005 of the optical depth of the accumulation mode derived from PARASOL are shown in Fig. 5.10.

Over land the evaluation of POLDER retrievals is made against the fine mode optical depth derived from AERONET measurements. The results show no significant bias and an RMS error on the order of 0.04 when dust-loaded atmospheres are excluded (i.e. validation in regions affected by biomass burning or pollution aerosols only).

5.5.2.2 Remote Sensing of Cloud Properties

POLDER provides an innovative method to derive the droplet radius. It is based on the analysis of cloud bows (Fig. 5.11) that can be observed in the polarized radiance measurements (Bréon and Goloub 1998). The method is very precise (error of <0.3 μm) with no known causes for error or biases, but it is applicable only in very specific conditions, i.e. extended cloud fields (150×150 km^2) with narrow size distributions at cloud top. The requirements are most often fulfilled over stratocumulus cloud fields on the Eastern part of oceanic basins, as well as over the Antarctic Ocean (Bréon and Colzy 2000). A comparison with MODIS estimates of cloud droplet radius show a high

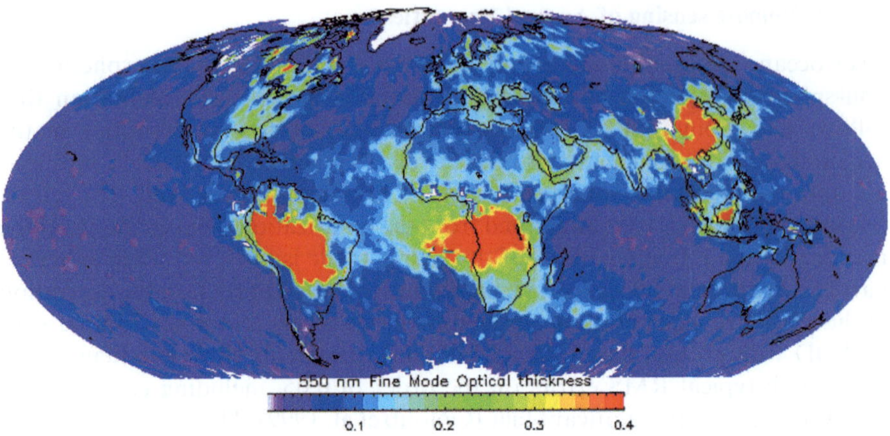

Fig. 5.10 POLDER aerosol monthly optical depth of the accumulation mode for Sep. 2005. The accumulation fraction of the aerosol loading is derived with different methods over land and ocean. *The continuity at the land/sea boundaries observed in most regions is a good indicator of the quality of the inversions.* Note the aerosols emitted from South America, Equatorial Africa and Asia (Tanré, personal communication)

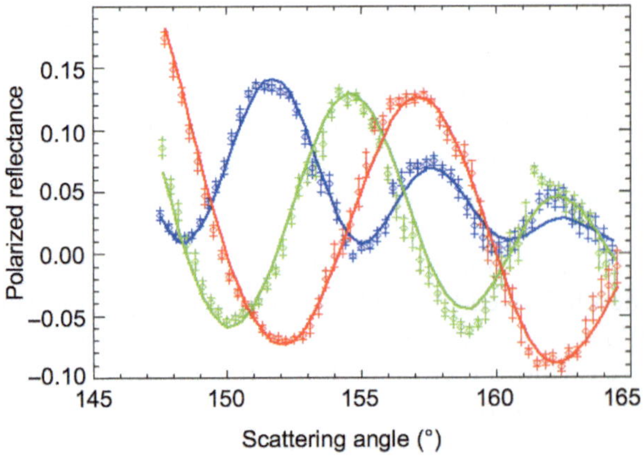

Fig. 5.11 Polarized reflectance from a cloud field as a function of the scattering angle in the three polarized channels of POLDER (*blue* – 443, *green* – 665 and *red* – 865 nm). The measurements over a 150×150 km^2 area have been averaged in 0.3 scattering angle bins. The figure shows the average (*diamond*) and the standard deviation (*crosses*) together with the result of a best fit (*line*). In this case, the best fit indicates a droplet effective radius of 8 μm. From Bréion and Goloub (1998) with permission of the American Geophysical Union

correlation over the oceans, but poor correlation over land surfaces. There is a bias in the two estimates that is not understood (Bréon and Doutriaux-Boucher 2005). The CDR from POLDER has been used together with aerosol estimates to assess their effect on cloud microphysics (Bréon et al. 2002).

The polarization capabilities of POLDER also permit an unambiguous identification of the cloud top phase. The method is based on the presence or the absence of a "cloud bow" for a scattering angle of 140° (Riédi et al. 2000).

The POLDER instruments provide two means of evaluating the cloud top pressure. One method is based on the measurement of the polarized radiance generated by molecular scattering (in the blue part of the spectrum where molecular scattering is very efficient), which decreases as a function of the cloud top pressure. The statistics show a negligible bias and a standard deviation of 65 hPa. The other method is based on the ratio of two radiance measurements centered on the oxygen A-band at 763 nm (Vanbauce et al. 1998). One channel is 10 nm wide while the other is 40 nm wide with the same central wavelength. As a consequence, the ratio of the narrow and wide channels provides an indication of the oxygen absorption between the reflector and the satellite, which is a direct proxy of the reflector pressure. A comparison with lidar measurements of cloud profiles has confirmed the validity of the two methods (Vanbauce et al. 2003).

5.5.3 *Remote Sensing of Aerosols and Clouds by MODIS*

The MODerate resolution Imaging Spectroradiometer (MODIS) began collecting data in February 2000 from the Terra (10:30 a.m. and 10:30 p.m. equatorial crossing time) spacecraft and June 2002 from the Aqua spacecraft (1:30 p.m. and 1:30 a.m. equatorial crossing time). Special emphasis is given to on-board calibration facilities, lunar observations, and detailed analysis of the calibration time series on the ground (Barnes et al. 1998). MODIS provides measurements of the spectral radiances from 0.41 to 14 μm in 36 spectral bands daily over most of the globe. MODIS measurements are used to characterize the global aerosol, clouds, water vapour, fires, vegetation, sea surface temperature, and ocean color. The aerosol characteristics are derived over the oceans (Tanré et al. 1997) and land (Kaufman et al. 1997) using independent algorithms

5.5.3.1 The Aerosol Algorithm

Over oceans the MODIS aerosol algorithm uses the measured 500 m resolution radiance from six MODIS bands (550–2100 nm) to retrieve aerosol information. Specifically, in cloud-free, glint-free ocean scenes (Martins et al. 2002), MODIS retrieves at a 10 km resolution: the aerosol optical depth at 550 nm, τ_{550}, the fraction of τ contributed by the fine (sub-micron size) mode aerosol, f, and the effective radius of the aerosol, r_{eff} (Tanré et al. 1997). Over land, the 2.1 μm channel is used to estimate the surface properties (Remer et al. 2005) and the 0.47 and 0.66 μm channels are used to derive f and τ. Aggregation of the

MODIS aerosol information from the 500 m pixels to the 10 km product allows for rigorous cloud screening, avoiding data gaps, and still generates large enough statistics for a stable and accurate product. The MODIS-derived aerosol properties have been validated using AERONET (Remer et al. 2005). In agreement with theoretical error analysis (Tanré et al. 1997; Kaufman et al. 1997), the aerosol optical depth is derived with an error of $\Delta\tau_{550} = \pm 0.03 \pm 0.05\tau$ over the oceans and $\Delta\tau_{550} = \pm 0.05 \pm 0.15\ \tau$ over the land. Over the oceans, the errors were found to be mostly random with very little bias. For aerosol dominated by dust, a bias of about +10% was observed. Figure 5.12 shows example of the MODIS observations over the globe. (MODIS monthly data can be accessed interactively from http://lake.nascom.nasa.gov/movas/. Information and images on the MODIS atmospheric products can be found at http://modis-atmos.gsfc.nasa.gov/).

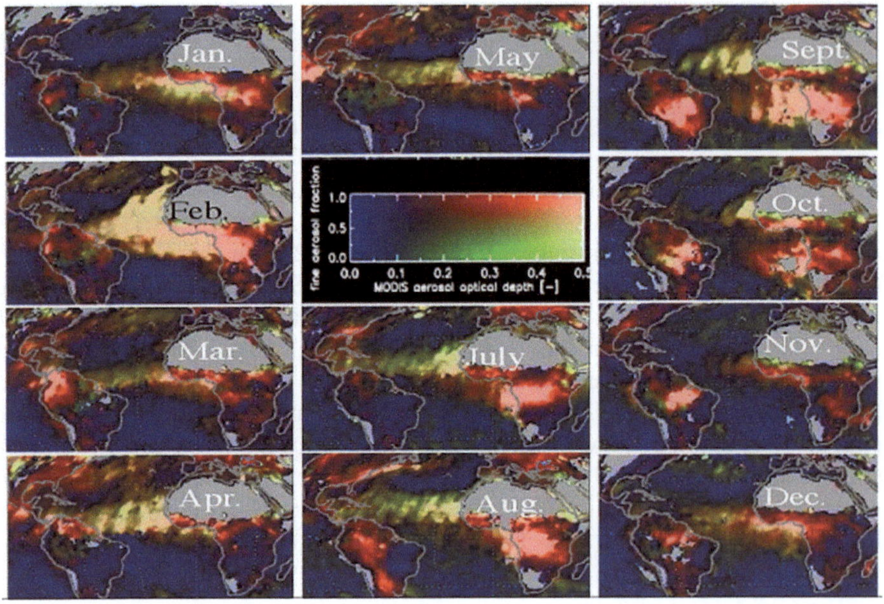

Fig. 5.12 MODIS aerosol monthly composites for 2001. Each composite is for the 15th of each month ± 5 days to find enough cloud free regions. Data for June are not shown since no MODIS data were available during the middle of the month. The color bar is located instead. The color bar was constructed so that blue represents clean conditions, aerosol optical depth <0.1, green and red show higher optical depth corresponding to the coarse (*green*) and fine (*red*) modes. Therefore, pure dust is shown as green and pure smoke or pollution red. Note that the color of the aerosol emitted from Africa changes from mixed red and green in January through April, to green in July–August. Biomass burning occurs in the Sahel during January–March, and moves to Southern Africa for July–August, when it is separated from the dust flow. From Kaufman et al. (2005a) with permission of the Americam Geophysical Union

5.5.3.2 Remote Sensing Measurements of Cloud Properties

The MODIS cloud product (Platnick et al. 2003; King et al. 2003) combines infrared and visible techniques to determine the physical, radiative, and microphysical properties of clouds. Cloud optical depth and effective radius are derived globally using six visible and near-infrared bands at 1-km spatial resolution. Cloud-top properties, including cloud-top temperature, cloud-top pressure, and effective emissivity are derived using the infrared split window and longwave CO_2 absorption bands (both day and night) at 5-km spatial resolution. Cloud thermodynamic phase is computed at 5-km resolution using a two-band algorithm that includes an additional thermal band at 8.55 μm, and also at 1-km resolution using a different technique based on results from the cloud mask tests followed by a bispectral threshold test, shortwave infrared tests, and finally cloud top temperature (King et al. 2004). Finally, the cloud product contains a cirrus reflectance product at a visible wavelength for use in removing cirrus scattering effects from the land surface. Hence, the cloud product contains many different cloud properties derived from 14 bands in total. The file size is different during the night (only cloud-top properties and thermodynamic phase at 5 km resolution) than during the day (when additional 1 km resolution products are included).

Figure 5.13 shows an example of the cloud optical depth, cloud top pressure, and effective radius for a daytime granule of Terra-MODIS data over the Western Pacific Ocean near the Kamchatka Peninsula on August 10, 2001 at 0025 UTC. The true color image in Fig. 5.13a shows extensive cloud cover over the Sea of Okhotsk, including mid-level and upper-level ice clouds, whereas the Bering Sea to the east of the Peninsula contains extensive marine stratocumulus clouds with numerous ship tracks in the south-eastern portion of the image. Figure 5.13b shows cloud optical depth, Fig. 5.13c shows cloud-top pressure, and Fig. 5.13d shows cloud effective radius, where different color scales for water and ice clouds were used in Fig. 5.13(b) and (d). The optically thick marine stratocumulus to the east of the peninsula is identified as water clouds with optical depths (0.65 μm) up to 25. The optically thick ice clouds over the Sea of Okhotsk and around the southern portion of the Kamchatka Peninsula have cloud optical depths approaching 40.

Ship tracks are not easily identified in the cloud optical depth image shown here, but they result in reduced effective radii in the microphysical retrievals shown in Fig. 5.13d. For this scene the thermal infrared and decision tree algorithms for deriving cloud thermodynamic phase, discussed by Platnick et al. (2003), are in quite good agreement. Finally, the cloud-top pressure for the cloud-filled pixels, shown in Fig. 5.13c, clearly show that the water clouds lie predominantly between 700 and 850 hPa, whereas the optically thick ice clouds over the Sea of Okhotsk lie at altitudes above the 500 hPa level.

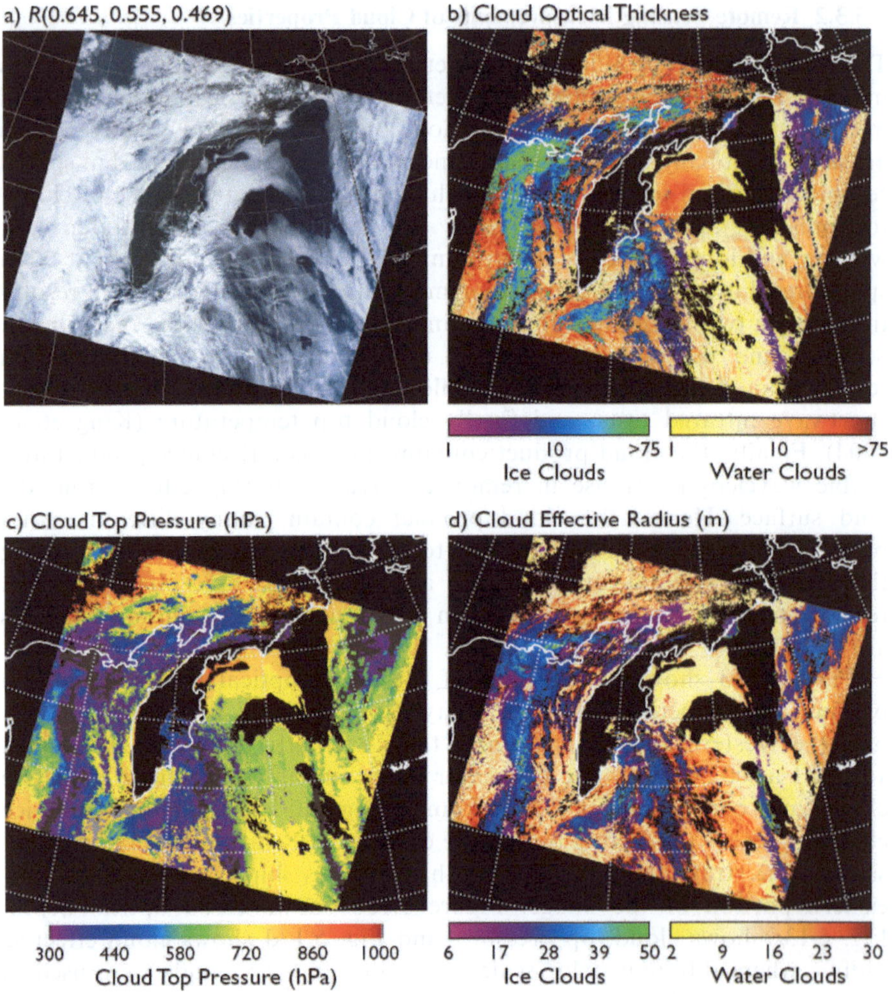

Fig. 5.13 Cloud properties over the western Pacific Ocean off the Kamchatka Peninsula on August 10, 2001. Panel (**a**) is a true color composite of one MODIS granule, showing marine stratocumulus clouds with ship tracks as well as upper level ice clouds. Panels (**b**) and (**d**) show the cloud optical depth and effective radius derived from all cloudy pixels, where a separate color bar is used to denote clouds processed as ice and water clouds. Panel (**c**) shows the cloud-top pressure for all clouds in this scene (King et al. 2003: © 2003 IEEE)

5.5.4 Measuring Aerosols with the Multi-angle Imaging Spectro Radiometer (MISR)

The Multi-angle Imaging SpectroRadiometer (MISR) instrument on Terra (Diner et al. 2002) observes the Earth globally at nine angles, ranging from 70° forward to 70° backward, along the spacecraft track. MISR aerosol, cloud, and surface products are available through the NASA Langley Atmospheric Sciences Data Center (http://eosweb.larc.nasa.gov); further information about the MISR Mission can be found at http://www-misr.jpl.nasa.gov.

Images are obtained in four spectral bands (446, 558, 672, and 866 nm) with a ground spatial sampling of 275 m to 1.1 km. Such data offers several advantages for aerosol retrieval. Oblique viewing provides high sensitivity to aerosol optical depth. Over major aerosol source regions, such as deserts and urban areas where the high surface reflectance presents a challenge, MISR algorithms make use of the systematically changing ratio of surface to atmospheric radiance with view angle to separate the surface from atmospheric signals (Martonchik et al. 1998, 2002). The resulting optical depths show good agreement with values derived from AERONET (Diner et al. 2001; Martonchik et al. 2004; Kahn et al. 2005; Abdou et al. 2005).

Over water MISR's scattering angle coverage (~60–160° in mid-latitudes) helps distinguish particle size and shape. Three to five particle size groupings between ~0.1–2.5 microns are identified in the data (Kahn et al. 2001, 2005), and recent work indicates that MISR is able to separate different mineral dust shape classes (Kalashnikova et al. 2005; Kalashnikova and Kahn 2006). Although detailed aerosol type validation opportunities are rare, retrievals from MISR data acquired during the CRYSTAL-FACE campaign identified three distinct aerosol modes within the column: maritime-type particles, cirrus, and Saharan dust, in proportions that agree with independent data to within about 20% (Kahn et al. 2001). On another occasion, retrievals over Galveston Bay, near Houston, TX, distinguished two aerosol modes, one with an effective radius near 0.1 μm and the other with an effective radius near 0.6 μm, showing sensitivity to differences between $>\sim 0.95$ and $<\sim 0.88$ in aerosol single scattering albedo, in agreement with field measurements upwind in the city, and consistent with pre-launch sensitivity expectations. Over water, MISR also generally has multiple glint-free cameras, making aerosol retrievals possible, even when nadir-viewing instruments are glint-contaminated.

MISR routinely obtains the heights of reflecting surfaces, such as cloud and aerosol plume tops, stereoscopically, provided there is some spatial contrast so that image pattern matching can be performed. This is often the case for aerosols near volcanic, forest fire, and dust source regions, where distinct plume features are present. The technique retrieves the altitude at which the reflectance contrast is greatest typically near the top of the plume. MISR algorithms perform pattern matching globally in an automated manner (Kahn et al. 2007), with 1.1 km horizontal resolution and about 0.5 km sensitivity in the vertical. An example for a wildfire is shown in Fig. 5.14. The stereo retrieval is successful in distinguishing the heights of two separate plumes associated with the fire complex.

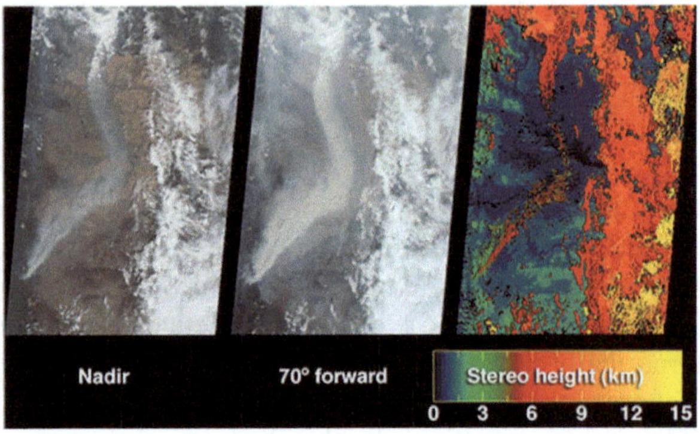

Fig. 5.14 MISR nadir camera image (left), 70° – camera image (*middle*), and standard stereo height retrieval product (*right*) for the B&B fire complex in Oregon, acquired September 4, 2003. The smoke snakes northward from several blazes located near the lower left. Based on Kahn et al, (2007) with permission of the American Geophysical Union

5.5.5 Meteosat Second Generation – MSG

The Meteosat series of geostationary satellites, at 0° longitude, are ideally located to follow the export of dust from the Saharan and Sahel dry areas and the biomass burning activity in Sahel and central Africa, two major sources of aerosols to the atmosphere. This opportunity has been exploited since the early 1990s to quantify the export of dust over the Atlantic and the Mediterranean (Jankoviak and Tanré 1992; Dulac et al. 1992; Moulin et al. 1997). Likewise, the thermal infrared channel of Meteosat was used to locate the main source areas of dust in Northern Africa through their impact on the apparent temperature at midday during dust storms, and a corresponding Infrared Difference Dust Index (IDDI) has been defined (Legrand and N'Doumé 2001).

The first generation of Meteosat satellites obtained images in only three wide spectral bands, in the visible band (0.45 to 1.0 μm), in the water vapour absorption band (5.7 to 7.1 μm) and in the thermal infrared (window) band (10.5 to 12.5 μm), so it is not possible, for instance, to distinguish between aerosol types like dust or biomass burning aerosols over the ocean. This limitation does not apply to the Meteosat Second Generation (MSG) that has three channels in the solar spectral range, from 0.6 to 1.6 μm, and two in the thermal infrared (window) band, 10.8 and 12.0 μm. It is expected that this operational satellite will permit an excellent monitoring of the aerosol sources and transport over the African continent. One major advantage of the geostationary satellites, compared to the polar orbiters, is their ability to provide images of a given area with high temporal sampling, every 15 min for MSG, which permits the monitoring of aerosol transport and impact on clouds and precipitation in different parts of the diurnal cycle.

5.5.6 Future Missions for Cloud and Aerosol Remote Sensing

The aerosol effect on the cloud albedo can be detected and quantified from space by means of long-term global measurements of the change in the number concentration of aerosol particles and the associated change in the cloud albedo and lifetime. The National Polar-orbiting Operational Environmental Satellite System (NPOESS) is being developed jointly by the National Ocean and Atmopsheric Administration (NOAA), the Department of Defense and NASA. The satellite system is planned for long term monitoring of the Earth and atmosphere, from the Visible/Infrared Imager/Radiometer Suite, VIIRS, a multi-spectral imager somewhat similar to MODIS.

Most CCN are smaller than the particles that can be detected with current passive remote sensing. The aerosol size spectra retrieval, together with information on the chemical composition, could be a good proxy for CCN estimation (Mishchenko et al. 1997). Although the chemical composition of aerosols cannot be determined directly from remote sensing measurements, the spectral refractive index and shape of aerosols provides strong constraints on aerosol composition. Other measurable manifestations of the indirect effect include modifications to the cloud droplet size distribution and number concentration and changing liquid water path (King et al. 1995; Schwartz et al. 1995; Brenguier et al. 2000, 2003). As previously described, single radiance-only measurements provide estimates of column optical thickness, Ångström exponent, fine mode fraction, and effective particle size of aerosols, as well as of optical depth, effective radius and phase (ice or water) of cloud droplets. Directional (MISR) and polarized radiances (POLDER) can better constrain the inversion to provide additional information on aerosol composition, single scattering albedo and shape, which also reduces the uncertainties on the retrieved aerosol and cloud parameters. This is of importance in understanding the different type of aerosols that play a role in modifying cloud microphysics. It is also essential that the cloud properties do not have biases that depend on the type of cloud or the season. Extension to the short infrared spectral (such as MODIS) of the present directional and polarization capabilities will provide a detailed estimate of aerosol and cloud properties. Such detailed information will result in better understanding and estimation of the aerosol effect on clouds and can explain the correlative studies being currently used.

In 2008 the NASA Glory mission will deploy a high precision multi-spectral photopolarimeter, called the Aerosol Polarimetry Sensor (APS). Since these measurements will only be available along the spacecraft ground track, other satellites are needed to obtain such data at a global scale on a daily basis. Applications of satellite data are constrained by the spatial resolution of the sensor. A spatial resolution of 500–1000 m is required for an efficient screening between aerosol plumes and clouds.

5.5.7 *Aerosol and Clouds Profiling from Space*

5.5.7.1 The GLAS Lidar Mission

The Geoscience Laser Altimeter System (GLAS) was launched in 2003. It is the sole scientific instrument on the Ice, Cloud, and land Elevation Satellite (ICESat). GLAS combines a precision surface lidar with a sensitive dual wavelength cloud and aerosol lidar. GLAS has 3 lasers, operating one at a time, that emit infrared and visible laser pulses at 1064 and 532 nm wavelengths. In orbit GLAS continuously emits laser pulses at a rate of 40 per second from the Earth facing (nadir) side of ICESat. GLAS measures precisely how long it takes for photons in a laser pulse to pass through the atmosphere to the Earth, reflect, and travel back to GLAS. Halving the total travel time, and applying corrections for the speed of light through the atmosphere, the distance from ICESat to the laser footprint on Earth can be calculated. ICESat collects data for calculating its position in space by using onboard GPS receivers augmented by a network of ground GPS receivers and satellite laser ranging stations. The angle of the laser beam relative to stars is measured precisely by GLAS with star-tracking cameras on the zenith side of ICESat. As ICESat orbits, GLAS takes data along ground tracks defined by the sequence of laser spots. GLAS produces a series of approximately 70 m diameter spots that are separated by nearly 170 m along track. Though the ICESat main mission is to measure polar ice, it also provides detailed information on the vertical distribution of clouds and aerosols illuminated by its laser spots (Fig. 5.15).

Fig. 5.15 GLAS profiles of aerosols on the region of the Himalayas. Image from http://icesat.gsfc.nasa.gov/images/movies/india_noclouds_dates.mpeg

5.5.7.2 The Calipso Satellite Mission

The Cloud-Aerosol Lidar and Infrared Pathfinder Satellite Observation (CALIPSO) satellite, was launched in April 2006. It continues the GLAS aerosol and cloud measurements, but with emphasis on atmospheric studies in coordination with other measurements on the A-train set of satellites. CALIPSO combines an active lidar instrument with passive infrared and visible imagers to probe the vertical structure and properties of thin clouds and aerosols over the globe (Winker et al. 2003). CALIPSO is a joint U.S. (NASA) and French (CNES) satellite mission that will fly as part of the A-train. It has an expected 3 year lifetime.

Retrieving the spatial and optical properties of clouds and aerosols from satellite lidar is confronted by a number of difficulties that are not faced in the analysis of ground-based lidars. Among these are the very large distance from the target, the high speed at which the satellite traverses the ground track, and the resulting low signal-to-noise ratios (Winker et al. 2004). To retrieve the aerosol extinction profile, the appropriate lidar ratio needs to be used in the optical analyses of aerosol layers. From field measurements, the aerosol lidar ratios show large variability, from 15 to 120 sr, hence it cannot be derived from the backscatter lidar but requires a Raman lidar or a high spectral resolution lidar (HSRL) for such measuremensts. CALIPSO therefore, determines the lidar ratios using a model-matching scheme, with the help of geophysical (e.g. latitude, longitude), and temporal (season) information. Six aerosol types are used, four of these (polluted continental, biomass burning, desert dust, and polluted dust) are derived from a comprehensive cluster analysis applied to AERONET data gathered from numerous sites around the globe (Winker et al. 2004).

Alternatively, the MODIS measurements can be used to constrain the inversion of the lidar data (Kaufman et al. 2003). The combined lidar and spectral measurements can be used to sense the vertical distributions of aerosol concentrations and properties. Figure 5.16 shows the results of inversions of combined aircraft lidar and MODIS data to study the properties of smoke off the southwest coast of Southern Africa. The inversion derives profiles of the aerosol extinction due to fine and coarse particles. Comparisons with three sets of airborne in situ measurements show excellent agreement of the aerosol extinction profiles; however, the lidar inversion derives smaller spectral dependence of the extinction than the in situ measurements.

5.5.7.3 The Cloudsat Satellite Mission

CloudSat (Stephens et al. 2002), launched in April 2006 with CALIPSO, provides from space the first global survey of the vertical structure of cloud systems using 94 GHz radar reflectivity measurements. The unique feature of this radar lies in its ability to observe jointly most of the cloud condensate and precipitation within its nadir field of view and its ability to provide profiles of

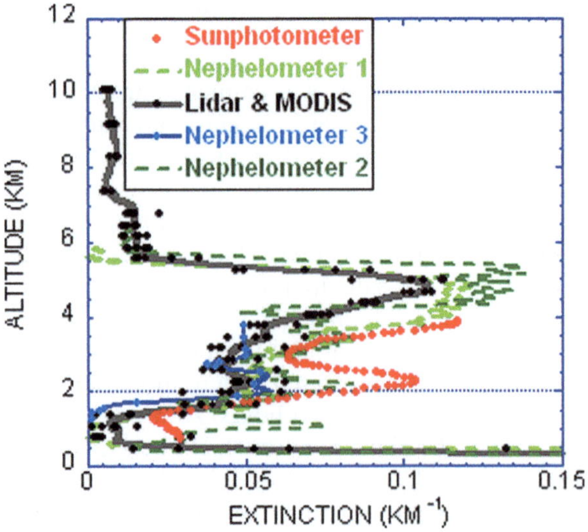

Fig. 5.16 Extinction profiles derived from the lidar & MODIS data (*black*), in situ nephelometer measurements (*blue*, *light* and *dark green*) converted to ambient extinction, airborne sunphotometer measured extinction profile (*red*), all for wavelength of 0.55 μm. The corrected in situ data reveal the smoke and boundary layer marine aerosol, shown in the inverted lidar-MODIS data. From Kaufman et al. (2003) with permission of the American Geophysical Union

these properties with a vertical resolution of 500 m. The primary product is the level-1B calibrated, range-resolved radar reflectivities and the essential level-2 products are the cloud profile properties derived from these radar data. It flies in on-orbit formation with the Aqua, PARASOL and CALIPSO satellites, providing a unique, multi-satellite observing system (the A-Train) particularly suited for studying the atmospheric processes of the hydrological cycle.

5.6 In Situ and Remote Sensing of Precipitation

Precipitation is a discontinuous geophysical field. It is raining over only ~4% of the earth's surface at any one time. Historical records indicate that rainfall has been measured for at least the last 2500 years. Simple rain gauges are based on catching rainfall in a bowl and measuring the contents at regular time intervals. Recent technological advances in rainfall measurement have focused on standardizing measurement in non-recording gauges, devising self-recording gauges, and on remote sensing methods.

The spatial variability of precipitation and its intermittency in time make areal rainfall difficult to measure accurately. The value of precipitation measurements depends on how well the measurement system is suited to the desired application. There is not a unique answer regarding the relative accuracy and precision of precipitation measurements, since the magnitude of errors are dependent on the spatial and time scales of interest, the storm structure, and the instrument system making the measurement. There is also currently no consensus on how to report rainfall measurement uncertainties, which makes comparison among different measurement systems and studies difficult.

5.6.1 In Situ Point Measurements of Precipitation

Rain gauge measurements are usually underestimates of the true precipitation at the location of the instrument (Groisman and Legates 1994). Wetting on the internal walls of the gauge and evaporation of liquid within the gauge decrease measured precipitation. Wind-related undercatch is usually the largest individual error source and is particularly large for snow. For objects such as precipitation gauges mounted above the surface, wind deviates around the object yielding a pattern of turbulence and accelerated and decelerated flow (Oke 1987). The divergence of flow acts to divert lighter particles, such as small drops and snow, away from the top orifice of the gauge, which decreases the precipitation falling on the gauge in comparison with what it would be at the surface if the instrument was not present (Folland 1988; Nešpor and Sevruk 1999). For a given particle size, undercatch increases with increasing wind speed (Sevruk 1982). Wind effects can be lessened by applying a wind correction based on the rain rate and wind speed (e.g. Sevruk 1982; Hasse et al. 1998), by deploying a wind shield around the instrument, by modifying the shape of the instrument (Folland 1988), and by placing the instrument in an area of low wind. Wind undercatch can be eliminated by placing the instrument in a pit so that the gauge orifice is at the ground surface. For typical rainfall events in the central plains of the USA, undercatch by above-ground tipping bucket and weighing gauges, compared to a pit gauge, is 4–5% (Duchon and Essenberg 2001). For an extreme event in the same region with very heavy rain (200 mm hr^{-1}) and high winds (12.5 m s^{-1}) undercatch was 15% (Duchon and Essenberg 2001). A separate study comparing measurements among above-ground gauges found that losses related to wind, wetting, and the common practice of counting trace precipitation (<0.127 mm) as 0 mm lead to average underestimates in rainfall of 20% for Barrow, Alaska (Yang et al. 1998).

The uncertainty in estimation of liquid water equivalent of snowfall is a function of wind speed, air temperature, whether a wind shield is present, and the type of wind shield and gauge (Goodison 1978; Groisman et al. 1991). Errors have been estimated to be as low as 10% at wind speeds $\leq$5.5 m s^{-1} for Canadian Nipher shielded snow gauges (Goodison 1978) to several hundred percent for unshielded 8-inch non-recording gauges historically used by the US weather service (Yang et al. 1998).

In situ rainfall measurements at sea vary in quality as a function of wind speed, instrument type and exposure, and to what degree the ship or buoy distorts the wind flow, in addition to the distortion created by the instrument itself (Yuter and Parker 2001).

5.6.1.1 Basic Types of In Situ Rain Instruments

Non-recording or manual rain gauges based on the bowl concept have hundreds of different designs and are still in widespread use. These instruments catch precipitation for recording by hand at regular time intervals. Measurement is performed either with the liquid in place or by pouring the contents of the gauge into a graduated cylinder. These methods yield slightly different estimates for the same catch. Depending on the length of time and the ambient environment, precipitation can be lost to evaporation before it is measured.

Different types of self-recording in situ rain instruments measure rainfall by myriad methods. Common methods for measuring rainfall include funneling rain into a bucket on a pivot that tips when it is full (tipping bucket), weighing collected water, measuring the capacitance of collected water, and the conversion of collected water into drops of fixed size and the counting of the formed drops (Strangeways 1996). Raindrop-induced optical scintillation is also used to estimate rain rate (Wang et al. 1979). Hot-plate sensors estimate rain and snow rates using cooling due to melting and evaporation of precipitation (Rasmussen et al. 2002a).

Disdrometers measure precipitation droplet size distributions by counting the number of drops within each of several size categories over a time interval. Disdrometers are a ground-based analog to aircraft precipitation particle probes (Sect. 5.3.2) since both yield size distributions of liquid and snow particles large enough to fall ($\geq\sim$0.2 mm diameter). The range of particles sizes detected and the distribution of bin sizes vary among different instruments. The minimum particle size detected can also vary as a function of rain rate, and with the ambient wind and noise environment of the instrument (Joss and Gori 1976; List 1988; Kruger and Krajewski 2002; Yuter et al. 2005). The size distributions are used to estimate bulk properties of the raindrop distribution including median diameter, rain rate, liquid water content, and equivalent radar reflectivity (Joss and Gori 1976). Some disdrometers use the kinetic energy of a raindrop impact to count natural drops in specified size ranges (Joss and Waldvogel 1967). Other disdrometers use optical methods including occultation and image processing (Löffler-Mang and Joss 2000; Fiser et al. 2002; Kruger and Krajewski 2002; Barthazy et al. 2004). Some optical disdrometers can measure particle size and velocity simultaneously and have the capability of distinguishing among particle types in mixed rain and snow (Löffler-Mang and Joss 2000; Kruger and Krajewski 2002; Yuter et al. 2006).

Vertically-pointing Doppler radar and profiler measurements are used to retrieve the drop spectra in a profile above the instrument (Gossard et al. 1990; Thomson and List 1995, 1996; Peters et al. 2002; Williams 2002, Nissen et al. 2005). Various methods are used to account for vertical air motion and turbulence, which if left uncorrected will reduce the quality of the retrieved drop size distribution (Joss and Dyer 1972). Methods to retrieve precipitation rate based on line-integrated attenuation between microwave transmitters and receivers, including cell phone towers, are under development (e.g. Messer et al. 2006).

Cloud radar (mm-wavelength) measurements attenuate in rain but can provide information on precipitation properties within shallow clouds (Wang and Geerts 2003; Bretherton et al. 2004; Li et al. 2004; Stevens et al. 2005). A combination of cloud radar and backscatter lidar measurements can be used to estimate mean particle size and to retrieve liquid water flux in stratocumulus clouds (e.g. O'Connor et al. 2005)

Self-recording precipitation measurement instruments can be ranked by their precision and latency. The smaller the detectable accumulation of rainfall, the better the instrument is at sensing low rain rates and the start and end of rainfall events. Latency is the time delay between when the rain passes through the plane of the sensing orifice and when it is registered by the instrument. Instruments such as impact or optical sensors have latency equal to their measurement time interval. Instruments with funnels to direct rainfall into the measuring mechanism have latency greater than or equal to their measurement time interval. For instruments with funnels, the percentage of rainfall lost to instrument wetting and to latency increases with decreasing rain rate. At low rain rates, it may take minutes to hours for sufficient rain to accumulate and be detected by an instrument with coarse precision. The impact of precision and latency on rainfall measurement is illustrated in Fig. 5.17 which shows 1 min rainfall data collected by an impact disdrometer (Joss and Waldvogel 1967) with precision <0.01 mm, a drop-forming rain gauge (Hasse et al. 1998) with

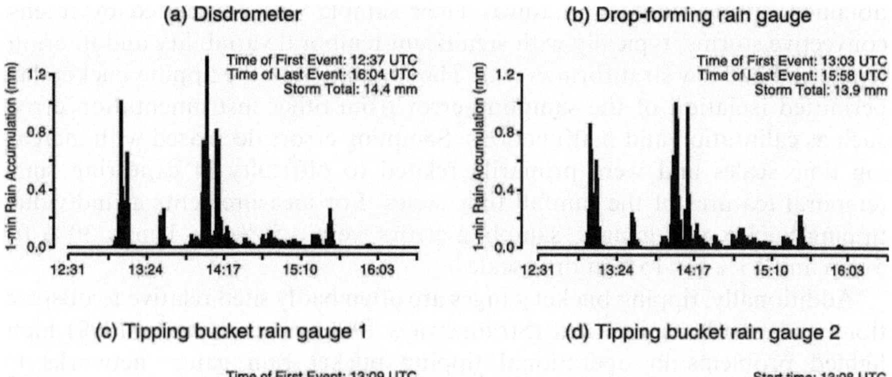

Fig. 5.17 Time series of recorded 1-min rain rates obtained from 1230 to 1630 UTC on 27 August 2003 from four instruments within 10 m of each other on Kwajalein Island in the Republic of the Marshall Islands. Instruments are (**a**) Distromet Inc. RD-69/ADA-90 impact disdrometer, (**b**) Eigenbrodt Inc. SRM-450 drop-form rain gauge, and (**c**)–(**d**) Qualimetrics Inc. 6011-A series 0.254 mm tipping bucket rain gauges. Based on Joss and Waldvogel (1967) with permission of S. Yuter

precision of 0.1 mm, and two tipping bucket rain gauges with a precision of 0.254 mm. While all four time-series show the same general temporal variation in precipitation, the 1 min rain rates, storm accumulations, and storm durations differ among the instruments. The highest precision instrument yields the highest accumulation and the longest storm duration, while the coarser precision instruments yield smaller accumulations and shorter storm durations. The latency at low rain rates is visible as gaps in the tipping bucket data, and to a lesser degree the drop-forming gauge data, when the disdrometer shows light rainfall. Relative errors between rain gauges and remote sensing estimates are addressed later under remote sensing systems.

5.6.1.2 Tipping Bucket Rain Gauge Networks

Tipping buckets rain gauges are one of the least expensive automated gauges available, and are often deployed in operational rain gauge networks. These instruments have sampling limitations related to their coarse precision (0.1 mm to 0.254 mm increments of rainfall). Habib et al. (2001) characterized sampling errors in 0.254 mm (0.01 in) tipping bucket gauges commonly used in the USA. They simulated tipping bucket data from a high resolution data sample obtained during summer in Iowa. Their sample was dominated by intense convective storms, typically with significant temporal variability and intermittency, and had few stratiform events. The simulation of the tipping bucket data permitted isolation of the sampling error from other instrumentation errors such as calibration and malfunctions. Sampling errors decreased with increasing time scales and were primarily related to difficulty in capturing small temporal features of the rainfall time series. For measurements at individual tipping bucket rain gauges, sampling errors were ~50% for 1 min, 30% for 5 min and 15% for 15 min time scales.

Additionally, tipping bucket gauges are often badly sited relative to obstructions and poorly maintained (Strangeways 1996). Steiner et al. (1999) highlighted problems in operational tipping bucket rain gauge networks by comparing nearby gauges to each other and to a radar-derived rainfall map. Over a 2 year study period, Steiner et al. (1999) reported that none of the 30 gauges in the network covering an area of 21.4 km^2 worked 100% of the time, but that for 80% of the storms investigated ~70% of the gauges worked reasonably well. Tipping bucket rain gauges were prone to malfunction due to mineral particulate and biological detritus accumulation such as dust, blown grass, pine needles, spider webs, dead insects, and bird droppings. Detritus malfunctions include partial or complete blockage of the gauge funnel leading to underestimates of rainfall.

5.6.1.3 Point to Point Spatial Variations of Precipitation

Thiessen (1911) was one of the first to discuss large differences in measured rainfall among nearby gauges. Huff and Shipp (1969) used statistical correlation methods to describe spatial variability and to define sampling requirements for rain gauge networks based on data obtained during four summer seasons. Spatial correlations decay with distance at rates that vary with rain type and synoptic pattern. Spatial correlations at a given scale decrease with increasing spatial variability. For thunderstorms they found $r^2 = 0.98$ at 1.6 km distance and 0.88 at 9.7 km distance compared to $r^2 = 1.00$ at 1.6 km and 0.98 at 9.7 km during passage of low pressure centers (Huff and Ship 1969). Spatial correlation increased for increasing storm durations up to 12 h and then decreased (Huff and Shipp 1969). Morin et al. (2003) obtained similar results based on analysis of 15 summer airmass thunderstorms observed in southeastern Arizona. To measure summer storms, Huff and Shipp's recommended rain gauge separations of 0.5 km for 1 min rain rates and 12 km for storm totals. In comparison, typical rain gauge densities in operational networks, such as the one in California, have gauge separations of ~10 km near population centers and ~20 km in rural areas. The Global Precipitation Climatology Project of the World Climate Research Program has estimated that between 5 and 20 rain gauges are required per 2.5° latitude box to meet a 10% criterion for relative sampling error for monthly precipitation (Rudolf et al. 1994). This recommended rain gauge density is met in most industrialized areas but is not met over the majority of the Earth's land surface (Fig. 5.18).

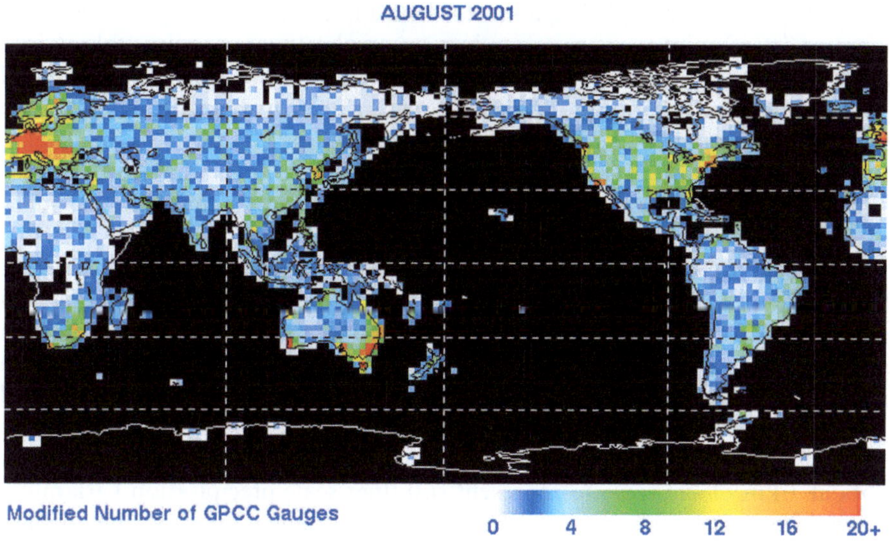

Fig. 5.18 Modified number of rain gauges used in GPCP monthly rainfall 2.5°×2.5° grid product. The modified number includes the gauges within an individual grid box and 0.25 gauges for each surrounding grid box with at least one gauge. Plots courtesy of G. Huffman, NASA

5.6.1.4 Point to Point Temporal Variation

The intermittence of rainfall describes its temporal variation between two points and provides information on the frequency of high spatial variability rainfall. Krajewski et al. (2003) examined the probability of the concurrent occurrence of rainfall accumulation and time fraction for rainy periods exceeding 0.5 mm hr^{-1} at two points separated by 1 km distance for five regions: central Florida, Oklahoma, and Iowa City, Iowa in the United States, the Island of Rota (near Guam in the west Pacific), and Ji Paraná, Rondonia, in Brazil. The fraction of total time raining was small, <3% in Florida, Oklahoma, Iowa, and the West Pacific, and <6% in Brazil. The joint occurrence of rainfall accumulations 1 km apart greater than or equal to 0.5 mm hr^{-1} for 5 min was lowest in the west Pacific location near Guam (79%) and slightly higher in Brazil (85%) and Florida (85%). Widespread precipitating clouds were more common in the mid-latitude locations, resulting in higher joint probabilities in Iowa (89%) and Oklahoma (89%) compared to the tropical sites. Joint occurrence of precipitation at gauges 1 km apart increased with increasing time interval.

5.6.1.5 Representativeness Error in Areal Precipitation Estimates

A good estimate of local precipitation (±10%) can be made at the location of a gauge when it is properly sited, calibrated, and maintained. However, most applications of rainfall mapping require areal estimates of precipitation, frequently at a hydrologic basin scale. The transformation of a point measurement or set of point measurements into an areal estimate requires taking into account the temporal and spatial variability of precipitation. Multi-gauge and gauge-radar studies of populations of storms can provide information on the typical distribution of temporal and spatial variability characteristics. Usually, the actual variability of rainfall around a gauge is unknown and yields an error source associated with transforming between spatial scales that is independent of the errors associated with the instrument itself.

Representativeness error occurs when observations at the native scale of the sensor are used to represent data at a different spatial scale (Tustison et al. 2001). For example, use of rain gauge data, a measurement made at a scale <1 m^2, to represent areal average rainfall for an area >>1 m^2 introduces a representativeness error. Analogously, use of satellite data with a sensor spatial resolution of 20 km×30 km to represent rainfall at 5 km×5 km scale also introduces a representativeness error, as does the use of operational radar data at 1 km×1 km scale to represent still finer scale precipitation variability. Representativeness error is independent of measurement errors at the intrinsic sensor spatial scale, such as calibration errors. Representativeness error between gauge measurements and areal estimates is dependent on the difference in spatial scales, the smoothness of the actual rainfall field, rain gauge spacing, and the method used to transform between the scales. A numerical experiment designed to isolate representativeness errors from other error sources found that representativeness error ranged from 25 to 95% of the hourly conditional mean rainfall and up to 50% of the hourly areal average rainfall for scales between 5 and 50 km (Tustison et al. 2001).

5.6.2 Radar Estimation of Surface Rainfall

Weather radar (Battan 1973; Doviak and Zrnic 1993; Meischner 2004; Rinehart 2004) can provide estimates of precipitating storm location, size, and intensity over large areas and at higher spatial resolutions and time resolutions than operational rain gauge networks. Ground-based scanning precipitation radars are used in short-term weather and flood forecasting, and to estimate the distribution and amount of cumulative rainfall over a region. The weather services of many countries have networks of operational radars that monitor precipitation near population centers, as well as in remote areas.

5.6.2.1 Methods and Sources of Error

Currently, radar reflectivity is the primary input to algorithms used to estimate surface rain rates. Both radar reflectivity (Z) and rain rate (R) are moments of the raindrop size distribution and can be related by a power-law of the form $Z = aR^b$ called the Z-R relation. Sources of error in radar-derived estimates of rainfall (Austin 1987; Joss et al. 1998; Krajewski and Smith 2002; Yuter 2002) include: non-meteorological echoes such as ground clutter, sea clutter, insects, birds, and anomalous propagation; instrument noise, absolute calibration, beam blockage, changes in the vertical profile of precipitation from the height of the radar beam to the surface, signal enhancement by melting particles or hail, attenuation, presence of downdrafts, inhomogeneities in rainfall within the radar resolution volume, and variations in the drop size distributions.

Inhomogenities within the radar resolution volume arise when precipitation comes in "sheets" of no more than a few hundred meters each (Thomson and List 1996). The large mismatch in spatial scales between fine-scale heavy precipitation features (often $<$200 m) producing the precipitation and the average reflectivity within a several kilometer wide resolution volume obtained by a scanning radar beam yields an incorrect representation of precipitation and potentially significant errors (Thomson and List 1996).

Different operational and research radar groups producing rain maps have different methods to mitigate these errors. Most apply some method of quality control to remove non-meteorological echo as an initial step. Some rain-map products use a mean field bias adjustment (Amitai et al. 2002; Gjertsen et al. 2004) to address multiple error sources simultaneously while others address individual sources of error independently (Joss et al. 1998; Zawadzki and Bellon

2003; Houze et al. 2004). The use of polarimetric radar variables for quantitative rainfall estimation mitigates some error sources such as attenuation and the presence of hail. Hybrid methods using a combination of reflectivity and polarimetric variables as input are being developed and evaluated (e.g. Bringi and Chandrasekar 2001; Illingworth 2004) and are planned for operational deployment in several countries later this decade.

5.6.2.2 Quantification of Errors in Radar-Rainfall Estimates

The weather service of Switzerland, MeteoSwiss, has a comprehensive program to quantify errors in surface precipitation estimates and, based on those empirical results, to refine quantitative rainfall products from their radar network (Joss et al. 1998; Germann and Joss 2004). Precipitation estimation in mountainous terrain has additional challenges compared to flat terrain. Gauge observations in mountains have large uncertainties associated with wind, snow-drift, and small-scale spatial variability. Radar observations are prone to both strong ground clutter and beam shielding by mountains. The historical data analysis and algorithm refinements by MeteoSwiss (Table 5.1) illustrate several important points. A particular algorithm refinement may improve some statistics but degrade others. To improve the statistics across the board, a combination of refinements was needed, including: rigorous elimination of clutter, correction for the vertical profile of precipitation between the altitude of radar measurement and the ground surface, hardware calibration to insure reproducibility of measurements, and long-term adjustment to independent data to minimize bias. These data also provide information on the residual

Table 5.1 Relative errors between MeteoSwiss RAIN product for daily radar-derived rainfall and daily rain gauge totals for summer season in Switzerland. The radar data are averaged over nine 1 km^2 pixels centered on 58 gauge locations including 19 gauges used for long-term adjustment. Bias and scatter statistics based on events with daily gauge rainfall $\geq$0.3 mm. Probability of Detection (POD, perfect score = 1) and False Alarm Rate (FAR, perfect score = 0) based on events with daily rainfall >0.3 mm. (adapted from Germann et al. (2006))

Year	Bias (dB)	Bias (%)	Scatter (dB)	Scatter (factor)	POD	FAR
1997	−2.4	−43	4.0	2.5	0.84	0.34
1998	−1.2	−23	4.0	2.5	0.87	0.30
Changes in operational clutter elimination						
1999	−4.2	62	4.8	3.0	0.73	0.08
2000	−3.2	52	4.4	2.8	0.74	0.11
Introduction of profile correction						
2001	−4.1	61	3.0	2.0	0.75	0.08
Long-term adjustment and modification of profile correction						
2003	0.2	5	3.4	2.2	0.89	0.18
Long-term adjustment modification						
2004	−0.5	−11	3.0	2.0	0.89	0.14

relative error for daily rainfall between rain gauge and the average of nine 1 km^2 radar pixels centered over the gauge when primary sources of error in radar rainfall mapping have been minimized. For 2004 storms with daily rainfall >0.3 mm, the probability of detection is 0.89 and false alarm rate is 0.14. For 2004 storms with daily gauge rainfall $\geq$0.3 mm, average bias for over all of Switzerland (41,290 km^2) is –11% and scatter is 3 dB(R). The scatter refers to the spread of daily radar/gauge ratios when pooling all rainy days and gauge stations over all of Switzerland together. For example, 3.0 dB(R) scatter in summer 2004 means that the radar-derived estimate of daily precipitation at a gauge location is within a factor of 2 of the gauge estimate for 68% of the rainy days, for the remaining 32% the uncertainty is larger. Uncertainties decrease closer to the radar. Within 70 km from a radar, scatter decreases to a factor of 1.4. For winter of 2004, average bias over Switzerland is –3% and scatter is 3.9 dB(R) or a factor of 2.5. These relative errors are a result of the superposition of gauge and radar uncertainties and hence the values in Table 5.1 are overestimates of the true uncertainty of the radar estimates (Germann et al. 2006).

Amitai et al. (2002) found average relative errors of 20% between monthly tipping bucket rain gauge accumulations and 4 km^2 radar-derived rainfall accumulations over the gauge locations for August–September 1998 in Melbourne, Florida. The 15 gauges were part of a dense rain gauge network of 10 km^2 area within 35 km range of the radar. They found that the average difference between the radar estimate over the gauge and the gauge accumulation was the same order as the differences between point rainfall measurements within the same 4 km^2 radar pixel.

Houze et al. (2004) used data from two rainy seasons at Kwajalein Atoll, Republic of the Marshall Islands, to estimate the magnitude of several independent components of radar-derived rain-map uncertainty under tropical oceanic conditions. Underlying assumptions regarding radar calibration, vertical profile correction, the Z-R relation, and missing data were varied within empirically determined bounds and their impact was assessed on the resulting monthly areal mean rainfall accumulation based on 4 km^2 pixels averaged over a $\sim 7 \times 10^4$ km^2 area (between 17 and 150 km range from the radar). The largest individual source of uncertainty was the absolute calibration of the radar, which was estimated at $\pm$30% based on the practical limit of $\pm$2 dB(Z) on reflectivity calibration using several methods in the Kwajalein setting. The next largest source of uncertainty ($\pm$10–15%) was associated with the correction for the vertical profile of precipitation from the altitude of the radar beam to surface. Since the radar beam altitude varies with range, this correction is largest for pixels at farthest ranges. Uncertainties in the Z-R relation based on disdrometer data accounted for only $\pm$5% uncertainty and the small number of data gaps when the radar was not operational had associated uncertainty of less than $\pm$2%. A total uncertainty in monthly areal average mean rainfall of 50% was estimated by adding the individual error sources.

5.6.3 Satellite Estimation of Surface Rainfall

Weather satellites in geosynchronous orbit (36,000 km altitude) and low earth orbit (typically 250–1000 km altitude) use remote sensing measurements to estimate precipitation at the Earth's surface. The primary methods are infrared (IR), passive microwave, and radar (active microwave). None of these methods measure rain rates directly. Rather, the satellite observed physical characteristics of precipitating clouds are used as inputs to algorithms of varying complexity to retrieve surface rainfall.

5.6.3.1 Satellite Precipitation Retrieval Algorithms

Methods to estimate precipitation from geosynchronous IR satellite data are based on empirical relations between area fraction of cold cloudiness and areal average rainfall. These methods decompose the areal average rainfall into a rainfall area estimated by satellite and a mean conditional rain rate estimated by independent measurements. In a series of studies (Arkin 1979; Richards and Arkin 1981; Arkin and Meisner 1987), Arkin and his collaborators refined empirical relations between IR cloud temperatures and areal-average surface rainfall accumulation estimated from an array of four shipborne radars using data sets collected in the eastern tropical Atlantic. The best correlations ($r^2 = 0.8$) were found for the largest spatial scale ($2.5° \times 2.5°$) and largest temporal scale (1 day). Correlations between rainfall area and IR cold cloudiness for deep tropical oceanic convection are very poor ($r^2 = 0.05$) for instantaneous data at 240 km scale (Yuter and Houze 1998). The GOES Precipitation Index (GPI) algorithm (Arkin and Meisner 1987) defines areal average precipitation as the product of the mean fractional cloudiness in a $2.5° \times 2.5°$ box lower than an IR temperature threshold, the time interval in hours, and a mean conditional rain rate in mm/hr. Based on data from the tropical eastern Atlantic and South America, they recommended a temperature threshold of 235 K, and a mean conditional rain rate of 3 mm/h for the tropics. A temperature threshold of 220 K was recommended for the extratropics based on comparisons over the continental USA. Comparisons between GPI estimates and independent estimates over land using rain gauges and radar show that GPI reproduces the occurrence of deep convective rainfall well but is less successful at estimating the intensity of rainfall (Arkin and Meisner 1987; Janowiak 1992).

Multi-spectral rainfall algorithms designed for geosynchronous satellites (e.g. Ba and Gruber 2001) often follow a similar methodology to GPI by decomposing rainfall into the probability of detectable rain and a mean rainfall rate but utilize IR as well as visible, near infrared, and water vapour channels. Multi-spectral methods exhibit improved detection of rainfall from lower altitude (higher temperature) cloud tops as compared to IR-only methods. A limitation of these algorithms is that the visible channel is available only during the day. Additionally, more fractional cloudiness is not always correlated with more precipitation. Subtropical marine stratocumulus clouds typically yield more drizzle from regions with broken cloud compared to regions of unbroken clouds that have higher areal average albedo (Comstock et al. 2005, Stevens et al. 2005).

Satellite passive microwave precipitation estimation techniques utilize the modification of surface upwelling radiation by precipitation (Spencer et al. 1983, Wilheit 1986). Microwave brightness temperatures are the result of the column scattering and absorption properties of precipitation, cloud, and water vapour. Observed brightness temperatures are dependent on the frequency of the sensor and the particular hydrometeor types and their mixing ratios within the column. Between 22 and 60 GHz both scattering and absorption are important and either may dominate depending on the specific situation. Below 22 GHz, absorption, primarily by liquid water, dominates. Above 60 GHz, ice scattering dominates. Scattering by precipitation is detectable over both water and land. Surface water has low emissivity at microwave frequencies and thus presents a cold background, which makes the small increase in brightness temperature associated with absorption by precipitation detectable. Land has larger and variable emissivities and presents a much warmer background against which the absorption signal is difficult to detect unambiguously (Wilheit 1986). As a result, passive microwave precipitation retrievals apply different weighting to the input frequency channels over oceans and land.

Passive microwave algorithms address the problem of relating brightness temperatures (column measurements) to surface rainfall (the flux at the bottom of the column) using some combination of "empirical" and "physically-based" methodologies (Smith et al. 1998). Empirical algorithms (e.g. Liu and Curry 1992; Ferriday and Avery 1994) derive a statistical regression between measured brightness temperatures at several frequencies and ground-based measurements of rainfall (radar and/or rain gauge network). Physically-based algorithms (e.g. Kummerow 1998; Bauer 2001) use cloud models to simulate the three-dimensional water vapour, cloud, and precipitation properties of storms, which are in turn used as input to forward radiative transfer calculations to estimate brightness temperatures. The cloud model and radiative transfer calculations yield a database associating sets of brightness temperatures with different precipitation structures and surface rainfall rates.

Error sources in both empirical and physically-based passive microwave estimates (Tesmer and Wilheit 1998; Bauer et al. 1999; Harris and Foufoula-Georgiou 2001; Olson et al. 2001) are related to the three-dimensional precipitation structure, and include uncertainties in the depth of the rain layer, mixing ratio profiles of cloud water, water vapour, and super-cooled precipitable water, the radiative characteristics of mixed phase layers such as wet aggregates in the melting layer, the natural variability of the profile of particle size distributions, and inhomogeneities of the rainfall pattern at scales smaller than the sensor field of view.

Intercomparison of different passive microwave precipitation retrieval algorithms using the same input data set revealed a bias uncertainty between estimates of ±30% (Smith et al. 1998). Differences among algorithms were partitioned into those associated with rainfall detection (rain area) and those associated with the conversion from brightness temperature to rain rate (rain intensity). For oceanic regions, detection differences mainly affected light rainfall regions that did not contribute significantly to area average rain rates. Over the ocean, the largest contribution to relative error among algorithms was the weighting of the input frequency channels, which impacts how intense rain rates were calculated and the maximum allowable rain rate. In contrast, over land rainfall detection was the dominant factor in producing differences among the algorithms.

The Tropical Rainfall Measuring Mission (TRMM) satellite, a collaborative effort between the U.S. and Japanese space agencies, has as one of its sensors the first space borne precipitation radar (PR) (Kummerow et al. 1998). The PR operates at K_u-band (2.3 cm) and yields reflectivity profiles at satellite-relative orientations from nadir (0°) to ±17°, yielding earth-relative vertical resolutions of 250 m to 1500 m. Unlike operational weather radars operating at S-band (10 cm) and C-band (5 cm), the K_u-band signal is strongly attenuated by rainfall. The signal attenuation is cumulative downward and is largest near the earth's surface. An attenuation correction must be applied to the near-surface reflectivity (Z) values before a meaningful transformation to surface rainfall (R) can be made. There are currently several methods to estimate total path-integrated attenuation and to distribute the attenuation correction in the vertical (Meneghini and Kozu 1990; Iguchi et al. 2000). Independent evaluation and error diagnosis of attenuation correction methods applied to satellite radar data has proven to be difficult. Error sources in the estimation of surface precipitation from an attenuated reflectivity profile include many of the same error sources in passive microwave algorithms. The set of passive microwave sensors on the TRMM satellite are referred to as the TRMM Microwave Imager (TMI). Zonal averages of areal average precipitation derived from TRMM TMI and PR at native sensor resolution exhibit relative differences of 20–40% (Kummerow et al. 2000).

Multi-satellite products attempt to combine the strengths of different types of satellite sensors to mitigate their individual weaknesses. Several multi-satellite products use variations of the GPI algorithm (Janowiak and Xie 1999; Huffman et al. 2001; Kuligowski 2002). IR cloud top temperatures from geosynchronous satellites are used to determine rain areas at high temporal resolution. The IR temperature threshold and average conditional rain rate are adjusted for designated time and spatial scales using some combination of global passive microwave data and rain gauge data over land. Comparisons between dense gauge networks over land to these types of multi-satellite products show similar results to those products using IR data alone. Multi-satellite products also perform better in deep convection typical of the warm season than in shallower convection typical of the cool season, and tend to perform poorly over complex terrain regardless of season (Arkin and Meisner 1987; Huffman et al. 2001).

Global satellite observations, rain gauge-derived areal estimates over land, and numerical model outputs are quality controlled and merged to create monthly analyses as part of the Global Precipitation Climatology Project (GPCP) of the Deutscher Wetter Dienst/WMO/WCRP Global Precipitation Climatology Centre (Huffman et al. 1997; Adler et al. 2003) and the NOAA Climate Prediction Center's Merged Analysis of Precipitation (CMAP) (Xie and Arkin 1997). GPCP is an element of the Global Energy and Water Cycle Experiment of the World Climate Research Program (WCRP). The GPCP and CMAP groups use different methods to combine rainfall data from multiple sources. Products include global maps of monthly precipitation on 2.5° by 2.5° and 1.0° by 1.0° latitude/longitude grids and monthly precipitation anomaly compared to a long-term average (see Chapter 1, Fig. 1.6). The value of merged products is based on the assumption that the combination of several estimates with unknown errors is closer to the truth than any single estimate. Analogous to ensemble numerical forecast modeling, this assumption holds if the individual estimates are distributed around the true value.

5.6.3.2 Uncertainties Associated with Satellite Temporal Sampling

Common to all satellite precipitation estimates are uncertainties related to estimating temporal average precipitation (e.g. daily, 5-day, monthly) from the nearly instantaneous samples for a given area obtained by satellite. Depending on storm duration, structure, and stage of evolution, a single snapshot of a storm has varying value as a representative sample of the storms temporal and spatial average. Steiner et al. (2003) utilized a set of operational radar network composites over the United States central plains during summer as a baseline data set to investigate errors associated with different satellite temporal sampling schemes. The radar composite had a spatial resolution of 2 km and a temporal resolution of 15 min. These data were sub-sampled at 1-, 3-, 6-, 8-, and 12-h intervals and averaged over 100-, 200-, and 500-km spatial scale domains. The uncertainty associated with discrete temporal sampling varies directly with sampling time interval and inversely with rainfall rate and space and time domain size. A set of low-earth orbit satellites with similar passive microwave sensors can be utilized as a group (Fig. 5.19) to obtain temporal sampling intervals of ~6 h in the tropics (30°S to 30°N) decreasing to ~3 h at latitudes above 60° (Turk et al. 2003).

Although summer-time rainfall in the United States central plains has different characteristics than rainfall in other seasons and at other locations, the uncertainty values in the Steiner et al. (2003) study provide guidance on the magnitude of errors due only to satellite temporal sampling. Based on Steiner et al.'s calculations for a 100 km scale domain with areal average rainfall of 0.1 mm hr^{-1}, median rainfall sampling uncertainty for daily rainfall is 10% for

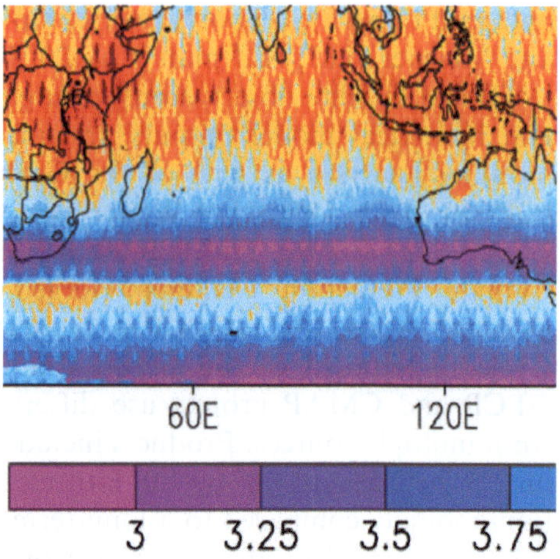

Fig. 5.19 Average revisit times in hours for the period 5–23 November 2005 from the set of satellite passive microwave radiometers on the following satellites: TRMM (TMI), AQUA (AMSR-E), DMSP (SSMI) F13, F14 and F15. The irregular pattern is a result of the combination of sun-synchronous (AQUA and DMSP) and precessing (TRMM) orbits. There were no samples (*white*) over northern hemisphere high latitude land areas with snow cover. Plots courtesy of R. Joyce, NOAA

1 h, 43% for 3 h, 92% for 6 h, and 154% for 12 h sampling time intervals. The uncertainties decrease for 500 km scale to 3% for 1 h, 11% for 3 h, 31% for 6 h and 61% for 12 h. Current geosynchronous satellites provide IR full-disk images at 1 to 3 h intervals. The temporal sampling of individual low-earth orbit satellites such as TRMM is dependent on the details of the orbit and latitude. TRMM Precipitation Radar revisit times over an area of 120 km radius vary from 78 h at the equator to 17 h at 35° latitude. A set of low-earth orbit satellites with similar passive microwave sensors can be utilized as a group to obtain temporal sampling intervals of ~6 h in the tropics (30°S to 30°N) decreasing to ~3 h at latitudes above 60° (Turk et al. 2003).

Since satellite algorithm retrieval errors are strongly dependent on the details of the three-dimensional precipitation structure, these errors vary with regional and seasonal changes in precipitation structure. It would be useful to isolate errors in the satellite retrieval algorithm from errors related to temporal sampling, differences in spatial scale, and differences in minimum detectable rain rate that are a function of the sensitivity of the sensor. In practice, this is difficult since the different types of sensors on the current generation of satellites were not designed to have matching spatial resolutions or sensitivity.

5.6.3.3 Future Satellite Precipitation Measurements

The proposed joint US and Japanese space agency Global Precipitation Measurement (GPM) Mission builds on the success of TRMM to address the critical need for a more comprehensive global precipitation measuring program. The GPM Mission consists of a Core satellite similar to TRMM and a constellation of internationally contributed multi-channel passive microwave satellites. The GPM Core satellite will be used as a calibration reference for moderate to heavy rainfall for the constellation satellites. Additionally, the replacement of the single frequency Ku radar, as used on TRMM, with a dual frequency radar (Ku/Ka) with improved sensitivity, gives the GPM Core satellite the ability to estimate bulk properties of the rain drop size distribution and to improve the attenuation-corrected vertical profile of reflectivity and rain rate using differential attenuation methods.

5.6.4 Discrepancies Among Different Methodologies

One of the most extensively analyzed data sets of recent years is from the Tropical Ocean Global Atmosphere Coupled Ocean Atmosphere Response Experiment (TOGA COARE) (Webster and Lukas 1992), which took place in the western Pacific warm pool region from November 1992 to February 1993. The TOGA COARE intensive flux array (IFA) had an area of $\sim 2.25 \times 10^5$ km^2. Areal average daily rainfall estimates for the IFA (Table 5.2) were obtained using many different methods for estimating surface rainfall (Johnson and Ciesielski 2000; Weller et al. 2004). Values varied from 5.4 mm/day for the shipborne radars to 9.3 mm/day for a merged gauge/satellite method. Discrepancies are a function of differences in subregion examined, temporal sampling, and spatial resolution of the rainfall estimates (Weller et al. 2004). These differences make it difficult to isolate errors associated with specific techniques or to determine the best estimate. For example, Johnson and Ciesielski (2000)

Table 5.2 TOGA COARE intensive flux array intensive observation period estimates of areal average daily rainfall using different techniques. "Sat/mixed" (Curry et al. 1999) utilized a combination of passive microwave brightness temperatures and visible and IR radiances from geosynchronous and low-earth orbiting satellites. "CMAP" refers to the CPC Merged Analysis of Precipitation of rain gauge, IR and microwave satellite estimates (Xie and Arkin 1997) (adapted from Johnson and Ciesielski (2000) and Weller et al. (2004))

Method	Rainfall estimate (mm/day)	Citation
Atmospheric moisture budget multiquadratic technique	8.2	Johnson and Ciesielski (2000)
Atmospheric moisture budget-barnes scheme	5.6	Johnson and Ciesielski (2000)
Ocean freshwater budget (near ship)	8.0	Feng et al. (2000)
Shipborne radar (2 km resolution within 145 km of ships)	5.4	Short et al. (1997)
Sat/Mixed	8.3	Curry et al. (1999)
CMAP	9.3	Johnson and Ciesielski (2000)
NCEP T62 reanalysis (210 km resolution)	8.4	Johnson and Ciesielski (2000)
ECMWF T106 reanalysis (125 km resolution)	6.7	Johnson and Ciesielski (2000)

showed that part of the reason the ship radar data estimate was lower than the other estimates was that the ships were not present for some of the heavier precipitation events during the study period, and that one of the two radar ships was stationed in a precipitation minimum.

Attempts to inter-compare monthly global estimates of precipitation (e.g. Adler et al. 2001) have suffered similar problems to the TOGA COARE rainfall comparisons. The impacts of differences in temporal sampling and spatial scales of the input sensor data are not usually removed. As a result, the factor of 2 to 3 differences among different satellite retrievals and ground-based observations (Adler et al. 2001) are difficult to diagnose in terms of specific algorithm or measurement system deficiencies. Estimates tend to cluster based on input data type, and hence similar temporal and spatial sampling (Ebert et al. 1996). Inclusion of all available operational data sets into blended or merged products limits opportunities for comparison to truly independent estimates.

The International Precipitation Working Group (IPWG) was established in 2001 as a permanent working group of the Coordination Group for Meteorological Satellites (CGMS) and is co-sponsored by CGMS and the WMO. The IPWG validation/intercomparison project provides standardized web-accessible intercomparisons over Australia, the continental US and Europe between daily precipitation estimates from rain gauge and radar networks and a variety of operational and semi-operational satellite algorithms and numerical weather prediction models (Ebert 2004; Ebert et al. 2007). Example daily statistics aggregated over two 3 month periods between the 0.25° operational rain gauge analysis for Australia (Weymouth et al. 1999) and a GPI algorithm, a merged multi-satellite algorithm, and the ECMWF forecast are shown in Table 5.3. Performance of both satellite algorithms is better during the December–February period corresponding to the wet season

Table 5.3 Aggregate daily statistics from IPWG intercomparison from 1 Dec 2003–29 Feb 2004 and 1 June 2004–31 August 2004. Statistics shown were compiled over land areas based on estimates of the daily accumulation from a GPI algorithm, a NASA TRMM 1 hourly multi-satellite product (3B41RT, Huffman et al. (2003)), and the ECMWF 36 h forecast for 24 h precipitation accumulation compared to the Australian national rain gauge network. Spatial scale of intercomparison is 0.25° for all products (data courtesy of E. Ebert)

	GPI Dec–Feb	TRMM multi-sat Dec–Feb	ECMWF forecast Dec–Feb	GPI June–August	TRMM multi-sat June–August	ECMWF forecast June–August
Ratio of estimated to observed average volume	1.18	1.08	0.74	0.38	0.55	0.75
Bias score	0.97	0.69	0.91	0.22	0.31	0.80
POD	0.71	0.56	0.73	0.13	0.16	0.69
FAR	0.26	0.19	0.19	0.41	0.49	0.13
Mean absolute error (mm/d)	4.0	3.9	2.8	0.66	0.89	0.39
RMS error (mm/d)	9.0	10.7	7.5	2.6	4.0	1.4

in tropical northern Australia as compared to the dryer June–August period. No one method performs best across all statistics for both the dry and wet seasons.

5.6.5 Testability of the Hypothesis that Aerosols Impact Precipitation

Uncertainties in precipitation measurement vary with the spatial scale, temporal scale, and storm type that are considered. For areal precipitation estimates derived from all types of sensors, storm total rainfall estimates from precipitating clouds covering larger areas with low spatial variability, of longer duration, and with moderate to heavy precipitation intensities, generally have smaller uncertainties than estimates from short-lived, isolated showers or light precipitation near the detection threshold of the measurement instrumentation. The few quantitative estimates of relative rainfall estimation error in the literature focus either on large areas and long time periods, such as a month or more, or on the subset of storms with moderate to heavy precipitation. These studies indicate that many of the independent sources of error have substantial biases and scatter. Representativeness error is estimated at 25–95% of hourly conditional mean rainfall and up to 50% for hourly areal average rainfall (Tustison et al. 2001). The daily radar/rain gauge ratio for storms with at least 0.3 mm daily precipitation is only within a factor of 2 for 68% of summer days in Switzerland, but these large daily uncertainties average to a summer seasonal bias of -11% (Germann et al. 2006). Comparison of radar derived rainfall over gauges (1 km^2) for storm totals with rainfall accumulations >10 mm had root mean square errors from 10 to 40% when using the subset of gauges passing quality control, and up to 80% when all gauges in the network were used (Steiner et al. 1999). Errors for storms with lighter precipitation were not quantified but are expected to be higher. Houze et al. (2004) estimated $\pm 50\%$ uncertainties in radar-derived monthly areal average rainrates over 7×10^4 km^2 tropical oceanic regions. Satellite temporal sampling errors, which are independent of the precipitation retrieval method, increase rapidly as temporal sampling decreases, and range from 10% for 1 hourly sampling to 154% for 12 hourly sampling of daily areal average rainfall over a 100 km scale domain (Steiner et al. 2003). These values are from studies that have better control of error sources, such as calibration and instrument maintenance, than is typical worldwide, and hence represent lower limits for rainfall measurement uncertainty. Uncertainties in snowfall measurement are larger than those for rain.

For long time scales, the impact of aerosols on precipitation is difficult to isolate from other sources of precipitation variability such as the location of storm tracks, variations in surface moisture and heat fluxes, urban boundary layer dynamics-thermodynamics, and interannual cycles. Analysis of time scales of storm duration or shorter and spatial scales sufficiently small to distinguish regions under clearly observed aerosol plumes ($\leq$ 50 km scale) would help to separate the impact of aerosols from other factors. At these short time and spatial scales, errors in precipitation measurement will typically be a factor of 2 or more. Given the large magnitude of these uncertainties, it will be infeasible to discern the impact of aerosols that modify rainfall by less than a factor of 2.

5.7 Conclusions

Even though we clearly see in measurements and in simulations the strong effect that aerosol particles have in cloud microphysics and development, we are not sure what is the magnitude or direction of the aerosol impact on precipitation and how it varies with meteorological conditions. The reasons behind this difficulty are the result of:

The Limited Accuracy of the Measurements: Quantitative radar measurements of precipitation can be off by up to a factor of 2 for daily rainfall and spatial scales $\leq$50 km.

Measurements of the cloud droplet size from satellites are biased by up to 30% for convective clouds, due to departure of the clouds from the plane parallel model. However, we can still measure the relative change of cloud properties with change in aerosol concentrations.

Aerosol column optical depth measured from satellites is not accurate enough: accuracy is $\pm$20% over the oceans, less accurate over vegetated land, deserts or urban areas. Ground based remote sensing of aerosol can have a high precision of $\pm$0.02 in units of aerosol optical depth, but with limited spatial coverage.

Translation of the optical thickness into the column aerosol concentration and into CCN is difficult due to variation in the aerosol size distribution and chemical composition, the vertical distribution of the aerosols and thermodynamic conditions, and the lack of satellite passive remote sensing sensitivity to most CCN-sized particles.

In situ aerosol measurements from aircraft provide more physical and especially chemical detail. However, the aircraft may influence the aerosol and cloud properties in some cases, and interpretation of the measurements must take account of limitations in the sampling of all aerosol and cloud droplets sizes. Surface based measurements offer highly accurate column aerosol optical depth and physical as well as chemical information about boundary layer particle properties, but the property measurements may not represent the total column, especially at cloud altitudes needed to study aerosol-cloud interactions.

Small Spatial and Temporal Coverage for Highly Variable Parameters: High spatial and temporal variability of precipitation and aerosols make it difficult to measure coincidental aerosol, clouds and precipitation. Therefore, we are not aware of studies that include direct statistical correlations of aerosol and precipitation.

Cloud properties have a diurnal cycle not well represented by satellite observations. Geostationary satellite measure the diurnal cycle but with limited ability to derive the cloud droplet size and aerosol amount and properties.

Cause and Effect: A considerable number of studies have revealed a correlation between higher aerosol concentrations and higher cloud droplet concentrations in stratocumulus clouds. However, the impact of aerosol-induced changes in droplet concentrations upon drizzle rates is not well understood, as different studies have yielded different results (Wood 2005). A confounding factor in determining causation is that once precipitation forms, aerosols are removed. It is still difficult to assess whether the changes in the precipitation patterns are primarily a result of the aerosol impact, or whether it is a result of changes of the local meteorological conditions, including boundary layer properties, that are the primary influences on changes in precipitation patterns near industrialized regions. Recent work (Ackerman et al. 2004; Wood 2007) shows that precipitation from stratocumulus clouds is strongly correlated with cloud thickness and that aerosol-induced changes in the cloud droplet spectra may lead to thicker or thinner clouds depending on the large scale meteorological conditions. These feedbacks on precipitation, both positive and negative, further complicate the unraveling of the relation between aerosols and precipitation. Even the most informative measurements so far on the effect of aerosols on precipitation do not include simultaneous quantitative measurements of aerosols, cloud properties, precipitation and the full set of meteorological parameters.

Field campaigns involving coordinated surface, aircraft, and spacecraft measurements, along with aerosol transport and cloud models, could aim at better understanding the physical mechanisms that must be represented in models, and at identifying key indicators in satellite observations that could be used to extrapolate the more detailed field measurements to regional and global scales.

Chapter 6
Effects of Pollution and Biomass Aerosols on Clouds and Precipitation: Observational Studies

Zev Levin and Jean-Louis Brenguier

6.1 Introduction

The impact of aerosols on cloud macrophysical (cloud extent, cloud thickness and liquid water path), cloud microphysical (droplet and ice crystals concentrations and size distributions), and precipitation has received a great deal of attention for over 50 years. The pioneering work of Gunn and Phillips (1957), Squires (1958), Squires and Twomey (1961), Warner (1968) and Warner and Twomey (1967), to mention just a few, pointed out that high concentrations of cloud condensation nuclei (CCN) from anthropogenic sources, such as from industrial pollution and from burning of sugarcane, can increase cloud droplet number concentration (CDNC), hence increasing cloud microphysical stability and potentially reducing precipitation efficiency. Although the connection between increased CCN and increased CDNC has been supported by many in situ observations, the impact of CCN on precipitation is less well established.

One of the obstacles in assessing such a cause and effect relationship is the fact that different aerosol types generally correspond to different air masses, hence to different vertical profiles of moisture and stability. The sensitivity of cloud and precipitation to these meteorological parameters is particularly high in the boundary layer. In convective clouds the ability to produce precipitation embryos depends to a large extend on the maximum liquid water conent (LWC), hence on the cloud depth, because the adiabatic liquid water content increases with altitude above cloud base (a rate of ~2 gm^{-3} of condensed water per km of ascent at 20°C). In subtropical boundary layer clouds, the altitude of the cloud base, hence the cloud depth and the maximum LWC slightly below cloud top, depend on the total water mixing ratio of the mixed layer. For example, in a boundary layer with a total water mixing ratio of 20 gm^{-3}, topped by a 200 m thick cloud, a very small drying of the boundary layer (1% decrease of the total water mixing ratio, at constant temperature), leads to a decrease of the cloud layer thickness by a factor of 2, which corresponds to a factor of 4 reduction of the liquid water path (LWP). Since the precipitation rate in

Z. Levin (✉)
Tel Aviv University, Tel Aviv, Israel

boundary layer clouds roughly scales with the LWP and the inverse of CDNC (Pawloswka et al. 2003, Comstock et al. 2004; Van Zanten et al. 2005), a 1% drying of the boundary layer might have the same impact on precipitation as a factor of four increase in CDNC due to pollution. The same argument can be made with a slight warming of the boundary layer, since a 0.6°C increase of the temperature, at constant humidity, leads to a reduction of the cloud depth by about 100 m.

Deep clouds which often produce more rain than boundary clouds, are less sensitive to small variations in the boundary layer moisture and temperature, but are highly sensitive to the stability profile and to the low level convergence of sensible and latent heat.

The following discussion illustrates how observational studies attempt to progressively overcome the obstacles to understanding by addressing each element of the long chain of physical processes connecting aerosols to precipitation.

6.2 Warm Clouds

6.2.1 Pollution Aerosol Impacts on CCN Properties and CDNC

There are numerous observational evidences that the CCN concentration is increased in polluted air masses and that their hygroscopic properties are modified.

Considerable effort on aerosol sources was expended in Australia in the 1970's and 1980's that paints a very clear picture on anthropogenic influence. Five years of automated CCN measurements at a rural site in New South Wales showed convincingly that cities and towns were copious sources of CCN, with levels of CCN qualitatively in proportion to size of town (Fig. 6.1) (Twomey et al. 1978).

Subsequently airborne methods for determining CN fluxes from individual anthropogenic sources were developed and applied to a very large, isolated, metal smelting complex at Mt Isa, then to towns and cities around Australia, demonstrating existence of well defined "pollution tracks" of aerosol CN and inferred CCN extending up to 1800 km from the source (Bigg and Turvey 1978; Bigg et al. 1978; Ayers et al. 1979; Carras and Williams 1981; Williams et al. 1981; Ayers et al. 1982; Carras et al. 1988). The extremely well defined scaling of town CN flux with population was shown theoretically to have a sound physical basis (Manton and Ayers 1982) (Fig. 6.2).

Radke and Hobbs (1976) measured CCN (S = 0.2%) concentrations of 1000–3500 cm^{-3} along the East Coast of the USA from Virginia to Long Island, compared to 100 cm^{-3} of CCN in clean maritime air and 300 cm^{-3} in continental air. They concluded that the global anthropogenic production rate of CCN might be comparable to the natural production rate.

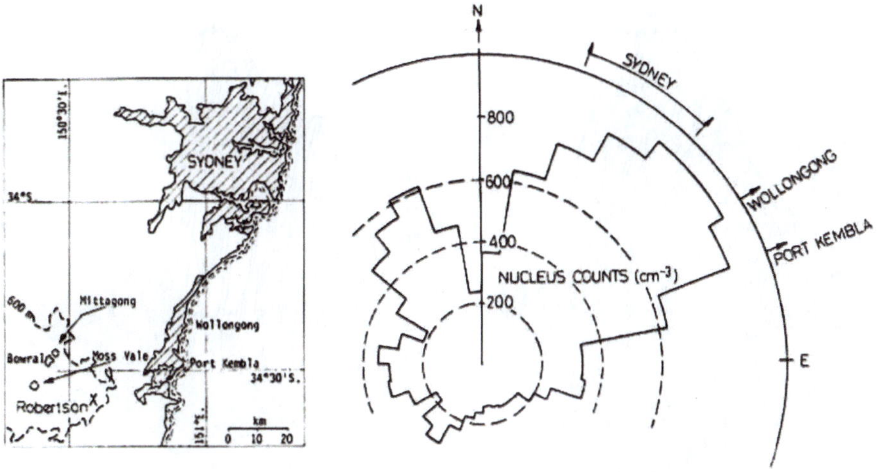

Local wind rose of nucleus counts over all seasons 1971–72.

Fig. 6.1 Five years CCN measurements at Robertson, showing evidence of CCN transport from cities/towns – Sydney to the NE, Mittagong, Bowral and Moss Vale to the NW and W. From Twomey et al. (1978) with permission of the American Meteorological Society

Mineral dust particles are known to be efficient ice nuclei (Pruppacher and Klett 1997), but not very efficient CCN. However, recent in situ measurements of dust from the Sahara over the Mediterranean (Levin et al. 1996), and of Asian dust over China and Japan during ACE-Asia (Trochkine et al. 2003), show that dust particles can get covered by sulphate from pollution during their transport through the atmosphere, which may increase their effectiveness as CCN. More recently, Levin et al. (2005) showed that low-lying dust storms over the sea could lead to sea salt coating on a large fraction of the dust particles. These types of coating could modify the particles into effective giant CCN (GCCN), as well as efficient ice nuclei. Indeed, sampling in clouds affected by particles coated with sulphate revealed the presence of relatively large drops close to cloud base (Fig. 6.3).

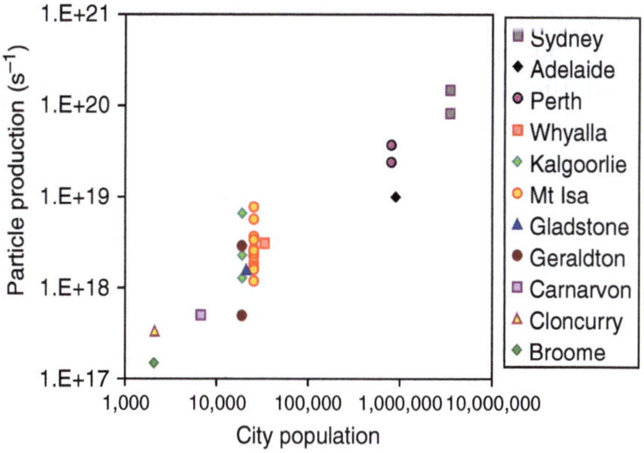

Fig. 6.2 CN flux measured downwind of Australian towns and cities plotted against population as a surrogate for source size (Ayers et al. 1982) with permission of Atmospheric Environment (@ Elsevier)

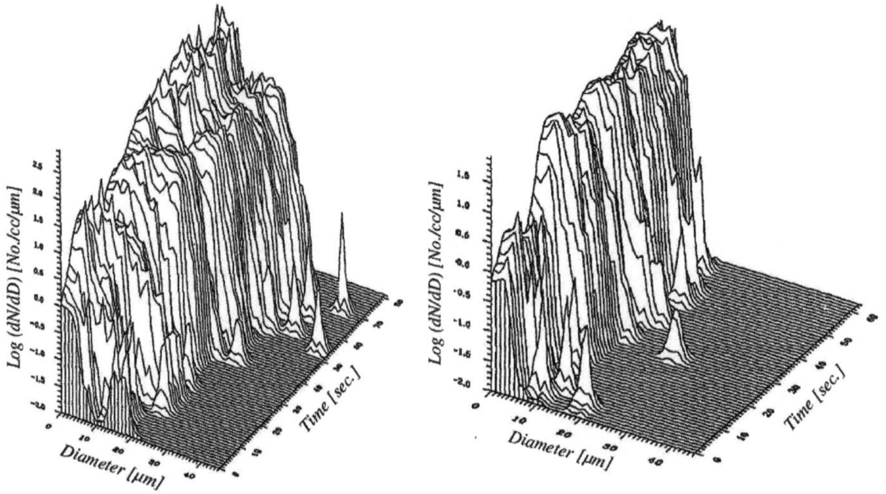

Fig. 6.3 Drop size spectra near the bases of clouds affected by large dust particles (*left*) as compared to the drop size spectra in cleaner clouds (*right*). The time axis represents the time of flight during the pass through the cloud. The airplane was flying at ~70 m s^{-1}. From Levin et al. (1996) with permission from the American Meteorological Society

At this point one may ask whether CDNC increases proportionally to the aerosol concentration. From co-located airborne measurements of condensation nuclei (CN), CCN and droplet concentrations, Leaitch et al. (1992), Martin et al. (1994), Taylor and McHaffie (1994), Gultepe et al. (1996), Chuang et al. (2000), Snider and Brenguier (2000), Ramanathan et al. (2001a), among others, have shown that regardless of location, increases in aerosol concentration lead to CDNC increases. However, these trends appear to taper off when the aerosol concentration increases, as shown in the compilation of these diverse results (Ramanathan et al. 2001b). The origin of the large discrepancies in this compilation is two fold: Firstly, CCN are only a fraction of the CN population, and that fraction varies significantly with the size distribution and chemical composition of the aerosol. Secondly, the fraction of available CCN that are actually activated in a cloud increases with the intensity of the updraft at cloud base (see Chapter 2). Stratocumulus clouds, with updraft speed of <1 ms^{-1} nucleate few of the available CCN particles (see the Atlantic Stratocumulus Transition Experiment (ASTEX) and the Aerosol Characterization Experiment-2 (ACE-2) case studies in Fig. 6.4. However, deeper convective clouds can develop updraft intensities of >10 ms^{-1}, hence activating most of the available CCN particles (see the North Atlantic case study in Fig. 6.4. The ratio of CDNC to the total aerosol concentration thus varies from 70% for North Atlantic convective systems to about 10% in ASTEX and ACE-2 sub-tropical stratocumuli. Note also that the total aerosol concentration varies significantly depending on whether the

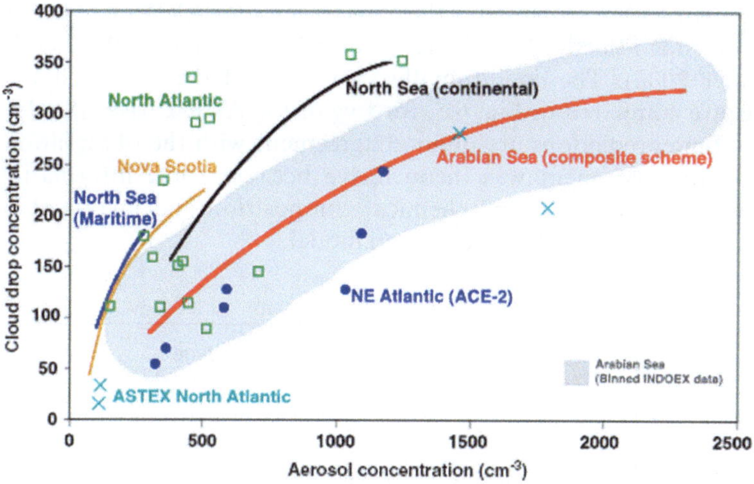

Fig. 6.4 Aircraft data illustrating the CDNC increase with the increase in aerosol number concentration. The thick red line is obtained from a composite theoretical parameterization that fits the INDOEX aircraft data for the Arabian Sea. From Ramanathan et al. (2001b) reprinted with permission of AAAS

instrument measures all the particles larger than 2.5, larger than 10 or larger than 20 nm. Although the concentrations of particles smaller than about 10–20 nm can be very high, their contribution to activation in clouds is negligible.

Observational studies of the CCN activation process therefore require a comprehensive understanding of the particle size distribution, chemical composition, mixing state and surface coating (see Chapters 3 and 4 for more details), all contributing to the activation properties of the observed aerosols (McFiggans et al. 2006). In actual clouds however, CCN activation is not uniform because the vertical velocity that governs the kinematics of the activation process varies in space and time, as does the resulting CDNC. Moreover, the droplet number concentration after activation can be significantly reduced by additional processes that are not directly linked to aerosol properties, such as entrainment-mixing, coalescence and precipitation scavenging.

A closure study on the CCN activation process has been performed during the ACE-2 experiment. Aerosol particle size distributions and chemical compositions were measured at the ground and on board research aircraft flying in the sub-cloud layer. The probability density function (PDF) of vertical velocity was also measured by an aircraft flying at the cloud base (Guibert et al. 2003). With these input parameters, a PDF of predicted CDNC was derived from the measured PDF of vertical velocity using different versions of an activation model. The actual values of droplet number concentration were measured in the cloud layer after selection of the samples that were not affected by mixing or drizzle scavenging (Pawlowska and Brenguier 2000). The conditionally sampled and the predicted PDFs of CDNC are compared in Fig. 6.5, for five of the ACE-2 cases (Snider et al. 2003). Some predictions are in good agreement with the observations, but the closure experiment was inconclusive because some information was lacking on the size resolved chemical composition and mixing state that would have constrained the activation model.

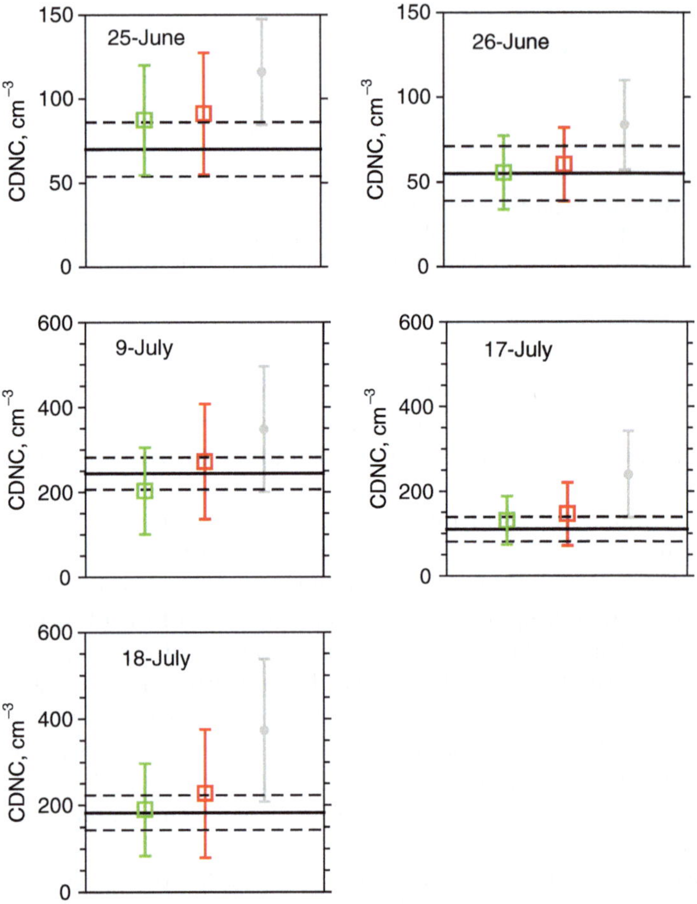

Fig. 6.5 CDNC values predicted by a CCN activation model for 5 ACE2 cases (colors correspond to different initialization modes). Error bars indicate the width (1 standard deviation) of the predicted CDNC probability distribution function corresponding to the measured frequency distribution of the vertical velocity at cloud base. Horizontal lines show the average and the plus or minus 1 standard deviation of the conditionally sampled CDNC probability density function. From Snider et al. (2003) with permission of the American Geophysical Union

6.2.2 *Aerosol Impact on Cloud Microphysics*

6.2.2.1 Deep Convective Clouds

Warner and Twomey (1967) and Warner (1968) reported that the *incorporation into clouds* of smoke particles *from sugar cane fires, leads to increases* in CDNC and *a* reduction in the sizes of cloud droplets (Fig. 6.6). *These effects* may impede the *growth of raindrops by* coalescence.

Recently, Andreae et al. (2004) reported on results from airborne and ground observations of cloud and precipitation development in the Amazon region. They divided their observations into four types of clouds: 1) Maritime shallow clouds over the ocean, which had low CCN concentrations and few cloud drops, including a few large drops. These clouds did not grow to be tall but developed precipitation very effectively, as can be seen by the widening of the drop size distribution with height. These were termed "Blue Ocean"; 2) The name "Green Ocean" was given to clouds that developed inland in the western Amazon in unpolluted conditions, mainly in the rainy season. These conditions were characterized by low CCN concentrations, due to washout by rain, and therefore resembled maritime clouds; 3) "Smoky Clouds" were those that were affected by high concentrations of CCN from smoke emitted from older fires. These particles remained in the atmosphere for a long time due to the lack of rain in the dry season. The authors reported that the high CCN concentrations

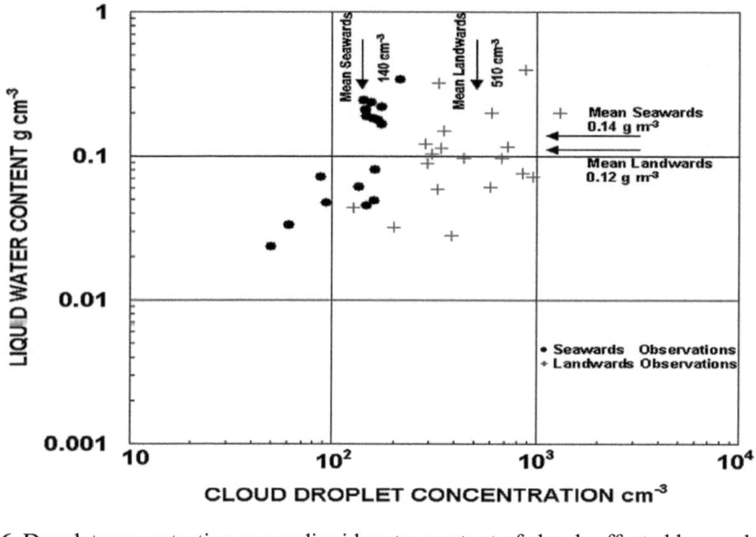

Fig. 6.6 Droplet concentration versus liquid water content of clouds affected by smoke from sugarcane fires in Australia. Note the CDNC increase from 140 to 510 cm^{-3} • (*black dots*) and the slight decrease in liquid water content from 0.14 to 0.12 g m^{-3}, in the clouds affected by smoke compared to those affected by clean on-shore airflow + (*gray pluses*). Adapted with modifications from Warner and Twomey (1967) with permission from the American Meteorological Society

produced high concentrations of cloud droplets, hence resulting in smaller sizes and narrower size distributions, which then reduced the efficiency for growth by collection. Therefore, these drops continued to grow by condensation and were surmised to reach higher altitudes and lower temperatures where ice could form. Since these clouds were deep, they could produce lightning, hail and heavy rain and 4) "Pyro-Clouds" were clouds fed directly with smoke and heat from biomass fires. The high concentrations of CCN in the smoke

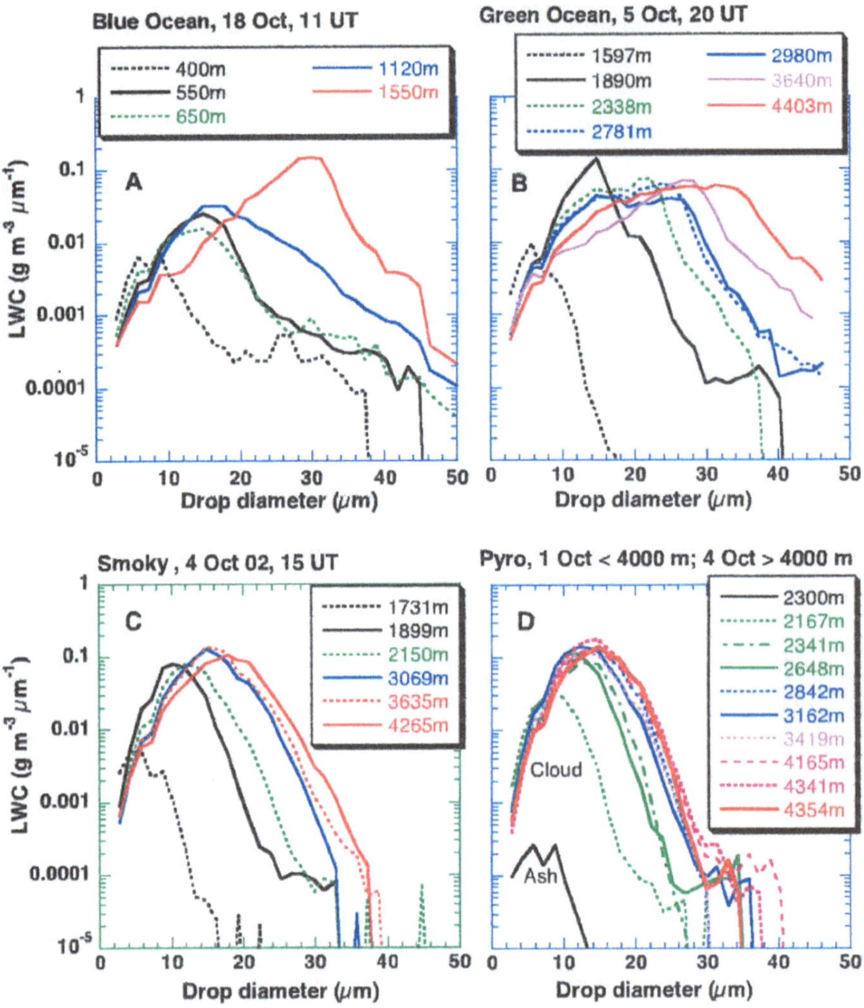

Fig. 6.7 The mass drop size distributions at different altitudes in a number of typical Amazonian clouds. Note the similarity and the wide drop size spectra in both the *blue* and *green ocean clouds*, compared to the narrow drop size spectra in the *smoky clouds*. Also note the large aerosol sizes in the ash and the resultant second peak in the drop size distribution in the *pyro-clouds*. From Andreae et al. (2004) reprinted with permission of AAAS

produced large concentrations of small drops, thus limiting the warm rain production (see also Fromm et al. 2006). Although some of the large ash particles in the smoke could produce large drops and enhance warm rain development, no raindrops were actually measured by the rain probe in the growing towers of these clouds. Figure 6.7 summarizes the drop size distributions in the four types of Amazonian clouds listed above and illustrates how the evolution of the droplet spectrum with altitude varies significantly depending on the type of aerosol.

6.2.2.2 Boundary Layer Clouds

Garrett and Hobbs (1995) studied the effects of long-range transport of particles from polluted and unpolluted sources on the characteristics of shallow, non-raining stratus clouds around the Azores. The aerosols originating from continental sources (Europe) had a monomodal particle size distribution at ~0.05 μm, while the clean maritime air contained a bimodal particle size distribution with modes at 0.02 and 0.08 μm. The clouds growing in polluted air had 160% higher concentrations of droplets, and a 27% smaller effective radius, than the clouds growing in clean air (Fig. 6.8).

Using Cape Grim observations of the CCN seasonal variability (see Sect. 6.2.4.3), the Southern Ocean Cloud Experiment (SOCEX) was designed to investigate whether coherent seasonal cycles extend to the microphysical characteristics of the maritime cloudy boundary layer. The experiment was conducted just west of Cape Grim, at the north-western tip of Tasmania, about 300 km south of the Australian continent. When the wind direction is from the

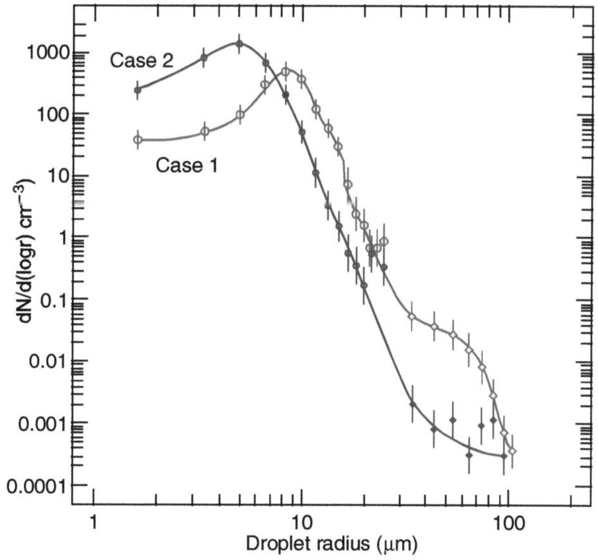

Fig. 6.8 Cloud droplet size distribution in clean (*case 1*) and polluted (*case 2*) warm clouds over the Azores Islands. Note the higher concentrations and the smaller sizes of the drops in the polluted clouds. Adapted from Garrett and Hobbs (1995) with permission from the American Meteorological Society

southwest (Baseline sector, between 190 and 280°) the air stream has a typical fetch of 4000 to 5000 km over the Southern Ocean and is virtually devoid of anthropogenic influences.

Two experiments were carried out: SOCEX I in July 1993 at the minimum of the seasonal cycles in Dimethyl Sulphide (DMS) flux and CCN concentration; and SOCEX II in February 1995 at the maximum of the seasonal cycle. The results (Boers et al. 1996; Boers et al. 1998) are summarized in Fig. 6.9. They show the average droplet concentration and the average effective radius as a function of pressure difference with respect to cloud base, for the summer experiment (circles) and winter experiment (pluses). Only droplet properties (Particle Measuring System- Forward scattering spectrometer probe, PMS-FSSP data) are shown, so the results exclude precipitation size droplets. The seasonal differences are large and provide a firm indication of the seasonality in the microphysical structure of clouds over the Southern Ocean. Near cloud base there is a factor of three differences in droplet concentration, ranging from 32 cm^{-3} in the winter to 94 cm^{-3} in summer.

These results are almost identical to the predictions by Boers et al. (1994) who found, using a parcel model of adiabatic ascent, that the cloud base droplet concentration in unpolluted conditions over the Southern Ocean would range from 35 cm^{-3} in winter to 92 cm^{-3} in summer.

The ACE-2 CLOUDY-COLUMN experiment in 1997 provided a very comprehensive data set on the impact of anthropogenic aerosol on the microphysics of stratocumulus (Brenguier et al. 2000a). Figure 6.10 compares vertical

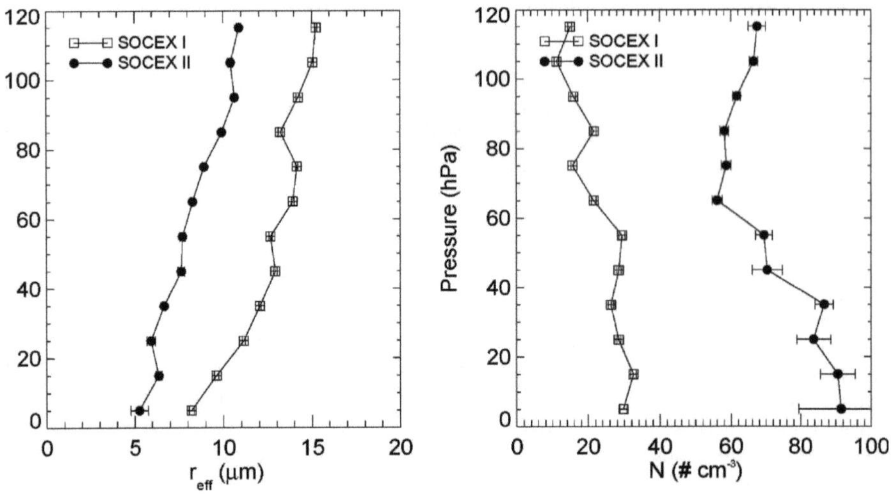

Fig. 6.9 In situ measured vertical profiles of droplet effective radius (*left*) and CDNC (*right*), for the SOCEX summer experiment (*circles*) and winter experiment (*pluses*). The ordinate is the pressure difference with respect to cloud base. From Boers et al. (1998) with permission from the Royal Meteorological Society

profiles of CDNC, mean volume droplet radius and liquid water mixing ratio for the most pristine case (26 June, CDNC = 51 cm^{-3}), and the most polluted one (9 July, CDNC = 256 cm^{-3}). Frequency distributions of these parameters are based on more than 400 km sampling along a 60 km^2 track, and stratified over 30 m thick layers from cloud base to cloud top. The average liquid water mixing ratio increases with altitude, as predicted by adiabatic convective models (Chapters 2 and 7). CDNC reaches its maximum value slightly above cloud base, because in the first 30 to 80 m all CCN are not fully activated, and some

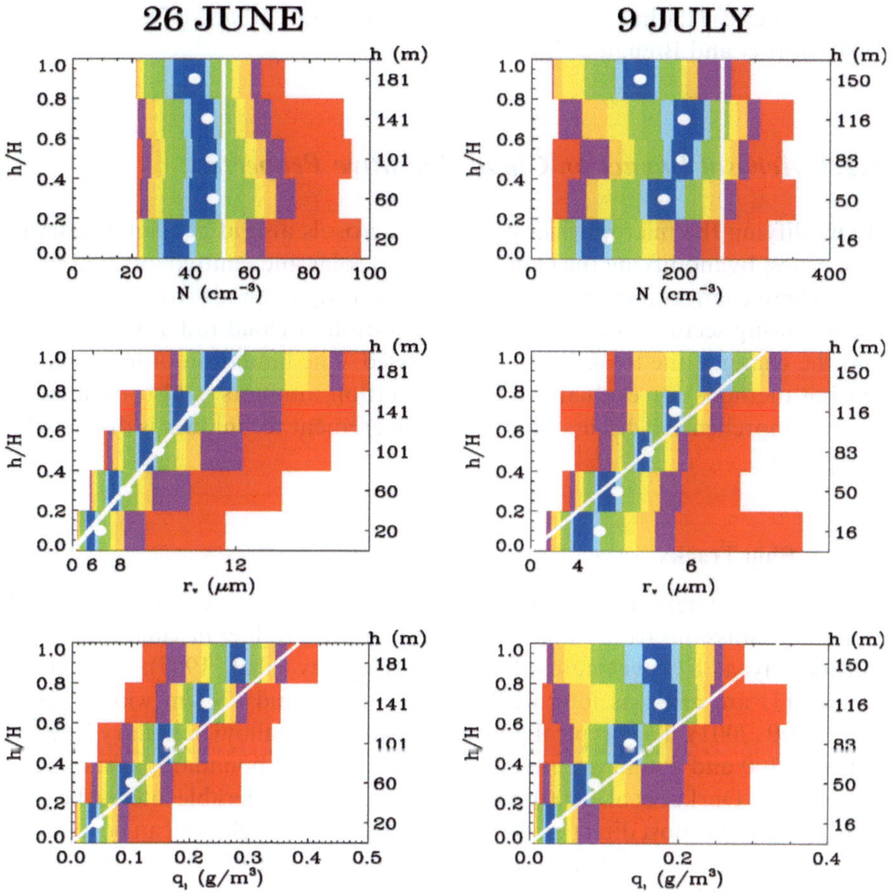

Fig. 6.10 Frequency distributions of N, r_v and q_l in five sub-layers, represented by their 5% percentiles, for the 26 June and 9 July ACE-2 cases. Statistics limited to cloud samples (N>20 cm^{-3}). The mean value in each sub-layer is indicated by a white circle. The vertical line in the top row corresponds to the characteristic CDNC value N_{act} (51 cm^{-3} for 26 June, and 256 cm^{-3} for 9 July). In the next two rows the lines indicate the adiabatic droplet size prediction, with $N = N_{act}$, in the r_v distribution, and the adiabatic LWC in the q_l distribution. From Brenguier et al. (2003a) with permission of the American Geophysical Union

droplets are too small to be detected by the airborne droplet spectrometer. Close to cloud top CDNC is substantially reduced by cloud-top entrainment-mixing. CDNC values that characterize CCN activation are derived by selecting samples with quasi-adiabatic LWC (>0.9 LWC_{ad}), and no precipitating particles (the white vertical lines in the top row). Figure 6.10 shows that the mean volume radius (second row) increases as predicted by the condensational growth theory, using the CDNC value N_{act} as defined above. It also shows that this diagnostic of the mean volume radius is still valid at cloud top despite the significant reduction of LWC following mixing with the overlying dry air. This feature is commonly referred to in the literature as inhomogeneous mixing (see Chapter 2) (Latham and Reed 1977; Baker and Latham 1979; Baker et al. 1980; Burnet and Brenguier 2007).

6.2.3 Aerosol Impact on Cloud Radiative Properties

By modifying the microphysics of clouds, aerosols also affect cloud radiative properties, by modifying the clouds' reflected solar and emitted infrared radiation. These changes are detected by remote sensing methods from satellites. In the following section, the issues of interpretation of cloud radiative properties and the changes due to aerosols are discussed. The limitation of using remote sensing methods for evaluation of precipitation amounts is also outlined. A more comprehensive discussion of the measurement techniques is discussed in Chapter 5.

6.2.3.1 Ship Tracks

Ship tracks are narrow lines of perturbed regions in marine stratiform clouds caused by emissions from ships. Ship tracks appear brighter in satellite imagery, particularly at a wavelength of 3.7 μm (Coakley et al. 1987). Ship tracks (Fig. 6.11) are often as long as 300 km or more and ~9 km wide (Durkee et al. 2000, 2001). They typically form in relatively shallow boundary layers of between 300 and 750 m deepth. They do not form in boundary layers thicker than 800 m (Durkee et al. 2000). Ship tracks are of considerable interest because they provide the most direct evidence that shallow, clean clouds can be modified by pollution sources that emit large concentrations of CCN.

Hobbs and Garrett (2000) reported that various types of ships, burning both low-grade marine fuel oil (MFO) and U.S. Navy distillate fuel (NDF), emitted from $\sim 4\times 10^{15}$–2×10^{16} total particles per kg of fuel burned ($\sim 4\times 10^{15}$–1.5×10^{16} particles per second). Diesel powered ships burning MFO emitted particles with a larger radius (~0.03–0.05 μm) than did ships powered by steam turbines using U.S. NDF. Consequently, for particles with similar chemical composition, those emitted by diesel ships burning MFO will serve as CCN at lower

Fig. 6.11 Ship tracks off France. From bands 1, 4 and 3 of MODIS on the Aqua Satellite (Courtesy of NASA)

supersaturations, and will therefore be more likely to produce ship tracks, than the particles emitted by ships powered by steam turbines using U.S. NDF.

The prevailing hypothesis is that ship trails appear brighter on satellite imagery because the effluent from ships is rich in CCN particles. The more numerous CCN particles create larger concentrations of cloud droplets, which reflect more solar energy than the surrounding clouds (King et al. 1993). Aircraft observations in ship tracks and surrounding clouds (Radke et al. 1989; Ferek et al. 1998) confirm that ship tracks exhibit higher droplet concentrations, smaller droplet sizes, and higher liquid water content than surrounding clouds. In some case ship tracks appear to be thicker than the ambient cloud (Ackerman et al. 2000a). Radke et al. (1989) found that the concentration of drizzle drops (droplets of diameter $\geq$200 μm) in ship tracks was only ~10% of that in surrounding clouds. This supports the laboratory results of Gunn and Phillips (1957) and Albrecht's (1989) hypothesis that a CDNC increase should reduce drizzle production. Another hypothesis is that ship track clouds exhibit higher liquid water content because heat and moisture emissions from ships create deeper and wetter (hence brighter) clouds (Porch et al. 1990). This hypothesis is supported by the observation of ship tracks in an apparently cloud-free sky (Conover 1966). Porch et al. examined this hypothesis and showed that ship tracks are not only characterized by greater brightness but also by clear bands on the edges of the cloud tracks. They speculated, and provided some modeling evidence, that the heat and moisture fluxes from ship effluents excite a dynamic mode of instability that, in some marine stratocumulus, leads to enhanced upward and downward motions associated with the cloud circulations. However, measurements during the Monterey Area Ship Track experiment (MAST) contradicted this hypothesis by observing that heat and moisture emissions from ships are much too small to have any impact on clouds (Hobbs et al. 2000). An alternate explanation for the formation of deeper and wetter clouds in ship tracks, and perhaps clearing to the sides of the tracks, is that when drizzle is suppressed, cloud top radiative cooling is enhanced. The enhanced cloud top cooling destabilizes the cloud layer, which results in strong ascending motions in the clouds that transports more moisture aloft, making the clouds wetter and deeper. In addition, in compensation for the enhanced upward motions, sinking motions surrounding the regions of enhanced ascent could cause clearing just outside of ship tracks. An extensive series of papers on ship tracks can be found in a special issue of the Journal of the Atmospheric Sciences (volume 57, 15 August 2000) devoted to the MAST.

Ship tracks nicely illustrate the complexity of the aerosol effects on clouds by showing that impacts on precipitation formation cannot be examined independently of concurrent dynamical and radiative feedback processes. Note however that ship tracks are an extreme realization of the aerosol impact on clouds, because the amount of CCN released and their narrow spatial extent may facilitate the development of very specific dynamical modes that are not representative of extended boundary layer clouds. Their narrow extent also facilitates detection through the radiance contrast with the surrounding cloud.

6.2.3.2 Extended Cloud Systems

During ACE2 CLOUDY-COLUMN, the aerosol properties and cloud microphysical properties, discussed above in Sect. 6.2.2.2, were measured by two instrumented aircraft flying below and inside the stratocumulus, respectively. Cloud radiative properties were measured simultaneously by a third aircraft equipped with a nadir pointing multispectral radiometer, flying above the cloud layer, and following the same track as the in situ aircraft. Figure 6.12 shows the vertical profiles of the mean volume droplet radius measured in situ and the conditional frequency distributions of the remotely sensed visible (VIS) versus Near IR (NIR) reflectances for the 26 June and 9 July ACE2 cases. The grids on the background reproduce radiative transfer simulations in adiabatic clouds for various values of CDNC and of the geometrical thickness (Schüller et al. 2003). They corroborate the Twomey hypothesis that polluted clouds have a higher reflectance than pristine clouds at constant liquid water path and that the optical thickness scales as $LWP^{5/6}N^{-1/3}$ (Brenguier et al. 2000b).

While the impact of aerosols on cloud microphysics has been clearly established since the sixties, the ACE2 closure experiment, relying on co-located and independent measurements of aerosol properties, cloud microphysics and cloud radiative properties, provided compelling observational evidence of the link between these microphysical changes and radiation at the scale of a cloud

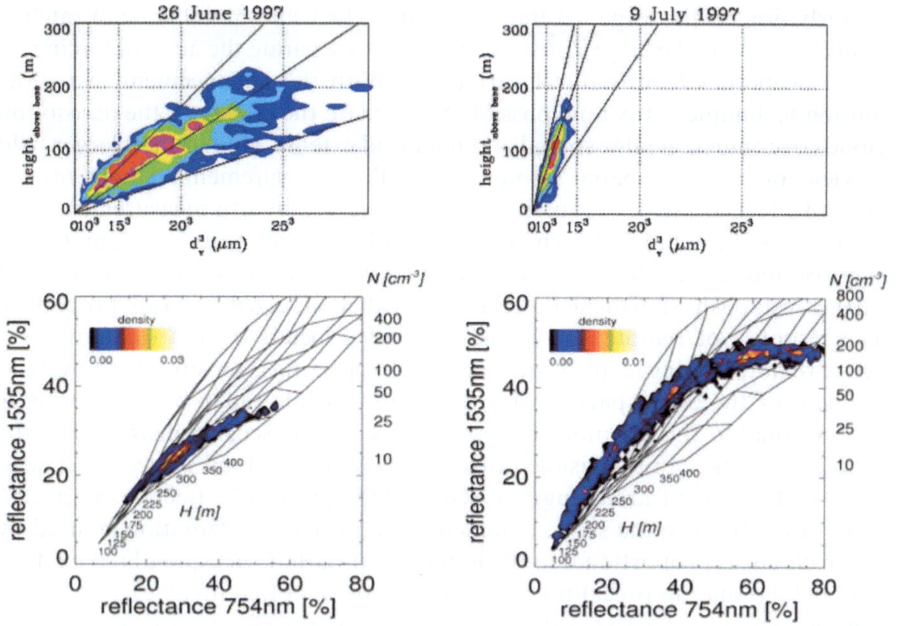

Fig. 6.12 Vertical profiles of the droplet mean volume diameter (*top*). The lines represent the expected droplet growth for a CDNC value equal to the mean, 0.5 and 1.5 of the mean for that day. Conditional frequency distributions of the remotely measured VIS versus NIR reflectances above the same cloud layer (*bottom*) for the ACE2 26 June pristine case (*left*) and 9 July polluted case (*right*). The background grid shows the expected radiances for a plane-parallel, vertically stratified and adiabatic cloud layer of diverse thicknesses (*H*) and CDNC (*N*) from Pawlowska and Brenguier (2000) (*top*) and Schüller et al. (2003) (*bottom*) with permission of the American Geophysical Union and of Tellus

system, i.e. at a much larger scale than the ship tracks. This quantitative assessment of the link between cloud microphysics and radiation is a crucial step because retrievals of cloud properties from ground or satellite remote sensing instruments that are discussed in the next sections and in more detail in Chapter 5 rely on similar measurements of cloud radiative properties.

6.2.4 Remote Sensing of the Effects of Aerosols on Cloud Properties

Field experiments are designed to address specific questions and test hypotheses in detail by focusing a comprehensive suite of in situ and remote sensing instruments on a selected atmospheric phenomenon. But they are limited in time and spatial coverage and only provide a few case studies. Long term observations of clouds, in contrast, rely on continuous remote sensing from the ground or on board satellites. Clouds vary widely in space and time due to variation of atmospheric conditions. The aerosol effect can be only a small perturbation on the large natural variability. The wide spatial scale of satellite data can provide the large statistics necessary to separate the aerosol forcing on cloud properties from natural variability. With the advancement in remote sensing techniques, it is now possible to estimate the effects of the aerosol on clouds over much greater spatial and temporal extents. Due to their larger field of view and coarser spatial resolution, satellite measurements cannot resolve aerosol effects on scales as small as can be observed by situ measurements.

Because of the link between cloud microphysics and radiation, remote sensing techniques can be used to retrieve cloud microphysical properties and further establish statistical correlations with co-located aerosol properties. The response of remote sensing instruments depends on the technique: for active remote sensing instruments, lidar backscattering is sensitive to the second moment of the cloud particle size distribution while radar backscattering is proportional to the 6th moment. For passive remote sensing instruments, the reflected radiance in the visible range depends on the cloud optical thickness, while in the near-infrared range it is sensitive to the particle effective radius (the ratio of the third to the second moment) and microwave radiation emitted by warm clouds is proportional to the liquid water path. Consequently, the retrieval of cloud microphysical properties from satellite often involves a combination of passive and active instruments operating at different wavelengths (see Chapter 5).

6.2.4.1 Retrievals of Aerosol Properties

Long term records of aerosol concentrations were obtained from TOMS and AVHRR that showed a systematic increase in the aerosol concentration in the developing world; e.g. 10% a year increase was reported for India (Massie et al. 2004). Global observations by TOMS and AVHRR show the impacts of volcanic eruptions and seasonal variability but no clear global trend (Torres et al. 2002; Geogdzhayev et al. 2005). The AVHRR study also showed evidence of reduction in the aerosol emissions in the former Soviet Union. AVHRR, METEOSAT and TOMS observations were used to track emission of dust from Africa (Carlson and Prospero 1971; Moulin 1997; Israelevich et al. 2002) and correlate them with changes in the air circulation,

These satellite observations used mainly the aerosol optical thickness (AOT) as a measure of the aerosol column concentration. There is an issue with that approach when applied to aerosol interaction with clouds. Indeed, the CCN particles that interact with cloud microphysics represent a small and highly variable fraction of the total aerosols that determine AOT. Nakajima et al. (2001) used the product of AOT with a measure of the aerosol size – the Angstrom exponent – to represent better the spatial variation of column CCN concentration. The usefulness of the aerosol index as a CCN proxy has recently been evaluated using in situ data by Kapustin et al. (2006), who pointed out that aerosol particle response to humidification may be a significant compounding factor.

Data from the new era of satellites can be better linked to CCN and cloud processes, by reporting not only the AOT but also separating it into fine and coarse aerosols. MODIS does the separation using spectral properties (Tanré et al. 1997), MISR uses the angular properties (Kahn 2001) and POLDER uses the strong polarization of the fine aerosols (Breon et al. 2002). The ability to separate fine from coarse aerosol from space enables an estimate of the anthropogenic contribution to the global aerosol system. It is estimated that 20% of the aerosol AOT over the ocean is of anthropogenic origin (Kaufman et al. 2005a,b).

Advance analysis of a combination of POLDER data from the PARASOL mission, together with the wide spectral MODIS data, shows the potential to retrieve not only higher accuracy optical thickness of the fine and coarse aerosols, but also to estimate the size of the fine aerosol, needed for precise conversion into number concentration of particles and the estimate of a CCN proxy (Gerard et al. 2005). These methods are yet to be applied to the study of aerosol-cloud interaction.

6.2.4.2 Retrievals of Cloud Optical Thickness and Droplet Effective Radius

The Twomey and Cocks (1989) and Nakajima and King (1990) technique described in Chapter 5 has been extensively used to retrieve cloud optical thickness and droplet effective radius from satellite multispectral radiances and to examine the aerosol impact on clouds. Kaufman and Nakajima (1993) used AVHRR 1 km resolution data to study well-developed clouds in the Amazon region during the dry-burning season. An increase in the smoke optical thickness (measured at a wavelength of 0.55 μm) from an average background value of 0.2 to 1.0 in very hazy conditions decreased the effective cloud droplet radius from 15 to 8 μm. Though the absolute droplet size is difficult to measure from satellites, the decrease in cloud droplet size by 40%, with five-fold increase in aerosol concentration (represented by the aerosol optical thickness measured in clear pixels) is a good indication of the overall effect of the aerosols on clouds. The increase in the smoke optical thickness and the corresponding decrease in the average drop size were surprisingly associated with a decrease in the cloud reflectance from 0.71 to 0.68. This may be due to absorption of sunlight by black carbon in the smoke, which reduces cloud brightness.

To detect if the aerosol affects cloud brightening due to increases in CCN in more susceptible (less reflective clouds), Kaufman and Fraser (1997) used a different cloud selection technique to highlight less developed clouds. The results showed that smoke increases cloud reflectance from 0.35 to 0.45, while reducing droplet size from 14 to 9 μm. However, Kaufman and Fraser found that the smoke effect varies systematically with the location of the measurements: in the moist tropical forest in the north of Brazil, the smoke had a maximum effect on clouds, while in the drier Cerrado in the south of Brazil the effects were minimal.

Further analysis of 2 years data over the Amazon region by Feingold et al. (2001) confirmed the latitudinal dependence of the effect of smoke on clouds, although they showed that it was unrelated to the humidity gradient. They also found a substantial difference between the 2 years they analyzed (1987 and 1995). Saturation of the response of cloud droplet size to aerosol occurred at an aerosol optical depth of 0.8 in 1987 and 0.4 in 1995.

Han et al. (1994) examined data from the International Satellite Cloud Climatology Project (ISCCP) and found "the expected systematic decreases of the droplet effective radius over land compared with over ocean and in the Northern Hemisphere compared with the Southern Hemisphere", but not the corresponding expected cloud albedo increase. Han et al. (1998) extended the statistical approach to the seasonal and geographic variability of cloud albedo and droplet effective radius, in relation with the LWP variability. They concluded "that cloud albedo increases with decreasing droplet size for most clouds over continental areas and for all optically thicker clouds, but that cloud albedo decreases with decreasing droplet size for optically thinner clouds over most oceans and the tropical forest". Observations using AVHRR satellite data over the oceans by Nakajima et al. (2001) however, do not show any systematic trend of liquid water path (derived from cloud optical depth and drop size) on column aerosol number concentration over the full range of column aerosol number concentrations.

Wetzel and Stowe (1999) used NOAA PATMOS data to evaluate the relationship between aerosol optical depth and cloud droplet size. POLDER satellite data were used by Breon et al. (2002) to derive aerosol concentrations and cloud droplet effective radii from 8 months of space borne measurements, and to explore the aerosol effect on cloud microphysics. It was found that the cloud droplet size decreased with increasing aerosol index (a measure of the aerosol column number concentration), showing that the effect of aerosols on cloud microphysics occurs on a global scale. However, the extent of the drop-size response to changes in aerosol was much lower than expected from theory. Rosenfeld and Feingold (2003) suggested that this low response was caused by the POLDER ability to measure only extensive homogeneous layer clouds and not convectively generated clouds.

Similar cloud optical thickness and droplet effective radius retrievals were examined against sulphate burden simulated with a chemical transport and transformation model (Harsvardhan et al. 2002; Schwartz et al. 2002). The analysis of two episodes of substantial influx of sulphate aerosol from industrial regions of Europe and North America to remote areas of the North Atlantic revealed "a decrease of the droplet effective radius concomitant with the increase in modeled sulphate burden", while "cloud optical thickness and albedo exhibit little evident systematic trend over the episodes". Han et al. (1998), Harsvardhan et al. (2002), and Schwartz et al. (2002) speculated that the variability of LWP was the most likely reason for this puzzling observation, but these studies were missing the independent LWP measurements required to corroborate their hypothesis.

These contradictory observations reflect the fact that cloud radiative properties are primarily determined by the cloud macrophysical properties, such as LWP. CDNC only modulates the optical thickness. Indeed, within the idealized framework of the Twomey hypothesis, the cloud optical thickness increases as $N^{1/3}LWP^{5/6}$. The two radiative parameters that are measured from satellite, optical thickness and droplet effective radius, are therefore not sufficient to distinguish between changes in the cloud macrostructure and possible impacts of aerosols on cloud microphysics and precipitation. To overcome the ambiguity due to the natural variability of LWP, two different approaches have been developed.

6.2.4.3 Retrievals of Liquid Water Path and Droplet Concentration

If the observed cloud field contains non-precipitating convective cells, optical thickness and droplet effective radius can be combined to derive information on CDNC and the cloud geometrical thickness or LWP. Brenguier et al. (2000b) developed the technique using a vertically stratified adiabatic cloud model to build up the neural network training database necessary for the processing of the measured radiances (see Chapter 5). In this scheme, LWP is proportional to H^2 (H is the cloud geometrical thickness). The technique was validated (Schüller et al. 2003) against the ACE2 data set, where cloud geometrical thickness and CDNC were directly measured in situ. Figure 6.13 shows the comparison between retrieved and in situ measured values of these two parameters. The CDNC differences between pristine (blue and purple) and polluted (yellow orange and red) are precisely captured. The cloud geometrical thickness, and hence LWP, are always overestimated. The origin of this bias, which is connected to the current overestimation of the droplet effective radius in most retrieval techniques, has not yet been identified. The technique has been extended to the retrieval of LWP (or cloud geometrical thickness) and CDNC from MODIS radiances (Schüller et al. 2005).

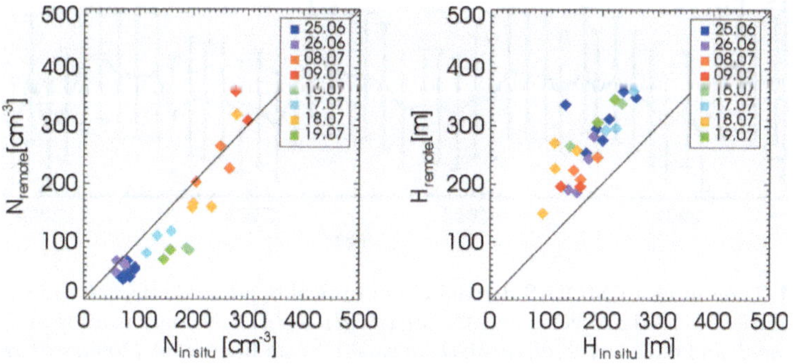

Fig. 6.13 Comparison between retrieved and in situ measured values of CDNC and cloud geometrical thickness in non-precipitating stratocumulus. From Schüller et al. (2005) with permission of the American Meteorological Society

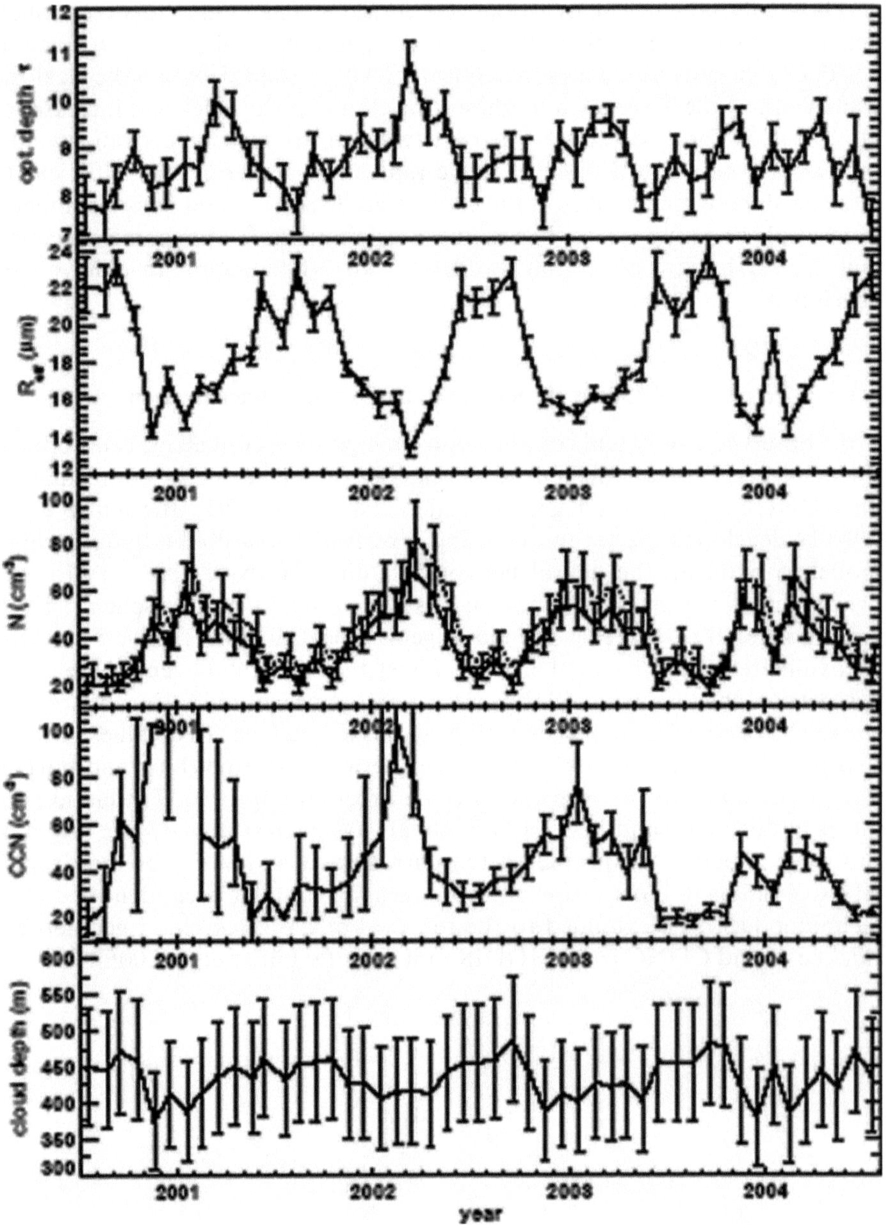

Fig. 6.14 Time series of MODIS-derived cloud optical depth and effective radius (*first two panels*), retrieved cloud droplet concentration (*third panel*), CCN concentration at Cape Grim (*fourth panel*) and retrieved cloud depth (*fifth panel*). From Boers et al. (2006) with permission from the American Geophysical Union

Recently Boers et al. (2005) [2006] have improved the technique by taking into account sub-adiabatic LWC values in the cloud layer and tested it against long term observations of the seasonal CCN variations recorded at Cape Grim. Over the 4 years, the retrieved CDNC closely follows the seasonal variations of the CCN concentration at the ground (Fig. 6.14).

Using these techniques thus allows discriminating among the observed variations of the cloud albedo, the respective contributions of the LWP natural variability (which is related to cloud depth, especially under adiabatic conditions) and of CDNC, that reflects the impact of the aerosol.

Note however that these techniques are only applicable to quasi adiabatic cloud layers, i.e. to non precipitating clouds that are not significantly affected by mixing with the environmental air. When mixing is prevalent, especially at the top of stratocumulus clouds, its impact on the droplet size distribution, hence on cloud radiative properties depend on the thermodynamic conditions at the mixing level. Either the droplet concentration is preserved while droplet sizes decrease (homogeneous mixing), or the concentration decreases while the sizes remain unchanged (heterogeneous mixing) (Latham and Reed 1977; Baker and Latham 1979; Baker et al. 1980; Burnet and Brenguier 2007). When mixing is of the inhomogeneous type (very dry inversion), the retrieved droplet concentration may be significantly underestimated and polluted clouds can be interpreted as pristine (Chosson et al. 2007).

6.2.4.4 Retrievals of Droplet Effective Radius Vertical Profile

Rosenfeld and Lensky (1998) applied a method based on retrieval of the droplet effective radius at cloud top from satellite observations. The method assumes that all the clouds in the field of view of the satellite behave the same way, namely, that they all have the same base altitude and the same droplet growth history. This assumption was validated by in situ aircraft measurements (Freud et al. 2005) and by microphysical analyses of sequences of 3-min rapid scans of METOSAT Second Generation satellite measurements (Lensky and Rosenfeld 2006). By looking at different cloud tops and their temperatures, they built profiles of effective radius and show that the effective radius increases as cloud depth increases. However, in continental clouds with higher droplet concentrations, the growth of the particles is slower than in maritime clouds. As a result, in continental clouds, drops commonly grow beyond an effective radius of 14 μm (assumed by Rosenfeld and Gutman 1994) to represent the demarcation between clouds that may precipitate and those that do not at much higher altitudes than in maritime clouds. In addition, the appearance of ice in the clouds appears as a sharp increase in effective radius with altitude. Rosenfeld and Lensky concluded that drops remain liquid to greater heights in continental clouds, leading to ice formation at much higher altitudes.

By analyzing the slope of the effective radius (r_e) versus temperature, Rosenfeld and Lensky (1998) defined five microphysical zones in convective clouds based on the temperature dependence of the r_e and therefore suggest that the method can diagnose changes in clouds microphysical properties in response to different aerosol types.

6.2.4.5 Ground Based Remote Sensing Retrievals of Aerosol and Cloud Properties

To overcome the limitations of satellite observations, surface-based remote sensing (lidar, radar, radiometer) data sets have recently been used to examine aerosol effects on clouds (Okamoto et al. 2002; Feingold et al. 2003). Although these methods can only be applied at a limited number of locations, they yield high temporal resolution data. They offer some interesting insights, and are complementary to the global satellite view. Feingold et al. (2003) used data collected at the Atmospheric Radiation Measurement (ARM) site to allow simultaneous retrieval of aerosol and cloud properties, with the combination of a Doppler cloud radar and a microwave radiometer, to retrieve cloud drop effective radius r_e profiles in non-precipitating (radar reflectivity $Z < -17$ dBZ), ice-free clouds. Simultaneously, sub-cloud aerosol extinction profiles were measured with a Lidar to quantify the response of drop sizes to changes in aerosol properties. The microwave radiometer made it possible to sort the cloud data according to LWP, consistent with Twomey's (1977) conceptual view of the aerosol impact on cloud microphysics. With high temporal/spatial resolution data (on the order of 20–100 m), realizations of aerosol-cloud interactions at the large eddy scale were obtained. Moreover, by examining updrafts only (using the radar Doppler signal), the role of updraft in determining the response of r_e to changes in aerosol (via changes in drop number concentration N_d) was examined. Analysis of data from 7 days showed that turbulence intensifies the aerosol impact on cloud microphysics.

In addition to radar/microwave radiometer retrievals of aerosol and cloud properties, surface based radiometers such as the Multi-filter Rotating Shadow-band Radiometer (MFRSR – Michalsky et al. 2001) have been used in combination with a microwave radiometer to measure an average value of r_e during daylight when the solar elevation angle is sufficiently high (Min and Harrison 1996). Using this retrieval, Kim et al. (2003) performed analyses of the r_e response to changes in aerosol at the same continental site, and instead of using extinction as a proxy for CCN, they used a surface measurement of the aerosol single scattering coefficient σ_s. Their analysis spanned much longer time periods and their data included a range of different aerosol conditions. A similar study was conducted by Garrett et al. (2004) at a location in the Arctic. The advantage of the MFRSR/microwave radiometer combination is that it derives r_e from cloud optical depth and LWP and it is not sensitive to large drops, Its drawback is that it can only be applied to clouds with extensive horizontal cover during daylight hours. It should be emphasized that none of these methods can directly address the effects of aerosols on precipitation.

6.2.5 Satellite Observations of Aerosol Impact on Precipitation Formation in Clouds

Since the 1960s, in-cloud measurements have clearly established that aerosols influence cloud microphysics. The overall feature is that increased CCN concentration leads to increased CDNC, although, as mentioned previously, the relationship also depends on vertical velocity at cloud base and on how CCN are distributed in size among the Aitken, accumulation and the coarse modes.

Since that time, many other field studies have revealed that the increased CDNC and decreased droplet sizes have an impact on the formation of precipitation embryos by droplet collection. For example, Boers et al. (1998), analyzing the SOCEX data set, found that the fraction of LWC contained in drizzle drops sharply increases when the droplet effective radius increases above 12 μm (Fig. 6.15).

Similarly, Pawlowska and Brenguier (2003) showed (Fig. 6.16) that for the 8 case studies of the ACE-2 experiment the averaged precipitation rate R, normalized by the cloud geometrical thickness H, scales with H^3/N. This feature, that may be useful for parameterizing precipitation in general circulation models, has further been corroborated with the data sets of the EPIC experiment (Comstock et al. 2005) and of the DYCOMS-II experiment (Van Zanten et al. 2005).

The effective radius profiling technique of Rosenfeld and Lensky (1998) described in Sect. 6.2.4.4 has been extensively used, in combination with TRMM remote sensing of precipitation, to examine the impact on precipitation of forest fires (Rosenfeld 1999), desert dust (Rosenfeld et al. 2001), sea spray (Rosenfeld et al. 2002), salt dust (Rudich et al. 2002), oil fires (Rudich et al. 2003), and the aerosol impact on the dynamics of convective clouds (Koren et al. 2005) and of storms (Fromm et al. 2006).

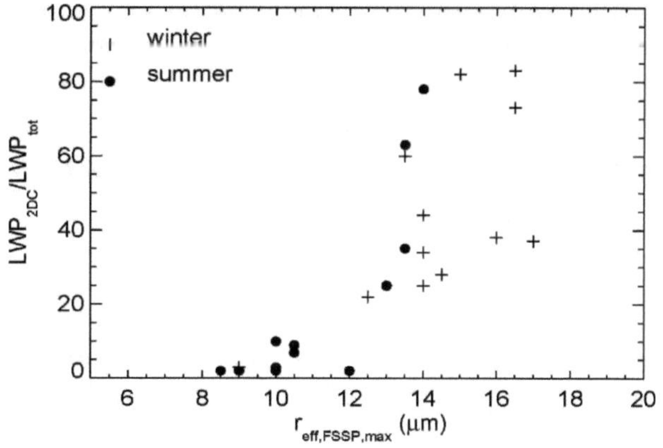

Fig. 6.15 Drizzle liquid water path fraction as a function of the maximum average effective radius observed in the case study profiles. Each data point is representative of individual stacks that were flown during the research flights. All *winter* and *summer* data have been included, except the 19 July 1993. From Boers et al. (1998) reprinted with permission from the Royal Meteorological Society

Fig. 6.16 Normalized precipitation rate (R/H) versus H^3/N. From Pawlowska and Brenguier (2003) with permission of the American Geophysical Union

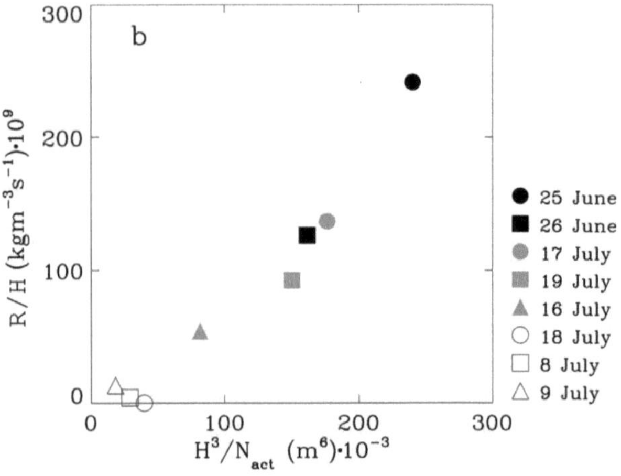

Overall, these satellite studies corroborate in situ measurements. Using TRMM data it was reported that precipitating particles were detected when the retrieved effective radius exceeded 14 μm threshold (Rosenfeld 1999; Rosenfeld et al. 2001, 2002). For example, Rosenfeld (2000), using AVHRR images, identified the signature of a pollution plume in a cloud layer (area 2 in Fig. 6.17a). Using TRMM measurements he pointed out that this polluted area did not develop any precipitation, unlike areas 1 and 3 in Fig. 6.17a,b. Ayers (2005) argued against the conclusion of Rosenfeld (2000) that rainfall was suppressed in the case shown in Fig. 6.17. Using radiosonde data he showed that the clouds depicted in the satellite retrievals were thin, located atop the boundary layer and were capped by a strong temperature inversion and low humidity. Therefore, they could not have produced much precipitation. This was borne out by 24 h rainfall that was measured by the Australian Bureau of Meteorology, showing that rainfall was isolated and almost non-existent everywhere across the region. In the debate over the interpretation of this study Ayers (2005) and Rosenfeld et al. (2006a) underscored the need for further examination of this issue using a large number of cases with differing methodologies.

6.2.6 Correlations Between Aerosol and Cloud Macrophysical Properties

In addition to its impact on droplet concentration, the aerosol also affects radiative transfer through clouds via absorption of solar light (semi-direct effect), which may accelerate cloud dissipation (cloud burning effect). Using satellite observations, Koren et al. (2004) reported that smoke reduced daytime boundary layer cloud cover over the Amazon forest, from 38% in clean

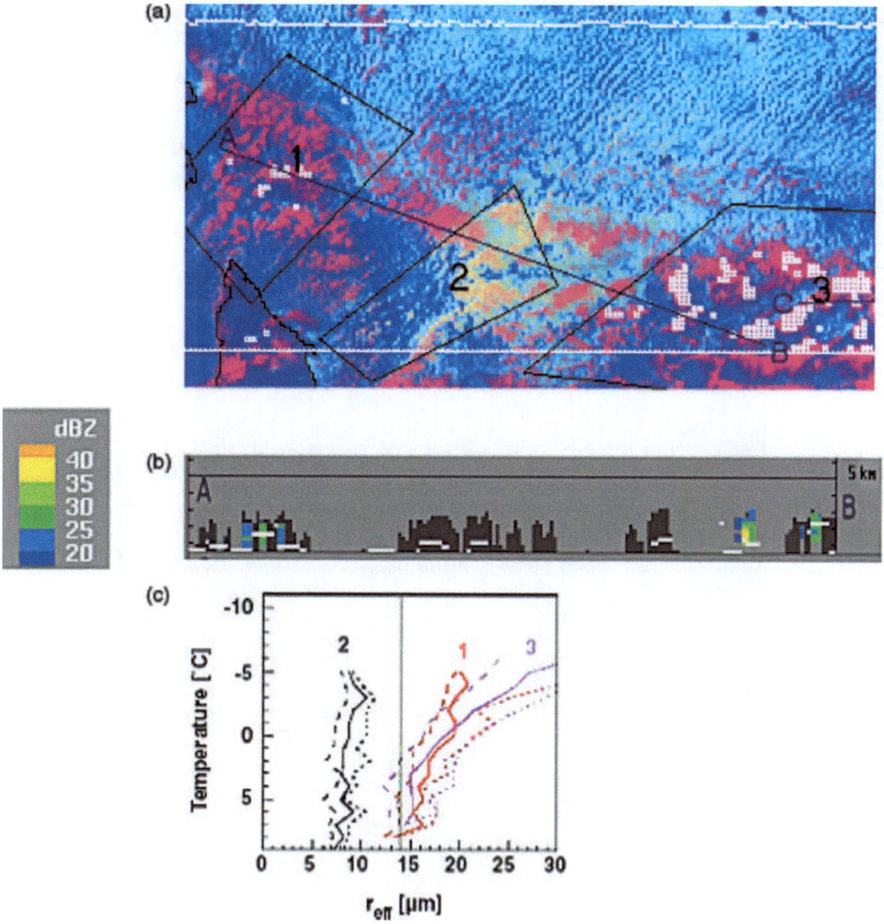

Fig. 6.17 (**a**) Pollution tracks (*yellow*) detected by AVHRR satellite imagery in clouds over South Australia. The higher reflectivity is due to reduced droplets sizes. (**b**) Radar shows precipitation as white patches outside pollution tracks, although these clouds have the same depth as adjacent non-polluted clouds, as shown in the vertical cross section. The gray clouds are the siluhets obtained from the TRMM VIRS. The colors represent the radar echo intensities. No radar echoes occur within the clouds with reduced effective radius in Area 2. (**c**) Average effective radius of particles (*probably water drops*) at cloud tops in these regions (1, 2, and 3). The dashed lines represent the 15 and 85 percentiles of the distributions (Rosenfeld 2000 function of the maximum average effective radius observed in the case study profiles. Each data point is representative of individual stacks that were flown during the research flights. All winter and summer data have been included, except the 19 July 1993. From Rosenfeld (2000) reprinted with permission of AASS

conditions to 0% in heavy smoke (optical thickness of 1.3). This response reverses the regional smoke instantaneous forcing of climate from –28 W m^{-2}, when only the direct effect is taken into consideration, to + 8 W m^{-2}, when the dissipation in cloud cover is accounted for (Fig. 6.18).

MODIS aerosol data over the ocean (Tanre et al. 1997; Remer et al. 2002a), and cloud data from King et al. (2003) and Platnick et al. (2003) from the Terra

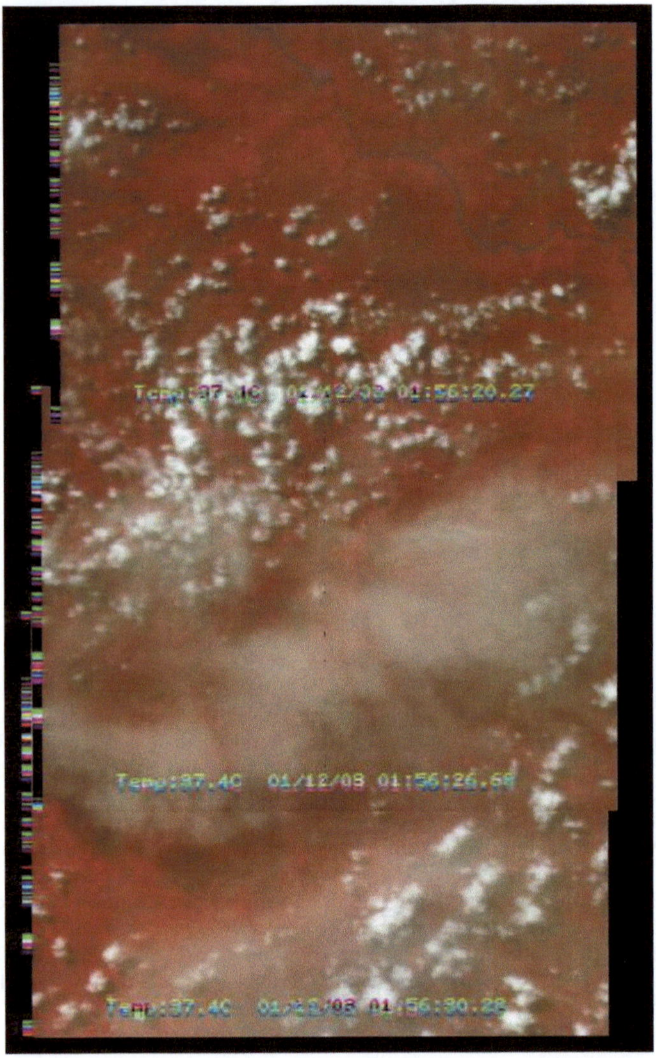

Fig. 6.18 Reduction of cloud cover due to absorption of solar radiation by particles from biomass burning in Brazil. Photograph taken from the NASA Columbia shuttle STS-107 tragic flight on January 2003. NASA image from the Columbia Shuttle, with permission of the MEIDEX team

satellite, have also been used for evaluating aerosol-cloud interactions for shallow, stratiform and trade-wind cumulus clouds and for deep convective clouds over the Atlantic Ocean. Kaufman et al. (2005b) showed [2005c] show that each of four aerosol types covering the Atlantic Ocean during June through August (maritime aerosol from 20 to 30°S; smoke from 20°S to 5°N; dust from 5 to 30°N; and, pollution aerosols from North America and Europe from 30 to 60°N) are correlated to a 20–40% cloud coverage variability of shallow stratiform and trade-wind cumulus clouds. Using co-variability analysis, they suggest that the aerosol particles might be responsible for ~40–100% of observed increases in cloud coverage. The increase in cloud cover attributed to aerosol is consistent with Albrecht's (1989) hypothesis that enhanced CCN concentration suppresses precipitation and thus might increase cloud lifetime. This hypothesis has been criticized recently (Ackerman et al. 2004) by showing, using large eddy simulations (LES) of boundary layer clouds, that increased CCN concentrations, and hence increased CDNC, might reinforce entrainment at cloud top, leading to a decrease in the LWP and in cloud cover.

Analysis of convective clouds by Koren et al. (2005) show systematic effects of pollution, desert dust and biomass burning aerosols on the development and coverage of convective clouds over the Atlantic. As expected, cloud droplet sizes are decreased by ~20% in high aerosol concentrations. This change has been accompanied by an increase in the coverage of convective and high clouds from ~30% for low aerosol concentrations to ~60% in hazy atmospheres. This sharp increase in cloud cover is associated with an increase in cloud top horizontal extent and altitude (the latter by as much as 1500 m), apparently through inhibition of early precipitation and the resultant strengthening of updrafts. These results are in agreement with the mechanism proposed in Williams at al. (2002) and Andreae et al. (2004) and simulated by Khain et al. (2005) and Teller and Levin (2006) (this last reference reported only relatively small increase in cloud top, about 500 m, but a large increase in horizontal extent).

In most of the papers discussed thus far, the effects of aerosols on clouds have been confined to warm non- or slightly raining clouds, and in a few cases to convective clouds that may have contained ice. Overall, these studies are lacking information on small changes in the moisture content, enthalpy and available convective instability that could also explain the observed changes in cloud cover between different air masses. Such thermodynamical fluctuations are presently not detectable from satellites, nor are they predictable from re-analyses of the meteorological fields.

Satellite monitoring of aerosol and cloud properties thus provides evidence of correlations between aerosol optical thickness and cloud macrophysical properties (cloud fraction, cloud extend, liquid water path). However, progressing from statistical correlations to cause and effect relationships remain a challenge. Satellite measurements of cloud optical thickness, as a proxy of CCN number concentration, does not allow discriminating between aerosol layers that may interact with clouds and those that do not interact because they are located higher up. Saharan dust for example is often exported towards the North-East Atlantic Ocean in the free troposphere, i.e. at an altitude much higher than the boundary layer where clouds develop. Such mineral dust is unlikely to interact with cloud microphysics but it may affect clouds in other ways. For example, a high dust layer may slightly decrease the incoming solar radiation, heat the atmospheric layer above the clouds and consequently increase the cloud cover of stratiform clouds (Johnson 2004), as measured off the coast of Africa (Kaufman et al. 2005b,c). Additionally, falling dust particles from above may nucleate ice particles near cloud top, thus modifying the development of precipitation via the ice phase. Quantifying the contribution of the meteorological circulation to cloud variability is also a serious obstacle because of the high sensitivity of the liquid water path to presently undetectable changes of the temperature and humidity in the boundary layer. Statistical analysis of the respective contributions of the aerosol and meteorology to cloud variability thus remains questionable because there are presently no robust meteorological predictors of shallow cloud cover.

The ACE-2 CLOUDY-COLUMN data base illustrates the paradox, with the 8 situations documented during the campaign. Figure 6.19 shows correlations between the cloud geometrical thickness H, the optical thickness τ, the droplet mean volume radius r_v at cloud top, as a proxy for the effective radius, and the droplet number concentration N, that characterize each case. Pristine cases are in blue, polluted cases in red, and intermediate cases in green. If H is constant, as in the Twomey hypothesis, or H/N correlations are restricted to the grey area corresponding in Fig. 6.19a to $H \propto N$ and $H \propto N^{-1/5}$, a negative correlation is expected between optical thickness and effective radius ($\tau \propto r_v^{-1}$, when H is constant). In fact, because polluted cases are thinner than the other cases, the data set shows a $H \propto N^{-1/2}$ correlation. In Fig. 6.19b,c,d, the dotted line corresponds to correlations anticipated by Twomey (H constant) while the solid line illustrates the expected correlations when $H \propto N^{-1/2}$. Figure 6.19d shows a positive correlation between optical thickness and effective radius, instead of the expected negative one.

This experiment demonstrates that the unexpected positive correlations observed from satellite between optical thickness and effective radius are not contradictory to the Twomey effect. They only reflect fortuitous correlations between pollution and geometrical thickness. In the ACE-2 specific case, back trajectories and meteorological analysis showed that the reduced thickness of the polluted cases was not due to their increased aerosol content, but only to the fact that these air masses were transported over the European continent for a few days before entering the maritime area of the experiment, north of Tenerife. As a consequence, the liquid water path, hence the optical thickness, of the polluted cloud systems sampled during ACE-2 was lower than the ones of the pristine or intermediate cases. In this case, the aerosol is not the cause of the cloud LWP variability, both being determined, rather, by the mesoscale circulation.

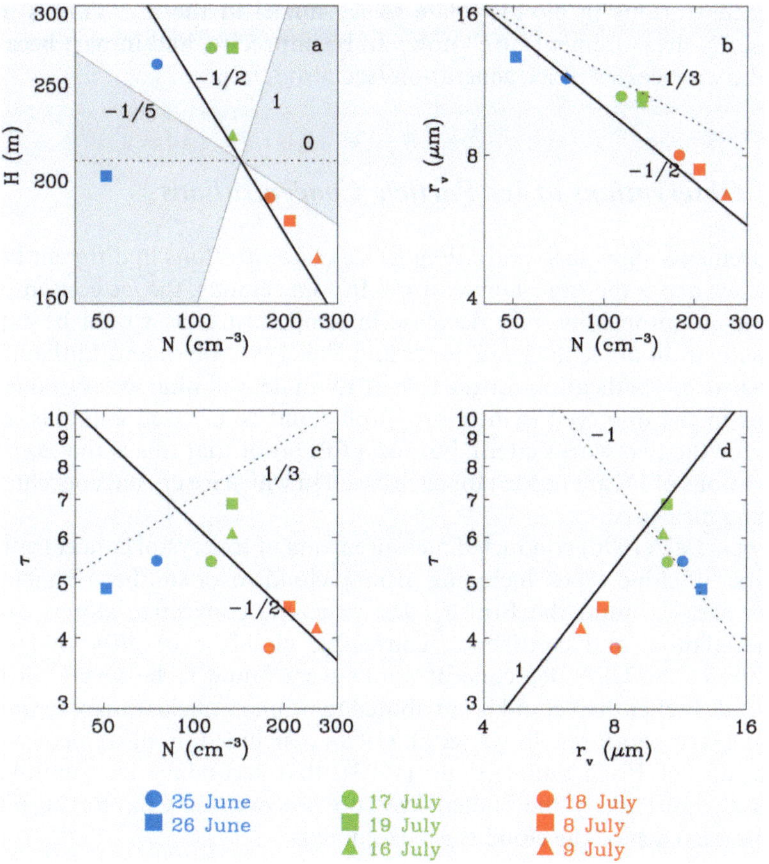

Fig. 6.19 Summary of the ACE-2 CLOUDY-COLUMN 8 case studies; (**a**) cloud geometrical thickness H as a function of droplet number concentration N; (**b**) droplet mean volume radius as a function of CDNC; (**c**) cloud optical thickness τ as a function of CDNC; and (**d**) optical thickness as a function of r_V. From Brenguier et al. (2003) with permission from the American Geophysical union

6.3 Field Studies of the Effects of Ice Nuclei on Clouds and Precipitation

Ice crystals may form in clouds whenever a cloud reaches an altitude where the temperature falls far below 0°C. However, at temperatures above –40°C, ice principally forms by what is called *heterogeneous nucleation* on particles that are effective as ice nuclei (IN). The role of the ice phase in the development of clouds and precipitation has been discussed in Chapter 2. Furthermore, Chapters 3 and 4 elaborate on the characteristics and the sources of IN. In spite of decades of research in this field it has not been clarified how observed ice concentrations in clouds relate to ice nuclei in the air. This is in part because IN measurements are known to be imprecise, and in part because of secondary processes of ice generation (see also Chapter 2).

6.3.1 Observations of Ice Particle Concentrations

Measurements show large variations in ice concentrations in different types of clouds, even at a the same temperature. In some clouds, the ice concentrations increase monotonically with decrease in temperature, as would be expected from activation of ice nuclei (Cooper and Vali 1981; Blyth and Latham 1993). This is a strong indication for the role of IN in determining ice crystal concentrations in the observed (winter orographic and cold-based cumulus, respectively, for the two works cited), but for a full proof that this is the case, direct observations of IN are needed for each cloud in which ice crystal concentrations are being measured.

Bower et al. (1996) conducted measurements of ice crystal concentrations in a number of cloud types, including: frontal clouds over southern England and the sea areas around the British Isles: maritime convective clouds over the North Atlantic; and continental convective clouds over New Mexico and Montana in the USA. Ice concentrations were found to be several orders of magnitude higher than could be attributed to primary nucleation of ice nuclei at cloud-top temperatures. Bower et al. (1996) concluded, in agreement with the conclusions of Blyth and Latham (1993), that secondary ice multiplication processes, most likely the Hallett-Mossop process (see Chapter 2), must be operating in each of the cloud types examined.

Our understanding of the processes of ice formation in clouds become even less clear following the reports of Gultepe et al. (2001) and Korolev et al. (2003), who showed that the average ice crystal concentration between ~–10 to –30°C in many stratiform clouds associated with frontal systems is almost constant (between ~2–5 cm^{-3}). The relatively constant concentrations with height are also much higher than those predicted by the parameterizations of Fletcher (1962) and Meyers et al. (1992). This suggests that processes other than nucleation by IN must play an important role in ice crystal formation in these clouds. Field et al. (2005) and Gayet et al. (2002) measured similar ice crystal concentrations as a function of temperature in different geographical areas in frontal and cirrus clouds, respectively.

Hobbs and Rangno (1985, 1990, 1998) and Rangno and Hobbs (1988, 1991, 1994, 2005) discussed ice particle initiation and multiplication in maritime and continental cumulus clouds in a number of geographic locations. They suggested that the ice particles originate in two-stages. During stage 1, initial ice particles seem to originate through the freezing of larger cloud droplets in concentrations comparable to that estimated with the formula of Meyers et al. (1992) derived from IN measurements. During stage 2, however, ice particle concentrations of 10's to 1000's per liter form in less than 10 min at temperatures as high as –13°C. Such high concentrations appear coincident with or soon after the formation of graupel particles. Supercooled raindrops or drizzle drops need not be present. In some cases, the time-scales for the formation of high ice particle concentrations are too short for the riming-splintering process to account for the observations.

It is recognized that ice crystal measurements in clouds suffer from the lack of adequate instrumentation for detecting small ice crystals (a few tens of microns), suggesting that ice crystal concentrations in clouds might be higher than reported (Korolev et al. (2003).

With regard to the vertical distribution of ice particles, Blyth and Latham (1993) reported that in the summertime convective clouds in New Mexico, primary ice nucleation was found to occur when the temperature within the cloud reached a value of between –10 and –12°C irrespective of whether this was in the updraught or downdraught. Drops with diameters of about 0.5 mm were often observed in concentrations of about 10 L^{-1} before the formation of ice, which suggests a nucleation mechanism involving large drops. There are some reports that for vigorous convective clouds, cloud drops sometimes remain unfrozen until they reach the homogeneous freezing levels (Rosenfeld and Woodley 2000; Rosenfeld et al. 2006b). The authors argue that pollution in clouds would result in smaller cloud drops many of which would remain unfrozen until they are lifted to the homogeneous freezing temperature of –38°C. However, if the pollution contains some particles that are effective IN, the formation of ice should start at higher temperatures.

The emission from steel mills in the Chicago area was used to at least partly explain the increase in rain and hail around La Porte, Indiana, downwind of Chicago (in what was named the La Porte anomaly. See a review by Changnon 1980). Furthermore, model results show (Yin et al. 2000a; Teller and Levin 2006) that if among the pollution particles large and giant CCN are present, the large drops that are formed enhance ice formation at warmer temperatures and lower altitudes. The presence of such ice crystals in regions with high liquid water content would expedite the process of riming and the formation of graupel particles.

6.4 Aerosol Impact on Rainfall on the Ground

As seen in the previous sections, aerosol impacts on cloud drop evolution are reasonably well understood, while the impact of aerosols on ice crystal formation is not yet clear. This makes it difficult to assess the impact of pollution on precipitation, especially from clouds where ice processes contribute to precipitation formation. Cloud and cloud system dynamics also have to be considered since they synergistically interact with microphysical changes.

Most measurement systems described in previous sections detect precipitation particles in clouds and do not provide direct measurements of precipitation rates and amounts at the surface. In this section, discussion is extended to field experiments in which rain gauge networks or remote sensing from satellites were used to assess the impact of aerosol on rainfall amounts.

6.4.1 The Effect on Precipitation from Paper Mill Smoke, Sugar Cane and Forest Fires

Warner (1968) summarized the potential effects of sugarcane smoke on rainfall, by looking at multi-decadal rainfall records from stations upwind and downwind of these prolific anthropogenic aerosol sources. His first summary, which appeared in 1968, suggested the possibility of rainfall suppression by the biomass burning aerosols affected cloud properties, (downwind versus upwind), as shown by Warner and Twomey (1967). However, a more detailed analysis (Warner 1971) led him to conclude that no such signal could be found in the data. In other words, his analysis gives no support to the idea that any association found between cane fires and rainfall at Bundaberg was due to inhibition of the coalescence process by a reduction in average cloud droplet size. He stated, "It would be surprising if the microphysics of a cloud played no part in determining its rainfall, but we must await further results if this is to be adequately demonstrated".

Hobbs et al. (1970) reported on an increase of up to 30% in precipitation from warm clouds downwind of paper mills in Washington State. Analyzing the same case through the use of a one dimensional numerical model Hindman et al. (1977a) concluded that the emitted GCCN from the paper mill could not by themselves account for the observed large increase in rainfall and that the combined effects of heat, water vapour and CCN from the paper mill may be responsible for the increased precipitation.

It is notable that Mather (1991) reported an increase in radar echo from clouds affected by particles emitted from paper mills in South Africa. This observation led to a large field experiment for rain enhancement using hygroscopic particles (see Chapter 8 for more detail).

Through the use of MODIS and TRMM satellite data, Lin et al. (2006) analyzed the effects of forest fires on precipitation in the dry season in the Amazon region. They report on increases in cloud heights and in precipitation with increases in aerosol optical depth. The increase cloud height led to enhanced growth of ice crystals, which culminated in heavier precipitation. However, in spite of the good correlation between these variables, the authors could not unequivocally establish causal links between aerosols and the observed changes in cloud height or with precipitation increases. The role of enhanced convection due to the heat from the fires and/or from heat due to absorption of solar radiation by the smoke itself could not be ruled out.

6.4.2 Effects of Urban Pollution on Rainfall

Extensive studies were conducted to explain the anomalous behavior of the precipitation around La Porte, downwind of Chicago. In that case, local records suggested an upward shift in warm season rainfall, thunderstorms and hail from the late 1930s to about 1965. The puzzling thing about this case is the fact that the anomaly appeared and then disappeared. Changnon (1980) reviewed the observations and concluded that the microphysical effects must have played a role, but without the larger scale dynamical or synoptic conditions this effect would not have occurred.

A large field experiment was carried out around St Louis Missouri, motivated by the examination of historical data that revealed summer increases in the immediate downwind area of the city (Fig. 6.20). The records show increases in: (1) rainfall (10–17%); (2) moderate rain days (11–23%); (3) heavy rainstorms (80%); (4) thunderstorms (21%); and (5) hailstorms (30%) (Changnon et al. 1971). The results of this experiment were reported in 1974 in the Bulletin of the American Meteorological Society. In his summary of METROMEX, Braham (1974) reported that the CCN production from the city was about $10^4\,m^{-2}\,s^{-1}$, much higher than the surrounding rural areas, accompanied by an increase in cloud drop concentrations and a decrease in drop size. However, the radar echoes from these clouds were usually lower than their counterparts in the rural surroundings. This seems to contradict our physical understanding of cloud growth, but it was concluded that one way to explain the observations is

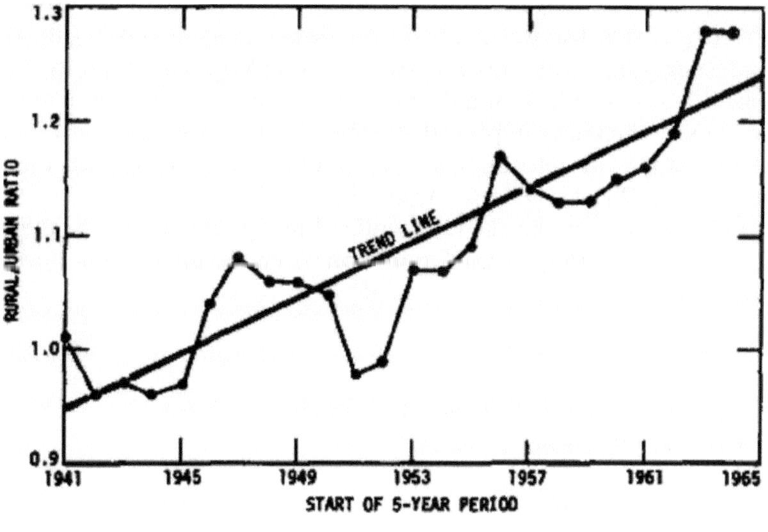

Fig. 6.20 Five year moving averages and time trend of Centerville (downwind of St. Louis) summer rainfall, 1941–1968. From Changnon et al. (1971) with permission from the American Meteorological Society

to assume that GCCN, which are not easily detected by the sampling methods, could be responsible for the increased precipitation.

Jin et al. (2005) analyzed diurnal, weekly, seasonal, and interannual variations of urban aerosols with an emphasis on summer months using 4 years of observations from the National Aeronautics and Space Administration (NASA) Earth Observing System (EOS) Moderate Resolution Imaging Spectroradiometer (MODIS), in situ data from AErosol RObotic NETwork (AERONET) observations, and in situ EPA PM2.5 data for one mid-latitude city (New York) and one sub-tropical city (Houston). The research reveals that spatial and temporal urban aerosol optical depth varies as a result of various parallel factors, such as human activity, land cover changes, cloud-aerosol interactions, and chemical processes. Diurnal, seasonal, and interannual variations of aerosol optical depths were examined and were found to be largely affected by weather conditions. On calm days, aerosols concentrations peak during the rush hours in the morning and evening. The weekly cycles of anthropogenic-induced aerosols and clouds are weak and are mixed with stronger natural weather variability. Analysis of monthly mean aerosol optical thickness and rainfall did not show strong relationships between aerosol and rainfall.

In their analysis, virtually no seasonality is observed for rainfall over Houston and New York City, suggesting that aerosols affect rainfall less than cloud microphysics. Around Houston, the TRMM satellite accumulated rainfall data show that the maxima monthly mean rainfall occurred in October 2000, May 2001, and September 2002. This is consistent with the transition between seasons in this region. In general, New York's rainfall had less month-to-month variation than Houston, with a maximum slightly above 200 mm/month in October 2002. The effective radius for water clouds was lower in New York City than in Houston, suggesting either more aerosols in New York City than in Houston, or thinner clouds. The lack of a direct relationship between rainfall and urban aerosol optical thickness implies that urban rainfall anomalies are not fully related to changes in aerosol. This observation is consistent with the earlier conclusions from METROMEX (Ackerman et al. 1978).

As can be seen from the above, in spite of many measurements, there is no conclusive evidence that aerosol pollution from urban regions does affect precipitation.

6.4.3 Effects of Air Pollution on Clouds and Rain from Orographic Clouds

Borys et al. (2000) and Borys et al. (2003) provided some evidence that pollution can delay precipitation in winter orographic clouds in the Rocky Mountains. Their analysis shows that pollution increases the concentration of CCN and therefore cloud drops, leading to the formation of smaller cloud drops. The

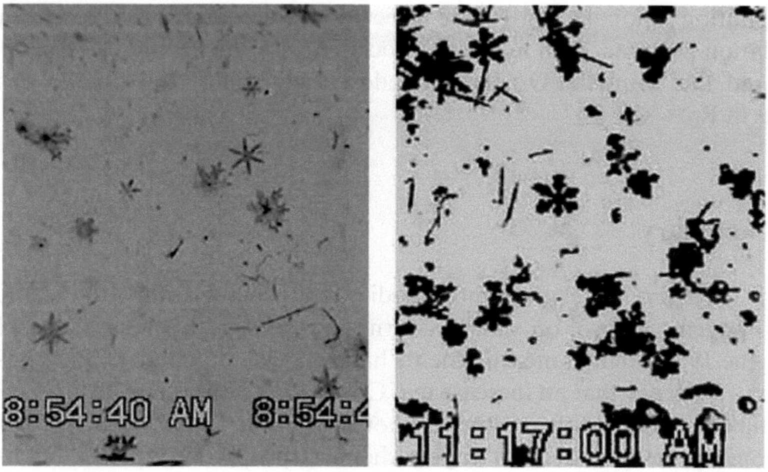

Fig. 6.21 Light riming of ice crystals in clouds affected by pollution (*left*) compared to heavier riming in non-polluted clouds (*right*). From Borys et al. (2003) with permission of the American Geophysical Union

reduced drop size leads to less efficient riming and therefore to smaller ice crystals (Fig. 6.21), smaller fall velocities, and less snowfall.

Givati and Rosenfeld (2004) analyzed about 100 years of precipitation records in regions downwind of pollution sources and compared them to precipitation in regions unaffected by these sources. In their study, Givati and Rosenfeld (2004) documented the precipitation trends in the orographic enhancement factor, R_0, which is defined as the ratio between precipitations over the hill with respect to the upwind lowland precipitation amount. Two geographical areas were chosen for this study: California and Israel. The topography in both regions is similar, although the mountains in Israel are much lower than the Sierra Nevada. The statistical results for both locations show that downwind of pollution sources, on the upslope of mountains and mountain tops, orographic precipitation is reduced by ~20 and ~7%, respectively. It was hypothesized that this decrease is due to an increase in droplet concentrations and a decrease in droplet size. Farther downwind on the lee side of mountains, the amount of precipitation is increased by ~14%. The authors postulate that this increase is due to smaller cloud particles taking longer time to grow, allowing the winds aloft to carry them over the mountain top (see an earlier study of similar effects, produced by deliberate over seeding with ice-producing particles, by Hobbs 1975a,b). However, they hypothesized that the integrated rainfall amount over the whole mountain range was reduced by the progressively increased pollution over the years. Subsequent studies show similar decreasing trends in R_0 over a few western States in the US (Griffith et al. 2005; Rosenfeld and Givati 2006) and the east slopes of the Colorado Rockies (Jirak and Cotton 2005). They argue that although absolute precipitation amounts and R_0 are affected by fluctuations in the atmospheric circulation patterns, such as those associated with the Pacific Decadal Oscillation and the Southern Oscillation Index, they cannot explain the observed trends in R_0.

6.5 Summary

Except for a very few papers, most studies to date have dealt with the effects of anthropogenic aerosol on shallow warm non- or slightly precipitating clouds. Since the 1960s, in situ measurements have provided numerous and consistent pieces of evidence that an increase in CCN concentrations results in an increase in droplet concentrations, although not in a linear fashion. The quantitative relationship between aerosol physio-chemical properties, updraft intensity at cloud base, and the resulting concentration of activated nuclei (i.e. the concentration of cloud droplets at cloud base) still needs to be improved for accurate evaluation of the aerosol impacts on clouds and climate.

In situ observations also corroborate the predictions of cloud microphysics theory, namely that clouds with more numerous, and hence smaller cloud droplets produce fewer precipitation embryos, than clouds with lower CDNC and a similar liquid water path.

Monitoring of these effects from satellite is less convincing because of the present uncertainties in remote sensing measurements of the liquid water path, which are required to stratify the observed data sets and separate the aerosol impact from the natural meteorological variability.

Beyond the LWP variability, there are additional cloud features that may significantly affect the retrieval of cloud microphysical properties from satellite. Important factors that determine cloud radiative properties include spatial heterogeneity, and the variability of mixing processes in the upper part of the cloud layer

In spite of the limitations, satellite retrieved T–r_e relations have been used to gain some insight into the cloud and precipitation forming processes, and possibly the impact of aerosols on them. Combined with space borne radar and passive microwave measurements, this methodology has been used in studies of the effects on precipitation of smoke from biomass burning, urban and industrial pollution, desert dust, and sea salt.

Recent remote sensing observations show that increased drop concentrations might lead to larger cloud fractions and deeper clouds. Similar results have been reported from numerical model simulations. However, there are indications from remote sensing studies that an increase in aerosol may decrease the cloud fraction of small warm cumulus clouds.

While numerous publications deal with the effects of pollution on clouds, there are few that treat the effects of pollution on the amount of precipitation on the ground. Some observations suggest that orographic precipitation may be reduced by increased aerosol pollution. The overall effects of cities on precipitation appear not to be clearly connected to aerosol pollution.

Large or GCCN in concentrations on the order of 1 per liter can significantly affect the early development of precipitation embryos in clouds. Their effect on precipitation on the ground is yet to be established.

Ice formation in clouds is not well understood. Certain effects of aerosol pollution on ice processes have been observed, but an overall link between aerosol pollution, ice nuclei, ice crystal concentrations and growth to precipitation has not been established.

To reach firmer conclusions on these effects, further observational studies and new techniques are needed on the following issues:

- For each aerosol type there is a need to characterize their CCN, IN and light absorption spectral properties (single scattering albedo).
- Measure the vertical profile of the aerosol particles in order to form a connection between them and the microphysical processes in the clouds.
- Evaluate the role of GCCN in the precipitation development.
- Improve the satellite retrieval techniques of the liquid water path, accounting for the cloud spatial heterogeneity and the mixing process biases.
- Improve measurements of the precipitation amount at the ground.
- Design statistical analyses that separate the effects of meteorology from the effects of aerosol pollution on precipitation on the ground.

Chapter 7
Effects of Pollution Aerosol and Biomass Burning on Clouds and Precipitation: Numerical Modeling Studies

Graham Feingold, William Cotton, Ulrike Lohmann and Zev Levin

7.1 Introduction

The history of numerical modeling of the effect of aerosols on clouds dates back at least 50 years to the work of Howell (1949) and Mordy (1959), who considered the growth of a population of aerosol particles in a rising parcel of air. Models such as these addressed the effects of both aerosol and dynamical parameters (i.e. updraft velocity) on the number and size distribution of cloud droplets. To this day similar models are in wide use to examine the effects of aerosol composition and atmospheric trace gases on droplet activation (e.g. Kulmala et al. 1993; Ghan et al. 1997; Feingold and Chuang 2002; Nenes et al. 2002).

From the early roots of cloud parcel models that simulated droplet growth by condensation, there have been two parallel and complementary foci (Fig. 7.1). The cloud physics community has pursued the modeling of precipitation formation in a variety of modeling frameworks, ranging from parcel models to 1-D, 2-D and 3-D models of both warm and cold clouds. These studies have been able to capture the salient features of cloud droplet activation and the timescales of growth to precipitation-sized drops in a reasonable manner. Simultaneously, the aerosol and chemistry communities have developed models that have placed more emphasis on the effect of aerosol composition on cloud microphysics, as well as on the role of clouds as processors of aerosol (e.g. via aqueous-phase chemistry). Although these efforts are depicted as parallel in Fig. 7.1, there has been some communication between the two communities with the result that the more recent modeling efforts include representation of the coupled aerosol-cloud-chemistry system (e.g. Barth et al. 1992; Respondek et al. 1995; Feingold and Kreidenweis 2002; Yin et al. 2005). The representation of all these components is required to model the effect of aerosols on precipitation because of the myriad feedbacks that can occur in the system. For example, aerosols that are modified by cloud processes, and are then released back into the atmosphere after cloud evaporation (most clouds evaporate and do not precipitate) could affect other clouds (e.g. Scott and Hobbs 1967; Hegg et al. 1980, 1996; Barth et al. 1992; Hobbs 1993; Wurzler et al.

G. Feingold (✉)
NOAA, Earth System Research Laboratory, Boulder, CO, USA

Z. Levin, W.R. Cotton (eds.), *Aerosol Pollution Impact on Precipitation*,

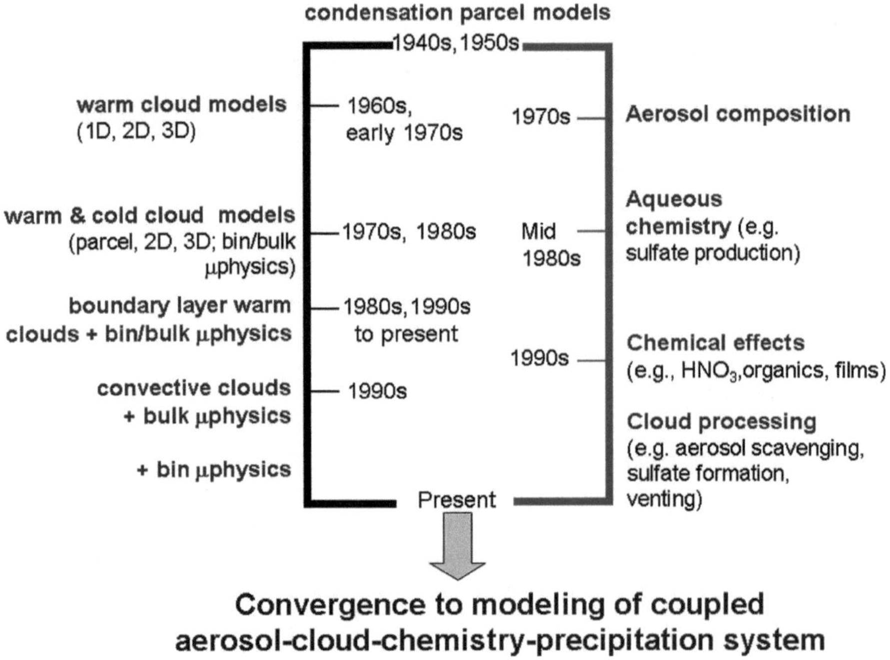

Fig. 7.1 Brief summary of parallel efforts in aerosol-cloud-precipitation studies over the past six decades. The *left branch* has focused on precipitation development, while the *right branch* has been concerned with aerosol composition and aqueous chemistry. Communication between these efforts over the past decade has stressed the importance of considering the aerosol-cloud-chemistry-precipitation system in a coupled manner

2000; Garrett et al. 2002; Tabazadeh et al. 2004). Therefore, not only must the effects of aerosols on clouds and precipitation be considered but also the effect of clouds and precipitation on aerosols.

The challenge to the cloud modeler is to represent the numerous, and sometimes poorly understood cloud processes depicted in Fig. 7.2 in a dynamical framework, and preferably in three dimensions (3-D). The lifetime of an individual cloud cell is on the order of an hour but, when considering precipitating cloud systems, multi-hour simulations need to be performed, with the result that computational costs quickly become prohibitive. The challenges therefore include adequate representation of (i) physics, (ii) chemistry, (iii) numerical methods for solving the equations, (iv) consideration of the most important processes and (v) consideration of temporal and spatial scales.

Below we describe some of the modeling techniques that have been used to represent precipitation, and parallel efforts at improving numerical methods, as well as the representation of the physical processes themselves. We then give examples of key results pertaining to aerosol-cloud-precipitation interactions. Finally, we discuss some of the challenges in quantifying aerosol effects on clouds and precipitation.

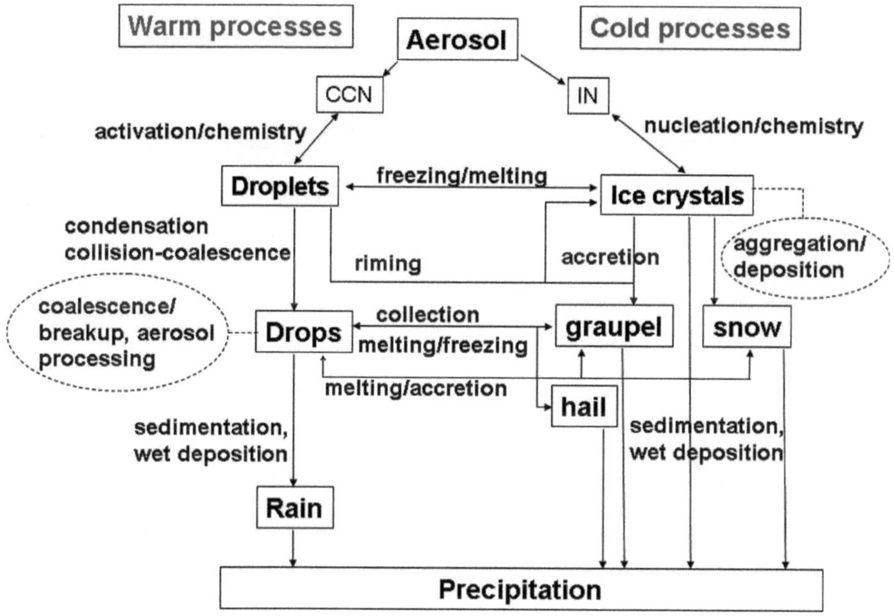

Fig. 7.2 Simplified schematic of *warm* and *cold* microphysical interactions and some of the processes that are represented in models. The multiple pathways are only a subset of the more complete description of processes shown in Fig. 2.14. From Reisin et al. (1996a) with permission of the American Meteorological Society

7.2 Numerical Methods for Modeling Cloud Microphysics

Cloud modeling requires a recognition that the aerosol cloud system comprises coupled components of dynamics, aerosol and cloud microphysics, radiation and even chemistry (Fig. 7.3), and that depending on the system being modeled, neglect of some of the components may have important consequences. The correct approach to modeling the effects of aerosol on precipitation requires an appropriate balance of treatment of these components. Earlier modeling efforts tended to focus on just parts of the system (e.g. the effect of aerosol size and composition on the drop size distribution with very detailed representation of the aerosol/drop size distribution (Mordy 1959), or the formation of precipitation in a 1-D model using simple representations of microphysics, frequently with little representation of aerosol (Silverman and Glass 1973). To this day, cloud models are often used to focus on just one aspect of the system, because the neglect of other components simplifies investigations and serves to elucidate specific processes. However, with the increase in computing power, and the evolution of numerical methods, the community has slowly moved towards representing both microphysics and dynamics with varying degrees of sophistication.

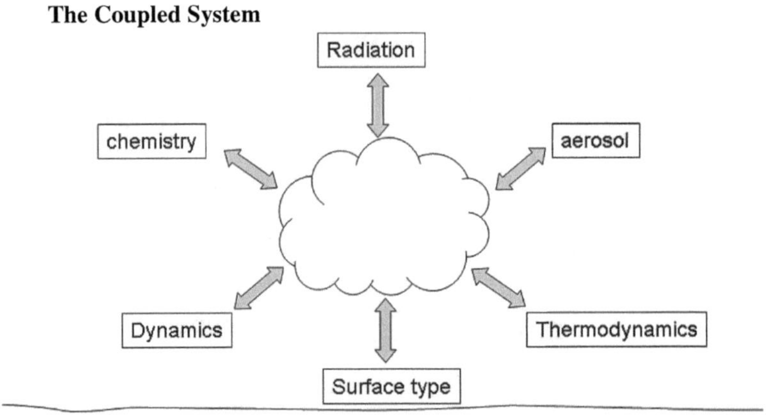

Fig. 7.3 Schematic indicating the coupled nature of the microphysics involving clouds and precipitation. Although great gains in knowledge have been acquired through consideration of a limited number of components in this figure, an appropriate balance between these components must be considered when modeling aerosol-cloud-precipitation interactions

In approaching numerical cloud modeling, the importance of numerical techniques that minimize spurious particle growth must be considered hand-in-hand with the representation of the physics inherent to cloud microphysical and dynamical processes. For example, the kernel function for collision and coalescence between different types of hydrometeors is based on both theory and laboratory experiments, in which the outcome of the interactions between pairs of hydrometeors of different sizes is calculated or measured. These are then parameterized into more general functions that can be embedded in the microphysical equations. Accurate parameterization of these processes is as important as consideration of numerical methods in determining our confidence in representing microphysical processes in cloud models. In fact it can be argued that the poor knowledge of collection kernels, particularly regarding ice-ice interactions, may limit the gains that can be realized by very detailed microphysical schemes.

7.2.1 Bulk Microphysics

When considering cloud microphysical processes, earlier models, and many current models, use a "bulk" representation of microphysics that represents a size distribution of hydrometeors by one or more of its moments (e.g. mass mixing ratio, number mixing ratio, surface area, radar reflectivity). In one approach drops and ice particles are represented by a simple exponential function (with fixed pre-exponent) and a single-moment of the hydrometeor spectra is predicted (Kessler 1969). In this case self-collection among cloud droplets is parameterized using an autoconversion formulation, and large hydrometeors such as raindrops are assumed to collect smaller drops and ice by continuous accretion. Moreover, all classes of large hydrometeors are assumed to fall with a constant fall speed, usually a water content-weighted fall speed. The autoconversion formulations have been either ad hoc (e.g. Kessler 1969; Manton and Cotton 1977; Cotton et al. 1986) or derived from parcel or box detailed bin-microphysics simulations (e.g. Berry 1967; Berry and Reinhardt 1974; Beheng 1994).

Multi-moment bulk schemes typically prescribe basis functions for the drop size distributions, such as gamma or log-normal distributions (Clark 1976; Clark and Hall 1983; Nickerson et al. 1986; Ferrier 1994; Meyers et al. 1997; Reisner et al. 1998; Milbrandt and Yau 2005a,b; Seifert and Beheng 2006a,b), and then explicitly predict the evolution of those basis functions (in terms of their moments) by vapour deposition/evaporation, stochastic collection, and sedimentation. The advantage of these multi-moment schemes is that they predict number concentration, mass mixing ratio (and sometimes higher order moments) and therefore are able to derive the broad features of the drop size distribution. In so doing they improve the representation of growth processes and precipitation formation. Multi-moment bulk methods have also been applied to representation of aerosol growth processes (e.g. McGraw 1997).

The two-moment scheme by Seifert and Beheng (2006a,b) predicts the evolution of mass as well as number densities of the five hydrometeor types, cloud droplets, raindrops, cloud ice, snow and graupel. It includes novel parameterizations of autoconversion, accretion and self-collection of water drops that have been derived by Seifert and Beheng (2001) directly from the stochastic collection equation. In contrast to most other schemes, the new autoconversion parameterization further considers aging of the cloud droplet size distribution with time by relying on a dynamic similarity theory. The parameters determining the shape of the droplet size distribution in this parameterization were chosen using droplet size distributions simulated within the dynamical model with bin microphysics model developed at the Hebrew University of Jerusalem (HUCM; Khain et al. 2004; Seifert et al. 2006). This calibration significantly increased the sensitivity of results to aerosol and allowed the authors to obtain similar results in bulk- and bin-microphysical simulations of maritime and continental clouds.

Another example of a multi-moment scheme implemented in a large scale model is the bin-emulating scheme, which is implemented in RAMS (Regional Atmospheric Modeling System, Pielke et al. 1992 and Cotton et al. 2003). The evolution of this modeling approach can be found in Verlinde et al. (1990), Walko et al. (1995), Meyers et al. (1997), Feingold et al. (1998), and Saleeby and Cotton (2004). Instead of using continuous accretion approximations, which has been common in cloud parameterizations, Feingold et al. (1998) showed that full stochastic collection solutions for self-collection among cloud droplets and for rain (drizzle) drop collection of cloud droplets can be obtained for realistic collection kernels by making use of look-up tables. This approach has been extended to all hydrometeor class interactions by collection, including the growth of graupel and hail by riming. The philosophy of bin representation of collection has been extended to calculations of drop sedimentation (Feingold et al. 1998). Bin sedimentation is simulated by dividing the basis function into discrete bins and then building look-up tables to calculate how much mass and number in a given grid cell fall into each cell beneath a given level in a given time step. Saleeby and Cotton (2004) refined this approach by the addition of a large cloud droplet mode from 40 to 80 μm diameter, and by predicting the number concentration of cloud droplets through explicit activation of CCN and giant-CCN. The large cloud droplet mode provides a better depiction of self-collection of cloud droplets (or autoconversion) and permits simulation of drizzle from fogs and marine stratocumulus clouds. The activation of CCN is parameterized through the use of a Lagrangian parcel model that considers ambient cloud conditions for the activation of cloud droplets from aerosol.

7.2.2 Size-resolved Microphysics

Size-resolved or explicit bin microphysical methods have been developed with the recognition that many cloud growth processes are highly sensitive to drop size. The first such class is the Lagrangian (or moving mass-grid) method, which represents particles at discrete sizes and allows each particle to grow by condensation on a moving mass grid. This approach eliminates numerical diffusion and allows for a smooth transition from aerosol to haze to cloud droplets without artificial distinctions between these classes. The cloud supersaturation is calculated based on source (cooling, which is related to updraft velocity) and sink (condensation) terms, enabling accurate determination of the number of activated droplets. These models typically focus on the initial growth phases from haze to droplet and include detailed representation of aerosol sizes and compositions (Mordy 1959; Fitzgerald 1974; Facchini et al. 1999; Feingold and Kreidenweis 2000; Feingold and Chuang 2002; Lohmann et al. 2004). They frequently also consider the effect of trace gases on activation (Kulmala et al. 1993). Aqueous production of sulphate has also been represented in studies that examine the effects of cloud processing on the aerosol size distribution (Hegg et al. 1991c; Bower and Choularton 1993; Feingold and Kreidenweis 2000). The Lagrangian method is used almost exclusively in kinematic cloud parcel models, where a parcel of air is moved either adiabatically or according to some known trajectory through a cloud. It is not easily adapted to the study of growth by processes such as collision-coalescence, and it is not suitable for general application in Eulerian dynamical models.

Fixed bin, or Eulerian (in size space), microphysical models have been developed to fill this need. Supersaturation calculations also follow from a balance equation for cooling (dynamical tendency) and condensation but accuracy depends on the grid resolution (Clark 1973, 1974). Activation is not represented as accurately because the Eulerian framework does not represent the transition from dry particle to haze to droplet as in the Lagrangian framework. Condensational growth is also subject to numerical diffusion associated with the mass-grid resolution. (This also feeds back to the accuracy of the supersaturation calculation.) The bin framework is most useful for collision-coalescence calculations. Early efforts by Telford (1955), Berry (1967), Berry and Reinhardt (1974), Kovetz and Olund (1969) and Bleck (1970) showed that great care must be taken in representing collision-coalescence to avoid numerical diffusion in the mass-transfer equations and rapid (spurious) acceleration of growth to precipitation-sized particles. More recent bin models, such as those used by Tzivion et al. (1987), Hounslow et al. (1988), and Chen and Lamb (1994), use a multi-moment representation of the cloud microphysics in each individual drop bin; this significantly reduces numerical diffusion and has the added benefit of conserving more than one moment of the size distribution. This has led to development of numerical methods that include a bin representation of aerosol in each individual hydrometeor size-bin (Bott et al. 1990; Chen and Lamb 1994; Kerkweg et al. 2003; Leroy et al. 2006). Such methods are very accurate since they maintain knowledge of the aerosol particle upon which the drop and/or ice particles form (Leroy et al. 2006) but are too computationally intensive to be included in 3-D models. Simpler methods that track dissolved aerosol within each hydrometeor bin are more commonly used (Flossmann et al. 1985; Toon et al. 1988; Respondek et al. 1995; Feingold et al. 1996; Yin et al. 2005). A review of aspects of bin microphysical modeling of both warm and cold cloud processes can be found in Khain et al. (2000).

Another approach to size-resolved microphysical modeling is the hybrid approach (Cooper et al. 1997; Jacobson 1999; Pinsky and Khain 2002) where the advantages of moving grids for condensational growth are combined with fixed grid approaches for collection processes.

While the focus here has been on numerical methods for solving microphysical interactions, it is important to consider numerical issues related to the dynamical aspects of cloud modeling. The advection equation is also susceptible to numerical diffusion (and dispersion). Modeling of the dynamical equations for hydrometeor species requires "positive-definiteness" for mass conservation. Numerical diffusion on the spatial grid has to be considered along with numerical diffusion on the mass grid; again, an appropriate level of balance is important (Clark 1973, 1974).

7.3 Modeling Aerosol-Cloud-Precipitation Interactions

7.3.1 Warm Clouds: Aerosol Effects on Drizzle in Stratocumulus Clouds

Stratocumulus clouds have been the focus of numerous intensive field campaigns, as well as detailed modeling efforts, primarily because of their importance to the Earth's radiation budget rather than because they produce precipitation in significant quantities. Their importance derives largely from their frequency of occurrence and extensive spatial coverage, as well as their high reflective contrast with the underlying surface (particularly over oceans). Over land and ocean, stratocumulus clouds have an annually averaged coverage of ~18% and 34%, respectively (Warren et al. 1986a,b). Stratocumulus clouds frequently produce drizzle (e.g. Paluch and Lenschow 1991; Stevens et al. 2003) and the initiation of drizzle has been the focus of much modeling effort. Modeling of aerosol-cloud interactions in stratocumulus clouds received attention in the early 1990s with the development of 1-D turbulence closure models coupled to bin microphysical models (Ackerman et al. 1995; Bott et al. 1996), and modeling at the large eddy scale in 2-D and 3-D simulations (Kogan et al. 1994; Liu et al. 2000; Feingold et al. 1994, 1996, 1997; Stevens et al. 1996, 1998; Khairoutdinov and Kogan 2000; Feingold and Kreidenweis 2002; Jiang et al. 2002).

Ackerman et al. (1995) used a 1-D turbulence closure model coupled to an aerosol and cloud particle size-resolved model to show that progressive aerosol scavenging by droplet activation, by drop collisions and drizzle depletion of liquid water in a stratocumulus cloud can reduce the cloud liquid water path (LWP) and optical depth to a point where there is insufficient cloud-top radiative cooling to maintain the cloud. Under such conditions the cloud collapses into a fog layer. The authors suggested that coalescence scavenging and drizzle could limit the lifetime of stratocumulus clouds. They contrasted this phenomenon with the more common hypothesis that increases in aerosol concentrations suppress drizzle and help to sustain clouds with higher LWPs (Albrecht 1989). Bott et al. (1996) used a similar 1-D model to examine dynamical-microphysical-radiation feedbacks. Recognizing the importance of better representation of boundary layer dynamics, Feingold et al. (1994) and Kogan et al. (1994) coupled similar bin microphysical models to 2-D and 3-D Eulerian models with grid sizes on the order of 50–100 m. Kogan et al. (1995) demonstrated the importance of microphysical-dynamical coupling, particularly through the supersaturation field. Feingold et al. (1996) showed that in-cloud residence time is an important factor in determining precipitation development in stratocumulus. Stevens et al. (1996) pointed out that air parcels that spend significant time at cloud top, where LWC is highest, are more likely to produce drizzle drops.

Feingold et al. (1997) showed that broadening of drop spectra via collision-coalescence tends to increase cloud susceptibility (defined by Twomey (1991) as the incremental change in cloud albedo for an incremental change in cloud drop concentration at constant cloud water). Khairoutdinov and Kogan (1999) compared their large eddy simulation (LES) model including bin microphysics to measurements from field experiments and showed generally good agreement in the dynamical and microphysical fields. Driven by the large computational expense of resolving drop size distributions in LES, Feingold et al. (1998) and Khairoutdinov and Kogan (2000) studied drizzle formation in stratocumulus using both bin and bulk (2 moment) schemes and showed that bulk schemes can reproduce many of the features of drizzle formation. Liu et al. (2000) studied ship tracks using LES and demonstrated the effects of boundary layer stability on both inclusion of the stack effluent into the clouds as well as maintenance of the ship track. Decoupling of the boundary layer suppresses transport of the effluent into cloud. Because boundary layer stability has a diurnal cycle, which modifies both cloud LWC and the amount of aerosol affecting the cloud, persistence of ship tracks is not easily predicted.

The LES/bin-microphysics simulations performed by Stevens et al. (1998) illustrate the importance of drizzle to the cloudy marine boundary layer. For strongly drizzling cases where the drizzle reaches the surface, evaporation results in cooling of the surface and stabilization of the sub-cloud layer (Paluch and Lenschow 1991). As a result, the cloud structure changes to a cumulus-under-stratus regime, in which lower-based cumulus clouds penetrate into the stratus layer supplying it with water vapour. The cumulus-under-stratus regime has very different optical properties than a pure solid stratocumulus field. However, in Jiang et al's. (2002) weakly drizzling LES model simulations, the response was quite different. The small amount of drizzle that was produced evaporated just below the cloud base, and had a destabilizing effect on the cloud layer. Addition of higher concentrations of aerosol suppressed this drizzle and consequently, the destabilization. This resulted in weaker penetrating cumulus and an overall reduction in the LWC and LWP of the clouds.

Other aspects of the effect of aerosol on cloud liquid water path (LWP) were considered by Ackerman et al. (2004). They also showed that increases in aerosol do not necessarily result in increases in LWP in stratocumulus clouds, as proposed by Albrecht (1989). A primary factor affecting the LWP response to aerosol changes appears to be the profile of humidity above the inversion. Only when the humidity above the inversion was high did increases in aerosol result in an increase in LWP. When dry air overlies the inversion, increases in aerosol tend to decrease LWP because of enhanced entrainment drying. Similar results were obtained by Lu and Seinfeld (2005). In large eddy simulations of marine trade cumulus clouds, Xue and Feingold (2006) showed that although aerosol has a strong effect on precipitation, aerosol effects on cloud LWP and cloud fraction are relatively small compared to the dynamical variability of the clouds at given aerosol concentrations. In fact their simulations showed a decrease in cloud fraction with increasing aerosol, apparently due to stronger evaporation of the smaller cloud droplets, and an evaporation entrainment feedback that reduced cloud size (see also Wang et al. 2003). Note that this result is for small cumulus clouds (order few 100 m) and that the result is contrary to the Albrecht hypothesis and the satellite-measured increase in cloud fraction with increasing aerosol for larger clouds (Kaufman et al. 2005b,c). The small aerosol effects on LWP in warm cumulus have also been modeled by Grabowski (2006) based on very long simulations (120 days) with a two-dimensional cloud model and bulk microphysics. The latter work stressed the importance of these extended simulations over large domains to obtain a statistically representative sample of aerosol-cloud interactions. It was shown that when viewed as a long time series, polluted clouds produced similar amounts of precipitation to clean clouds. This result is influenced by the fact that sea surface temperature was held fixed in the model, which predetermines the surface fluxes and precipitation rates. The author stressed the importance of results for ensembles of clouds versus those for single clouds. We will return to this theme in Sect. 7.3.4.2.

7.3.2 Role of Giant Cloud Condensation Nuclei

An early hypothesis for droplet spectral broadening and the development of precipitation in warm clouds is that the aerosol population includes giant cloud condensation nuclei (GCCN) and/or ultragiant aerosol particles that can act as the embryos for initiating coalescence growth (Chapter 2). Johnson (1979, 1982), using a parcel model representing droplet growth on a population of aerosol particles via condensation and coalescence calculated that the addition of GCCN at concentrations on the order of 1 per liter (i.e. about 1 particle in 10^5 or 10^6; typical measured concentrations) can account for rapid development of precipitation-sized particles, even in colloidally stable, continental clouds. Woodcock et al. (1971) and Takahashi (1976) found that giant salt nuclei do not contribute substantially to warm rain initiation in maritime cumulus clouds because drizzle is active anyway under clean conditions. Feingold et al. (1999) used several models, including parcel models and 3-D large eddy simulations of marine stratocumulus clouds with detailed bin microphysics, to study the effect of GCCN on drizzle formation in stratocumulus. They found that observed GCCN concentrations of 10^{-4} cm^{-3} to 10^{-2} cm^{-3} were sufficient to transform a non-precipitating stratocumulus into a precipitating one when CCN concentrations were in the range 50–250 cm^{-3}. At lower CCN concentrations, addition of GCCN has little impact on drizzle formation, in agreement with earlier work in clean cumulus clouds. Note however, that although the relative importance of GCCN increases with increasing pollution, there is a point of diminishing returns since precipitation amounts tend to decrease steadily with increasing pollution (all else being equal).

Similar results were obtained by Yin et al. (2000a) for warm convective clouds (using a 2-D slab symmetric cloud model). Yin et al. found the intriguing result that a polluted cloud with CCN concentrations of ~1700 cm^{-3} and GCCN of 0.02 cm^{-3} can precipitate more readily than a cleaner cloud (~1000 cm^{-3} CCN) with no GCCN. This modeling result is supported by the observations of Eagan et al. (1974a,b) and Hindman et al. (1977b) on the effects of effluents from paper mills on cloud micro-structures and of Rudich et al. (2002) on the effects of salt on polluted clouds near the Aral Sea. Wurzler et al. (2000) showed that cloud-processed dust particles coated with a soluble material could enhance precipitation. Rosenfeld et al. (2002) suggested that GCCN could be an effective mechanism for controlling precipitation, in agreement with earlier studies. Segal et al. (2004) used a high resolution (2000-bin) Lagrangian parcel model describing diffusional and collisional growth of aerosols and drops to investigate the comparable effects of small and large CCN on the droplet spectrum and raindrop formation. It was shown that when updrafts increase in strength above cloud base, bimodal spectra may form as a result of activation of CCN associated with the increasing supersaturation (see also Erlick et al. 2005). As in prior studies, it was shown that the collision-coalescence process begins to play an important role even at the stage of condensational growth and GCCN play an important role at higher CCN concentrations.

In addition to influencing the warm rain process, GCCN also affect cold clouds by increasing the concentration of larger graupel particles (e.g. Yin et al. 2000a; Teller and Levin 2006). Yin et al. (2000a) showed that by enhancing collision-coalescence in polluted clouds (CCN $\sim$1700 cm^{-3}), the inclusion of GCCN decreased both drop and graupel concentrations; however, graupel mass was substantially increased. Precipitation was initiated earlier, rain intensities were higher, and the total accumulated precipitation as much as 4 times higher. Moderately polluted clouds (CCN $\sim$1000 cm^{-3}) exhibited weaker, but still significant enhancements in rain due to GCCN (increase in the order of 60%).

The importance of GCCN requires consideration of the body of modeling studies of hygroscopic seeding that consider the addition of giant salt particles into clouds as a means of artificially enhancing precipitation (Hindman et al. 1977b; Cooper et al. 1997; Segal et al. 2004). The studies of Cooper et al. (1997), Yin et al. (2000b), and Segal et al. (2004) are of special interest because they elucidate the role not only of the giant seeded particles, but also of the other hygroscopic particles present in flares that have been used for artificial seeding. (See Chapter 8, Sect. 8.4 for a more detailed discussion of these modeling results in the context of deliberate cloud seeding.)

The relative sensitivity of a cloud to GCCN has to be considered in the context of the non-linear response of the collection process to cloud liquid water content, the sensitivity to cloud drop concentration, and the time available for collection. Quantification of precipitation in terms of GCCN alone is therefore not feasible; no study to date has quantified the effect of GCCN in terms of all the pertinent parameters.

7.3.3 Cloud Processing – Cloud Effects on Aerosol

The recognition that aerosols not only affect clouds but that clouds affect aerosols (e.g. Easter and Hobbs 1974; Hegg et al. 1980) provided challenges to the modeling community by requiring that CCN be treated as size-resolved prognostic species and be tracked as soluble material inside drops (Flossmann et al. 1985; Toon et al. 1988; Bott et al. 1990; Chen and Lamb 1994; Feingold et al. 1996). Modeling of aqueous sulphate production, and the resultant modification of the aerosol size distribution, has most often been simulated in Lagrangian parcel models (e.g. Easter and Hobbs 1974; Hegg and Larson 1990; Roelofs 1992; Roelofs 1993; Bower and Choularton 1993; Wurzler et al. 2000).

These models have reproduced the creation of a bimodal aerosol size distribution as observed by Hoppel et al. (1990). The implications of this processing range from increases in light scattering (Hegg et al. 1996) to effects on CCN and drop number concentration (Bower and Choularton 1993; Feingold and Kreidenweis 2000; Yin et al. 2002). Eulerian models have also been used to study cloud processing. Bott (2000) used a 1-D model, with a very detailed 2-D representation of the joint aerosol-drop distribution function, to study the modification of aerosol size distributions by aqueous chemistry for a range of aerosol and gas phase conditions. Feingold et al. (1996) incorporated bin microphysics for aerosols and drops in 2-D simulations of stratocumulus and studied the rate of aerosol scavenging via drop collection. Drop collection reduces the total drop number concentration, so that on subsequent evaporation the available solute mass is redistributed among fewer, and therefore larger, particles. These authors compared their results to aqueous chemistry processing and suggested that at high LWC, coalescence processing of aerosol would likely dominate aqueous chemistry processing. Similar inferences were made by Hatzianastassiou et al. (1998) and Wurzler et al. (2000). Feingold and Kreidenweis (2002) used a 3-D LES model with size-resolved aerosol and cloud drops, as well as aqueous sulphate production. They showed that the addition of sulphate mass could increase or decrease drizzle formation in stratocumulus depending on the size distribution of the background aerosol. When mass addition occurred at sizes significantly larger than the background CCN, drizzle was increased; conversely, when mass addition occurred close to the mode of the background CCN, drizzle was reduced.

In precipitating clouds, the most direct form of cloud processing of aerosol is through wet deposition to the Earth's surface. It has been shown that 50–90% of aerosol number concentration is removed by in-cloud nucleation (activation) processes (Flossmann et al. 1985; Flossmann and Pruppacher 1988; Taylor 1989). Brownian motion is effective at removing the smaller and more numerous particles. For example, Jacobson (2003) found that 42% of the number concentration of aerosol was scavenged through this process, but with little effect on aerosol mass. The reduction in aerosol mass concentration by in-cloud activation is >99%.

Below cloud, impaction scavenging (washout) has a negligible effect on particle concentration since it affects the larger particles that exist in very low concentrations. The reduction in mass is also much smaller than in-cloud losses (5–10%) (Flossmann and Pruppacher 1988; Taylor 1989). In spite of this, both of these studies suggest similar contributions of in-cloud and below-cloud scavenging to the mass of aerosol deposited to the surface by rain. This is a result of the low precipitation efficiencies (~10–15%) for these simulated clouds and the fact that much of the in-cloud scavenged aerosol mass does not reach the surface as rain. Note that impaction scavenging removes both hydrophobic and hydrophilic aerosols, while nucleation scavenging removes only hydrophilic aerosols.

These results have been confirmed by later studies such as those by Flossmann et al. (1998) and Yin et al. (2005). The latter study coupled a bin microphysical model to an aqueous sulphate chemistry model. Their simulations showed that the addition of sulphate mass had an insignificant effect on cloud evolution, even for polluted conditions (10 ppbv SO_2 and 1 ppbv H_2O_2).

Convective clouds also act to transport boundary layer aerosol to the middle and upper troposphere (Cotton et al. 1995; Flossmann 1998). This can have important implications for precipitation, as well as for upper tropospheric heterogeneous reactions and the life cycle of aerosols (Clarke et al. 1999). Additionally, Teller and Levin (2006) showed that polluted clouds containing high concentrations of CCN (with about 1000 droplets cm^{-3} as compared 100 cm^{-3}) not only lead to reduced precipitation but also leave behind as much as three times more water vapour from the evaporation of cloud drops and ice crystals at higher altitudes after precipitation stops. This could affect radiation transfer as well as various chemical reactions. See Wang (2003) for discussion on the role of convective storms in redistributing water vapour.

Cloud processing of aerosol also has important consequences on regional scales. Aerosol is processed through cloud during long-range transport, with the extent of this processing dependent on cloud contact time, cloud liquid water content, and the availability of SO_2 and oxidants.

7.3.4 Mixed-phase Clouds: Effects of Aerosols on Precipitation

7.3.4.1 Single Clouds

A number of early studies have shown that the speed of glaciation of a cloud is highly dependent on the presence of large supercooled raindrops or on the concentrations of CCN (Cotton 1972a,b; Koenig and Murray 1976; Scott and Hobbs 1977). When CCN concentrations are high, warm-cloud collection processes are suppressed and supercooled raindrops are few in number, small ice crystals must first grow to several 100 micrometers in diameter before they begin collecting cloud droplets. Then the riming process proceeds slowly until the ice particles have grown to millimeter-size. Furthermore, since riming is suppressed in clouds forming in air masses with high CCN concentrations, secondary ice particle production by the riming-splintering process (Hallett and Mossop 1974; Mossop and Hallett 1974) is suppressed as well.

Reisin et al. (1996a,b) studied the effects of changes in CCN on precipitation in an axisymmetric cloud model with a comprehensive bin microphysical representation of water and ice hydrometeors. They showed that a simulation with relatively low CCN (100 cm^{-3} at 1% supersaturation) produced precipitation efficiently through the freezing of large droplets interacting with ice crystals. With increasing CCN concentrations and decreasing drop sizes, graupel growth was suppressed and the precipitation efficiency decreased (for CCN = 900 cm^{-3}, precipitation was reduced by 85%). This study also showed that for the sounding considered, increases in ice nucleus (IN) concentrations tended to produce higher precipitation amounts, but the changes were very small for a three order of magnitude change in IN in a moderately polluted cloud (CCN = 600 cm^{-3} at 1% supersaturation). They found a greater sensitivity to IN concentrations when CCN concentrations were high (1100 cm^{-3}).

Khain et al. (1999) used a 2-D (slab-symmetric) model with bin microphysics to simulate aerosol effects on precipitation in an Eastern Mediterranean coastal setting. They too found that warm rain was significantly reduced by an increase in the aerosol concentrations. However, they noted that in the case of polluted clouds, the lofting and downwind transport of smaller precipitation particles could produce more rain a few tens of kilometers from the convective region than was produced by the cleaner clouds with their efficient warm rain process. The result points to the importance of both dynamics and microphysics through particle "size sorting" and spatial redistribution of ice and water phases in the spatial domain.

In a recent study, Teller and Levin (2006) using a 2-D slab symmetric model with detailed treatment of cloud microphysics showed that increases in CCN due to pollution in Mediterranean type winter convective clouds decrease rain amounts on the ground. When the ingested air into the cloud also contained GCCN in concentrations similar to those found in the atmosphere (a few per liter) the amount of rain increased. They showed that increasing CCN concentration from 600 to 1000 cm^{-3} resulted in a decrease in rainfall by a factor of 3, as compared to a decrease by a factor of only 1.7 when GCCN were present (Fig. 7.4). Furthermore, they showed that increasing ice concentrations above those produced through the use of Meyers et al. (1992) parameterization resulted in a decrease in rainfall. This decrease was more pronounced in the clean than the polluted clouds.

Khain et al. (2001) simulated the effect of varying the concentration of CCN on ice-crystal concentrations in deep convective clouds. Clean clouds (CCN~100 cm^{-3}) produced large drops and rapid freezing at ~–15°C, resulting in significant depletion of cloud water. In a polluted cloud, (CCN = 1260 cm^{-3}), freezing was delayed until much lower temperatures (~–35°C) because the small drop sizes rendered mechanisms that convert water droplets to ice inefficient, until the level of homogeneous freezing was reached. These simulations compare well with in situ measurements exhibiting high LWC at temperatures as low as –38°C under polluted conditions (Rosenfeld and Woodley 2000). Precipitation amounts were about half of those in the clean case. Khain et al. (2001) suggested that the general assumption that droplet-graupel collision efficiencies are equal to droplet-droplet collision efficiencies strongly overestimates the rate of collection of small droplets by small graupel; a sensitivity test showed that cloud LWC is depleted at much higher temperatures if this assumption is made. This result points to the importance of quantifying collision efficiencies and the importance of commensurate efforts in both modeling methodology and laboratory studies.

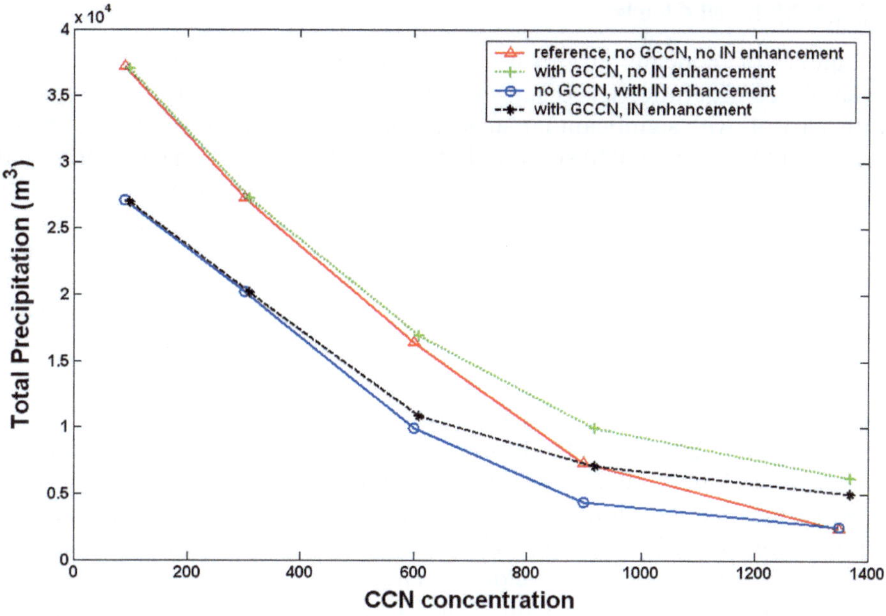

Fig. 7.4 Dependence of total precipitation on CCN concentration based on 2-D single cloud simulations including bin microphysical representation of water and ice. The inclusion of GCCN increases precipitation under polluted conditions but the trend of decreasing precipitation with increasing CCN is robust (Teller and Levin 2006)

The importance of treating aerosol and cloud microphysics as a coupled system has been highlighted in recent papers by Fridlind et al. (2004), Yin et al. (2005), and Seifert et al. (2006). Fridlind et al. (2004) combined measurements from Florida cumulonimbus with 3-D modeling (with bin-microphysics for aerosol and hydrometeors) to infer that most anvil ice-crystals form on mid-tropospheric aerosol and not boundary layer aerosol. Lateral entrainment incorporates these particles into the updrafts where they are exposed to supersaturations and activated. Similar conclusions were drawn by Yin et al. (2005) with a 2-D axisymmetrical model employing bin microphysics for aerosol and cloud microphysics, as well as aqueous sulphate chemistry. They showed that aerosol regeneration, which returns aerosol to the mid-troposphere upon evaporation of hydrometeors, has a significant effect on microphysical pathways; when this process was neglected, graupel concentrations were nearly doubled and surface precipitation increased by $\sim$50%. Another example of the importance of tracking aerosol budgets can be found in Seifert et al. (2006). Squall line simulations that excluded budgets on the aerosol evolved quite differently from those that did not. When aerosol depletion was allowed, the simulation resulted in cleaner conditions and suppression of secondary convection; when aerosol depletion was not allowed, secondary convection did occur.

7.3.4.2 Multicell Clouds

A number of modeling groups using very different microphysics models and dynamic configurations are suggesting complicated dynamical interactions with aerosols with significant impacts on simulated precipitation. For example, Seifert and Beheng (2006b) showed that the effect of changes in CCN on mixed phase convective clouds is quite dependent on cloud type. They found that for small convective storms, an increase in CCN decreases precipitation and the maximum updraft velocities. For multicellular storms, the increase in CCN has the opposite effect – namely, promoting secondary convection, and increasing maximum updrafts and total precipitation. Supercell storms were the least sensitive to CCN. Their study also showed that the most important pathway for feedbacks from microphysics to dynamics is via the release of latent heat of freezing. Likewise, Lynn et al. (2005a) performed mesoscale simulations with bin-resolving microphysics for Florida deep convective clouds and found that for shallow clouds, high concentrations of CCN delayed the formation of precipitation in initial clouds relative to clean clouds. Thus clouds in a maritime airmass precipitated sooner than clouds forming in a continental airmass but this led to weaker secondary clouds than those formed in the continental airmass. As a consequence clouds that formed in the continental airmass exhibited stronger updrafts, higher cloud tops, greater peak rainwater amounts and heavier precipitation rates.

When the environment can support deeper convective clouds, cloud modeling studies with bin-resolving microphysics by Khain et al. (2004) found that pollution-induced smaller cloud droplets reduce the production of drizzle drops. When these droplets froze, the associated latent heat release resulted in more vigorous convection. In contrast, in a clean cloud, drizzle depleted the cloud liquid water so that less latent heat was released when the cloud glaciated, resulting in less vigorous convection. Thus, they found that a squall line developed under continental aerosol conditions and produced more precipitation after 2 hours whereas this did not happen with clean aerosol conditions. Zhang et al. (2005b) came to similar conclusions in their model simulations for different three-week periods over the ARM site in Oklahoma.

Similarly, mesoscale simulations of deep convection over Florida Lynn et al. (2005b) showed that higher CCN concentrations delayed the onset of precipitation but led to more intense convective storms with higher peak precipitation rates. However, the accumulated precipitation was largest for the cleaner atmosphere.

In mesoscale simulations of entrainment of Saharan dust into Florida thunderstorms with bin-emulating bulk microphysics, Van den Heever et al. (2006) found that dust not only impacts cloud microphysical processes but also the dynamical characteristics of convective storms. Dust may serve as CCN, GCCN, and IN. The effect of dust on cloud microstructure and storm dynamics in turn alters the accumulated surface precipitation and the radiative properties of anvils. These results suggest that the dynamic structure of the storms is influenced by varying dust concentrations. In particular, the updrafts are consistently stronger and more numerous when Saharan dust is present compared with a clean airmass. Like Seifert and Beheng (2006b) they found that dust results in enhanced glaciation of convective clouds, which then leads to dynamical invigoration of the clouds, larger amounts of processed water, and thereby enhanced rainfall at the ground. However, Van den Heever et al. simulations suggested that rainfall is enhanced by dust ingestion during the first 2 hours of the formation of deep convective cells, but it is reduced on the ground later in the day. Thus the clean aerosol simulations produced the largest surface rain volume at the end of the day. This is a result of complex dynamical responses of clouds to aerosol changes associated with subcloud evaporation of rain in which low-level cold pools influence storm propagation and to scavenging of dust, so that few GCCN and IN remained late in the day.

Another study that illustrates the complexity of aerosol interactions with convective storms is van den Heever and Cotton's (2007) examination of the impacts of urban-enhanced aerosol concentrations on convective storm development and precipitation over and downwind of St. Louis, MO. In the van den Heever and Cotton (2007) study RAMS was set up as a cloud-resolving, mesoscale model with both sophisticated land-use processes and aerosol microphysics using a bin-model emulation approach. The results indicate that urban land-use forced convergence downwind of the city, rather than the presence of greater aerosol concentrations, is the dominant control on the locations and amounts of precipitation in the vicinity of an urban complex. Once convection is initiated, urban-enhanced aerosols can exert a significant effect on the dynamics, microphysics and precipitation produced by these storms. The model results indicate, however, that the response to urban-enhanced aerosol depends on the background concentrations of aerosols; a weaker response occurs with increasing background aerosol concentrations.

It was found that when aerosol concentrations were enhanced that cloud water and rain formed more rapidly than in the clean control simulation in which only observed rural aerosol concentrations were utilized. The updrafts were also stronger initially, and the downdrafts developed more quickly. The larger amounts of supercooled liquid water available, together with the stronger updrafts, led to the generation of greater ice mixing ratios earlier in the storm development. Greater amounts of surface precipitation were also produced in this case during the first hour and a quarter to hour and a half of convective storm formation. However, the greater and more rapid production of surface precipitation generates stronger downdrafts and more intense cold pools earlier in the storm lifecycle than in the clean control simulation. This is detrimental to the updraft development and strength, the evidence of which is the earlier demise of the storm closest to the urban region following storm splitting.

In the clean control simulation, the updrafts develop later in association with the delayed hydrometeor development, but they are eventually stronger than those in the simulation in which aerosol concentrations are enhanced. The storms last longer following storm splitting, and new storm development occurs downwind of the city later on in the simulation. This results in increased amounts of accumulated surface precipitation during this time. The variations in storm dynamics in response to variations in aerosol concentrations, result in the greatest accumulated surface precipitation when aerosols are increased early in the afternoon. However as the simulation progresses, this trend reverses, and later in the afternoon, the largest accumulated precipitation occurred in the clean control case (Fig. 7.5). The extremely complex, non-linear relationships between the microphysics and dynamics therefore make it difficult to make absolute statements regarding the impacts of urban-enhanced aerosol on downwind convection and precipitation.

It is worth pointing out at this stage that there is very little modeling work on aerosol effects on hail, with the exception of Danielsen et al. (1977) 1D parcel model simulations. He concluded that hail formation was sensitive to input CCN and IN concentrations. (See Chapter 2 for discussion on hail.) It is possible that if pollution invigorates storm updraft velocities, as discussed above, then the size of hailstones will likely be increased, since hail size correlates with the strength of storm updraft velocities.

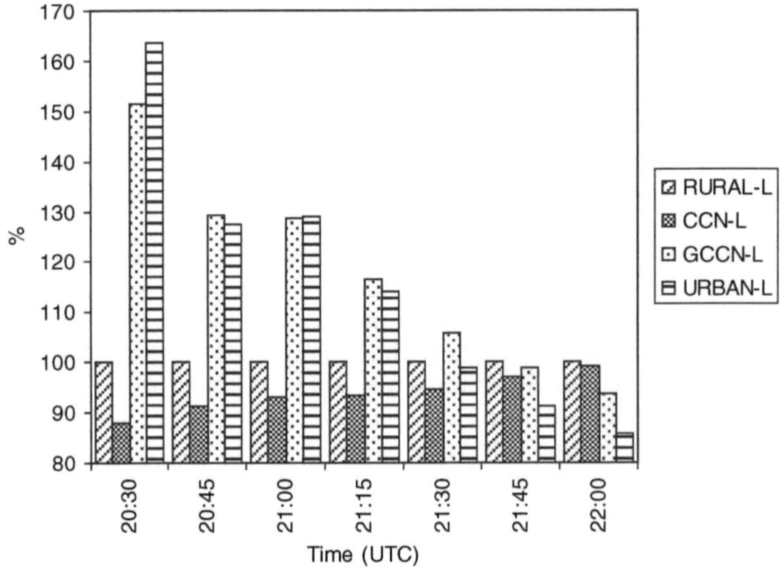

Fig. 7.5 Time series of the accumulated volumetric precipitation in the downwind region expressed as a percentage of the RURAL-Clean background simulation. *Rural-L* represents the simulation with clean background aerosols; *CCN-L* represents the affects of increasing CCN concentrations only over a clean background concentration; *GCCN-L* represents the affects of pollution serving as giant CCN only relative to a clean background; *Urban-L* represents the affects of pollution acting to enhance CCN and GCCN concentrations relative to a clean background aerosol. From van den Heever and Cotton (2007) with permission of the American Meteorological Society

In summary, all the above reviewed studies on aerosol impacts on convective storm precipitation illustrate that the ultimate determining factor controlling whether pollution increases or decreases precipitation is the dynamic response of the storm to enhanced or delayed precipitation formation. Moreover, the particular responses depend on the unique properties of the storm environment (i.e. static stability, wind shear, regional convergence) and many of those complex responses occur through the formation of secondary cells and cold-pool dynamics, often many hours following initial convective cell formation.

7.3.5 Orographic Clouds

The influence of aerosol pollution on orographic precipitation has been investigated in three dimensions with bin-emulating microphysics and in two-dimensions with full bin microphysics.

Using the three dimensional version of RAMS with bin-emulating microphysics, Saleeby and Cotton (2005) examined the influence of varying concentrations of CCN and GCCN on simulated wintertime orographic clouds and precipitation over the Park Range of Colorado. They found that higher CCN concentrations lead to the formation of smaller, more numerous droplets and reduced riming consistent with the observations by Borys et al. (2000, 2003) as discussed in Chapter 6. Total precipitation on the ground was therefore reduced. Moreover, with higher CCN concentrations the smaller droplets evaporate more readily when ice crystals grow at the expense of cloud droplets (the Bergeron-Findeisen processs) thus depleting cloud liquid water contents and further decreasing riming growth. (See also Teller and Levin 2006.) Higher concentrations of GCCN resulted in enhanced surface precipitation when CCN concentrations were high. When CCN concentrations were low, however, higher GCCN concentrations caused greater vapour competition between ice particles and droplet activation and vapour deposition growth, which led to a reduction in surface precipitation. Those simulations were performed using the Meyers et al. (1992) formula for IN activation. Observations of IN concentrations over the Park Range using a continuous flow diffusion chamber by DeMott et al. (2003b) suggest that the Meyers formula overestimates IN concentrations. Using the lower IN concentrations produced very small changes in cloud mixing ratios, but lowered ice water contents, and reduced surface precipitation. Recent observations in the Arctic basin in the fall by Prenni et al. (2007) also suggest lower IN concentrations than given by the Meyers formula. They found that when these IN observations are inserted in a mesoscale model the predictions of cloud structure are consistent with observations.

Rasmussen et al. (2002b) investigated the formation of freezing drizzle in the mesoscale model MM5 coupled to a bin microphysics model fashioned after Reisin et al. (1996a). They showed that the formation of freezing drizzle is significantly enhanced in conditions of low aerosol concentrations, high LWC (or cloud depth), long in-cloud residence times, and a dearth of ice crystals ($<0.08\ L^{-1}$). The latter condition was supported by both model results and observations.

Recently, Lynn et al. (2007) presented results designed to interpret the study of Givati and Rosenfeld (2004) (Chapter 6). Using the Weather and Research Forecast model (WRF) coupled to their bin microphysics model they presented a series of two-dimensional simulations of a Sierra mountain (California) case study, in which they varied CCN input, horizontal wind velocity, and relative humidity to examine the effect of these factors on the nature and distribution of condensate and precipitation relative to the mountain slope. In the control case, a five-fold increase in CCN (at 1% supersaturation) resulted in greater amounts of suspended ice condensate, and significant downwind displacement of this condensate, to the lee of the mountain. The associated decrease in surface precipitation was 27% (consistent with the analysis of precipitation trends in the Sierra Mountains by Givati and Rosenfeld 2004), although there is no observational evidence to support the assumed five-fold increase in CCN. Interestingly, the downwind shift in the surface precipitation associated with the highest topographical feature was small ($\sim$10 km). Sensitivity studies showed that under high relative humidity the surface precipitation amounts were a factor of about 10 higher than for the control (drier) conditions, and the differences in precipitation between clean and polluted orographic clouds were negligible. A decrease in horizontal wind speed to 0.75 of that in the control case resulted in very significant reduction in precipitation and a stronger difference between clean and polluted conditions. This study places the aerosol effects on precipitation in the context of some important dynamical controlling factors such as horizontal wind speed and relative humidity. These factors had a stronger effect on precipitation than the (very large) five-fold increase in CCN.

7.3.6 Arctic Stratus Clouds

A recent modeling study suggests that Arctic stratus clouds may also be susceptible to aerosol pollution. This should not be too surprising as they have many properties similar to orographic clouds, especially low liquid water contents and long lifetimes. Cloud-resolving simulations of Arctic boundary layer clouds using RAMS (Cotton et al. 2003; Saleeby and Cotton 2004) were carried out for a particular day (Carrió et al. 2005a) and for an entire spring season (Carrió et al. 2005b). The model was initialized either with clean sub-cloud aerosol concentrations throughout the boundary layer or with observed moderately polluted aerosol concentrations above the inversion and clean below. During the spring season simulations were performed with the model coupled to a sea-ice model. The multi-month simulations were performed using 2–3 daily soundings nudged into the cloud-resolving model to represent daily variations in the synoptic atmosphere. Mixed-phase clouds prevailed during the first 2 months of simulation, while predominately liquid clouds were simulated during the last month. The effects of IN entrainment in the presence of mixed-phase clouds decreased LWC, while ice water paths increased. Even though IN concentrations above the inversion were much lower than estimates from mid-latitudes using the Meyers et al. (1992) formula, the clouds were essentially over-seeded (e.g. ice crystal concentrations were so high that the crystals were very small). As a result crystal fall speeds were reduced, as were precipitation rates. This resulted in longer residence times of the ice particles and increased total condensate paths. This produced enhanced downward longwave radiation. Enhanced albedo associated with enhanced CCN concentrations (hence droplet concentrations) also occurred, but net surface radiation was still higher. As a consequence, sea-ice melting rates were greater. Changes in surface fluxes of the opposite sign were simulated for short time periods when predominately liquid clouds were present, and during the last month. The effect on melting rates associated with the presence of enhanced CCN concentrations above the boundary layer is opposite although less important than that of IN entrainment. Overall these model simulations suggest that the entrainment of a polluted layer of air overriding an inversion enhances sea-ice melting rates. Melting rates were approximately 4% higher when the air above the boundary layer was polluted than when the entire layer was composed of clean sub-cloud air.

7.3.7 Semi-direct Aerosol Effects

Heating of the air by absorbing aerosols can result in stabilization of a moist moderately stable layer and weaken convection and precipitation. This could have important climatic implications through the hydrological cycle. In regions where there is insufficient moisture or instability to support deep convection, aerosol impacts should be less. This effect of aerosols has been termed the semi-direct effect (Grassl 1975; Hansen et al. 1997). The reduction in cloud cover associated with this effect can alter the surface energy budget significantly. If the aerosol contains a large fraction of soot, such as the south Asian haze, then warming of the aerosol layer can desiccate stratocumulus cloud layers and alter the properties of the trade-wind cumulus layer (Ackerman et al. 2000b). The influence of black carbon dominates via its absorption of solar radiation within the atmosphere, which leads to lower surface temperatures (Ramanathan et al. 2001; Lohmann and Feichter 2001), and reduced outgoing fluxes. General circulation model simulations by Menon et al. (2002b) suggest that black carbon emissions over China may be producing changes in the general circulation, which contribute to the observed increases in summer flooding in south China and drought in north China. Thus, in spite of the fact that anthropogenic aerosol emissions are regionally concentrated, their potential for global impacts is great. Wang (2004) studied the radiative effects of black carbon in a coupled aerosol-climate model and also showed that black carbon modifies precipitation patterns. Interestingly, the model shows an increase in low clouds and a decrease in mid-level clouds as a result of black carbon.

Johnson et al. (2004) conducted large eddy simulations (with bulk microphysics) of stratocumulus clouds and imposed a range of heating rate profiles. They found that the location of the absorbing aerosol was important in determining cloud response. When the absorbing aerosol was located within the well mixed boundary layer, cloud LWP and cloud fraction were reduced. However, when the absorbing aerosol resided above the cloud layer, LWP was increased due to a strengthening of the inversion and reduction in entrainment of dry air. Using the same model, Johnson et al. (2005) also considered the semi-direct effect for trade cumulus clouds with cloud fractions that were either high (0.1–0.5) or low (0.05–0.20). Their results indicated significant semi-direct effects for the higher cloud fraction but rather insignificant effects for the low cloud fraction. Note however, that their absorbing aerosol was considered to be a separate class of aerosol from the CCN (external mixture), and fixed in space and time.

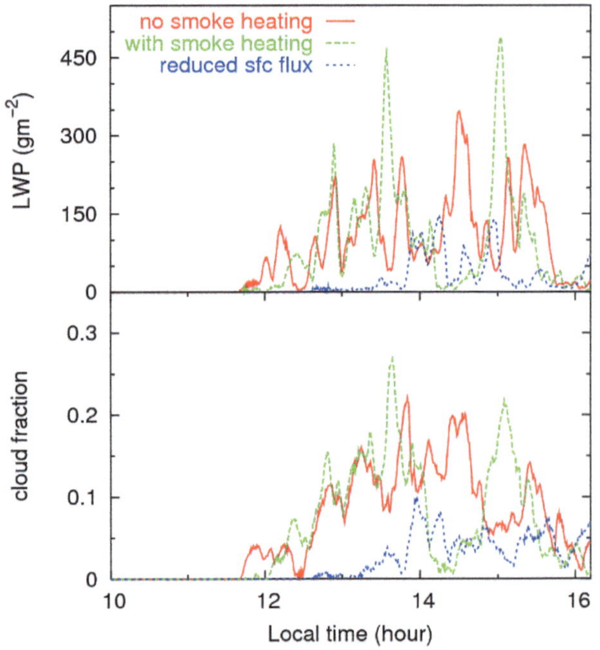

Fig. 7.6 Simulations of smoke aerosol effects on LWP and cloud fraction for warm cumulus clouds, where the smoke is mostly confined to the boundary layer. The control simulation ignores the absorption and heating due to the smoke (*no smoke heating*). The first sensitivity test includes absorption and heating and shows that LWP and cloud fraction are modified but not significantly reduced. The second sensitivity test (*reduced surface flux*) tests the importance of the reduction of the surface latent and sensible heat fluxes due to the presence of the absorbing aerosol and shows that this is the primary reason for the reduction in LWP and cloud fraction. From Feingold et al. (2005) with permission of the American Geophysical Union

In response to Koren et al. (2004) observations that smoke in biomass burning regions significantly reduces cloud fraction, Feingold et al. (2005) modeled the semi-direct effect using a LES model with bin microphysics for aerosol and cloud drops. Predictive equations included aerosol advection and therefore heating rates evolved with the boundary layer and cloud fields. (Recall, previous work had imposed fixed heating profiles.) Through a series of numerical experiments, they established that in addition to the importance of the location of the smoke, the reduction in net surface radiation, and therefore land-surface latent and sensible heat fluxes associated with the smoke was likely the most important factor in suppressing cloud formation (see Fig. 7.6). This reduction in surface fluxes was also a robust feature in Wang (2004). Note that the fact that the presence of aerosol, and particularly absorbing aerosol, modifies the sensible and latent heat fluxes is of great importance, since it points to possible non-monotonic responses of LWP to aerosol. As aerosol increases, it may act to increase LWP and suppress precipitation, but with increasing aerosol loading a point may be reached where the associated reduction in surface fluxes will act to suppress convection and therefore reduce LWP (Jiang and Feingold 2006). This result can be broadened to stress the importance of land-surface characteristics and land management for cloud development. For example, as large tracts of forest are converted to pasture in the Amazon, clouds and precipitation are expected to be modified (Fisch et al. 2004).

7.4 Aerosol Effects on Precipitation and the Hydrological Cycle from a GCM Perspective

General circulation models (GCMs) represent not only the primary tool for simulation of global climate change, but they also play an important role in evaluating the regional effects of pollution on modifying precipitation distribution and amounts. By integrating atmospheric, radiative, oceanic, and land-surface processes on the global scale they provide an indication of expected changes in the coupled system, including possible consequences of coupled greenhouse gas warming and aerosol on the climate system.

GCMs have been used to examine the influence on global climate of widespread sources of CCN. First it must be recognized that GCMs have grid spacings of 150 km to 250 km and that clouds often cover a small fraction of the grid-cell area of a GCM, and the average vertical velocities in a grid cell are very small ($\sim$0.01 m s^{-1}) whereas actual cloud-scale vertical velocities are $\sim$1 m s^{-1} or greater. The poor representation of convection is likely a major source of the disparity in model-predicted LWP amongst GCMs (Fig. 7.7). Space-based measurements of LWP are also difficult, although uncertainties in observations are smaller than the inter-model differences in Fig. 7.7; nevertheless, models are not well constrained by observations. Model predictions of total precipitation show less variability, reflecting the fact that models are heavily

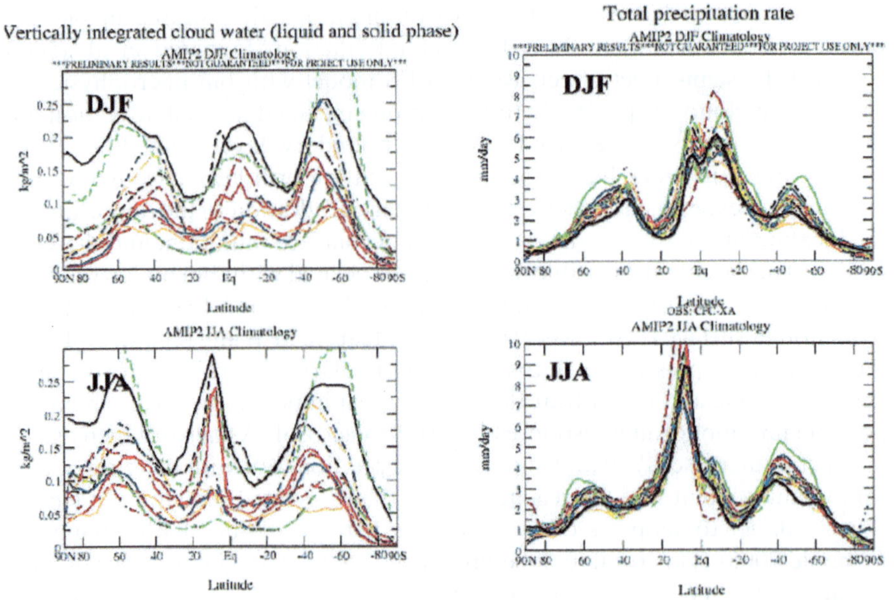

Fig. 7.7 GCM calculations of (*left panel*) LWP (*liquid* and *solid*) and (*right panel*) total precipitation rate from a variety of model simulations as part of the Atmospheric Model Intercomparison Project (AMIP). From Stephens et al. (2002) with permission of the American Meteorological Society

tuned to observations (Stephens et al. 2002) and because the global annual mean amount of precipitation is constrained by the global mean evaporation.

The problem of prediction of condensate is not limited to GCMs; much of the ensuing discussion applies to mesoscale and regional scale models as well. For example, comparisons between total liquid water content (ice plus water) generated by three different cloud schemes in a forecast model show very large variability amongst themselves, and as compared to in situ aircraft data (Guan et al. 2002). Similar discrepancies in condensate between single column models (SCMs) were obtained by Menon et al. (2003) in their simulations of clouds observed during the ACE-2 campaign.

Cloud-scale velocities generate the peak supersaturations in clouds, which together with the number of CCN available determine the number of activated drops. It is therefore necessary to estimate cloud-scale vertical velocities to predict cloud drop concentrations in all models that do not resolve these cloud scales. Some modelers have assumed an empirical relationship between predicted sulphate mass concentrations and droplet concentrations (Martin et al. 1994; Boucher and Lohmann 1995; Kiehl et al. 2000), which is equivalent to assuming there is only a single-value of cloud updraft velocity for all clouds in the model. Others have estimated vertical velocity based on predicted turbulent kinetic energy from boundary layer models (Lohmann et al. 1999). This is a step in the right direction, but it does not take into account the fact that cloudy updrafts are at the tail of the probability density function (PDF) of vertical velocity. Ghan et al. (1997) and Chuang et al. (1997) assumed a normal distribution of vertical velocity with a mean value given by the GCM grid point mean. They then determined the velocity-weighted mean droplet concentration that takes into account the tails of their assumed PDF of vertical velocity. However, observed PDF's of vertical velocity in clouds in the boundary layer are multimodal and are better represented by double-Gaussian PDF's (Larson et al. 2001) with a mean that is a function of the RMS vertical velocity rather than a GCM grid point mean (Peng et al. 2005). Moreover, the complications of precipitation or drizzle processes on cloud lifetime, cloud water contents, and cloud radiative properties discussed above cannot be simulated well in GCM cloud parameterization schemes. For example, precipitation processes are nonlinear functions of total condensate water contents. As a result, the mean LWC from a GCM model grid-box is essentially meaningless for the representation of precipitation production (Stevens et al. 1998; Pincus and Klein 2000). A PDF approach to subgrid modeling may be the optimum approach to resolving these deficiencies. PDFs of subgrid quantities, such as vertical velocity and LWP, are determined from prescribed basis functions in which various moments of the basis functions are predicted in the models (Pincus and Klein 2000; Golaz et al. 2002a,b; Larson et al. 2005).

Another option, though considerably more computationally expensive, is to use what has been called "super-parameterizations", in which cloud resolving models are activated at model grid points (Grabowski et al. 1999; Randall et al. 2003). These models have the capability of predicting cloud-scale vertical velocities and LWC and thus explicitly representing precipitation processes but have yet to be applied to aerosol effects on precipitation. If the representation of aerosol and clouds in such models, or others using new and innovative techniques for representing subgrid processes can be improved, they should improve predictions of the role of aerosols in changing clouds and precipitation spatial distribution and amounts as well as climate. With these caveats in mind, we will now describe some of the results of GCM simulations of the effects of aerosol on clouds and precipitation.

7.4.1 Aerosol Impacts on Clouds; the Twomey and Cloud Lifetime Effects in GCMs

Aerosol-cloud interactions in GCMs are usually divided into the first indirect effect, (also referred to as Twomey effect (Twomey 1977) or cloud albedo effect), and the second indirect effect, or "lifetime effect". The albedo effect is posed for constant cloud water, which is met in GCMs by doing two radiation calculations each timestep. One radiative transfer calculation is conducted with the cloud droplet number concentration that is obtained from present-day aerosol concentrations and one with the cloud droplet number concentration that is obtained from pre-industrial aerosol concentrations keeping all other variables the same. The cloud lifetime effect is based on Albrecht's (1989) hypothesis that higher concentrations of cloud droplets resulting from higher concentrations of CCN will decrease droplet size, suppress drizzle, and increase liquid water contents and thereby result in clouds with greater lifetimes. As noted in Sect. 7.3.1, however, there is recent evidence from large eddy simulations that under certain conditions higher concentrations of cloud droplets can result in lower liquid water content clouds and reduced cloud liquid water path. (Jiang et al. 2006) analyzed hundreds of clouds generated by large eddy simulations, and showed that aerosol effects on cloud lifetime are minimal compared to the variability in cloud lifetime associated with convective variability. Single cloud models even suggest a decrease in cloud lifetime. It is also noted that there are only a few studies (cloud-scale modeling or observational) of cloud lifetime and that the spatial/temporal resolution of GCMs makes them ill-suited to calculation of cloud lifetime. The "cloud lifetime effect" in GCMs is therefore just an attempt to study the effect of autoconversion parameterizations on climate simulations.

There is considerable difference amongst models in predicted forcing due to the albedo effect, and even more significant disagreement in the lifetime effect. The global mean magnitude of the cloud albedo effect from different climate models since pre-industrial times is estimated to be between –0.5 and –1.9 W m^{-2} from different climate models and the cloud lifetime effect to be between –0.3 and –1.4 W m^{-2} (Lohmann and Feichter 2005). The importance of the cloud albedo effect versus the cloud lifetime effect varies even when the models use the same aerosol fields (Penner et al. 2006). In most simulations shown in Fig. 7.8 the total aerosol effect is restricted to warm clouds, except for the simulations by Jacobson (2006) and Lohmann and Diehl (2006) who also include aerosol effects on mixed-phase and ice clouds. The radiation effect at the top of the atmosphere ranges from –0.2 W m^{-2} in the combined GCM + satellite simulations (Quaas et al. 2005) to –1.8 W m^{-2} in the simulations by Ming et al. (2007) and Kristjansson (2002) with an average forcing of –1.1 W m^{-2}. The total aerosol effect is larger on the Northern Hemisphere than on the Southern Hemisphere in all models (Fig. 7.8). The models, however, disagree on the dominance of the indirect effect over ocean or land, and also the two combined GCM + satellite simulations provide conflicting answers (Fig. 7.8). Although most model estimates also include the direct and semi-direct effects, their contribution to the top ot the atmoshere (TOA) radiation is generally small compared with the indirect effect ranging from +0.1 to –0.5 W m^{-2} due to variations of the different locations of black carbon with respect to the cloud (Lohmann and Feichter 2005). Other differences among the simulations include an empirical relationship between aerosol mass and cloud droplet number concentration versus a mechanistic relationship, the dependence of the indirect aerosol effect on the assumed background aerosol or cloud droplet number concentration, the competition between natural and anthropogenic aerosols as CCN, and differences in the cloud microphysics schemes, especially in the autoconversion rate.

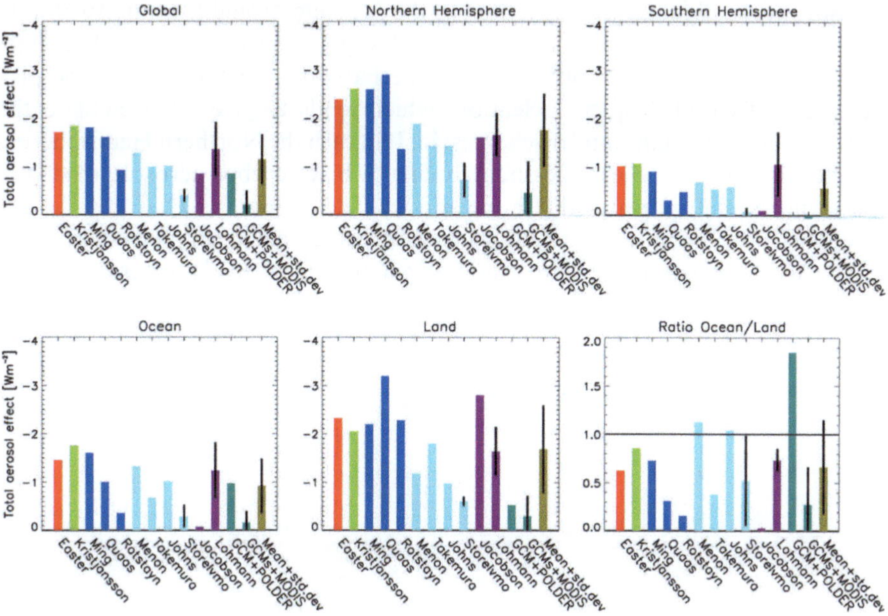

Fig. 7.8 Global mean total anthropogenic aerosol effect (*direct, semi-direct and indirect cloud albedo* and *lifetime effects*) defined as the response in net radiation at the top-of-the-atmosphere from pre-industrial times to present-day and its contribution over the Northern Hemisphere (NH), Southern Hemisphere (SH), over oceans and over land, and the ratio over oceans/land. *Red bars* refer to anthropogenic sulphate (Easter et al. 2004), *green bars* refer to anthropogenic sulphate and *black carbon* (Kristjánsson 2002[*,+]), *blue bars* to anthropogenic sulphate and organic carbon (Quaas et al. 2004[*,+]; Rotstayn and Liu 2005[+]; Ming et al. 2007[+]), *turquoise bars* to anthropogenic sulphate, black, and organic carbon (Takemura et al. 2005; Johns et al. 2006; Storelvmo et al. 2006; Menon and Del Genio 2007), *dark purple bars* to the mean and standard deviations of anthropogenic sulphate, black, and organic carbon effects on water and ice clouds (Jacobson 2006; Lohmann and Diehl 2006), *teal bars* refer to a combination of GCM and satellite results (ECHAM + POLDER, Lohmann and Lesins 2002; LMDZ/ECHAM + - MODIS, Quaas et al. 2006) and *olive bars* to the mean plus standard deviation from all simulations. Vertical black lines for individual results refer to ± one standard deviation in case of multiple simulations/results. This figure is an updated version of a similar figure in the IPCC AR4 2007 report

* refers to estimates of the aerosol effect deduced from the shortwave radiative flux only.

+ refers to estimates solely from the indirect effects.

7.4.2 Aerosol Influences on Precipitation in GCMs

Several studies have considered the response of a GCM with a mixed-layer ocean to indirect aerosol effects (Rotstayn et al. 2000; Williams et al. 2001; Rotstayn and Lohmann 2002) or to a combination of direct and indirect aerosol effects (Feichter et al. 2004; Takemura et al. 2005; Kristjansson et al. 2005). All of these found a substantial cooling that was strongest in the Northern Hemisphere, with a consequent southward shift of the Intertropical Convergence Zone (ITCZ) and the associated tropical rainfall belt. Rotstayn and Lohmann (2002) suggested that aerosol effects might have contributed to the Sahelian droughts of the 1970s and 1980s. The southward shift of the ITCZ was less pronounced in the Feichter et al. (2004) study than in the other studies, perhaps due to their more complex treatment of cloud droplet nucleation, which tends to give less weight to the effects of sulphate than simpler schemes do. If in turn the Northern Hemisphere is warmed, for instance due to the direct forcing by black carbon aerosols, the ITCZ was found to shift northward (Chung and Seinfeld 2005).

Global climate model estimates of the change in global mean precipitation due to the total aerosol effects are summarized in Fig. 7.9. Consistent with the

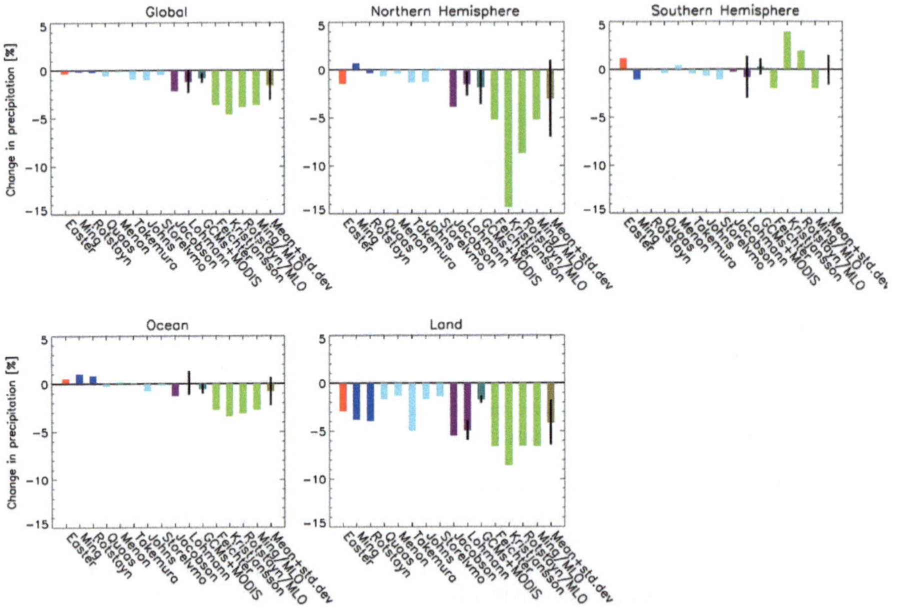

Fig. 7.9 Global mean change in precipitation ((present-day–pre-industrial)/present-day) in % due to the total anthropogenic aerosol effect (direct, semi-direct and indirect cloud albedo and lifetime effects) from pre-industrial times to present-day and its contribution over the Northern Hemisphere (NH), Southern Hemisphere (SH) and over oceans and over land. *Red bars* refer to anthropogenic sulphate (Easter et al. 2004$^+$), *blue bars* to anthropogenic sulphate and organic carbon (Ming et al. 2007$^+$; Rotstayn and Liu 2005$^+$), *turquoise bars* to anthropogenic sulphate, black, and organic carbon (Takemura et al. 2005; Johns et al. 2006; Storelvmo et al. 2006; Quaas et al. 2006; Menon and Del Genio 2007), *dark purple bars* to the mean and standard deviations of anthropogenic sulphate, black, and organic carbon effects on water and ice clouds (Jacobson 2006; Lohmann and Diehl 2006), *teal bars* refer to a combination of GCM and satellite results (LMDZ/ECHAM + MODIS, Quaas et al. 2006), *green bars* refer to results from coupled atmosphere/mixed-layer ocean (MLO) experiments (Feichter et al. (2004) – sulphate, black, and organic carbon; Kristjansson et al. (2005) – sulphate and black carbon; Rotstayn and Lohmann, (2002)$^+$ – sulphate only; Ming et al. personal communication – sulphate and organic carbon) and *olive bars* to the mean plus standard deviation from all simulations. Vertical black lines refer to ± one standard deviation. This figure is an updated version of a similar figure in the IPCC AR4 2007 report
$^+$ refers to estimates solely from the indirect effects.

conflicting results from detailed cloud system studies, the global mean precipitation decreases between 0.1 and 4.6% from pre-industrial times to the present-day. These differences are amplified over the Northern Hemisphere, ranging from +0.7 to –14.3% of their present-day values. The decreases in precipitation are larger when the atmospheric GCMs are coupled to mixed-layer ocean models (green bars), where the sea surface temperature and, hence, evaporation is allowed to vary. When aerosol effects on warm convective clouds are included in addition to their effect on warm stratiform clouds, the overall indirect aerosol effect and the change in surface precipitation can be larger or smaller than if just the aerosol effect on stratiform clouds is considered (Nober et al. 2003; Menon and Rotstayn 2006). Besides changes in the distribution of precipitation the frequency of extreme events may also be reduced by the presence of aerosols (Paeth and Feichter 2006).

Observations by Borys et al. (2003) in midlatitude orographic clouds show that for a given supercooled liquid water content, both the riming and the snowfall rates are reduced if the supercooled cloud has more cloud droplets (see also Chapter 6). Examination of this effect in global climate model simulations with pre-industrial and present-day aerosol concentrations showed that while the riming rate in stratiform clouds has indeed decreased due to the smaller cloud droplets in polluted clouds, the snowfall rate has actually increased because of feedbacks within the climate system. The pollution induced increase in aerosol and cloud optical thickness reduces the solar radiation at the surface and causes a cooling that favors precipitation formation via the ice phase (Lohmann 2004).

7.4.3 Coupling of the Hydrological Cycle and Greenhouse Gas Warming

Greenhouse warming, primarily as a result of enhanced CO_2 concentrations, is amplified when the global hydrological cycle is enhanced and greater amounts of water vapour are evaporated into the air, principally over the oceans but also over land. Because water vapour is a much more powerful greenhouse gas than CO_2, the increased amount of water vapour in the air, in turn, results in a strong positive feedback to CO_2 warming. Recent GCM simulations of greenhouse warming and direct and indirect aerosol effects (Liepert et al. 2004) show that in accord with the discussion in Sect. 7.3.7, the indirect and direct cooling effects of aerosols reduce surface latent and sensible heat transfer (Fig. 7.10) and, as a consequence, act to spin-down the hydrological cycle and thereby substantially weaken greenhouse gas warming. This is important since most investigators compare top of the atmosphere radiative differences for greenhouse gas warming and aerosol direct and indirect effects separately. Since greenhouse warming causes a spin-up of the hydrological cycle, and aerosol direct and indirect cooling counteracts the spin-up, the potential influence of aerosols on global climate could be far more significant than previously thought. The simulated decrease in global mean precipitation from pre-industrial times to the present is

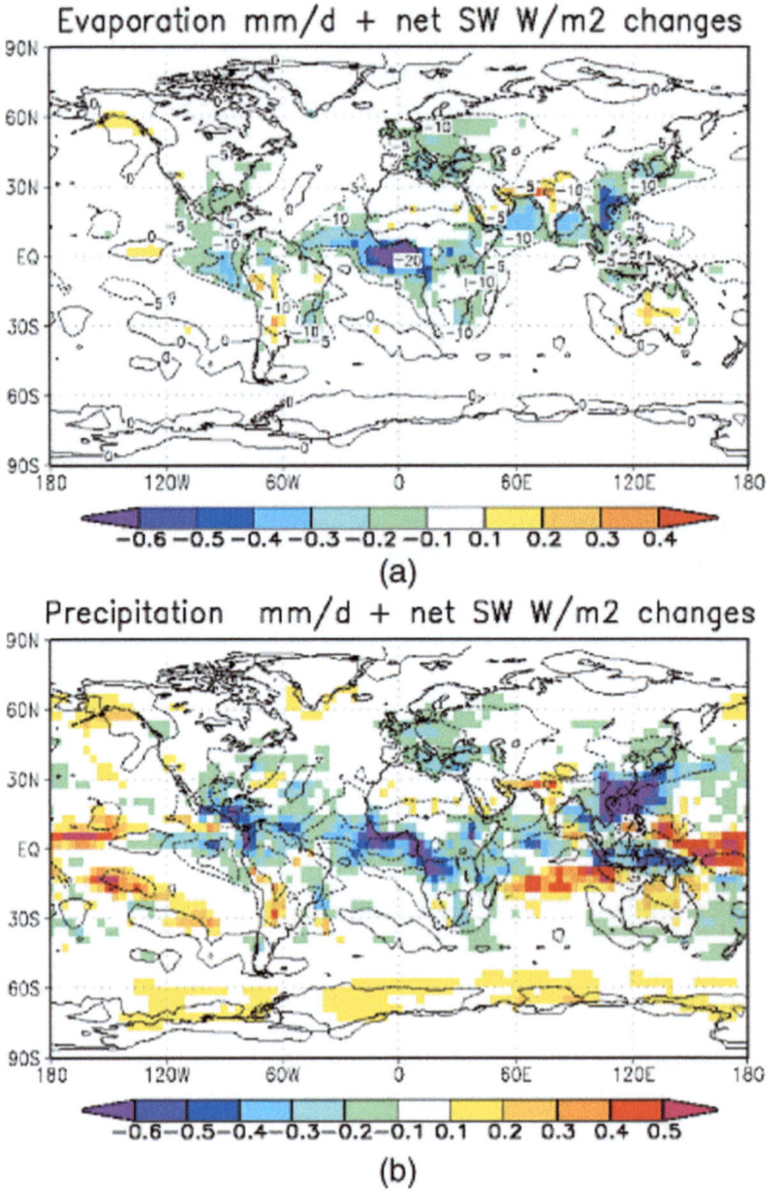

Fig. 7.10 Simulated changes in (**a**) evaporation (*colors*) and net surface solar radiation (*contours*) and (**b**) precipitation (*colors*) and net surface solar radiation (*contours*). Model results are differences of present-day minus pre-industrial climate simulations with modified anthropogenic aerosol and greenhouse gas concentrations. From Liepert et al. (2004) with permission of the American Geophysical Union

in contrast to the observed precipitation in the last century and points to an overestimation of aerosol influences on precipitation in these simulations. Nevertheless, this decrease in precipitation changes sign to an increase of about 1% in simulations to 2031–2050 as compared to 1981–2000, because the increased warming due to black carbon and greenhouse gases then dominates over the sulphate cooling (Roeckner et al. 2006).

We summarize by stressing that the results of these GCM simulations should not be viewed as quantitative forecasts of the effects of aerosols on patterns and amounts of regional precipitation. As noted previously, there are many uncertainties in the distribution and concentrations of aerosols and condensate. In addition, there are many simplifications in the models that limit their ability to realistically simulate the indirect effects of aerosols. However, these model simulations demonstrate the potential effects of direct and indirect aerosol forcing on clouds and precipitation in a coupled global system.

7.5 Summary

- Cloud-resolving models are able to capture the salient features of cloud droplet activation and the timescales of growth to precipitation-sized drops in a reasonable fashion (warm clouds).
- There are still unresolved questions regarding the nature and causes of droplet growth to precipitation embryos. Convective clouds entrain clear environmental air, which results in dilution of the cloud water, modification of cloud buoyancy, and introduction of aerosol particles into the cloud. Cloud models require meter-scale resolution to adequately represent the details of entraining convective clouds; some of these details may have important consequences for formation of precipitation embryos.
- Size-resolved microphysical treatment of cloud microphysics is the preferred approach to representing aerosol effects on precipitation because it provides more flexibility for representing the tail of precipitation embryos than bulk schemes.
- Uncertainties in collection kernels, especially those involving ice particles, are a major source of uncertainty in modeling of development of precipitation. This limits the gains that can be realized by size resolved microphysical schemes.
- Aerosol effects on precipitation must be considered together with cloud and precipitation effects on aerosol. These include precipitation scavenging, aqueous chemistry, new particle formation and cloud processing. Models that include these processes produce different amounts of precipitation than those that do not. This places strong demands on models because of added computational overhead.
- Evaluation of aerosol effects on clouds and precipitation must be considered in a dynamical/meteorological context. Aerosol-cloud interactions occur in a

tightly coupled system and it is important that models couple dynamics, microphysics, radiation, chemistry, and surface characteristics. Some model results suggest that even the sign of aerosol effects on precipitation can change depending on the dynamical context.

- Simulations of individual clouds show differences from a field of interacting clouds, e.g. as a result of cold pool outflow and convergence zones. Historically the cloud modeling community has focused on aerosol effects on single clouds. While these studies are still of great value there is need to pursue more studies of cloud ensembles;
- There are major challenges in modeling precipitation at the appropriate range of spatial/temporal scales, primarily due to limitations in computing power.
- Mesoscale, regional scale and global scale models do not always resolve the appropriate spatial and temporal scales of the physical processes represented, and their use in predicting the effects of aerosol on clouds and precipitation should be regarded with caution. The scaling-up of bulk parameterizations developed for cloud-scale models to regional and global scale models does not have a solid scientific basis.
- Because of these uncertainties and limitation in simulating aerosol effects on precipitation, climate model estimates of aerosol-induced changes in precipitation since pre-industrial times vary tremendously.

7.6 Recommendations for Future Modeling Studies

To place our understanding of aerosol-precipitation interactions on a firmer footing, we recommend the following for future modeling efforts.

7.6.1 Improve Model Physics

There is a strong need for an improvement in the representation of cloud microphysical processes in models. This includes both the representation of the *physics of hydrometeor behavior, hydrometeor interactions* (e.g. laboratory studies of collision and coalescence efficiencies for various combinations of hydrometeor types), *the parameterization of these processes* so that they can be applied in numerical algorithms, and *numerical methods.* Advances in representation of hydrometeor collection kernels have generally lagged the development of new bin and bulk microphysical methods, specific recommendations include:

- Simulations of aerosol-cloud interactions should be done at the *appropriate scale* so that the development of cloud supersaturations, droplet activation, and growth processes are resolved. This will yield greater confidence in the models' ability to represent actual cloud responses to variable aerosol inputs. Because of the coarse grid spacing of mesoscale, regional and global models,

clouds and cloud properties essentially have to *be parameterized as sub-grid-scale processes*. Parameterizations at the full range of scales should be pursued.
- Aerosol effects on clouds must be considered within the *appropriate dynamical and meteorological context*. It is likely that the response of clouds to aerosol will vary depending on input thermodynamic sounding, surface fluxes, subsidence, large-scale advective tendencies etc. *Simulations of these interactions should be extended to larger scales to enable dynamical interactions between neighboring clouds.*

Land surface characteristics determine the surface latent and sensible heat fluxes and drive boundary layer and cloud development. Current and future changes in vegetation and soil moisture are therefore an important consideration. Moreover, because the presence of atmospheric aerosol (particularly absorbing aerosol) modifies the surface radiation budget, it also changes the surface fluxes, and these changes are likely to differ from one land-surface regime to another. Consideration of the coupled aerosol-cloud-land-surface system is therefore of great importance and may have consequences at the range of spatial scales.

- Neither GCMs nor regional-scale models *resolve topography* sufficiently well to depict cloud-scale vertical motions responsible for the activation of aerosol and the formation of precipitation. Solving this problem goes beyond traditional thinking for cloud parameterization schemes.
- Model evaluation should be performed through a series of case study intercomparisons of modeled and observed aerosol-cloud-precipitation interactions. These should occur at a large range of spatial and temporal scales.

7.6.2 Need for Model Input Data

- To model aerosol-cloud-precipitation interactions *input data on aerosol that have the potential to act as CCN, GCCN,* and IN are needed. Ideally these data should include not only surface measurements but also *vertical profiles*. Long-term data sets should form the backbone, with intensive field campaigns being supplementary. Where possible, CCN proxies such as accumulation mode aerosol concentration together with composition information should be considered.
- *Ice nucleation* deserves specific mention. New methods of measuring ice nuclei and analyzing their composition need to be developed. Current field-deployable ice nucelei counters represent only a few of the known modes of ice nucleation. The representation of ice formation in cloud models is still rudimentary and relies on a very limited number of measurements and parameterizations.
- Analyze satellite data to establish constraints on aerosol effects on clouds and precipitation. This may require combination of passive radiometry to

characterize aerosol and cloud drop sizes, and radar to characterize precipitation. The launch of CLOUDSAT and CALIPSO in the spring of 2006 as part of the A-Train (see Chapter 5) promises to provide important data on aerosol, clouds, and precipitation.

7.6.3 Assessment, Attribution of Cause-and-Effect

- Models are particularly useful for providing quantified estimates of parameters such as precipitation amounts for a given simulation. However, the skill level of models in so doing should be quantified statistically. Models should be used to *isolate the role of aerosols from the influence of the meteorology and to quantify the relative importance of each process.*
- Models should be used to develop metrics to quantify aerosol effects on precipitation efficiency (Chapter 2).
- Models should consider the role of *clouds as aerosol scavengers and the scavenging efficiency of clouds and rain.* Cloud scavenging, both through coalescence scavenging and precipitation washout are important aerosol sinks. The links between precipitation scavenging of aerosol and precipitation efficiency should also be explored further. Consideration of these processes should be included in clouds of all scales.
- Identify *cause-and-effect.* Models are very effective tools for isolating cause-and-effect in a physical system. It is important to *correctly ascribe the reasons for agreement or disagreement with observations.* As models become more complex, it becomes increasingly difficult to isolate causal relations. Effort should be put into applying existing and new methodologies for quantifying the sensitivity of a given parameter to a change in an input parameter or change in a combination of inputs.

Chapter 8
Parallels and Contrasts Between Deliberate Cloud Seeding and Aerosol Pollution Effects

William R. Cotton

8.1 Introduction

Deliberate cloud seeding, with the goal of increasing precipitation by the injection of specific types of particles into clouds, has been pursued for over 50 years. Efforts to understand the processes involved have led to a significant body of knowledge about clouds and about the effects of the seeding aerosol. A number of projects focused on the statistical evaluation of whether a seeding effect can be distinguished in the presence of considerable natural variability. Both the knowledge gained from these experiments, and the awareness of the limitations in that understanding, are relevant to the general question of aerosol effects on precipitation. Definite proof from the seeding projects for an induced increase in precipitation as a result of the addition of seeding material to the clouds would represent a powerful demonstration of at least one type of dominant aerosol-precipitation link in the clouds involved. Therefore, in this chapter we review the fundamental concepts of cloud seeding and overview the parallels and contrasts between evaluations of deliberate and inadvertent modification of precipitation by aerosols. It is not our intent to provide a comprehensive assessment of the current status of cloud seeding research. We direct the reader to more comprehensive weather modification assessments in NRC (2003), Cotton and Pielke (2007), Silverman (2001, 2003), and Garstang et al. (2005).

Deliberate cloud seeding experiments can be divided into two broad categories: *glaciogenic seeding* and *hygroscopic seeding*. *Glaciogenic seeding* occurs when ice-producing materials (e.g. dry ice (solid CO_2), silver iodide, liquid propane etc.) are injected into a supercooled cloud for the purpose of stimulating precipitation by the ice particle mechanism (see Sect. 2.2). The underlying hypothesis for glaciogenic seeding is that there is commonly a deficiency of natural ice nuclei and therefore insufficient ice particles for the cloud to produce precipitation as efficiently as it would in the absence of seeding.

The second category of artificial seeding experiments is referred to as *hygroscopic seeding*. In the past this type of seeding was usually used for rain enhancement from

W.R. Cotton (✉)
Colorado State University, Fort Collins, CO, USA

Z. Levin, W.R. Cotton (eds.), *Aerosol Pollution Impact on Precipitation*,

warm clouds (see Cotton 1982 for a review of early hygroscopic seeding research). However, more recently this type of seeding has been applied to mixed phase clouds as well. The goal of this type of seeding is to increase the concentration of *collector drops* that can grow efficiently into raindrops by collecting smaller droplets and by enhancing the formation of frozen raindrops and graupel particles. This is done by injecting into a cloud (generally at cloud base) large or giant hygroscopic particles (e.g. salt powders) that can grow rapidly by the condensation of water vapour to produce collector drops (see Sect. 2.3).

8.2 Static Glaciogenic Cloud Seeding

Static cloud seeding refers to the use of glaciogenic materials to modify the microstructures of supercooled clouds and precipitation. Many hundreds of such experiments have been carried over the past 50 years or so. Some are operational cloud seeding experiments (many of which are still being carried out around the world) which rarely provide sufficient information to decide whether or not they modified either clouds or precipitation. Others are well designed scientific experiments that provide extensive measurements and modeling studies that permit an assessment of whether artificial seeding modified cloud structures and, if the seeding was randomized, the effects of the seeding on precipitation. While there still is some debate of what constitutes firm "proof" (see NRC 2003; Garstang et al. 2005) that seeding affects precipitation, generally it is required that *both* strong physical evidence of appropriate modifications to cloud structures and highly significant statistical evidence be obtained.

8.2.1 Glaciogenic Seeding of Cumulus Clouds

The static seeding concept has been applied to supercooled cumulus clouds and tested in a variety of regions. Two landmark experiments (Israeli I and Israeli II), carried out in Israel, were described in the peer-reviewed literature. The experiments were carried out by researchers at the Hebrew University of Jerusalem (HUJ), hereafter the experimenters. These two experiments were the foundation for the general view that under appropriate conditions, cloud seeding increases precipitation (e.g. N.R.C. 1973; Sax et al. 1975; Tukey et al. 1978a,b; Simpson 1979; Dennis 1980; Mason 1980, 1982; Kerr 1982; Silverman 1986; Braham 1986; Cotton 1986a,b; Cotton and Pielke 1992, 1997; Young 1993).

Nonetheless, reanalysis of those experiments by Rangno and Hobbs (1993) suggested that the appearance of seeding-caused increases in rainfall in the Israel I experiment was due to "lucky draws" or a Type I statistical error. Furthermore, they argued that during Israel II naturally heavy rainfall over a wide region encompassing the north target area gave the appearance that seeding caused increases in rainfall over the north target area. At the same time, lower natural rainfall in the region encompassing the south target area gave the appearance that seeding decreased rainfall over that target area. But this speculation could not explain the positive effect when the north target area was evaluated against the north upwind control area. Details of this controversy can be found in the March 1997 issue of the Journal of Applied Meteorology (Rosenfeld 1997; Rangno and Hobbs 1997a; Dennis and Orville 1997; Rangno and Hobbs 1997b; Woodley 1997; Rangno and Hobbs 1997c; Ben-Zvi 1997; Rangno and Hobbs 1997d; Rangno and Hobbs 1997e). Some of these responses clarified issues; others have left a number of questions unanswered.

It is interesting to note that in the Israeli experiments the effects of artificial seeding with silver iodide appeared to be an increase in the duration of precipitation, with little if any effect on the intensity of precipitation (Gagin 1986; Gagin and Gabriel 1987), a finding compatible with the "static" seeding hypothesis.

Givati and Rosenfeld (2005) wrote, "that cloud seeding with sliver iodide enhances precipitation especially where the orographic enhancement factor (see Chapter 6) was the largest. Likewise, the pollution effects reduced precipitation by the greatest amount at the same regions". They suggested that this is because the shallow and short-living orographic clouds are particularly susceptible to such impacts. This suggests that attempts to alter winter precipitation should be concentrated on orographic clouds. Or interpreted in terms of inadvertent modification of clouds; winter orographic clouds may be the most susceptible to precipitation modification by pollution.

This also suggests that the conceptual model on which the Israeli cloud seeding experiments was based is not exactly as postulated. The seeding was originally aimed at the convective clouds that formed over the narrow coastal plain, with the intent of nucleating ice crystals and forming graupel earlier in the cloud lifecycle (Gagin and Neumann 1974), thus leading to increased rainfall in the catchment basin of the Sea of Galilee to the east of the Galilee Mountains. However, the report of Givati and Rosenfeld (2005) concluded that "cloud seeding did not enhance the convective precipitation, but rather increased the orographic precipitation on the upwind side of the Mountains, probably by the Bergeron-Findeisen process".

The lack of enhancement of the convective clouds in Israel might be explained by their tendency to mature and dissipate inland during the winter storms. Seeding of mature convective clouds cannot affect them much. The lack of enhancement is also consistent with the microphysically maritime nature of the convective clouds. This appears to be caused mainly due to the natural hygroscopic seeding by sea spray or mineral dust particles coated with soluble material (Levin et al. 1996, 2005) in the winter storms that enhance the warm precipitation (Rosenfeld et al. 2001) as well as promoting the formation of ice hydrometeors that is followed by ice multiplication (Hallett and Mossop 1974). These suggestions are supported by the results of glaciogenic cloud seeding in Tasmania, which targeted a hilly area by seeding along an upwind coastline. The seeding in Tasmania was shown to enhance precipitation from the stratiform orographic clouds, but not from the convective clouds (Ryan and King 1997). This is consistent with the microphysical conclusions of Rangno and Hobbs (1993), who asserted that cloud seeding as done in Israel could not have possibly caused the statistically documented rain enhancement from the convective clouds there.

Recently, Givati and Rosenfeld (2005) carried out a study in which the effects of pollution on rainfall suppression in orographic clouds were separated from the effects of cloud seeding in Israel. They concluded that the two effects have the opposite influence on rainfall, demonstrating the sensitivity of clouds to anthropogenic aerosols of different kinds. By analyzing the rainfall amounts in northern Israel during the last 53 years during days in which no seeding was carried out, they observed a decreasing trend of the orographic factor R_0 (discussed in Chapter 6) with time from the beginning of the study. They associated this decrease with the increase in aerosol pollution. The same trend, but shifted upward by 12–14%, was observed for days in which seeding was carried out. Thus, it appears that the opposing effects of air pollution and seeding appear to have nearly canceled each other.

Another noteworthy experiment was carried out in the high plains of the U.S. (High Plains Experiment (HIPLEX-1) Smith et al. (1984). Analysis of this experiment revealed the important result that after just 5 min, there was no statistically significant difference in the precipitation between seeded and non-seeded clouds, (Mielke et al. 1984). Cooper and Lawson (1984) found that while high ice crystal concentrations were produced in the clouds by seeding, the cloud droplet region where the crystals formed evaporated too quickly for the incipient artificially produced ice crystals to grow to appreciable sizes. Instead they formed low density, unrimed aggregates having the water equivalent of only drizzle drops, which were too small to reach the ground before evaporating. Schemenauer and Tsonis (1985) affirmed the findings of Cooper and Lawson in a reanalysis of the HIPLEX data emphasizing their own earlier findings (Isaac et al. 1982) that cloud lifetimes were too short in the HIPLEX domain for seeding to have been effective in the clouds targeted for seeding (i.e. those with tops warmer than –12°C). Although the experiment failed to demonstrate statistically all the hypothesized steps, the problems could be traced to the physical short lifetimes of the clouds (Cooper and Lawson 1984; Schemenauer and Tsonis 1985). This in itself is a significant result that shows the ability of physical measurements and studies to provide an understanding of the underlying processes in each experiment. The results suggested that a more limited window of opportunity exists for precipitation enhancement than was thought previously. Cotton and Pielke (1995) summarized this window of opportunity as being limited to: Clouds that are relatively cold-based and continental; Clouds with top temperatures in the range –10 to –25°C, and a timescale confined to the availability of significant supercooled water before depletion by entrainment and natural precipitation processes.

Today, this window would even be viewed as too large, since many cold based continental clouds with tops >–25°C have copious ice particle concentrations (e.g. Hobbs and Rangno 1985; Rangno and Hobbs 1988). The HIPLEX results also indicated that small clouds make little contribution to rainfall.

This begs the question, should we expect a similar *window of effectiveness* for inadvertent IN pollution?

8.2.2 Seeding Winter Orographic Clouds

The static mode of cloud seeding has also been applied to orographic clouds. Precipitation enhancement of orographic clouds by cloud seeding has several advantages over cumulus clouds. The clouds are persistent features that produce precipitation even in the absence of large-scale meteorological disturbances. Much of the precipitation is spatially confined to high mountainous regions thus making it easier to set up dense ground based seeding and observational networks. Moreover, orographic clouds are less susceptible to a "time window" as they are steady clouds that offer a greater opportunity for successful precipitation enhancement than cumulus clouds. A time window of a different type does exist for orographic clouds, which are related to the time it takes a parcel of air to condense to form supercooled liquid water and ascend to the mountain crest. If winds are weak, then there may be sufficient time for natural precipitation processes to occur efficiently. Stronger winds may not allow efficient natural precipitation processes but seeding may speed up precipitation formation. Stronger winds may not provide enough time for seeded ice crystals to grow to precipitation before being blown over the mountain crest and evaporating in the sinking subsaturated air to the lee of the mountain. A time window related to the ambient winds, however, is much easier to assess in a field setting than the time window in cumulus clouds.

The land-mark randomized cloud seeding experiments at Climax, near Fremont Pass, Colorado (referred to as *Climax I* and *Climax II*), Colorado, reported by Grant and Mielke (1967) and Mielke et al. (1970, 1971) suggested increases in precipitation of 50% and more on favorable days (e.g. Grant and Mielke 1967; Mielke et al. 1970, 1971), and the results were widely viewed as demonstrating the efficacy of cloud seeding (e.g. NRC 1973; Sax et al. 1975; Tukey et al. 1978a,b, American Meteorological Society 1984), even by those most skeptical of cloud seeding claims (e.g. Mason 1980, 1982). Nonetheless, Hobbs and Rangno (1979), Rangno and Hobbs (1987, 1993) question both the randomization techniques and the quality of data collected during those experiments and conclude that the Climax II experiment failed to confirm that precipitation can be increased by cloud seeding in the Colorado Rockies. Even so, in their reanalysis, Rangno and Hobbs (1993) did show that precipitation increased by about 10% in the combined Climax I and II experiments. This should be compared, however, to the original analyses by Grant and Mielke (1967), Grant and Kahan (1974), Grant and Elliott (1974), Mielke et al. (1971), Mielke et al. (1976) and Mielke et al. (1981) that indicated greater than 100% increase in precipitation on seeded days for Climax I and 24% for Climax II.

Two other randomized orographic cloud seeing experiments, the Lake Almanor Experiment (Mooney and Lunn 1969) and the Bridger Range Experiment (BRE) as reported by Super and Heimbach (1983) and Super (1986) suggested positive results. However, these particular experiments used high elevation AgI generators, which increase the chance that the AgI plumes get into the supercooled clouds. Moreover, both experiments provided physical measurements that support the statistical results (Super and Heimbach 1983, 1988).

There have been a few attempts to use mesoscale models to evaluate cloud seeding programs. Cotton et al. (2006) applied the Colorado State University Regional Atmospheric Modeling System (RAMS) to the simulation of operational cloud seeding in the central Colorado Mountains in the 2003–2004 winter season. The model included explicit representation of surface generator production of AgI at the locations, burn rates, and times supplied by the seeding operator. Moreover, the model explicitly represented the transport and diffusion of the seeding material, its activation, growth of ice crystals and snow, and precipitation to the surface. Detailed evaluation of model forecast orographic precipitation was performed for 30 selected operational seeding days. It was shown that the model could be a useful forecasting aid in support of the seeding operations. But the model over-predicted natural precipitation, particularly on moist southwest flow days. The model also exhibited virtually no enhancement in precipitation due to glaciogenic seeding. There are a number of possible causes for the lack of response to seeding, such as over prediction of natural precipitation, which prevented the effects of seeding from being seen. In addition, the background CCN and IN concentrations are unknown, therefore lower CCN concentrations than occurred would make the clouds more efficient in precipitation production, thus reduceing seeding effectiveness.

Finally, Ryan and King (1997) reviewed over 14 cloud seeding experiments covering much of southeastern, western, and central Australia, as well as the island of Tasmania. They concluded that static seeding over the plains of Australia is not effective. They argue that for orographic stratiform clouds, there is strong statistical evidence that cloud seeding increased rainfall, perhaps by as much as 30% over Tasmania when cloud top temperatures are between –10 and –12°C in southwesterly airflow. The evidence that cloud seeding had similar effects in orographic clouds over the mainland of southeastern Australia is much weaker. Note that the Tasmanian experiment had both strong statistical and physical measurement components and thus meets, or at least comes close to meeting, the NRC (2003) criteria for scientific "proof." Cost/benefit analysis of the Tasmanian experiments suggests that seeding has a gain of about 13:1. This is viewed as a real gain to hydrologic energy production.

A complication revealed in the analysis of some of the Australian seeding experiments is that precipitation increases were inferred one to three weeks following seeding in several seeding projects (e.g. Bigg and Turton 1988). Bigg and Turton (1988) and Bigg (1988, 1990, 1995) suggested that silver iodide seeding can trigger biogenic production of additional ice nuclei. The latter research suggests that fields sprayed with silver iodide release secondary ice nuclei particles at intervals of up to ten days.

In summary, the "static" mode of cloud seeding has been shown to cause the expected alterations in cloud microstructure including increased concentrations of ice crystals, reductions of supercooled liquid water content, and more rapid production of precipitation elements in both cumuli (Isaac et al. 1982; Cooper and Lawson 1984) and orographic clouds (Reynolds and Dennis 1986; Reynolds 1988; Super and Boe 1988; Super et al. 1988; Super and Heimbach 1988). The documentation of increases in precipitation on the ground due to static seeding of cumuli, however, has been far more elusive, with the Israeli experiment (Gagin and Neumann 1981) providing the strongest evidence that static seeding of cold-based, continental cumuli can cause significant increases of precipitation on the ground. The evidence that orographic clouds can cause significant increases in snowpack is far more compelling, particularly in the more continental and cold-based orographic clouds (Mielke et al. 1981; Super and Heimbach 1988).

Perhaps, however, the most challenging obstacle to evaluating cloud seeding experiments to enhance precipitation, is the inherent natural variability of precipitation in space and time, and the inability to increase precipitation amounts to better than ~10%. This last obstacle puts great demands on the measuring accuracy and the duration of the experiments. Shouldn't we expect similar obstacles in evaluating inadvertent effects of IN pollution on precipitation?

8.3 Dynamic Glaciogenic Seeding

So far we have considered only *static seeding*, in which the principle thrust is to modify the microstructures of clouds generally for the purpose of enhancing precipitation. There is, however, another glaciogenic seeding hypothesis in which the cloud-scale dynamics of a cloud is enhanced by stimulating buoyancy and upward motions of air. This is referred to as *dynamic cloud seeding*. In principal, this can be done by glaciating convective clouds so that large quantities of latent heat are released by the freezing of copious liquid water, to invigorate updrafts in the cloud. This can be particularly effective if prior to seeding, the tops of the clouds are restricted by a shallow stable layer produced by a temperature inversion. In this case, the sudden release of a large quantity of latent heat might provide enough buoyancy to push the top of the cloud through the stable layer and into a region where the air is naturally unstable. The cloud might then rise to much greater heights than it would have done naturally.

In a series of randomized experiments carried out in Florida in 1968 and 1970–1973 (called the Florida Area Cumulus Experiment or FACE), it was found that precipitation (measured by radar) from isolated cumulus clouds ~5 km in diameter, which were artificially seeded to induce explosive growth, was about twice that from the unseeded control clouds (e.g. Simpson and

Woodley 1975; Woodley et al. 1982). The seeded clouds rained more than the control clouds because they were bigger and lasted longer, rather than their rainfall rates being significantly greater. In FACE II, the attempt was made to confirm and replicate the results of FACE I by going the additional step of specifying the manner in which clouds would respond to seeding based on what was understood to have been the response in FACE I. While there were several suggestions of seeding effects on some clouds and some days (e.g. Woodley et al. 1983), the overall experiment officially failed to confirm the results of FACE I (Nickerson 1979, 1981; Flueck et al. 1981).

In recent years the dynamic seeding strategy has been applied in Thailand and West Texas. Results from exploratory dynamic seeding experiments over west Texas have been reported by Rosenfeld and Woodley (1989, 1993). Analysis of the seeding of 183 convective cells suggests that seeding increased the maximum height of the clouds by 7%, the areas of the cells by 43%, the durations by 36%, and the rain volumes of the cells by 130%. The results are encouraging but such small increases in vertical development of the clouds are hardly consistent with earlier exploratory seeding experiments.

As a result of their experience in Texas, Rosenfeld and Woodley (1993) proposed an altered conceptual model of dynamic seeding in which explosive vertical development of seeded clouds is not emphasized. As pointed out by Silverman (2001), however, application of the revised hypothesis in Thailand (Woodley et al. 2003a,b) indicated rainfall enhancement, but the results did not reach statistical significance. Moreover the enhanced downdraft presumably produced by it did not appear to be delayed (Woodley et al. 1999b).

In summary, the concept of dynamic seeding is a physically plausible hypothesis that offers the opportunity to increase rainfall by much larger amounts than simply enhancing the precipitation efficiency of a cloud. It is a much more complex hypothesis, however, requiring greater quantitative understanding of the behavior of cumulus clouds and their interaction with each other, with the boundary layer, and with larger-scale weather systems. As discussed in Chapter 7, similar complex dynamical responses to pollution aerosols is found in modeling studies, thus requiring greater quantitative understanding of the behavior of cumulus clouds and their interaction with each other, with the boundary layer, and with larger-scale weather systems as well.

8.4 Hygroscopic Cloud Seeding

Hygroscopic seeding was mainly used in the past in warm clouds without any ice in them. However, more recently, this type of seeding methods have been tried in mixed-phase clouds. The aim in seeding warm clouds is to enhance drop growth by coalescence and thus improving the efficiency of rainfall formation. On the other hand, seeding mixed-phase clouds seems to affect both drop growth and ice formation, probably through the efficient formation of graupel particles. Appropriately sized salt particles, water droplets from sprays of either water or saline solution (Bowen 1952; Biswas and Dennis 1971; Cotton 1982; Murty et al. 2000; Silverman and Sukarnjanasat 2000), and hygroscopic flares (Mather et al. 1997; WMO 2000) have been used for seeding. Essentially hygroscopic seeding is similar in concept to inadvertently infecting clouds with higher than normal GCCN and recent enthusiasm for the concept was motivated by Mather's (1991) study showing the effects of paper mill effluent on precipitation. Statistical results, observations and modeling results have provided some evidence that under certain conditions and with optimal seed drop size spectra, precipitation may be enhanced (Farley and Chen 1975; Rokicki and Young 1978; Young 1996; Reisin et al. 1996c; Yin et al. 2000a,b). Hygroscopic flare particle seeding experiments have provided statistical support for rainfall increases (Fig. 8.1) due to seeding based on single cloud analyses (Mather et al. 1997; Bigg 1997; WMO 2000; Silverman 2003). Model simulations suggest that the increase in rainfall amounts stems from the increase in graupel numbers and masses, which are generated by the increased concentrations of large drops (Yin et al. 2000a,b). Such increases could generate more

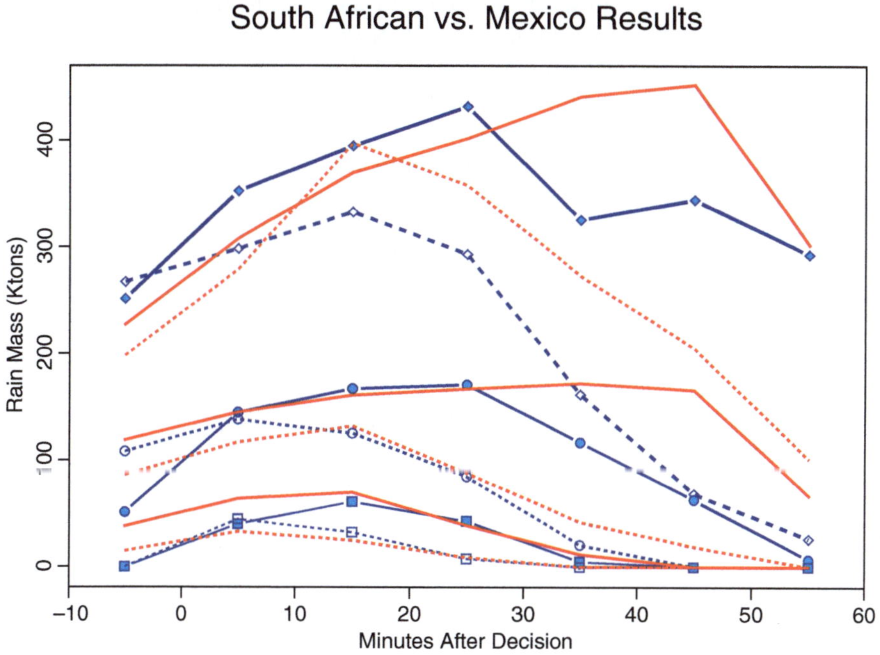

Fig. 8.1 Quartile values of radar derived rain mass (Precipitation Flux integrated over time) versus time after "decision time" for seeded and non-seeded cases for the South African (*red lines*) and Mexican (*blue lines*) experiments. Rain Mass is the Precipitation Flux integrated over 5-min intervals. 1st quartile is the value of Rain Mass that is larger than the value for 25% of the storms; 2nd quartile value exceeds the value for 50% of the other storms; and 3rd quartile value exceeds the value for 75% of the storms, with permission of R. Bruintjes

rain but it is not clear how these procedures can affect the clouds for such a length of time as some of the measurements suggest (e.g. Silverman 2003).

The principle of enhancing the coalescence process via hygroscopic seeding is dependent on three important parameters: the chemistry (hygroscopicity), size and concentration of the particles (CCN) produced from the flares or large particle salt seeding. In addition, the effectiveness of seeding will depend on the natural background particles and their characteristics with regard to the same three parameters. The principle of flare seeding is to have the flares produce effective CCN (usually salts such as sodium chloride, potassium chloride, or calcium chloride) particles in larger sizes (large or giant nuclei) than occur in the natural environment.

The majority of the hygroscopic cloud seeding flares, including the flares used in South Africa, are based on the formula described by Hindman (1978) that was developed to initiate fog cover for military vessels. The flares used in the South African experiment (Mather et al. 1997) provide large CCN (>0.3 μm diameter) to a growing cloud, influencing the initial condensation process and allowing fewer CCN to activate to cloud droplets as shown in the modeling study by Cooper et al. (1997). If the CCN that are introduced into the cloud from the flare are larger in size than the natural CCN, the introduced CCN will activate preferentially over the natural CCN and change the character of the cloud drop size distribution to favor the collision-coalescence process and the formation of rain. The larger artificial CCN inhibit the smaller natural CCN from nucleating, resulting in a broader droplet spectrum at cloud base. The fewer cloud droplets grow to larger sizes, and are often able to start growing by collision-coalescence with other cloud droplets within $\sim$15 min, initiating the rain process earlier within a typical cumulus cloud lifetime of $\sim$30 min.

Model simulations (Reisin et al. 1996c; Yin et al. 2000a,b; Caro et al. 2002; Segal et al. 2004) suggest that in addition to the size of the hygroscopic particles the time of seeding is a crucial parameter. Seeding too early or too late can lead to reduced rain amounts or to be ineffective at all, respectively. The models suggest that the window of opportunity for effective hygroscopic seeding is small, although it is longer than that of glaciogenic seeding.

In both South Africa (Mather et al. 1997) and Mexico (WMO 2000), hygroscopic flares have been applied to mixed-phase convective cloud systems in limited physical and statistical experiments. Aircraft microphysical measurements were made to verify some of the processes involved. Radar measured 30 dBZ volumes produced by the convective complexes were tracked by automated software and various storm and track properties were calculated. These two sets of experiments produced remarkably similar results in terms of the difference in radar-estimated rainfall between the seeded and non-seeded groups. The South African data have been re-evaluated independently by Bigg (1997) and Silverman (2000); both concluded that there is statistically significant evidence of an increase in radar-estimated rainfall from seeded convective cloud systems.

Mather et al. (1997), Bigg (1997), and Silverman (2000) all allude to apparent dynamic effects of seeding clouds, manifest in the seeded cloud systems being longer-lived. It was speculated that the relation between precipitation loading and evaporation, and the characteristics of the downdraft that is generated, and between the downdraft and the storm organization, evolution, and lifetime, determines the dynamic effect of seeding on rainfall. Another factor not mentioned is the possible consequences of altered raindrop size distributions. If seeding shifts the raindrop size distribution to smaller raindrops, then greater sub-cloud evaporation would ensue, which would alter cold-pool dynamical effects. If seeding shifts the raindrop spectrum to larger drops the opposite response would be expected (Yin et al. 2001).

The individual storms selected for the experiment, almost without exception, extended well above the freezing level. For the exploratory analyses on the South African data, marked differences were found in storm properties above 6 km (Mather et al. 1997). The 6 km level generally corresponds to the –5 to –10°C level, which points to probable ice-phase processes being part of the apparent seeding effect. Some indication of how the microphysical changes of broadening the droplet spectrum can be brought about by hygroscopic flare particles, as well as supporting measurements, are given by Cooper et al. (1997), Reisin et al. (1996c) and Yin et al. (2000b). Although the modeled and observed mixed-phase cloud responses to hygroscopic seeding are not fully understood, they are consistent with earlier modeling and observational studies, which indicated:

- High concentrations of large ice particles or graupel particles are found in maritime clouds (in which collision-coalescence is active) that extend above the freezing level (Koenig 1963; Cotton 1972a,b; Koenig and Murray 1976; Scott and Hobbs 1977), compared to the relatively low concentrations of large ice particles (graupel) in continental clouds (in which coalescence is not active);
- Freezing temperatures increase with an increase in droplet size, due to the higher probability that larger droplets will contain (or come in contact with) ice nuclei and riming is thereby enhanced (Johnson 1987); and,
- Secondary ice particle formation by ice splinter formation during riming is enhanced in the presence of relatively large cloud droplets (Hallett and Mossop 1974).

It appears that continental convective storms are remarkably sensitive to changes in the CCN ingested at cloud base. For example, both the South African and Mexican experiments with hygroscopic flares show very strong signals in terms of increased storm lifetime in seeded storms, increases in reflectivity aloft, and increases in storm densities. Thus, these hygroscopic flare seeding experiments suggest that it is possible, under appropriate conditions, to produce large differences in cloud properties by injection of hygroscopic particles into cloud bases.

The calculations of Reisin et al. (1996a,b,c), Cooper et al. (1997), Yin et al. (2000a,b), Caro et al. (2002) and Segal et al. (2004) show that for clouds with maritime cloud droplet spectra, hygroscopic seeding should have no effect, since the collision-coalescence acts very efficiently in such clouds. Relatively high concentrations of large natural CCN prevent the seeded particles from dominating the growth. Model calculations show that clouds with natural CCN concentrations $<500\ cm^{-3}$ should not respond favorably to hygroscopic seeding.

It has been postulated that the initial spreading of the seeding effect through a cloud occurs through the formation of drizzle drops. The modeling study by Cooper et al. (1997) indicated that the formation of drizzle is highly dependent on the particle size spectra produced by the flares, and that the concentrations of drizzle drops produced can vary by several orders of magnitude depending on the particle size spectra. Cooper et al. found that the optimum particle size for seeding with hygroscopic flares in order to obtain higher concentrations of drizzle size drops is $\sim 1\ \mu m$ diameter. Reisin et al. (1996c), Yin et al. (2000a,b) and Segal et al. (2004) showed that particles larger than 1 μm radius are the most effective for seeding. All these modeling studies also indicate that background CCN (size and concentration) properties are crucial for determining the effectiveness of the seeded particles because the seeded nuclei compete with the background aerosols for the available water vapour.

The NRC (2003) assessment of weather modification research concluded that the South African and Mexican experiments have demonstrated responses in clouds to treatment in accordance with understanding of the chain of physical reactions leading to precipitation. But since the analyzed statistical results are for radar-defined floating targets, they still do not prove that rainfall can be increased by hygroscopic seeding on the ground for specific watersheds. Moreover, since seeding may alter the size-spectrum of raindrops, which alters the radar return, uncertainties exist in the evaluation of actual rain amounts for seeded versus not-seeded floating targets. Finally, since the main response to seeding found in the South African, Mexican, and Thailand experiments is delayed in time for as much as 1 and 6 h, respectively, following the cessation of seeding, we lack a clear understanding of the actual processes that can lead to such a physical response.

Thus while the results of the hygroscopic seeding experiments are quite promising they still do not constitute a "proof" that hygroscopic seeding can enhance rainfall on the ground over an extended area. The areas affected by cloud seeding remain an open question. In after-the-fact analyses several rain enhancement projects have reported evidence for physical effects outside the area or timing originally designated as the target, or beyond the time interval when seeding effects were anticipated. For example, in recent large particle hygroscopic and glaciogenic seeding trials involving warm-base convective clouds in Thailand and Texas, increases in rain were reported three to 12 h after seeding was conducted, well beyond the time at which direct effects of seeding were expected and possibly outside the target area. In Project Whitetop the seeding appears to have decreased rain in the area immediately downwind of the seeding release line. This was followed by apparent rainfall increases well downwind in space and time. Does this mean that the scientists misjudged where seeding materials were actually reaching receptive cloud conditions or does it mean that the primary effects of seeding were followed by secondary effects well beyond the original target? Such secondary effects could occur, for instance, if seeding materials become entrained in a downdraft and then are carried outward into the updraft of other clouds. Or are seeding produced cold pools beneath the clouds colder than unseeded clouds thereby forming new cells on the boundaries of the cold pools? In the case of the hygroscopic seeding experiments the postulated dynamic effects due to microphysical and dynamical interactions in the cloud and sub-cloud region and with the environment could result in longer-lived or progeny clouds. Another related uncertainty in seeding convective systems is whether a positive effect on some individual clouds (or cloud complexes) will aggregate to result in increased area rainfall. As more is learned about the global water balance and as new tools enable the cloud scientist to better understand clouds and their response to seeding, the question of extended area affects will likely become better defined and understood.

8.4.1 Hail Suppression

In addition to studies using cloud seeding principles to enhance rainfall, cloud seeding with aerosols have been used to reduce hail damage. The main concepts for suppressing hail by cloud seeding are as follows:

8.4.1.1 The Glaciation Concept

The aim of hail suppression by glaciation is to introduce so many ice crystals via seeding that the ice crystals consume all the available supercooled liquid water as they grow by vapour deposition and riming of cloud droplets. To be effective this technique requires the insertion of very large amounts of seeding materials in the storm updrafts. Modeling studies (Weickmann 1964; Dennis and Musil 1973; English 1973; Young 1977) have suggested that unless very large amounts of seeding material are used, the strongest updrafts remain all liquid and hail growth is not substantially affected. Therefore, the glaciation concept is generally thought not to be a feasible approach to hail suppression. The glaciation concept is also not popular because many scientists think that it may result in a reduction in rainfall along with hail. Since most hail-prone areas are semi-arid, the loss of rainfall can have a greater adverse impact on agriculture than economic gains from hail suppression.

8.4.1.2 The Embryo Competition Concept

The competing embryo concept, first introduced by Iribarne and dePena (1962), involves the introduction of modest concentrations of hailstone embryos (on the order of 10 m^{-3}) in the regions of major hailstone growth. The idea is that millimeter-sized ice particles will then compete "beneficially" for the available supercooled water and result in numerous small hailstones or graupel particles rather than a few large, damaging hailstones. Because it is not economically feasible to introduce hailstone embryos directly in the cloud, one must use a seeding strategy, which utilizes the storm's natural hailstone embryo manufacturing process. For example, the Soviet concept of hail suppression (Bibilashvili et al. 1974) can be considered an embryo competition strategy. In their case the hypothesized hailstone embryos are frozen supercooled raindrops. By dispersing seeding material into a region containing supercooled raindrops, the raindrops readily freeze and immediately become millimeter-sized hailstone embryos. The numerous hailstone embryos then beneficially compete for the available supercooled water resulting in the formation of numerous small hailstones, many of which would melt before reaching the ground.

Now consider a cloud in which supercooled raindrops are not present. In such clouds millimeter-sized ice particles first must form by vapour deposition until ice crystals of the size of a few hundred micrometers (0.1 mm) form. This takes a significant amount of time, on the order of 5 to 10 min. The larger vapour-grown ice crystals can then settle through a population of cloud droplets and grow rapidly by riming those droplets to form graupel particles. The larger ice crystals can also collide with each other to form aggregates. Both the graupel particles and aggregates can serve as hailstone embryos since they have significant fall velocities and cross-sectional areas to enable them to grow rapidly by accreting supercooled cloud droplets to form hailstones.

As a result of the significant amount of time for hailstone embryos to form, seeding intense updrafts, such as exist in supercell storms and the mature cell of severe multicell storms with weak echo regions, is unlikely to have any significant effect on hail growth. The ice crystals formed from seeding would probably be swept aloft into the anvil before becoming large enough to serve as embryos of hailstones. In the case of multicell storms, the recommended approach is to seed in the flanking towering cumulus clouds where updrafts are weaker and transient. If the cell is a daughter cell or a cell that eventually becomes a mature cell, it may be laden with numerous artificially produced hailstone embryos. Likewise, if the flanking cell is in the right location to serve as a feeder cell, then the natural and artificially produced hailstone embryos will be entrained into the mature cell to beneficially compete for the supercooled liquid water and reduce the size of hailstones.

8.4.1.3 Early Rainout and/or Trajectory Lowering Liquid Water Depletion

The idea behind early rainout is to initiate ice-phase precipitation lower into the feeder or daughter cell clouds where temperatures are in the range –5 to –15°C. If the prematurely initiated precipitation settles below an otherwise rain-free base, it could fall into the inflow into the storm and impede the flow of moisture into the storm, which, in turn, would reduce supercooled liquid water contents deeper in the storm. In addition, initiation of ice lower in the smaller turrets has the potential of reducing supercooled liquid water available for hail growth in the larger turrets, where updrafts are stronger and more conducive to the growth of larger hailstones.

Essentially the same seeding strategy can be used to achieve early rainout and trajectory lowering, since if the precipitation initiated lower in the cloud does not precipitate from the rain-free cloud base, the number of hail embryos is increased causing beneficial competition.

Another proposed approach to hail reduction is to seed the base of clouds with hygroscopic materials and thereby initiate a warm rain process in the lower levels of the cloud. The concept behind this trajectory-lowering technique is that the precipitation settling out of the lower part of the cloud will deplete the liquid water in the cloud and therefore limit hailstone growth. It is based on the hygroscopic seeding strategies discussed above in which salt particles or some other hygroscopic material is introduced into the base of flanking clouds, thereby initiating a more vigorous warm rain process in the lower levels of the cloud. Some early cloud modeling studies (Young 1977) suggest that this may be a feasible approach in regions such as the High Plains of the United States or Canada where cloud base temperatures are cold and cloud droplet concentrations are large.

It has also been proposed to use a combination of salt seeding and ice phase seeding to both deplete supercooled liquid water and to promote beneficial embryo competition (Dennis and Musil 1973). Only limited exploratory field studies have been performed to examine the feasibility of this approach.

These concepts are illustrated in Fig. 8.2. Developing flanking line cells with weaker updrafts are shown on the left of the figure and the mature cell with strong updrafts on the right. One can also interpret the figure as being a time-height cross section of a storm with zero time on the left and the time of the dissipating cells on the far right. We emphasize here that it is the weaker updraft regions of developing cumulus congestus, rather than the main cumulonimbus cell that is the preferred region for seeding.

In general, operational hail suppression operations follow the embryo competition concept. While operational hail suppression programs are carried out in many countries in Europe, Canada, China, the U.S. and elsewhere, scientific field confirmation of the concepts have been unsuccessful (Atlas 1977; Federer et al. 1986).

Only long-term statistical analyses of non-randomized, operational programs have provided more convincing evidence suggested that seeding could significantly reduce hail frequency (Mesinger and Mesinger 1992; Rudolph et al. 1994; Smith et al. 1997; Dessens 1998; Eklund et al. 1999;). An advantage of evaluating an operational program is that often one can work with long-period records, such as 40 years in Yugoslavia, whereas randomized research programs typically cannot get funding for more than 5 years or so. The disadvantage is that one cannot totally eliminate concerns about natural variability in the climate (see comments in NRC 2003).

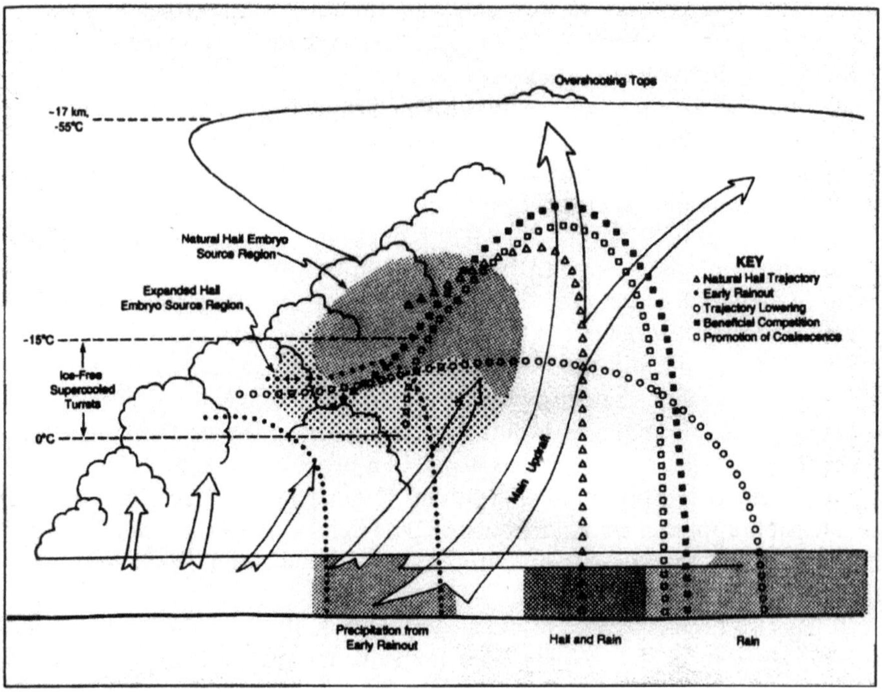

Fig. 8.2 Hail suppression concepts. (World Meteorological Organization 1996)

Overall the scientific basis of hail suppression remains not fully resolved. Similarly there is no sound basis on which to base an assessment of the potential impact of pollution hailstorms. Nonetheless owing to the persistence of pollution aerosols and potential for pollution to serve as CCN, GCCN, and IN, pollution may have a more demonstratable effect on hailstorms than deliberate hail suppression.

8.5 Implications of Cloud Seeding Research to Investigations of Aerosol Pollution Effects on Precipitation

In view of the findings summarized in the foregoing sections, it is clear that aerosol from pollution and biomass burning will have certain effects on clouds and on the conditions that govern precipitation development (ice concentrations, drop sizes etc.). It is less clear whether precipitation on the ground can be significantly altered. However, in drawing parallels between deliberate and inadvertent seeding of clouds some important differences must be kept in mind. In some respects, the situation with inadvertent seeding is more complex, but there are also simplifying factors.

Some of the complicating factors in evaluating the inadvertent effects of aerosol pollution on clouds are as follows:

- At any specific location in the atmosphere the particles that may affect cloud structures and precipitation derive from both natural and anthropogenic sources (biomass burning, industrial, cities, agricultural practices and others). In most cases as opposed to cloud seeding experiments, it is difficult, if not impossible, to determine and differentiate the type of particles (Chapters 2, 3, 6 and 7).
- Particles with different physical and chemical properties are emitted by most pollutant sources (cf. the variety of emissions from a biomass burning, city, industrial etc.). These various particles may affect clouds in different ways, some of which could be self-canceling (See Chapters 2, 6 and 7).
- In addition to any cloud-active particles emitted directly by an anthropogenic source, various gases (e.g. SO_2, NO_x) emitted by the same source may be converted into cloud-active particles downwind (e.g. into sulphates and nitrates). One consequence of such gas-to-particle conversion is that an anthropogenic source may affect cloud structures and precipitation well downwind of the source (see Chapters 2 and 3). This is particularly true once the anthropogenic source material is processed through clouds downwind of the source.

In general, it is more difficult to carry out controlled randomized experiments using anthropogenic emissions as a source of particles. Alternate methods of evaluation are needed. On the other hand, there are some advantages in utilizing anthropogenic sources to investigate the modification of clouds and precipitation by particles.

- In some cases concentrations of cloud-active particles emitted by certain industries (e.g. cloud condensation nuclei from paper mills) have been measured. Moreover, these particles are generally emitted continuously, day and night. Consequently, cloud structures and precipitation downwind of such sources may differ significantly from regions upwind of the source.
- In those cases where a new industrial site is established in what was a previously pristine location, the structures of clouds and precipitation in the vicinity of the site can be compared before and after the establishment of the industry. Also, in the case of an isolated industry, it should be possible to characterize rather well the nature of the emissions. However, a complicating factor is the potential for climate variations during the period prior to and after the construction of the new industrial site.
- Variations in pollution output diurnally and for weekdays versus weekends may provide a means of assessing cause and affect

In summary, the history of evaluating the effects of cloud seeding on precipitation has taught us a number of lessons that are useful for evaluating inadvertent modification of clouds and precipitation by aerosols.

The scientific community has established a set of criteria for determining that there is "proof" that seeding has enhanced precipitation. For firm "proof" (see

NRC 2003; Garstang et al. 2005) that seeding affects precipitation, *both* strong physical evidence of appropriate modifications to cloud structures and highly significant statistical evidence is required. Likewise, for firm “proof” that human production of aerosol in the atmosphere is altering precipitation, both strong physical evidence of appropriate modifications to cloud structures and significant statistical evidence is required. Unfortunately, as noted in Cotton and Pielke (2007), there is a human tendency to accept the results of assessments of inadvertent modification of clouds and precipitation even though those studies do not meet the standards of “proof”, while at the same time requiring that cloud seeding evaluations meet such “proof” criteria.

Another lesson from evaluating cloud seeding experiments is that natural variability of clouds and precipitation can be quite large and thus can inhibit conclusive evaluation of even the best designed statistical experiments. The same can be said for evaluating the effects of inadvertent introduction of aerosols into the climate system. If the signal is not strong, then to evaluate if human activity has produced some observed effect (cause and effect), one requires much longer time records than is available for most if not all data sets. Thus we often do not even have a measure of the natural variability of the climate effecting clouds and precipitation.

Chapter 9
Summary

Aerosols impact global climate in a number of ways. First they directly affect the Earth's radiation budget by absorbing and reflecting solar radiation and to a lesser extent altering the profile of IR absorption in the atmosphere. Second, by serving as cloud condensation nuclei (CCN) and ice nuclei (IN), they determine cloud microphysics, the formation of precipitating particles and cloud radiative properties, particularly cloud albedo and emission. As a consequence, these properties influence the local radiation budget, atmospheric temperatures, land-surface and ocean temperatures. Aerosols can therefore affect regional cloud properties and may affect precipitation amounts. If these effects are large enough, they could adversely affect water availability, and will impact global climate by changing land-surface radiative properties and surface energy budgets, which can alter the general circulation of the atmosphere and climate.

This report summarizes our present knowledge of the effects of pollution on clouds and precipitation from regional to global scales. Although a great deal of progress has been made in the past few years regarding our understanding of the effects of pollution on cloud processes, the effects on precipitation on the level of a single storm, on a regional and on a global scale are not yet understood. Below we summarize some of the effects that have been observed and some of the gaps in our knowledge and the difficulties in assessing the effects of pollution on rainfall.

Although our ability to forecast precipitation based on climate models is still limited, we can already conclude that some changes in precipitation would take place. Global climate model estimates of the change in global mean precipitation due to the total aerosol effects vary between 0 and –4.5%. These differences are amplified over land, ranging from –1.5 to –8.5%. Recent coupled ocean mixed layer GCM simulations show that the cooling due to the direct and indirect effects by aerosols reduce surface latent and sensible heat transfer and, as a consequence, act to reduce the intensity of the hydrological cycle. Thus the potential influence of aerosols on climate could be far more significant than previously thought. Estimates by GCMs of the consequences of combined greenhouse gas warming and aerosol cooling in the future depend on uncertain estimates of future pollution emissions and greenhouse

Z. Levin, W.R. Cotton (eds.), *Aerosol Pollution Impact on Precipitation*,

gas releases into the atmosphere, as well as the uncertainties associated with the parameterizations of cloud-aerosol interactions in the models.

9.1 Specific Points

- The mere presence of aerosols in the atmosphere modifies the radiation balance by warming or cooling the atmosphere and thus affecting convection and precipitation.
- The role of aerosols in clouds containing both water and ice is complex because precipitation can grow in a number of ways: by the collision-coalescence process, by ice growth by deposition, by the riming of ice particles, or by all of the above.
- Ice plays an important role in precipitation development in many mid-to high latitude clouds. The role of ice in more temperate climates is probably less dominant and precipitation development by the warm cloud processes (condensation and collection) often dominates.
- There are recent satellites measurements showing positive correlations between cloud cover (cloud fraction) and the loading of scattering aerosol (aerosols from pollution and from natural sources, e, g. dust), but the reasons for this have yet to be established. Similar results from model simulations have been obtained.
- There are indications from both observations and modeling studies that a predominantly absorbing fine particle aerosol can reduce cloud fraction.

9.1.1 The Effects of CCN on Clouds (Chapters 2, 6 and 7)

- Cloud droplets are formed by nucleation on cloud condensation nuclei from natural or polluted sources.
- The CCN are a subset of the total aerosols present in the atmosphere. Most of the pollution aerosols above a certain size threshold are effective CCN (e.g. sulphate, nitrates and certain organics).
- Many particles that are not very efficient CCN become effective CCN due to various chemical reactions (gas and aqueous phase) that take place during their transport in the atmosphere, especially when they move through clouds.
- Field measurements of CCN and GCCN concentrations in dusty atmospheres exhibit higher concentrations than in corresponding dust-free environments. Moreover, dust particles coated with soluble material (sulphate and sea salt) have been analyzed in the residual materials from sampled ice crystals, further indicating their role in cloud and ice formation in the atmosphere.
- The number of cloud drops formed depends on these aerosol properties and the strength of the updraft. For a given aerosol, stronger updrafts produce

higher supersaturations leading to a larger fraction of the aerosols being activated, thus higher supersaturation leads to the activation of more CCN, when they are available.

- Field measurements and modeling studies show that increasing CCN concentrations (from pollution or from natural sources) increases cloud droplet concentrations.
- At given liquid water content the increase in droplet concentrations results in smaller cloud drops.
- Very small concentrations ($\sim$1 L^{-1}) of large or giant CCN (GCCN) can serve as precipitation embryos and significantly affect the growth of precipitation.

9.1.2 The Role of Ice Nuclei in Cloud Processes

- While it is well established that ice particles can have a major role in the formation of precipitation globally, our understanding of the details of the processes are critically short of what would be required for valid representation of these processes in cloud and climate models.
- Ice formation in the atmosphere can occur through nucleation on ice nuclei (IN) ("primary"), freezing of liquid droplets (homogeneous freezing) or by secondary processes (e.g. multiplication processes).
- Ice nuclei are a very small subset of atmospheric aerosols and are generally smaller than the CCN concentrations by up to five to six orders of magnitude. The concentration of activated nuclei increases with decreasing temperature and with supersaturation with respect to ice.
- Many measurements suggest that the most effective IN in the atmosphere originate from mineral dust. Some particles from industrial sources and biogenic materials have also been shown to be effective IN.
- There have been many measurements showing that the number of ice crystals in clouds exceeds the concentrations of IN by several orders of magnitude under certain conditions. To explain the discrepancy a number of ice multiplication mechanisms have been proposed.
- Recent observations suggest that in midlatitude frontal clouds the average concentrations of ice crystals seem to be independent of temperature over a large range of temperatures (−5 to −30°C).

9.1.3 The Effects of Aerosol Pollution on Precipitation

- Obtaining quantitative estimation of the effects of aerosol pollution on precipitation necessitates the design of new analysis methods and new experiments that could separate the aerosol effects from meteorological effects.

- There are only a few observational studies of aerosol effects on precipitation and they lack quantitative estimates of the aerosol and precipitation changes.
- Measurements in the Amazon region show that clouds affected by biomass burning have higher concentrations of cloud drops, narrower drop size spectra and slower development of precipitation embryos.
- When clouds form directly above the concentrated intense plumes of the fires, they developed strong updrafts, taller clouds with small drops, thus delaying precipitation formation.
- Precipitation downwind from paper mills has been shown to increase due to the emission of effective GCCN. Motivated by these studies, GCCN have been used to artificially seed clouds for rain enhancement (see chapter 8).
- Precipitation downwind from sugar cane fires in Australia was found to decrease, however the role of meteorology in this reduction could not be ruled out.
- Precipitation around urban regions (such as St Louis, New York and Houston) has been shown to increase in some cases, but the connection to increased aerosol pollution could not be established. A more convincing connection was found between changes in precipitation and urban-land use and other meteorological factors.
- Modeling studies of single mixed phase clouds show both decreases and increases in precipitation. Extensive modeling of single mixed phase clouds show that precipitation from clean clouds is most susceptible to aerosol pollution. In these cases tripling the CCN concentrations decreases precipitation amounts by about the same factor. Polluted clouds, on the other hand, are less susceptible to increases in CCN, and beyond about 1500 CCN per cm^3(active at 1% supersaturation) very little change in precipitation is shown.
- Both observational and modeling studies suggest that orographic clouds are the most susceptible to modification in precipitation by pollution owing to the modest liquid water contents in them, the relatively short time the drops and ice crystals spend in the clouds and the large areal coverage. Some observational studies suggest that aerosol pollution may substantially decrease orographic precipitation. Although these studies needs to be validated by more experiments, they have major ramifications in watersheds which depend upon orographic precipitation as the major water resource,
- Recent statistical analysis of the ratio (R_0) of orographic rainfall downwind to upwind of polluted sources shows an inverse trend between pollution and rainfall on the upslope of the mountains. Increases in pollution leads to a decrease in R_0. Smaller decreases were reported at the mountain top and a small increase was reported on the downwind side of mountains.
- A few giant CCN per liter of air enhance precipitation development. These effects are more pronounced in polluted clouds where the GCCN grow rapidly to precipitation size particles, both large drops and graupel. Model simulations suggest that the increase in precipitation due to GCCN pollution only moderates the larger decrease in rainfall due to small pollution aerosols.

- Models show that since pollution alters droplet distributions, this can influence glaciation and precipitation development, but only limited observational studies have quantified this process at this point.
- Models also show that the changes in cloud and precipitation development due to pollution have effects on the dynamics of clouds. Enhanced downdrafts create cold pools below clouds, which in some cases can transform individual clouds into long-lived squall lines with enhanced precipitation, and in other cases lead to weakening of storms and reduced precipitation.
- Some models of cloud fields and some cloud seeding experiments suggest that the response of the clouds to aerosol changes may take place over periods of several hours and multiple cloud lifecycles; the reasons for these lasting effects are not clearly understood.

9.1.4 Gaps in Our Knowledge

- Ice formation in clouds is still not fully understood.
- The impact of pollution on ice formation has not been clearly demonstrated.
- Although models suggest that GCCN enhance precipitation formation, there are relatively few observations that confirm the enhanced effects on precipitation on the ground.
- Evaluation of the effects of pollution or aerosols from natural sources on precipitation has to rely on large amounts of data. Although case studies of individual storms can shed some light on the role of various processes, they cannot determine the long-term climatological effects of pollution aerosol on precipitation. At the same time, long term measurements also suffer from difficulties stemming from climatic fluctuations that take place during the experiment. Researchers in weather modification have for years faced similar problems in their evaluation of cloud seeding experiments.
- A method to assess the effect of aerosols on precipitation can include direct measurements of the correlation between precipitation efficiency and aerosol concentration, whether from the surface, aircraft or satellites. Precipitation efficiency involves not only the measurement of surface precipitation but also horizontal water vapour flux and vertical air motion within clouds. Even though measurements and model simulations show the strong pollution interference with cloud microphysics, the magnitude or direction of the effect on precipitation and its variations with meteorological conditions have not been clearly identified.
- The expected changes in precipitation due to enhanced pollution or biomass burning could be smaller than the errors of the measurements. Therefore, in order to increase signal to noise ratio, the measurements of all variables must be accurate enough to be able to resolve the expected changes. Unfortunately, many of the instruments listed in Chapter 5 fall short of this requirement. Reasons for *the limited accuracy of the measurements include:*

  - Radar measurements of precipitation can be off by up to 100%.
  - Retrievals of the cloud droplet size (effective radius) from satellites are biased by up to 30% for convective clouds, due to departure of the clouds from the plane parallel model. However we can still measure the relative change of cloud properties with change in aerosol concentrations. Moreover, retrievals of effective radius are biased heavily by the presence of ice in a pixel. Therefore, effective radii are not necessarily representative of droplet sizes that affect precipitation formation.
  - Aerosol column optical thicknesses measured from satellites are not yet accurate enough (±20% over the oceans, but less accurate over vegetated land and over the deserts).
  - Ground based remote sensing of aerosol can have a precision of ±0.02 in units of optical depth.
  - Translating the optical thickness into the column aerosol concentration and into CCN is difficult due to variation in the aerosol size distribution and chemical composition, water uptake and the vertical distribution of the aerosols. These are critical parameters needed for interpreting aerosols ingestion and effects on clouds.
  - In situ measurements from aircraft have limitations in correctly sampling all aerosol sizes. In addition, such measurements only represent small volumes that may not always correctly represent the processes that affect clouds. Ground based measurements may not always represent the properties of the aerosols and the clouds above.

- Cloud properties have a diurnal cycle not well represented by polar satellite observations. Geostationary satellite measure the diurnal cycle but with limited ability to derive the cloud droplet size.
- Once a statistical connection is established between the aerosol concentration, the cloud properties, the large scale water vapour flux, and precipitation, it is still difficult to assess whether the changes in the precipitation patterns are a results of the aerosols impact, whether it is changing of the precipitation patterns that affect the aerosol properties or that both phenomena are concomitant, reflecting properties of the air mass.
- Most informative measurements so far on the effect of aerosols on precipitation do not include simultaneous quantitative measurements of both aerosols and precipitation.
- Because of the coarse grid spacing of GCMs, clouds and cloud properties essentially have to be parameterized as sub-grid-scale processes. The challenge is to parameterize aerosol activation, the influence of aerosols on precipitation processes, the precipitation processes and their feedbacks on cloud and storm dynamics, and cloud processing of aerosols all in sub-grid-scale parameterizations.

Chapter 10
Recommendations

- It is recommended that a series of international projects targeted toward unraveling the complex interactions among aerosols, clouds, and precipitation be implemented. A series of international workshops and field studies are needed to address the impacts on clouds and precipitation of aerosols from a range of sources including biomass burning, dust, and industrial pollution within different regional weather regimes in the tropics, middle latitudes, and the polar regions. All of these studies need to have an adequate scientific component, but should also deal with specific situations occurring in the developing world by addressing the impact of pollution/precipitation on economic, sociological, environmental and health issues.
- It is recommended that WMO/IUGG take the lead in such projects together with other UN and International Organizations. Some of these projects could be sponsored and financially supported by the countries involved. For example, the effects of an evolving industrial economy, such as China, on precipitation should be studied. Similarly, a study of the effects of biomass burning and dust in some of the African regions would be highly valuable for our understanding of these complex issues.
- It is also proposed to consider a project in a country where the relationship between pollution and precipitation can be explored with the a large arsenal of the newest scientific instrumentation. The WMO/IUGG can play a key coordination role in making sure the following recommendations are implemented:

10.1 Better Characterization of Aerosols

10.1.1 Emission Inventories

- To be useful as a reference for studies of pollution effects on cloud and precipitation, size-resolved measurements of the numbers of particles emitted is required for the various primary aerosol sources, especially sea salt, mineral dust, pyrogenic aerosol and primary biogenic aerosol.

- The abundance and rate of production of submicron sea salt particles is still under dispute. Since these particles can play a significant role in cloud microphysics and precipitation in remote marine regions, observational constraints to their source parameterizations are urgently needed.
- Due to the complexity of carbonaceous aerosol composition and emission processes, a critical concern is to define an adequate classification scheme for these aerosols based on observable characteristics.
- There is a need to build a reliable inventory of emission parameterization of carbonaceous aerosols (BC, PPOC and SOA) that includes size distributions and number fluxes.
- In building an inventory of emission the injection height should be included since changes in injection height due to shifts in agricultural or technical practices will have important consequences on the lifetime and fate of particles.

10.1.2 Chemical Processes, Physical Properties and Instrumentation

- Accurate knowledge of the rates of the processes leading from SO_2 to sulphate aerosols is needed for clear and cloudy atmospheres.
- The ability of different types of particles (e.g. mineral dust particles, biomass smoke, biogenic aerosols, carbonaceous aerosols) to act as CCN, GCCN, and IN as a function of aerosol size, origin, and air mass history needs to be determined.
- In order to assess the human impact on cloud physics, we need to know the aerosol, CCN and IN distributions in the pre-anthropogenic atmosphere. Novel methods of inferring or backing-out pre-industrial cloud nucleating aerosol distributions are needed. The sources, characteristics, and fluxes of natural aerosol types must be investigated by carefully conducted field and laboratory studies.
- New instruments should be developed and measurements should be carried out to determine CCN, GCCN and IN concentrations as a function of particle size, composition and supersaturation. Uncertainties are greatest for IN because only a few of the known modes of ice nucleation are measured in current field-deployable devices. These measurements should be conducted in the context of closure studies.
- A program for validation and intercomparison of CCN, GCCN, and IN measurements should be carried out. This should be in the form of workshops in which instruments are compared using the same air samples.
- Global coordination of observational networks is needed for more complete coverage of global aerosols. These could combine in situ and ground-based remote sensing methods (e.g. AERONET).

- Continuity of the observations and suitable locations of the observation sites (depending on the scale of phenomena) is very important. Aerosol type classification methods using depolarization ratio, wavelength dependence, and the lidar ratio (if available) are useful for the comparison.
- More accurate assessment from satellites of the aerosol column concentration, vertical and size distribution, refractive index and single scattering albedo is needed.
- It is important to extend the measurements to small (100 nm particle diameter) particles in the retrievals, since many of these could be efficient CCN.

10.2 The Effects on Clouds and Precipitation

- Vertical profiles of cloud microphysical processes, size, composition and phase of hydrometeors should be carried out by profiling from aircraft and from satellites and ground measurements. This is especially important in the tropics in areas affected by biomass burning and highly populated areas. Such data could be used to validate model simulations and to shed more light on the main processes leading to precipitation under clear and polluted conditions.
- More accurate measurements of cloud droplet size, liquid water path and total condensate through remote sensing should be developed.
- Most optical spectrometers today cannot accurately distinguish small droplets from ice crystals. In light of recent measurements showing high concentrations of very small (<50 μm) ice crystals it is recommended that new in situ instruments for measuring very small (5–30 micrometer diameter) ice crystals be developed.
- Experiments should be designed to better understand the role of ice in precipitation development and the connection between IN and ice crystals in clouds.
- Multi-year measurements from space of precipitation patterns, along with retrievals of cloud nucleating aerosols, are needed to assess both regional and global impacts of aerosol pollution on precipitation. These space borne studies should be supplemented with intensive field campaigns in which coordinated detailed observations of clouds and aerosols are obtained.
- Because both observational and modeling studies suggest that orographic clouds are the most susceptible to precipitation modifications due to pollution aerosols, and because the ramifications on water resources are large, it is recommended that focused, coordinated observational and modeling campaigns be implemented to study aerosol precipitation interactions for selected watersheds where pollution effects are likely to be large. Examples include the 2006 Suppression of Precipitation (SUPRECIP) study in California, the planned Convective and Orographically-induced Precipitation Study (COPS) experiment in Germany in 2007 and the Colorado River

Basin in the western U.S. It is recommended that similar studies be conducted in regions such as China and India, where the effects of mega-cities on precipitation could be evaluated. These studies should be basin wide and include ground-based, airborne and satellite measurements of aerosol properties, cloud microphysical measurements, surface and radar (particularly cloud radar) measurements of cloud structures and precipitation, basin-wide hydrological measurements of liquid precipitation, snowpack and runoff etc.

- Because of the indications of strong influences of urban land-use and aerosol pollution on precipitation and lightning, it is recommended that a coordinated modeling and observational campaign be established in a number of large metropolitan regions where convection is prevalent. Crucial to site selection is the importance of finding locations where the local physiography (i.e. topography, land-water interfaces) is relatively simple and where generally widespread aerosol pollution is minimal. Measurements should include the list given above plus greater attention to surface measurements of fluxes and land-use properties.
- Because cloud-resolving models suggest a significant dynamic response of cumulus clouds to pollution aerosols which then modifies simulated precipitation, particularly through secondary dynamic responses of clouds via cold pools and gravity waves, it is recommended that a coordinated observational and modeling campaign be organized to investigate the response of cloud systems to varying amounts and characteristics of pollution. These studies should include ground-based, airborne, and satellite measurements of aerosol properties, cloud microphysics measurements, surface mesonets, ground-based multiple Doppler radars, and airborne Doppler radars. These campaigns should be done in regions of significant biomass burning or urban aerosol emissions, and in other locations where the meteorology varies such that different cloud responses can be expected.
- Modeling studies that extend beyond single-cloud responses and also consider the cloud ensemble response to variations in cloud nucleating aerosols are needed. These types of simulations should be combined with a number of runs using random variations in initial conditions, in order to determine the statistical variations of the final product, namely the precipitation on the ground.
- Models are particularly useful for providing quantified estimates of parameters such as precipitation amounts for a given simulation. However, the skill level of models in so doing should be quantified statistically.
- Models should be used to isolate the role of aerosols from the influence of the meteorology and to give a quantitative answer as to the relative importance of each process.
- Models are very effective tools for isolating *cause-and-effect* in a physical system. As models become more complex, it becomes increasingly difficult to isolate causal relations. Effort should be put into new methodologies for quantifying the sensitivity and response of a parameter such as precipitation to changes in input parameters such as pollution or changes in a combination

of inputs. Possibly the method of factor separation should be more frequently used to determine the relative importance of each parameter.

- Because of the coarse grid spacing of GCMs, clouds and cloud properties essentially have to be parameterized as sub-grid-scale processes. The challenge is to parameterize aerosol activation, the influence of aerosols on cloud microphysics and precipitation processes, the precipitation processes and their feedbacks on cloud and storm dynamics, and cloud processing of aerosols, all in sub-grid-scale parameterizations. There is a need to develop new and novel methods of parameterizing aerosol interactions with clouds in large-scale models. It is urged that aerosol-cloud-precipitation specialists in observations and modeling collaborate with GCM developers to refine the representation (parameterization) of aerosol-cloud-precipitation processes in GCMs. Such a strategy of combining observations, cloud resolving models and climate models has been put forward in the GEWEX Cloud System Study context.

10.3 Immediate Action Items

- It is recommended that aerosol monitoring stations be equipped with instruments for measuring CCN activation spectra in order to be able to establish empirical relationships between measured aerosol properties and their potential effects on clouds.
- A Workshop should be held on developing improved instrumentation for measuring IN, small ice particles, and precipitation. Because of the complexities involved, it might be necessary to develop an overall strategy to address these issues.
- A Workshop should be held to determine a strategy for a focused, coordinated observational and modeling campaign that could address the effects that aerosols have on orographic clouds. This might include additions to existing planned measurement campaigns or a new campaign(s).
- It is recommended that aerosol cloud precipitation cloud resolving model intercomparison studies be implemented to serve as a stimulus for model refinement research and evaluations of model performance. These should be done for a variety of regimes including warm and cold-season orographic cloud models, shallow convection and deep convection, and stratiform cloud systems. These studies could be linked to some of the campaigns suggested above or to other research campaigns where implementation of an aerosol component could be accomplished easily and economically.
- There is a strong need to assemble data sets or climatology that can be used for the assessment of climate simulations with GCMs, including aerosol-cloud-precipitation parameterizations. Some of the parameters that need evaluation include seasonal and annual precipitation amounts, regional precipitation climatology, global and regional aerosol distributions, and

TOA radiation budgets. This could be done in the context of GCM inter-comparison studies much like that recommended for cloud resolving models.

- It is recommended that existing and new statistical methods be applied to current and future data sets to distinguish aerosol effects on precipitation from meteorological influences. Numerical models should play an integral role in this process where they can serve as "predictors" or "covariates" in the statistical evaluations.
- Finally it is recommended that the WMO creates a web site to serve as a central index to archives of public domain measurements of aerosol, cloud, and precipitation parameters. Individual agencies/countries are to be encouraged to maintain long term web-accessible archives of research (including short-term field experiments) and operational measurements. Archives are encouraged to include key measured instrument parameters as well as retrieved parameters.

Appendix A
List of Acronyms Used in the Report

ABC	Atmospheric Brown Cloud
ACE	Aerosol Characterization Experiments
ACE-2 CLOUDY-COLUMN	CLOUDY-COLUMN is one of ACE-2 projects, which took place in June–July 1997, between Portugal and the Canary Islands
AD-Net	Asian Dust Network
AEROCE	Atmospheric Ocean Chemistry Experiment
AEROCOM	Aerosol model intercomparison
AERONET	Aerosol Robotic Network http://aeronet.gsfc.nasa.gov
AEROSIBNET	Siberian system for aerosol research http://sibrad.iao.ru/
AGCM	atmospheric general circulation model
AMS	American Meteorological Society
AOD	Aerosol optical depth
AOT	Aerosol Optical Thickness
APS	Aerosol Mass Spectrometer
ARM	Atmospheric Radiation Measurement www.**arm**.gov/
ASTEX	Atlantic Stratocumulus Transition Experiment
ATSR	Along Track Scanning Radiometer
AVHRR	Advanced Very High Resolution Radiometer
BASE	Beaufort and Arctic Storms Experiment
BC	Black Carbon
BRE	Bridger Range Experiment
BSRN	Baseline Surface Radiation Network/World Radiation Monitoring Center
CALIPSO	Cloud-Aerosol Lidar and Infrared Pathfinder Satellite Observation
CAPMoN	Canadian Air and Precipitation Monitoring Network
CAPS	Cloud Aerosol Precipitation Spectrometer
CASTnet	Clean Air Status and Trends Network
CCM3	Community Climate Model
CCN	Cloud Condensation Nuclei
CDNC	Cloud Droplet Number Concentration
CDP	Cloud Droplet Probe
CFDC	Continuous-Flow Diffusion Chamber
CFDE (I and III)	Canadian Atlantic Storms Program (I and III)
CFS	Continuous Flow Spectrometers
CGMS	Coordination Group for Meteorological Satellites
CIS-LiNet	Commonwealth of Independent States Lidar Network

Climax	Cloud seeding experiment for rain enhancement near Climax Colorado
CLIVAR	Climate Variability and Predictability
CMAP	NOAA Climate Predication Center's Merged Analysis of Precipitation
CMDL	Climate Monitoring & Diagnostics Laboratory.
CN	Condensation Nuclei
CNES	French Space Agency
COS	Carbonyl Sulphide
COSAM	Comparison of Large Scale Sulphate Aerosol Models
CPC	Condensation Particle Counter
CVI	Counterflow Virtual Impactor
DISORT	Discrete Ordinate Radiative Transfer Model
DMPS	Differential Mobility Particle Sizer
DMS	Dimethyl Sulphide
DWD	Deutscher Wetterdienst
DSD	Drop Size Distribution
DYCOMS-II	Dynamics and Chemistry of Marine Stratocumulus
EARLINET	European Aerosol Research Lidar Network
EC	Elemental Carbon
ECHAM5	A general circulation model of the atmosphere, developed at the Max Planck Institute, Hamburg.
EF	Emission Factors
EI	Electron Impact Ionization
EMEP	European Monitoring and Evaluation Program
EPIC	Eastern Pacific Investigation of Climatexperiment
EUROTRAC	Project on the Transport and Chemical Transformation of Environmentally Relevant Trace Constituents in the Troposphere over Europe
FACE	Florida Area Cumulus Experiment
FAO	Food and Agriculture Organization of the United Nations
FAR	False Alarm Rate
FIRE	First ISCCP Regional Experiment
FIRE.ACE	First (ISCCP) Regional Experiment – Arctic Cloud Experiment
FPD	Flame Photometric Detectors
GAW	Global Atmospheric Watch
GAW-PFR	Global Atmospheric Watch Precision Filter Radiometer Network
GCCN	Giant CCN
GCM	Global Climate Model
GEIA	Global Emission Inventory
GLAS	Geoscience Laser Altimeter System
GOES	Geostationary Operational Environmental Satellites
GPCP	Global Precipitation Climatology Project
GPI	GOES Precipitation Index algorithm
GPS	Global Positioning System
HIPLEX	High Plains Experiment
HSRL	High-spectral resolution lidar
HTDMA	Humidified Tandem Differential Mobility Analyzers
HUJ	The Hebrew University of Jerusalem

HULIS	Humic-like substances
IC	Ion chromatography
ICESat	Ice, Cloud, and land Elevation Satellite
ICP-MS	Inductively Coupled Plasma-Mass Spectrometry
IDDI	Infrared Difference Dust Index
IFA	Intensive Flux Array
IMAGES	Intermediate Model of Global Evolution of Species SCAPE2
IMPROVE	Interagency Monitoring of Protected Visual Environments
INAA	Instrumental Neutron Activation Analysis
INDOEX	Indian Ocean Experiment
IN	Ice Nuclei
IPCC	TAR Intergovernmental Panel for Climate Change
IPWG	International Precipitation Working Group
ISCCP	International Satellite Cloud Climatology Project
ITCZ	Inter Tropical Convergence Zone
KNMI	Koninklijk Nederlands Meteorologisch Instituut http://www.knmi.nl
LAMMA	Laser Microprobe Mass Analysis
LBA-SMOCC	Large Scale Biosphere-Atmosphere Experiment in Amazonia-Smoke, Aerosols, Clouds, Rainfall and Climate
LCF	Longwave Cloud Forcing
LCL	Lifting Condensation Level
LFC	Level of Free Convection
LITE	Lidar In-Space Technology Experiment
LMDzT INCA	A model, coupling the general circulation model LMDzT of the Laboratoire de Météorologie Dynamique to the INCA (Interaction with chemistry and aerosols) module of CEA-CNRS,
LWC	Liquid Water Content
LWP	Liquid Water Path
MAAP	Multi Angle Absorption Photometer
MAST	Monterey Area Ship Track experiment
METEOSAT	Meteorological Satellite
METROMEX	An Investigation of Inadvertent Weather Modification downwind of St Louis,
MFO	Low-grade Marine Fuel Oil
MFRSR	Multi-filter Rotating Shadow-band Radiometer
MISR	Multi-angle Imaging Spectro Radiometer
MMD	Mean Mass Diameter
MODIS	MODerate resolution Imaging Spectroradiometer
MOUDI	Micro Orifice Uniform Deposit Impactor
MPLNET	Micro-Pulse Lidar Network of NASA
MSA	Methane Sulphonic Acid
MSG	Meteosat Second Generation
NAPS	National Air Pollution Surveillance
HASA	The US National Aeronautic and Space Administration
ND	Not Determined
NDF	U.S. Navy Distillate Fuel

NIES	National Institute for Environmental Studies
NOAA PATMOS	A project to derive atmospheric and surface climate records from AVHRR data.
NPOESS	National Polar-orbiting Operational Environmental Satellite System
NRC	National Research Council
NWP	Numerical Weather Prediction model
OPC	Optical Particle Counter
PARASOL	Polarization and Anisotropy of Reflectances for Atmospheric Sciences coupled with Observations from a Lidar
PBAP	Primary Biogenic Aerosol Particles
PBL	Planetary Boundary Layer
PCASP	Passive Cavity Aerosol Spectrometer Probe
PCTE	Polycarbonate Track Etched
PE	Precipitation Efficiency
PDF	Probability Density Function
PFR	Precision Filter Radiometers
PHOTONS	Sunphotometer network operated by the University of Lille www-loa.univ-lille1.fr/photons
PILS	Particle-Into-Liquid Sampler
PM	Particulate Matter
PMOD/WRC	Physikalisch-Meteorologisches Observatorium Davos.
PMS FSSP	Particle Measuring System- Forward scattering spectrometer probe.
POC	Particulate Organic Carbon
POD	Probability Of Detection
POLDER	Polarization and Directionality of the Earth Reflectances
POM	Particulate Organic Matter
PPOC	Primary Particulate Organic Carbon
PPOM	Primary Particulate Organic Matter
PSA	Particle Surface Area
PSAP	Particle Soot Photometer
QMS	Quadrupole Mass Spectrometry
RAMS	Colorado State University Regional Atmospheric Modeling System
RCFM	Reconstructed Fine Mass
REALM	Regional East Atmospheric Lidar Mesonet
RH	Relative Humidity
SAFARI-2000	Southern African Fire-Atmosphere Research Initiative in 2000.
SCAR	Sulphates, Clouds and Radiation Experiments (A in America and B in Brazil)
SDCC	Static thermal gradient Diffusion Cloud Chamber
SFU	Stacked Filter Units
SKYNET	Atmospheric radiation and weather observation network in Japan http://atmos.cr.chiba-u.ac.jp/
SMPS	Scanning Mobility Particle Spectrometer
SOA	Secondary Organic Aerosols
SOCEX	Southern Ocean Cloud Experiment
SPOT	Satellite Pour l'Observation de la Terre
SRES	IPCC Special Report on Emissions Scenarios

SURFRAD	Surface Radiation Network
SVOC	Semi-volatile organic Carbon
TEOM	Tapered Element Oscillating Microbalance
TIM	TRMM Microwave Imager
TM-3	3-D Transport Model
TOA	Top Of the Atmosphere
TOGA COARE	Tropical Ocean Global Atmosphere Coupled Ocean Atmosphere Response Experiment
TOMS	Total Ozone Measuring Spectrometer
TRACE-P	Transport and Chemical Evolution over the Pacific
TRMM	Tropical Rainfall Measuring Mission
TSP	Total Suspended Particles
UNFCCC	United Nations Framework Convention on Climate Change
USDA	United States Department of Agriculture
VIIRS	Visible/Infrared Imager/Radiometer Suite
VMD	Volume Medium Diameter
VOC	Volatile Organic Compounds
WINSIC	Water Insoluble Inorganic Carbon
WINSOC	Water Insoluble Organic Compounds
WMO	World Meteorological Organization
WORCC	World Optical Depth Research and Calibration Center
WRC	World Radiation Center
WRCP	World Climate Reference Programme
WSOC	Water Soluble Organic Compounds
XRF	X-Ray Fluorescence
ZSR	Zdanovskii-Stokes-Robinson

References

Abbatt, J.P.D., K. Broekhuizen, and P.P. Kumal, Cloud condensation nucleus activity of internally mixed ammonium sulfate/organic acid aerosol particles, *Atmos. Environ.*, **39**, 4767–4778, 2005.

Abdou, W.A., D.J. Diner, J.V. Martonchik, C.J. Bruegge, R.A. Kahn, B.J. Gaitley, K.A. Crean, L.A. Remer, and B. Holben, Comparison of coincident multiangle imaging spectroradiometer and moderate resolution imaging spectroradiometer aerosol optical depths over land and ocean scenes containing aerosol robotic network sites, *J. Geophys. Res.*, **110**, D10S07, doi:10.1029/2004JD004693, 2005.

Abdul-Razzak, H., and S.J. Ghan, A parameterization of aerosol activation. Part 2: Multiple aerosol types, *J. Geophys. Res.*, **105**, 6837–6844, 2000.

Abdul-Razzak, H., and S.J. Ghan, A parameterization of aerosol activation. Part 3: Sectional representation, *J. Geophys. Res.*, **107**, 4026, doi:10.1029/2001JD000483, 2002.

Ackerman, B., S.A. Changnon, G.L. Dzurisin, D.F. Gatz, R.C. Grosh, S.D. Hilberg, F.A. Huff, J.W. Mansell, H.T. Ochs III, M.E. Peden, P.T. Schickedanz, R.G. Semonin, and J.L. Vogel, Summary of METROMEX. Volume 2: Causes of Precipitation Anomalies, Illinois State Water Survey, Champaign, IL, pp. 395, 1978.

Ackerman, A.S., P.V. Hobbs, and O.B. Toon, A model for particle microphysics, turbulent mixing, and radioactive transfer in the stratocumulus-topped marine boundary layer and comparisons with measurements, *J. Atmos. Sci.*, **52**, 1204–1236, 1995.

Ackerman, A.S., O.B. Toon, J.P.J. Taylor, W. Doug, P.V. Hobbs, and R.J. Ferek, Effects of aerosols on cloud albedo: Evaluation of twomey's parameterization of cloud susceptibility using measurements of ship tracks, *J. Atmos. Sci.*, **57**, 2684–2695, 2000a.

Ackerman, A.S., O.B. Toon, D.E. Stevens, A.J. Heymsfield, V. Ramanathan, and E.J. Welton, Reduction of tropical cloudiness by soot, *Science*, **288**, 1042–1047, 2000b.

Ackerman, A.S., M.P. Kirkpatrick, D.E. Stevens, and O.B. Toon, The impact of humidity above stratiform clouds on indirect aerosol climate forcing, *Nature*, **432**, 1014–1017, 2004.

Adams, P.J., and J.H. Seinfeld, Disproportionate impact of particulate emissions on global cloud condensation nuclei concentrations, *Geophys. Res. Lett.*, **30**, 1239, doi:10.1029/2002GL016303, 2003.

Adams, P.J., J.H. Seinfeld, and D.M. Koch, Global concentrations of tropospheric sulfate, nitrate, and ammonium aerosol simulated in a general circulation model, *J. Geophys. Res.*, **104**, 13,791–13,823, 1999.

Adams, P.J., J.H. Seinfeld, D. Koch, L. Mickley, and D. Jacob, General circulation model assessment of direct radiative forcing by the sulfate-nitrate-ammonium-water inorganic aerosol system, *J. Geophys. Res.*, **106**, 1097–1111, 2001.

Adler, R.F., C. Kidd, G. Petty, M. Morrissey, and H.M. Goodman, Intercomparison of global precipitation products: The third Precipitation Intercomparison Project (PIP-3), *Bull. Amer. Meteor. Soc.*, **82**, 1377–1396, 2001.

Adler, R.F., J. Susskind, G.J. Huffman, D. Bolvin, E. Nelkin, A. Chang, R. Ferraro, A. Gruber, P.P. Xie, J. Janowiak, B. Rudolf, U. Schneider, S. Curtis, and P. Arkin, The version-2 Global Precipitation Climatology Project (GPCP) monthly precipitation analysis, (1979–present), *J. Hydrometeor.*, **4**, 1147–1167, 2003.

Albrecht, B.A., Aerosols, cloud microphysics, and fractional cloudiness, *Science*, **245**, 1227–1230, 1989.

Al-Naimi, R., and C.P.R. Saunders, Measurements of natural deposition and condensation-freezing ice nuclei with a continuous flow chamber, *Atmos. Environ.*, **19**, 1871–1882, 1985.

American Meteorological Society, Statement on planned and inadvertent weather modification, *Bull. Amer. Meteor. Soc.*, **66**, 447–448, 1984.

Amitai, E., D.B. Wolff, D.A. Marks, and D.S. Silberstein, Radar rainfall estimation, Lessons learned from the NASA/TRMM validation program. Second European Conference on Radar Meteorology (ERAD), Delft, The Netherlands. ERAD Publication Series, **1**, 255–260 (Copernicus GmbH peer reviewed publication, ISBN:3-936586-04-7), 2002.

Andreae, M.O., Ocean-atmosphere interactions in the global biogeochemical sulfur cycle, *Mar. Chem.*, **30**, 1–29, 1990.

Andreae, M.O., Climatic effects of changing atmospheric aerosol levels, in *World Survey of Climatology. Volume 16: Future Climates of the World*, edited by A. Henderson-Sellers, pp. 341–392, Elsevier, Amsterdam, 1995.

Andreae, M.O., and P.J. Crutzen, Atmospheric aerosols: Biogeochemical sources and role in atmospheric chemistry, *Science*, **276**, 1052–1056, 1997.

Andreae, M.O., and A. Gelencsér, Black carbon or brown carbon? The nature of light-absorbing carbonaceous aerosols, *Atmos. Chem. Phys.*, **6**, 3131–3148, 2006.

Andreae, M.O., and P. Merlet, Emission of trace gases and aerosols from biomass burning, *Global Biogeochem. Cycles*, **15**, 955–966, 2001.

Andreae, M.O., R.J. Charlson, F. Bruynseels, H. Storms, R.E. van Grieken, and W. Maenhaut, Internal mixture of sea salt, silicates and excess sulfate in marine aerosols, *Science*, **232**, 1620–1623, 1986.

Andreae, M.O., H. Berresheim, T.W. Andreae, M.A. Kritz, T.S. Bates, and J.T. Merrill, Vertical distribution of dimethylsulfide, sulfur dioxide, aerosol ions, and radon over the northeast Pacific Ocean, *J. Atmos. Chem.*, **6**, 149–173, 1988.

Andreae, M.O., R. Talbot, H. Berresheim, and K.M. Beecher, Precipitation chemistry of Central Amazonia, *J. Geophy. Res.*, **95**, 16,987–16,999, 1990.

Andreae, M.O., E. Atlas, H. Cachier, W.R. Cofer, III, G.W. Harris, G. Helas, R. Koppmann, J.P. Lacaux, and D.E. Ward, Trace gas and aerosol emissions from savanna fires, in *Biomass Burning and Global Change*, edited by J.S. Levine, pp. 278–295, MIT Press, Cambridge, MA, 1996.

Andreae, M.O., W. Elbert, Y. Cai, T.W. Andreae, and J. Gras, Non-seasalt sulfate, methanesulfonate, and nitrate aerosol concentrations and size distributions at Cape Grim, Tasmania, *J. Geophys. Res.*, **104**, 21,695–21,706, 1999.

Andreae, M.O., W. Elbert, R. Gabriel, D.W. Johnson, S. Osborne, and R. Wood, Soluble ion chemistry of the atmospheric aerosol and SO_2 concentrations over the eastern North Atlantic during ACE-2, *Tellus*, **52B**, 1066–1087, 2000.

Andreae, M.O., P. Artaxo, H. Fischer, S.R. Freitas, J.M. Gregoire, A. Hansel, P. Hoor, R. Kormann, R. Krejci, L. Lange, J. Lelieveld, W. Lindinger, K. Longo, W. Peters, M. deReus, B. Scheeren, M.A.F.S. Dias, J. Strom, P.F.J. vanVelthoven, and J. Williams., Transport of biomass burning smoke to the upper troposphere by deep convection in the equatorial region, *Geophys. Res. Lett.*, **28**, 951–954, 2001.

Andreae, T.W., M.O. Andreae, C. Ichoku, W. Maenhaut, J. Cafmeyer, A. Karnieli, and L. Orlovsky, Light scattering by dust and anthropogenic aerosol at a remote site in the Negev desert, Israel, *J. Geophys. Res.*, **107**, 4008, doi:10.1029/2001JD900252, 2002.

Andreae, M.O., D. Rosenfeld, P. Artaxo, A.A. Costa, G.P. Frank, K.M. Longo, and M.A.F. Silva-Dias, Smoking rain clouds over the Amazon, *Science*, **303**, 1337–1342, 2004.

Andres, R.J., G. Marland, I. Fung, and E. Matthews, A 1°x1° distribution of carbon dioxide emissions from fossil fuel consumption and cement manufacture, 1950–1990, *Global Biogeochem. Cycles*, **10**, 419–429, 1996.

Anttila, T., and V.M. Kerminen, Influence of organic compounds on the cloud droplet activation: A model investigation considering the volatility, water solubility, and surface activity of organic matter, *J. Geophys. Res.*, **107**, 4662, doi:10.1029/2001JD001482, 2002.

Arimoto, R., B.J. Ray, N.F. Lewis, U. Tomza, and R.A. Duce, Mass-particle size distributions of atmospheric dust and the dry deposition of dust to the remote ocean, *J. Geophys. Res.*, **102**, 15,867–15,874, 1997.

Arimoto, R., Y.J. Kim, Y.P. Kim, P.K. Quinn, T.S. Bates, T.L. Anderson, S.L. Gong, M.C.I. Uno, B.J. Huebert, A.D. Clarke, Y. Shinozuka, R.J. Weber, J.R. Anderson, S.A. Guazzotti, R.C. Sullivan, D.A. Sodeman, K.A. Prather, and I. Sokolik, Characterization of Asian Dust during ACE-Asia, *Global Planet. Change*, **52**, 23–56, 2006.

Arkin, P.A., The relationship between fractional coverage of high cloud and rainfall accumulations during GATE over the B-scale array, *Mon. Wea. Rev.*, **107**, 1382–1387, 1979.

Arkin, P.A., and B.M. Meisner, The relationship between large-scale convective rainfall and cold cloud over the western hemisphere during 1982–1984, *Mon. Wea. Rev.*, **115**, 51–74, 1987.

Artaxo, P., Aerosol sampling and analysis, in *Environmental Monitoring Handbook*, edited by A.Guenther, McGraw-Hill Professional, New York, USA, ISBN:0-07-135176-0, 2002.

Artaxo, P., H. Storms, F. Bruynseels, R.V. Grieken, and W. Maenhaut, Composition and sources of aerosols from the Amazon basin, *J. Geophys. Res.*, **93**, 1605–1615, 1988.

Artaxo, P., W. Maenhaut, H. Storms, and R.V. Grieken, Aerosol characteristics and sources for the Amazon Basin during the wet season, *J. Geophys. Res.*, **95**, 16,971–16,986, 1990.

Artaxo, P., M.L.C. Rabello, W. Maenhaut, and R. Van Grieken, Trace elements and individual particle analysis of aerosol particles from the Antarctic peninsula, *Tellus*, **44B**, 318–334, 1992.

Artaxo, P., F. Gerab, and M.L.C. Rabello, Elemental composition of aerosol particles from two background monitoring stations in the Amazon Basin, *Nuc. Instru. and Meth. in Phys. Res.*, **B75**, 277–281, 1993.

Artaxo, P., F. Gerab, M.A. Yamasoe, and J.V. Martins, Fine mode aerosol composition at three long-term atmospheric monitoring sites in the Amazon Basin, *J. Geophys. Res.*, **99**, 22,857–22,868, 1994.

Artaxo, P., P. Oyola, and R. Martinez, Aerosol composition and source apportionment in Santiago de Chile, *Nuc. Instru. and Meth. in Phys. Res.*, **B150**, 409–416, 1999.

Artaxo, P., J.V. Martins, M.A. Yamasoe, A.S. Procópio, T.M. Pauliquevis, M.O. Andreae, P. Guyon, L.V. Gatti, and A.M.C. Leal, Physical and chemical properties of aerosols in the wet and dry season in Rondonia, Amazonia, *J. Geophys. Res.*, **107**, 8081, doi:10.1029/2001JD000666, 2002.

Atlas, D., The paradox of hail suppression, *Science*, **195**, 139–145, 1977.

Auer, Jr., A.H., D.L. Veal, and J.D. Marwitz, Observations of ice crystal and ice nuclei concentrations in stable cap clouds, *J. Atmos. Sci.*, **26**, 1342–1343, 1969.

Austin, P.M., Relation between measured radar reflectivity and surface rainfall, *Mon. Wea. Rev.*, **115**, 1053–1070, 1987.

Austin, P., Y. Wang, R. Pincus, and V. Kujala, Precipitation in stratocumulus clouds: Observational and modeling results, *J. Atmos. Sci.*, **52**, 2329–2352, 1995.

Ayers, G.P., Air pollution and climate change: Has air pollution suppressed rainfall over Australia? *Clean Air and Environ. Quality.*, **39**(2), 51–57, 2005.

Ayers, G.P., and J.L. Gras, Seasonal relationship between cloud condensation nuclei and aerosol methanesulfonate in marine air, *Nature*, **351**, 834–835, 1991.

Ayers, G.P., E.K. Bigg, and D.E. Turvey, Aitken particle and cloud condensation nucleus fluxes in the plume from an isolated industrial source, *J. Appl. Meteoro.*, **18**, 449–459, 1979.

Ayers, G.P., E.K. Bigg, D.E. Turvey, and M.J. Manton, Urban influences on condensation nuclei over a continent, *Atmos. Environ.*, **16**, 951–954, 1982.

Ayers, G.P., J.P. Ivey, and R.W. Gillett, Coherence between seasonal cycles of dimethyl sulfide, methansulphonate and sulfate in marine air, *Nature*, **349**, 404–406, 1991.

Ba, M.B., and A. Gruber, GOES Multispectral Rainfall Algorithm (GMSRA), *J. Appl. Meteor.*, **40**, 1500–1514, 2001.

Baker, M.B., and J. Latham, The evolution of droplet spectra and the rate of production of embryonic raindrops in small cumulus clouds, *J. Atmos. Sci.*, **36**, 1612–1615, 1979.

Baker, M.B., R.G. Corbin, and J. Latham, The influence of entrainment on the evolution of cloud droplet spectra: I. A model of inhomogeneous mixing, *Q. J. Roy. Meteor. Soc.*, **106**, 581–598, 1980.

Baltensperger, U., E. Weingartner, H. Burtscher, and J. Keskinen, Dynamic mass and surface area measurements, in *Aerosol Measurement, Principles, Techniques, and Applications*, Second Edition, edited by A.B. Paul, pp. 387–418, Wiley-Interscience, 2001.

Baltensperger, U., N. Streit, E. Weingartner, S. Nyeki, A.S.H. Prevot, R. Van Dingenen, A. Virkkula, J.P. Putaud, A. Even, H. ten Brink, A. Blatter, A. Neftel, and H.W. Gaggeler, Urban and rural aerosol characterization of summer smog events during the PIPAPO field campaign in Milan, Italy, *J. Geophys. Res.*, **107**, 8193, doi:10.1029/2001JD001292, 2002.

Baltensperger, U., M. Kalberer, J. Dommen, D. Paulsen, M. Alfarra, H. Coe, R. Fisseha, A. Gascho, M. Gysel, S. Nyeki, M. Sax, M. Steinbacher, A. Prevot, S. Sjoren, E. Weingartner, and R. Zenobi, Secondary organic aerosols from anthropogenic and biogenic precursors, *Faraday Discuss.*, **130**, 265–278, 2005.

Barger, W.R., and W.D. Garrett, Surface-active organic material in air over Mediterranean and over eastern Equatorial Pacific, *J. Geophys. Res.*, **81**, 3151–3157, 1976.

Barkstrom, B.R., Some effects of 8–12 um radiant energy transfer on the mass and heat budgets of cloud droplets, *J. Atmos. Sci.*, **35**, 665–673, 1978.

Barnes, W.L., T.S. Pagano, and V.V. Salomonson, Pre-launch characteristics of MODIS on EOS-AM1, *IEEE Trans. Geosci. Remote Sens.*, **36**, 1088–1100, 1998.

Baron, P.A., K. Willeke, *Aerosol Measurement: Principles, Techniques, and Applications*, Second Edition, Wiley-Interscience, NY, ISBN:0471356360, 2001.

Barrie, L.A., Arctic air pollution: An overview of current knowledge, *Atmos. Environ.*, **20**, 643–663, 1986.

Barrie, L.A, Arctic air pollution: A case study of continent-to-ocean-to-continent transport, in *The Long-Range Atmospheric Transport of Natural and Contaminant Substances*, edited by A.H. Knap, and M.S. Kaiser, pp. 137–148, Kluwer Academic Publishers, Dordrecht, Holland, 1990.

Barrie, L.A., Y.Yi, W.R. Leattch, U. Lohmann, P. Kasibhatla, G.J. Roelofs, J. Wilson, F. McGovern, C. Benkovitz, M.A. Mélières, K. Law, J. Prospero, M. Kritz, D. Bergmann, C. Bridgeman, M. Chin, J. Christensen, R. Easter, J. Feichter, C. Land, A. Jeuken, and jellström, A Comparison of Large-scale Atmospheric Sulphate Aerosol Models (COSAM): Overview and highlights, *Tellus*, **53B**, 615–645, 2001.

Barth, M.C., and A.T. Church, Regional and global distributions and lifetimes of sulfate aerosols from Mexico city and southeast China, *J. Geophys. Res.*, **104**, 30,231–30,239, 1999.

Barth, M.C., D.A. Hegg, and P.V. Hobbs, Numerical modeling of cloud and precipitation chemistry associated with two rainbands and some comparisons with observations, *J. Geophys. Res.*, **97**, 5825–5845, 1992.

Barthazy, E., S. Göke, R. Schefold, and D. Högl, An optical array instrument for shape and fall velocity measurements of hydrometeors, *J. Atmos. Ocean. Tech.*, **21**, 1519–1531, 2004.

Bates, T.S., B.J. Huebert, J.L. Gras, F.B. Griffiths, and P.A. Durkee, The International Global Atmospheric Chemistry (IGAC) Project's First Aerosol Characterization Experiment (ACE-1) – Overview, *J. Geophys. Res.*, **103**, 16,279–16,318, 1998.

Bates, T.S., P.K. Quinn, D.S. Covert, D.J. Coffman, J.E. Johnson, and A. Wiedensohler, Aerosol physical properties and processes in the lower marine boundary layer: A comparison of shipboard sub-micron data from ACE-1 and ACE-2, *Tellus*, **52B**, 258–272, 2000.

Bates, T.S., P.K. Quinn, D.J. Coffman, J.E. Johnson, T.L. Miller, D.S. Covert, A. Wiedensohler, S. Leinert, A. Nowak, and C. Neusuess, Regional physical and chemical properties of the marine boundary layer aerosol across the Atlantic during Aerosols 99: An overview, *J. Geophys. Res.*, **106**, 20,767–20,782, 2001.

Bates, T.S., D.J. Coffman, D.S. Covert, and P.K. Quinn, Regional marine boundary layer aerosol size distributions in the Indian, Atlantic, and Pacific Oceans: A comparison of INDOEX measurements with ACE-1, ACE-2, and Aerosols99, *J. Geophys. Res.*, **107**, 8026, doi:10.1029/2001JD001174, 2002.

Battan, L.J., *Radar Observation of the Atmosphere*, 324 pp., University of Chicago Press, Chicago, Ill, 1973.

Bauer, P., Over-ocean rainfall retrieval from multisensor data of the Tropical Rainfall Measuring Mission. Part I: Design and evaluation of inversion databases, *J. Atmos. Ocean. Tech.*, **18**, 1315–1300, 2001.

Bauer, P., J.P.V. Poiares Baptisa, and M. De Iulis, The effect of the melting layer on the microwave emission of clouds over the ocean, *J. Atmos. Sci.*, **56**, 852–867, 1999.

Bauer, H., H. Giebl, R. Hitzenberger, A. Kasper-Giebl, G. Reischl, F. Zibuschka, and H. Puxbaum, Airborne bacteria as cloud condensation nuclei, *J. Geophys. Res.*, **108**, 4658, doi:10.1029/2003JD003545, 2003.

Baumgardner, D., H. Chepfer, G.B. Raga, G.L. Kok, The shapes of very small cirrus particles derived from in situ measurements, *Geophys. Res. Lett.*, **32**, L01806, doi:10.1029/2004GL021300, 2005.

Beard, K.V., and H.T. Ochs, III, Warm-rain initiation: An overview of microphysical mechanisms, *J. Appl. Meteor.*, **32**, 608–625, 1993.

Beheng, K.D., A parameterization of warm cloud microphysical conversion processes, *Atmos. Res.*, **33**, 193–206, 1994.

Ben-Zvi, A., Comments on "A new look at the Israeli cloud seeding experiments", *J. Appl. Meteor.*, **36**, 255–256, 1997.

Bergeron, T., On the physics of clouds and precipitation, in *Proc. 5th Assembly UGGI*, Volume 2, Lisbon, 1933.

Berglen, T.F., T.K. Berntsen, I.S.A. Isaksen, and J.K. Sundet, A global model of the coupled sulfur/oxidant chemistry in the troposphere: The sulfur cycle, *J. Geophys. Res.*, **109**, D19310, doi:10.1029/2003JD003948, 2004.

Berry, E.X., Cloud droplet growth by coalescence, *J. Atmos. Sci.*, **24**, 688–701, 1967.

Berry, E.X., and R.L. Reinhardt, An analysis of cloud drop growth by collection. Part I: Double distributions, *J. Atmos. Sci.*, **31**, 1814–1824, 1974.

Bey, I., D.J. Jacob, J.A. Logan, and R.M. Yantosca, Asian chemical outflow to the Pacific: Origins, pathways and budgets, *J. Geophys. Res.*, **106**, 23,097–23,114, 2001.

Bibilashvili, N. Sh., I.I. Gaivoronski, G.G. Godorage, A.I. Kartsivadze, and R.N. Stankov, *Proc. WMO/IAMAP Scientific Conf. on Weather Modification*, Tashkent, 1973, p. 333, 1974.

Bigg, E.K., Discrepancy between observation and prediction of concentrations of cloud condensation nuclei, *Atmos. Res.*, **20**, 81–86, 1986.

Bigg, E.K., Secondary ice nucleus generation by silver iodide applied to the ground, *J. Appl. Meteor.*, **27**, 453–457, 1988.

Bigg, E.K., Aerosol over the southern ocean, *Atmos. Res.*, **25**, 583–600, 1990.

Bigg, E.K., Tests for persistent effects of cloud seeding in a recent Australian experiment, *J. Appl. Meteor.*, **34**, 2406–2411, 1995.

Bigg, E.K., An independent evaluation of a South African hygroscopic cloud seeding experiment, 1991–1995, *Atmos. Res.*, **43**, 111–127, 1997.

Bigg E.K., and D.E. Turvey, Sources of atmospheric particles over Australia, *Atmos. Environ.*, **12**, 1643–1655, 1978.

Bigg, E.K., and E. Turton, Persistent effects of cloud seeding with silver iodide, *J. Appl. Meteor.*, **27**, 505–514, 1988.

Bigg E.K., Ayers, G.P, and D.E. Turvey, Measurements of the dispersion of a smoke plume at large distances from the source, *Atmos. Environ.*, **12**, 1815–1818, 1978.

Bilde, M., and B. Svenningsson, CCN activation of slightly soluble organics: The importance of small amounts of inorganic salt and particle phase, *Tellus*, **56B**, 128–134, 2004.

Birmili, W., B. Yuskiewicz, A. Wiedensohler, F. Stratmann, T.W. Choularton, and K.N. Bower, Climate-relevant modification of the aerosol size distribution by processes associated with orographic clouds, *Atmos. Res.*, **50**, 241–263, 1999.

Birmili, W., A. Wiedensohler, J. Heintzenberg, and K. Lehmann, Atmospheric particle number size distribution in central Europe: Statistical relations to air masses and meteorology, *J. Geophys. Res.*, **106**, 32,005–32,018, 2001.

Biswas, K.R., and A.S. Dennis, Formation of rain shower by salt seeding, *J. Appl. Meteor.*, **10**, 780–784, 1971.

Bleck, R., A fast approximative method for integrating the stochastic coalescence equation, *J. Geophys. Res.*, **75**, 5165–5171, 1970.

Blyth, A.M. and J. Latham, Development of ice and precipitation in New Mexican summertime cumulus clouds, *Q. J. Roy. Meteor. Soc.*, **119**, 91–120, 1993.

Boers, R., G.P. Ayers, and J.L. Gras, Coherence between seasonal cycles in satellite observed cloud optical depth ad boundary layer CCN concentration at a mid-latitude Southern Hemispheric site, *Tellus*, **46**B, 123–131, 1994.

Boers, R., J.B. Jensen, P.B. Krummel, and H. Gerber, Microphysical and radiative structure of wintertime stratocumulus clouds over the Southern Ocean, *Q. J. Roy. Meteor. Soc.*, **122**, 1307–1339, 1996.

Boers, R., J.B. Jensen, P.B. Krummel, Microphysical and radiative structure of marine stratocumulus clouds over the Southern Ocean: Summer results and seasonal differences, *Q. J. Roy. Meteor. Soc.*, **124**, 151–168, 1998.

Boers, R., J.R. Acarreta, J.L Gras, Satellite monitoring of the first indirect aerosol effect: Retrieval of the droplet concentration of water clouds, *J. Geophys. Res.*, **111**, D22208, doi:10.1029/2005JD006838, 2006.

Bond, T.C., D.S. Covert, J.C. Kramlich, T.V. Larson, and R.J. Charlson, Primary particle emissions from residential coal burning: Optical properties and size distributions, *J. Geophys. Res.*, **107**, 8347, doi:10.1029/2001JD000571, 2002.

Bond, T.C., D.G. Streets, K.F. Yarber, S.M. Nelson, J.H. Woo, and Z. Klimont, A technology-based global inventory of black and organic carbon emissions from combustion, *J. Geophys. Res.*, **109**, D14203, doi:10.1029/2003JD003697, 2004.

Borys, R.D., D.H. Lowenthal, and D.L. Mitchel, The relationship among cloud physics, chemistry and precipitation rate in cold mountain clouds, *Atmos. Environ.*, **34**, 2593–2602, 2000.

Borys, R.D., D.H. Lowenthal, S.A. Cohn, and W.O.J. Brown, Mountaintop and radar measurements of anthropogenic aerosol effects on snow growth and snow rate, *Geophys. Res. Lett.*, **30**, 1538, doi:10.1029/2002GL016855, 2003.

Bott, A., A flux method for the numerical solution of the stochastic collection equation: Extension to two dimensional particle distributions, *J. Atmos. Sci.*, **57**, 284–294, 2000.

Bott, A., U. Sievers, and W. Zdunkowski, A radiation fog model with a detailed treatment of the interaction between radiative transfer and fog microphysics, *J. Atmos. Sci.*, **47**, 2153–2166, 1990.

Bott, A., T. Trautmann, and W. Zdunkowski, A numerical model of the cloud-topped planetary boundary-layer: Radiation, turbulence and spectral microphysics in marine stratus, *Q. J. Roy. Meteorol. Soc.*, **122**, 635–667, 1996.

Boucher, O., and U. Lohmann, The sulfate-CCN-cloud albedo effect, *Tellus*, **47B**, 281–300, 1995.

Bowen, E.G., A new method of stimulating convective clouds to produce rain and hail, *Q. J. Roy. Meteor. Soc.*, **78**, 37–45, 1952.

Bower, K.N., and T.W. Choularton, Cloud processing of the cloud condensation nucleus spectrum and its climatological consequences, *Q. J. R. Meteorol. Soc.*, **119**, 655–679, 1993.

Braham, R.R., What is the role of ice in summer rain-showers? *J. Atmos. Sci.*, **21**, 640–646, 1964.

Braham, R.R., Meteorological bases for precipitation development, *Bull. Amer. Met. Soc.*, **49**, 343–353, 1968.

Braham, R.R. Cloud Physics of urban weather modification – A preliminary report, *Bull. Amer. Meteor. Soc.*, **55**, 100–106, 1974.

Braham, R.R., Rainfall enhancement – A scientific challenge, in *Rainfall Enhancement – A Scientific Challenge, AMS Meteor. Monogr.*, **21**(43), 1–5, Amer. Meteor. Soc., Boston, MA, 1986.

Braham, R.R., and P. Spyers-Duran, Ice nucleus measurements in an urban atmosphere, *J. Appl. Meteor.*, **13**, 940–945, 1974.

Brechtel, F.J., and S.M. Kreidenweis, Predicting particle critical supersaturation from hygroscopic growth measurements in the humidified TDMA. Part I: Theory and sensitivity studies, *J. Atmos. Sci.*, **57**, 1854–1871, 2000a.

Brechtel, F.J., and S.M. Kreidenweis, Predicting particle critical supersaturation from hygroscopic growth measurements in the humidified TDMA. Part II: Laboratory and ambient studies, *J. Atmos. Sci.*, **57**, 1872–1887, 2000b.

Brechtel, F.J., S.M. Kreidenweis, and H.B. Swan, Air mass characteristics, aerosol particle number concentrations, and number size distributions at Macquarie Island during the First Aerosol Characterization Experiment (ACE 1), *J. Geophys. Res.*, **103**, 16,351–16,367, 1998.

Brenguier, J.L., and L. Chaumat, Droplet spectra broadening in cumulus clouds. Part I: Broadening in adiabatic cores, *J. Atmos. Sci.*, **58**, 628–641, 2001.

Brenguier, J.L., T. Bourrianne, A. Coelho, J. Isbert, R. Peytavi, D. Trevarin, and P. Wechsler, Improvements of droplet size distribution measurements with the Fast-FSSP, *J. Atmos. Oceanic. Technol.*, **15**, 1077–1090, 1998.

Brenguier, J.L., P.Y. Chuang, Y. Fouquart, D.W. Johnson, F. Parol, H. Pawlowska, J. Pelon, L. Schuller, F. Schroder, and J. Snider, An overview of the ACE-2 CLOUDYCOLUMN closure experiment, *Tellus B.*, **52**, 815–827, 2000a.

Brenguier, J.L., H. Pawlowska, L. Schüller, R. Preusker, J. Fischer, and Y. Fouquart, Radiative properties of boundary layer clouds: Droplet effective radius versus number concentration, *J. Atmos. Sci.*, **57**, 803–821, 2000b.

Brenguier, J.L., H. Pawlowska, and L.J. Schüller, Cloud microphysical and radiative properties for parameterization and satellite monitoring of the indirect effect of aerosol on climate, *J. Geophys. Res.*, **108**, 8632, doi:10.1029/2002JD002682, 2003.

Bréon, F.M., and S. Colzy, Global distribution of cloud droplet effective radius from POLDER polarization measurements, *Geophys. Res. Lett.*, **27**, 4065–4068, 2000.

Bréon, F.M. and M. Doutriaux-Boucher: A comparison of cloud droplet radii measured from space, *IEEE Trans. Geos. Rem. Sens.*, **43**(8), 1796–1805, 2005.

Bréon, F.M., and P. Goloub, Cloud droplet effective radius from spaceborne polarization measurements, *Geophys. Res. Lett.*, **25**, 1879–1882, 1998.

Bréon, F.M., D. Tanre, and S. Generoso, Aerosol effect on cloud droplet size monitored from satellite, *Science*, **295**, 834–838, 2002.

Bretherton, C.S., T. Uttal, C.W. Fairall, S. Yuter, R. Weller, D. Baumgardner, K. Comstock, and R. Wood, The EPIC 2001 stratocumulus study, *Bull. Amer. Meteor. Soc.*, **85**, 967–977, 2004.

Bringi, V.N., and V. Chandrasekar, *Polarimetric Doppler Weather Radar: Principles and Applications*, 636 pp., Cambridge University Press, Cambridge, UK, 2001.

Broekhuizen, K., P.P. Kumar, and J.P.D. Abbatt, Partially soluble organics as cloud condensation nuclei: Role of trace soluble and surface active species, *Geophys. Res. Lett.*, **31**, 2004.

Broekhuizen, K., R.Y.W. Chang, W.R. Leaitch, S.M. Li, and J.P.D. Abbatt, Closure between measured and modeled Cloud Condensation Nuclei (CCN) using size-resolved aerosol compositions in downtown Toronto, *Atmos. Chem. Phys.*, **6**, 2513–2524, 2006.

Brook, J.R., T.F. Dann, and R.T. Burnett, The relationship among TSP, PM10, PM2.5, and Inorganic constituents of atmospheric particulate matter at multiple Canadian locations, *J. Air and Waste Manage. Assoc.*, **47**, 2–19, 1997.

Brooks, S.D., P.J. DeMott, and S.M. Kreidenweis, Water uptake by particles containing humic materials and mixtures of humic materials with ammonium sulfate, *Atmos. Environ.*, **38**, 1859–1868, 2004.

Browning, K.A., and F.H. Ludlam, Airflow in convective storms, *Q. J. Roy. Meteor. Soc.*, **88**, 117–135, 1962.

Bukowiecki, N., M. Hill, R. Gehrig, C.N. Zwicky, P. Lienemann, F. Hegedüs, G. Falkenberg, E. Weingartner, U. Baltensperger, Trace metals in ambient air: Hourly size-segregated mass concentrations determined by Synchrotron-XRF, *Environ. Sci. Tech.*, **39**, 5754–5762, 2005.

Buriez, J.C., C. Vanbauce, F. Parol, P. Goloub, M. Herman, B. Bonnel, Y. Fouquart, P. Couvert, and G. Seze, Cloud detection and derivation of cloud properties from POLDER, *Int. J. Remote.*, **18**, 2785–2813, 1997.

Burnet, F., and J.L. Brenguier, Observational study of the entrainment-mixing process in warm convective clouds, *J. Atmos. Sci.*, **64**, 1995–2011, 2007.

Cachier, H., P. Buat-Ménard, M. Fontugne, and R. Chesselet, Long-range transport of continentally-derived particulate carbon in the marine atmosphere: Evidence from stable carbon isotope studies, *Tellus*, **35B**, 161–177, 1983.

Cachier, H., C. Liousse, M.H. Pertuisot, A. Gaudichet, F. Echalar, and J.P. Lacaux, African fire particulate emission and atmospheric influence, in *Biomass Burning and Global Change*, edited by J.S. Levine, pp. 428–440, MIT Press, Cambridge, MA, 1996.

Caminade, C., L. Terray, and E. Maisonnave, West African monsoon response to greenhouse gas and sulphate aerosol forcing under two emission scenarios, *Climate Dynamics.*, **26**, 531–547, 2006.

Campuzano-Jost, P., C.D. Clark, H. Maring, D.S. Covert, S. Howell, V. Kapustin, K.A. Clarke, E.S. Saltzman, and A.J. Hynes, Near-real-time measurement of sea salt aerosol during the SEAS campaign: Comparison of emission-based sodium detection with an aerosol volatility technique, *J. Atmos. Ocean. Technol.*, **20**, 1421–1430, 2003.

Cantrell, W., and A. Heymsfield, Production of ice in tropospheric clouds, *Bull. Amer. Meteor. Soc.*, **86**(6), 795–807, 2005.

Cantrell, W., G. Shaw, C. Leck, L. Granat, and H. Cachier, Relationships between cloud condensation nuclei spectra and aerosol particles on a south-north transect of the Indian Ocean, *J. Geophys. Res.*, **105**, 15,313–15,320, 2000.

Cantrell, W., G. Shaw, G.R. Cass, Z. Chowdhury, L.S. Hughes, K.A. Prather, S.A. Guazzotti, and K.R. Coffee, Closure between aerosol particles and cloud condensation nuclei at Kaashidhoo Climate Observatory, *J. Geophys. Res.*, **106**, 28,711–28,718, 2001.

Carlson, T.N., and J.M. Prospero, Large-scale movements of Saharan air impulses over western tropical Atlantic, *Bull. Amer. Meteor. Soc.*, **52**(8), 779, 1971.

Carmichael, G.R., D.G. Streets, G. Calori, M. Amann, M.Z. Jacobson, J. Hansen, and H. Ueda, Changing trends in sulfur emissions in Asia: Implications for acid deposition, air pollution, and climate, *Environ. Sci. Technol.*, **36**, 4707–4713, 2002.

Caro, D., W. Wobrock, and A.I. Flossmann, A numerical study on the impact of hygroscopic seeding on the development of cloud particle spectra, *J. Appl. Meteor.*, **41**, 333–350, 2002.

Carras, J.N., and D.J. Williams, The long-range dispersion of a plume from an isolated point source, *Atmos. Environ.*, **15**, 2205–2217, 1981.

Carras, J.N., and D.J. Williams, Measurements of relative σy up to 1800 km from a single source, *Atmos. Environ.*, **22**, 1060–1069, 1988.

Carrió, G.G., H. Jiang, and W.R. Cotton, Impact of aerosol intrusions on the Arctic boundary layer an on sea-ice melting rates. Part I: May 4, 1998 case, *J. Atmos. Sci.*, **62**, 3082–3093, 2005a.

Carrió, G.G., H. Jiang, and W.R. Cotton, Impact of aerosol intrusions on the Arctic boundary layer an on sea-ice melting rates. Part II: Sea ice melting rates, *J. Atmos. Sci*, **62**, 3094–3105, 2005b.

Castro, A., J.L. Marcos, J. Dessens, J.L. Sanchez, and R. Fraile, Concentration of ice nuclei in continental and maritime air masses in Leon (Spain), *Atmos. Res.*, **48**, 155–167, 1998.

Cavalli, F., M.C. Facchini, S. Decesari, M. Mircea, L. Emblico, S. Fuzzi, D. Ceburnis, Y.J. Yoon, C.D. O'Dowd, J.P. Putaud, and A. Dell'Acqua, Advances in characterization of size-resolved organic matter in marine aerosol over the North Atlantic, *J. Geophys. Res.*, **109**, D24215, doi:10.1029/2004JD005137, 2004.

Chan, M.N., and C.K. Chan, Hygroscopic properties of two model humic-like substances and their mixtures with inorganics of atmospheric importance, *Environ. Sci. Technol.*, **37**, 5109–5115, 2003.

Chan, M.N., M.Y. Choi, N.L. Ng, and C.K. Chan, Hygroscopicity of water-soluble organic compounds in atmospheric aerosols: Amino acids and biomass burning derived organic species, *Environ. Sci. Technol.*, **39**, 1555–1562, 2005.

Chand, D., O. Schmid, P. Gwaze, R. Parmar, G. Helas, K. Zeromskiene, A. Wiedensohler, A. Massling, and M. Andreae, Laboratory measurements of smoke optical properties from the burning of Indonesian peat and other types of biomass, *Geophys. Res. Lett.*, **32**, Art. No. L12819, 2005.

Changnon, Jr., S.A., More on the LaPorte anomaly: A review, *Bull. Amer. Meteor. Soc.*, **61**, 702–711, 1980.

Changnon, Jr. , S.A., F.A. Huff and R.G. Semonin, METROMEX: An investigation of inadvertent weather modification, *Bull. Amer. Meteor. Soc.*, **52**, 958–968, 1971.

Charlson, R.J., J.E. Lovelock, M.O. Andreae, and S.G. Warren, Oceanic phytoplankton, atmospheric sulphur, cloud albedo, and climate, *Nature*, **326**, 655–661, 1987.

Charlson, R.J., J.H. Seinfeld, A. Nenes, M. Kulmala, A. Laaksonen, and M.C. Facchini, Reshaping the theory of cloud formation, *Science*, **292**, 2025–2026, 2001.

Chate, D.M., Study of scavenging of submicron-sized aerosol particles by thunderstorm rain events, *Atmos. Environ.*, **39**, 6608–6619, 2005.

Chate, D.M., and T.S. Pranesha, Field measurements of sub-micron aerosol concentration during cold season in India, *Current Science*, **86**, 1610–1613, 2004.

Chaumat, L., and J.L. Brenguier, Droplet spectra broadening in cumulus clouds. Part II: Micro-scale droplet concentration heterogeneities, *J. Atmos. Sci.*, **58**, 642–654, 2001.

Che, H.Z., G.Y. Shi, X.Y. Zhang, R. Arimoto, J.Q. Zhao, L. Xu, B. Wang, and Z.H. Chen, Analysis of 40 years of solar radiation data from China, 1961–2000, *Geophys. Res. Lett.*, **32**, L06803, doi:06810.01029/02004GL022322, 2005.

Chen, J.P., and D. Lamb, Simulation of cloud microphysical and chemical processes using a multicomponent framework. Part I: Description of the microphysical model, *J. Atmos. Sci.*, **51**, 2613–2630, 1994.

Chen, Y.Z., N. Shah, F.E. Huggins, G.P. Huffman, W.P. Linak, and C.A. Miller, Investigation of primary fine particulate matter from coal combustion by computer-controlled scanning electron microscopy, *Fuel Process. Technol.*, **85**, 743–761, 2004.

Chiapello, I. et al., Understanding the long-term variability of African dust transport across the Atlantic as recorded in both Barbados surface concentrations and large-scale Total Ozone Mapping Spectrometer (TOMS) optical thickness, *J. Geophys. Res.*, **110**, D18S10, doi:10.1029/2004JD005132, 2005.

Chin, M., P. Ginoux, S. Kinne, O. Torres, B.N. Holben, B.N. Duncan, R.V. Martin, J.A. Logan, A. Higurashi, and T. Nakajima, Tropospheric aerosol optical thickness from the GOCART model and comparisons with satellite and sun photometer measurements, *J. Atmos. Sci.*, **59**, 461–483, 2002.

Choi, M.Y., and C.K. Chan, Continuous measurements of the water activities of aqueous droplets of water-soluble organic compounds, *J. Phys. Chem. A.*, **106**, 4566–4572, 2002.

Chosson, F., J.L. Brenguier, and L. Schüller, Entrainment-mixing and radiative transfer simulation in boundary-layer clouds, *J. Atmos. Sci.*, **64**, 2670–2682, 2007.

Chowdhary, J., B. Cairns, and L.D. Travis, Case studies of aerosol retrieval over the ocean from multiangle, multispectral photopolarimetric remote sensing data, *J. Atmos. Sci.*, **59**, 383–397, 2002.

Chowdhury, Z., L.S. Hughes, and L.G. Salmon, Atmospheric particle size and composition measurements to support light extinction calculations over the Indian Ocean, *J. Geophys. Res.*, **106**, 28,597–28,605, 2001.

Christensen, J.H., The Danish Eulerian Hemispheric ModelAa three-dimensional air pollution model used for the Arctic, *Atmos. Environ.*, **31**, 4169–4191, 1997.

Chu, D.A., Y.J. Kaufman, C. Ichoku, L.A. Remer, D. Tanré, and B.N Holben, Validation of MODIS aerosol optical depth retrieval over land, *Geophys. Res. Lett.*, **29**, doi:10.1029/2001GL013205, 2002.

Chuang, C.C., J.E. Penner, K.E. Taylor, A.S. Grossman, and J.J. Walton, An assessment of the radiative effects of anthropogenic sulfate, *J. Geophys. Res.*, **102**, 3761–3778, 1997.

Chuang, P.Y., D.R. Collins, H. Pawlowska, J.R. Snider, H.H. Jonsson, J.L. Brenguier, R.L. Flagan, and J.H. Seinfeld, CCN measurements during ACE-2 and their relationship to cloud microphysical properties, *Tellus*, **52B**, 843–867, 2000.

Chung, S.H., and J.H. Seinfeld, Global distribution and climate forcing of carbonaceous aerosols, *J. Geophys. Res.*, **107**, 4407, doi:10.1029/2001JD001397, 2002.

Chung, S.H., and J.H. Seinfeld, Climate response of direct radiative forcing of anthropogenic black carbon, *J. Geophys. Res.*, **110**, D11102, doi:10.1029/2004JD005441, 2005.

Chylek, P., M.K. Dubey, U. Lohmann, V. Ramanathan, Y.J. Kaufman, G. Lesins, J. Hudson, G. Altmann, and S.C. Olsen, Aerosol indirect effect over the Indian Ocean, *Geophys. Res. Lett.*, **33**, L06806, doi:10.1029/2005GL025397, 2006

Claeys, M., W. Wang, A.C. Ion, I. Kourtchev, A. Gelencser, and W. Maenhaut, Formation of secondary organic aerosols from isoprene and its gas-phase oxidation products through reaction with hydrogen peroxide, *Atmos. Environ.*, **38**, 4093–4098, 2004.

Clark, T.L., Numerical modelling of the dynamics and microphysics of warm cumulus convection, *J. Atmos. Sci.*, **30**, 857–878, 1973.

Clark, T.L., On modeling nucleation and condensation theory in Eulerian spatial domain, *J. Atmos. Sci.*, **31**, 2099–2117, 1974.

Clark, T.L., Use of log-normal distributions for numerical calculations of condensation and collection, *J. Atmos. Sci.*, **33**, 810–821, 1976.

Clark, T.L., and W.D. Hall, A cloud physical parameterization method using movable basis functions: Stochastic coalescence parcel calculations, *J. Atmos. Sci.*, **40**, 1709–1728, 1983.

Clarke, A.D., F. Eisele, V.N. Kapustin, K. Moore, D. Tanner, L. Mauldin, M. Litchy, B. Lienert, M.A. Carroll, and G. Albercook, Nucleation in the equatorial free troposphere: Favorable environments during PEM-Tropics, *J. Geophys. Res.*, **104**, 5735–5744, doi:10.1029/98JD02303, 1999.

Clarke, A.D., V. Kapustin, S. Howell, K. Moore, B. Lienert, S. Masonis, T. Anderson, and D. Covert, Sea-salt size distributions from breaking waves: Implications for marine aerosol production and optical extinction measurements during SEAS, *J. Atmos. Ocean. Technol.*, **20**, 1362–1374, 2003.

Clarke, A.D., Y. Shinozuka, V.N. Kapustin, S. Howell, B. Huebert, S. Doherty, T. Anderson, D. Covert, J. Anderson, X. Hua, M.K.G. II, C. McNaughton, G. Carmichael, and R. Weber, Size distributions and mixtures of dust and black carbon aerosol in Asian outflow: Physiochemistry and optical properties, *J. Geophys. Res.*, **109**, D15S09, doi:10.1029/2003JD004378, 2004.

Clegg, S.L., and J.H. Seinfeld, Improvement of the Zdanovskii-Stokes-Robinson model for mixtures containing solutes of different charge types, *J. Phys. Chem. A.*, **108**, 1008–1017, 2004.

Clegg, S.L., P. Brimblecombe, and A.S. Wexler, Thermodynamic model of the system H^+-NH_4^+-SO_4^{2-}-NO_3^--H_2O at tropospheric temperatures, *J. Phys. Chem. A.*, **102**, 2137–2154, 1998.
Clegg, S.L., J.H. Seinfeld, and P. Brimblecombe, Thermodynamic modelling of aqueous aerosols containing electrolytes and dissolved organic compounds, *J. Aerosol Sci.*, **32**, 713–738, 2001.
Coakley, J.A., R.L. Bernstein, and P.A. Durkee, Effects of ship-track effluents on cloud reflectivity, *Science*, **255**, 423–430, 1987.
Coakley, J.A., M.A. Friedman, and W.R. Tahnk, Retrieval of cloud properties for partly cloudy imager pixels, *J. Atmos. Ocean Tech.*, **22**, pp. 3–17, 2005.
Cofala, J., M. Amann, and R. Mechler, *Scenarios of World Anthropogenic Emissions of Air Pollutants and Methane upto 2030*, International Institute for Applied Systems Analysis (IIASA), 2005.
Collins, D.R., H.H. Jonsson, J.H. Seinfeld, R.C. Flagan, S. Gasso, D.A. Hegg, P.B. Russell, B. Schmid, J.M. Livingston, E. Ostrom, K.J. Noone, L.M. Russell, and J.P. Putaud, In situ aerosol-size distributions and clear-column radiative closure during ACE-2, *Tellus*, **52B**, 498–525, 2000.
Comstock, K.K., C.S. Bretherton, and S.E. Yuter, Mesoscale variability and drizzle in southeast Pacific stratocumulus, *J. Atmos. Sci.*, **62**, 3792–3807, 2005.
Conover, J.H., Anomalous cloud lines, *J. Atmos. Sci.*, **23**, 778–785, 1966.
Cooke, W.F., C. Liousse, H. Cachier, and J. Feichter, Construction of a 1°x1° fossil fuel emission data set for carbonaceous aerosol and implementation and radiative impact in the ECHAM-4 model, *J. Geophys. Res.*, **104**, 22,137–22,162, 1999.
Cooper, W.A., Ice formation in wave clouds: Observed enhancement during evaporation, *Proc. Conf. Cloud Physics*, Dallas, *Amer. Meteor. Soc.*, 147–152, 1995.
Cooper, W.A., and R.P. Lawson, Physical interpretation of results from the HIPLEX-1 experiment, *J. Clim. Appl. Meteor.*, **23**, 523–540, 1984.
Cooper, W.A. and G. Vali, The origin of ice in mountain cap clouds, *J. Atmos. Sci.*, **38**, 1244–1259, 1981.
Cooper, W.A., R.T. Bruintjes, and G.K. Mather, Calculations pertaining to hygroscopic seeding with flares, *J. Appl. Meteor.*, **36**, 1449–1469, 1997.
Corrigan, C.E., and T. Novakov, Cloud condensation nucleus activity of organic compounds: A laboratory study, *Atmos. Environ.*, **33**, 2661–2668, 1999.
Cotton, W.R., Numerical simulation of precipitation development in supercooled cumuli. Part I: *Mon. Wea. Rev.*, **11**, 757–763, 1972a.
Cotton, W.R., Numerical simulation of precipitation development in supercooled cumuli. Part II: *Mon. Wea. Rev.*, **11**, 764–784, 1972b.
Cotton, W.R., Modification of precipitation from warm clouds – A review, *Bull. Amer. Meteor. Soc.*, **63**, 146–160, 1982.
Cotton, W.R., Testing, implementation and evolution of seeding concepts – A review, *Meteor. Monogr.* **21**, 63–70, 1986a.
Cotton, W.R., Testing, implementation, and evolution of seeding concepts – A review, in *Precipitation Enhancement – A Scientific Challenge*, edited by R.R. Braham, Jr., *AMS Meteorol. Monogr. Ser.*, **43**, 139–149, Amer. Meteor. Soc., Boston, MA, 1986b.
Cotton, W.R., *Storms*, ASTeR Press, Fort Collins, CO, 158 pp., 1990.
Cotton, W.R., and R.A. Anthes, *Storm and Cloud Dynamics*, Academic Press, San Diego, CA, 1989.
Cotton, R.J., and P.R. Field, Ice nucleation characteristics of an isolated wave cloud, *Q.J. Roy. Met. Soc.*, **128**, 2417–2437, 2002.
Cotton, W.R., and R.A. Pielke, *Human Impacts on Weather and Climate*, 288 pp., ASTeR Press, Ft. Collins, Colo., 1992.
Cotton, W.R., and R.A. Pielke, *Human Impacts on Weather and Climate*, Second Edition, 288 pp., Cambridge University Press, New York, 1995.

Cotton, W.R., and R.A. Pielke, Sr., *Human Impacts on Weather and Climate*, Cambridge University Press, New York, 2007.

Cotton, W.R., G.J. Tripoli, R.M. Rauber, and E.A. Mulvihill, Numerical simulation of the effects of varying ice crystal nucleation rates and aggregation processes on orographic snowfall, *J. Clim. Appl. Meteorol.*, **25**, 1658–1680, 1986.

Cotton, W.R., G.D. Alexander, R. Hertenstein, R.L. Walko, R.L. McAnelly, and M. Nicholls, Cloud venting-A review and some new global annual estimates, *Earth Sci. Rev.*, **39**, 169–206, 1995.

Cotton, W.R., R.A. Pielke, Sr., R.L. Walko, G.E. Liston, C.J. Tremback, H. Jiang, R.L. McAnelly, J.Y. Harrington, M.E. Nicholls, G.G. Carrió, and J.P. McFadden, RAMS 2001: Current status and future directions. *Meteor. Atmos Physics*, **82**, 5–29, 2003.

Cotton, W.R., R. McAnelly, G. Carrió, P. Mielke, and C. Hartzell, Simulations of Snowpack Augmentation in the Colorado Rocky Mountains, *J. Weather Modification Association.*, **38**, 58–65, 2006.

Couch, R.H., C.W. Rowland, K.S. Ellis, M.P. Blythe, C.R. Regan, M.R. Koch, C.W. Antill, W.L. Kitchen, J.W. Cox, J.F. DeLorme, S.K. Crockett, R.W. Remus, J.C. Casas, and W.H. Hunt, lidar In-space Technology Experiment (LITE): NASA's first in-space lidar system for atmospheric research, *Opt. Eng.*, **30**, 88–95, 1991.

Covert, D.S., J.L. Gras, A. Wiedensohler, and F. Stratmann, Comparison of directly measured CCN with CCN modeled from the number-size distribution in the marine boundary layer during ACE 1 at Cape Grim, Tasmania, *J. Geophys. Res.*, **103**, 16,597–16,608, 1998.

Croft, B., U. Lohmann, and K. von Salzen, Black carbon ageing in the Canadian Center for Climate modeling and analysis atmospheric general circulation model, *Atmos. Chem. Phys.*, **5**, 1931–1949, SRef-ID: 1680-7324/acp/2005-1935-1931, 2005.

Crutzen, P.J., and M.O. Andreae, Biomass burning in the tropics: Impact on atmospheric chemistry and biogeochemical cycles, *Science*, **250**, 1669–1678, 1990.

Cruz, C.N., and S.N. Pandis, A study of the ability of pure secondary organic aerosol to act as cloud condensation nuclei, *Atmos. Environ.*, **31**, 2205–2214, 1997.

Cruz, C.N., and S.N. Pandis, The effect of organic coatings on the cloud condensation nuclei activation of inorganic atmospheric aerosol, *J. Geophys. Res.*, **103**, 13,111–13,123, 1998.

Curry, J.A., W.B. Rossow, D. Randall, J.L. Schramm, Overview of Arctic cloud and radiation characteristics, *J. Climate.*, **9**, 1731–1764, 1996.

Curry, J.A., C.A. Clayson, W.B. Rossow, R. Reeder, Y.C. Zhang, P.J. Webster, G.Liu, and R.-S. Sheu, High resolution satellite-derived data set of the surface fluxes of heat, freshwater, and momentum for the TOGA COARE IOP, *Bull. Amer. Meteor. Soc.*, **80**, 2059–2080, 1999.

Cziczo, D.J., D.M. Murphy, P.K. Hudson, and D.S. Thomson, Single particle measurements of the chemical composition of cirrus ice residue during CRYSTAL-FACE, *J. Geophys. Res.*, **109**, D04201, doi:10.1029/2003JD004032, 2004a.

Cziczo, D.J., P.J. DeMott, S.D. Brooks, A.J. Prenni, D.S. Thomson, D. Baumgardner, J.C. Wilson, S.M. Kreidenweis, and D.M. Murphy, Observations of organic species and atmospheric ice formation-art, *Geophys. Res. Lett.*, **31**, L12116, doi:10.1029/2004GL019822, 2004b.

Danielsen, E.F., Inherent difficulties in hail probability prediction, in *Meteor Monographs*, 16(38), *Amer. Met. Soc.*, edited by G.B. Foote, and C.A. Knight:, pp. 135–143, Boston, MA, 1977.

Davenport, H.M., and L.K. Peters, Field studies of atmospheric particulate concentration changes during precipitation, *Atmos. Environ.*, **12**, 997–1008, 1978.

Davison, P.S., D.L. Roberts, R.T. Arnold, and R.N. Colvile, Estimating the direct radiative forcing due to haze from the 1997 forest fires in Indonesia, *J. Geophys. Res.*, **109**, D10207, doi:10210.11029/12003JD004264, 2004.

Decesari, S., S. Fuzzi, M.C. Facchini, M. Mircea, L. Emblico, F. Cavalli, W. Maenhaut, X. Chi, G. Schkolnik, A. Falkovich, Y. Rudich, M. Claeys, V. Pashynska, G. Vas, I. Kourtchev, R. Vermeylen, A. Hoffer, M.O. Andreae, E. Tagliavini, F. Moretti, and P. Artaxo, Characterization of the organic composition of aerosols from Rondônia, Brazil, during the LBA-SMOCC 2002 experiment and its representation through model compounds, *Atmos. Chem. Phys.*, **6**, 375–402, 2006.

DEH, *State of the Air: National Ambient Air Quality Status and Trends Report 1991–2001.* Department of Environment and Heritage, Australian Government, ISBN:0-642-54990-7, April, 2004

Delene, D.J., and T. Deshler, Calibration of a photometric cloud condensation nucleus counter designed for deployment on a balloon package, *J. Atmos. Oceanic Technol.*, **17**, 459–467, 2000.

Delene, D.J., and T. Deshler, Vertical profiles of cloud condensation nuclei above Wyoming, *J. Geophys. Res.*, **106**, 12,579–12,588, 2001.

DeMott, P.J., Laboratory studies of cirrus cloud processes, Chp 5, in *Cirrus*, edited by D.K. Lynch, K, Sassen, D.O.C. Starr, and G. Stephens, Oxford University Press, New York, 2002.

DeMott, P.J., M.P. Meyers, and W.R. Cotton, Parameterization and impact of ice initiation processes relevant to numerical model simulations of cirrus clouds, *J. Atmos. Sci.*, **41**, 77–90, 1994.

DeMott, P.J., K. Sassen, M.R. Poellet, D. Baumgardner, D.C. Rogers, S.D. Brooks, A.J. Prenni, and S.M. Kreidenweis, African dust aerosols as atmospheric ice nuclei, *Geophys. Res. Lett,*. **30**(14), 1732, doi:10.1029/2003GL017410, 2003a.

DeMott P.J., D.J. Cziczo, A.J. Prenni, D.M. Murphy, S.M. Kreidenweis, D.S. Thomson, R. Borys, and D.C. Rogers, Measurements of the concentration and composition of nuclei for cirrus formation, *Proc. Natnl. Acad. Sci.*, **100**(25), 14,655–14,660, 2003b

Dennis, A.S., *Weather Modification by Cloud Seeding*, 267 pp., Academic Press, New York, 1980.

Dennis, A.S., and D.J. Musil, Calculations of hailstone growth and trajectories in a simple cloud model, *J. Atmos. Sci.*, **30**, 278–288, 1973.

Dennis, A.S., and H.D. Orville, Comments on "A new look at the Israeli cloud seeding experiments," *J. Appl. Meteor.*, **36**, 277–278, 1997.

Dentener, F.J., G.R. Carmichael, Y. Zhang, J. Lelieveld, and P.J. Crutzen, Role of mineral aerosol as a reactive surface in the global troposphere, *J. Geophys. Res.*, **101**, 22,869–22,889, 1996.

Dentener, F.J., S. Kinne, T. Bond, O. Boucher, J. Cofala, S. Generoso, P. Ginoux, S. Gong, J.J. Hoelzemann, A. Ito, L. Marelli, J.E. Penner, J.P. Putaud, C. Textor, M.Schulz, G.R. van der Werf, and J. Wilson, Emissions of primary aerosol and precursor gases in the years 2000 and 1750 prescribed data sets for AeroCom, *Atmos. Chem. Phys.*, **6**, 4321–4344, 2006.

Derwent, R.G., M.E. Jenkin, C.E. Johnson, and D.S. Stevenson, The global distribution of secondary particulate matter in a 3-D Lagrangian chemistry transport model, *J. Atmos. Chem.*, **44**, 57–95, 2003.

Deschamps, P.Y., F.M. Bréon, M. Leroy, A. Podaire, A. Bricaud, J.C. Buriez, and G. Seze, The POLDER mission: Instrument characteristics and scientific objectives, *IEEE Trans. Geosci. Remote Sens.*, **32**, 598–615, 1994.

Dessens, J., A physical evaluation of a hail suppression project with silver iodide ground burners in southwestern France, *J. Appl. Met.*, **37**, 1588–1599, 1998.

Deuzé, J.L., F.M. Bréon, C. Devaux, P. Goloub, M. Herman, B. Lafrance, F. Maignan, A. Marchand, F. Nadal, G. Perry, and D. Tanré, Remote sensing of aerosols over land surfaces from POLDER-ADEOS-1 polarized measurements, *J Geophys. Res.*, **106**, 4913–4926, 2001.

Diehl, K., C. Quick, S. Matthias-Maser, S.K. Mitra, and R. Jaenicke, The ice nucleating ability of pollen. Part I: Laboratory studies in deposition and condensation freezing modes, *Atmos. Res.*, **58**, 75–87, 2001.

DiGirolamo, L., T.C. Bond. D. Bramer. D.J. Diner, F. Fettinger, R.A. Kahn, J.V. Martonchik, M.V. Ramana, V. Ramanathan, and P.J. Rasch, Analysis of Multi-angle Imaging SpectroRadiometer (MISR) aerosol optical depths over greater India during winter 2001–2004, *Geophys. Res. Let.*, **31**, L23115, doi:10.1029/2004GL021273, 2004.

Diner, D.J., J.C. Beckert, T.H. Reilly, C.J. Bruegge, J.E. Conel, R. Kahn, J.V. Martonchik, T.P. Ackerman, R. Davies, S.A.W. Gerstl, H.R. Gordon, J.P. Muller, R. Myneni, R.J. Sellers, B. Pinty, and M.M. Verstraete, Multiangle Imaging SptectroRadiometer (MISR) description and experiment overview, *IEEE Trans. Geosci. Remt. Sensing*, **36**, 1072–1087, 1998.

Diner, D.J., W.A. Abdou, C.J. Bruegge et al., MISR aerosol retrievals over southern Africa during the SAFARI-2000 dry season campaign, *Geophys. Res. Lett.*, **28**, 3127–3130, 2001.

Diner, D.J., J.C. Beckert, G.W. Bothwell, and J.I. Rodriguez, Performance of the MISR instrument during its first 20 months in Earth orbit, *IEEE Trans. Geosci. Remote Sens.*, **40**, 1449–1466, 2002.

Diner, D.J., T.P. Ackerman, T.L. Anderson, et al., PARAGON - An integrated approach for characterizing aerosol climate impacts and environmental interactions, *Bullet. Amer. Meteor. Soc.*, **85**, 1491–1501, 2004.

Dinger, J.E., H.B. Howell, and T.A. Wojciechowski, On the source and composition of cloud nuclei in a subsident air mass over the North Atlantic, *J. Atmos. Sci.*, **27**, 791–797, 1970.

Dong Y.Y., R.G. Oraltay, and J. Hallett, ice particle generation during evaporation, *Atmos. Res.* **32**, 45–53, 1994.

Doswell, C.A., H.E. Brooks, and R.A. Maddox, Flash flood forecasting: An ingredients-based methodology, *Wea. Forecasting*, **11**, 560–581, 1996.

Doviak, R.J., and D.S. Zrnic, *Doppler Radar and Weather Observations*, 562 pp., Academic Press, New York, NY, 1993.

Dubovik, O., and M.D. King, A flexible inversion algorithm for retrieval of aerosol optical properties from Sun and sky radiance measurements, *J. Geophys. Res.*, **105**, 20,673–20,696, 2000.

Dubovik, O., A. Smirnov, B.N. Holben, M.D. King, Y.J. Kaufman, T.F. Eck, and I. Slutsker, Accuracy assessments of aerosol optical properties retrieved from AERONET sun and sky-radiance measurements, *J. Geophys. Res.*, **105**, 9791–9806, 2000.

Dubovik, O., B.N. Holben, T.F. Eck, A. Smirnov, Y.J. Kaufman, M.D. King, D. Tanré, and I. Slutsker, Variability of absorption and optical properties of key aerosol types observed in worldwide locations, *J. Atmos. Sci*, **59**, 590–608, 2002.

Duce, R., Distributions and fluxes of mineral aerosol, in *Aerosol Forcing of Climate*, edited by R.J. Charlson, and J. Heintzenberg, pp. 43–72, Wiley, Chichester, UK, 1995.

Duchon, C.E., and G.R. Essenberg, Comparative rainfall observations from pit and aboveground rain gauges with and without wind shields, *Water Resour. Res.*, **37**, 3253–3263, 2001.

Dulac, F., Tanré, D., Bergametti G, Buat-Menard, P., Desbois, M., Sutton, D., Assessment of the african airborne dust mass over the western mediterranean sea using Meteosat data, *J. Geophys. Res.*, **97**, 2489–2506, 1992.

Durant, A.J. and R.A. Shaw. Evaporation freezing by contact nucleation inside-out, *Geophys. Res. Lett.*, **32**, L20814, doi:10.1029/2005GL024175, 2005.

Durkee, P.A., R.E. Chartier, A. Brown, E.J. Trehubenko, S.D. Rogerson, C. Skupniewicz, and K.E. Nielsen, S. Platnick and M.D. King, Composite Ship Track Characteristics, *J. Atmos. Sci.*, **57**, 2542–2553, 2000.

Durkee, P.A., K.J. Noone, R.J. Ferek, D.W. Johnson, J.P. Taylor, T.J. Garrett, P.V. Hobbs, J.G. Hudson, C.S. Bretherton, G. Innis, G.M. Frick, W.A. Hoppel, C.D. O'Dowd, L.M. Russell, R. Gasparovic, K.E. Nielsen, S.A. Tessmer, E. Öström, S.R. Osborne, R.C. Flagan, J.H. Seinfeld and H. Rand, The Impact of Ship-Produced Aerosols on the Microstructure and Albedo of Warm Marine Stratocumulus Clouds: A Test of MAST Hypotheses 1i and 1ii, *J. Atmos. Sci.*, **57**, 2554–2569, 2001.

Dusek, U., D.S. Covert, A. Wiedensohler, C. Neususs, D. Weise, and W. Cantrell, Cloud condensation nuclei spectra derived from size distributions and hygroscopic properties of the aerosol in coastal south-west Portugal during ACE-2, *Tellus B.*, **55**, 35–53, 2003.

Dusek, U., G.P. Frank, L. Hildebrandt, J. Curtius, J. Schneider, S. Walter, D. Chand, F. Drewnick, S. Hings, D. Jung, S. Borrmann, M.O. Andreae, Size matters more than chemistry in controlling which aerosol particles can nucleate cloud droplets, *Science*, **312**, 1375–1378, 2006.

Eagan, R.C., P.V. Hobbs, and L.F. Radke, Measurements of cloud condensation nuclei and cloud droplet size distributions in the vicinity of forest fires, *J. Appl. Meteor.*, **13**, 553–557, 1974a.

Eagan, R.C., P.V. Hobbs, and L.F. Radke, Measurements of cloud condensation nuclei and cloud droplet size distributions in the vicinity of forest fires, *J. Appl. Meteor.*, **13**, 553–557, 1974b.

Easter, R.C., and P.V. Hobbs, The formation of sulfates and the enhancement of cloud condensation nuclei in clouds, *J. Atmos. Sci.*, **31**, 1586–1594, 1974.

Easter, R.C., S.J. Ghan, Y. Zhang, R.D. Saylor, E.G. Chapman, N.S. Laulainen, H. Abdul-Razzak, L.R. Leung, X.D. Bian, and R.A. Zaveri, MIRAGE: Model description and evaluation of aerosols and trace gases, *J. Geophys. Res.*, 109, D20210, doi:10.1029/2004JD004571, 2004.

Ebert, E.E., Monitoring the quality of operational and semi-operational satellite precipitation estimates – The IPWG Validation/intercomparison study. 2nd International Precipitation Working Group (IPWG) Workshop, Monterey, CA, 25–28 October 2004, 9 pp., 2004.

Ebert, E.E., M.J. Manton, P.A. Arkin, R.J. Allam, G.E. Holpin, and A. Gruber, Results from the GPCP Algorithm Intercomparison Programme, *Bull. Amer. Meteor. Soc.*, **77**, 2875–2887, 1996.

Ebert, E.E., J.E. Janowiak, and C. Kidd, Comparison of Near Real Time Precipitation Estimates from Satellite Observations and Numerical Models, *Bull. Amer. Meteor. Soc.*, **88**, 47–64, 2007.

Eck, T.F., B.N. Holben, J.S. Reid, O. Dubovik, A. Smirnov, N.T. O'Neill, I. Slutsker, and S. Kinne, Wavelength dependence of the optical depth of biomass burning, urban and desert dust aerosols, *J. Geophys. Res.*, **104**, 31,333–31,350, 1999.

Eck, T.F., B.N. Holben, D.E. Ward, M.M. Mukelabai, O. Dubovik, A. Smirnov, J.S. Schafer, N.C. Ilsu, S.J. Piketh, A. Queface, J. LeRoux, R.J. Swap, and I. Slutsker, Variability of biomass burning aerosol optical characteristics in southern Africa during the SAFARI 2000 dry season campaign and a comparison of single scattering albedo estimates from radiometric measurements – Art. No. 8477, *J. Geophys. Res. Atmos.*, **108**, 8477, 2003.

Eklund, D.L., D.S. Jawa, and T.K. Rajala, Evaluation of the western Kansas weather modification program, *J. Wea. Mod.*, **31**, 91–101, 1999.

Eleftheriadis, K., and I. Colbeck, Coarse atmospheric aerosol: Size distributions of trace elements, *Atmos. Environ.*, **35**, 5321–5330, 2001.

Endresen, O., E. Sorgard, J.K. Sundet, S.B. Dalsoren, I.S.A. Isaksen, T.F. Berglen, and G. Gravir, Emission from international sea transportation and environmental impact, *J. Geophys. Res.*, **108**, 4560, doi:10.1029/2002JD002898, 2003.

Endresen, O., J. Bakke, E. Sorgard, T.F. Berglen, and P. Holmvang, Improved modelling of ship SO_2 emissions – A fuel-based approach, *Atmos. Environ.*, **39**, 3621–3628, 2005.

English, M., Alberta hailstorms. Part II: Growth of large hail in the Storm, *Meteor. Monogr.*, **14**, 37–98, 1973.

Erlick, C., A.P. Khain, M. Pinsky and Y. Segal, The effect of turbulent velocity fluctuations on drop spectrum broadening in stratiform clouds, *Atmos. Res.* **75**, 15–45, 2005.

Ervens, B., G. Feingold, and S.M. Kreidenweis, The influence of water soluble organic carbon on cloud drop number concentration, *J. Geophys. Res.*, **110**, D18211, doi:10.1029/2004JD005634, 2005.

Ervens, B., M. Cubison, E. Andrews, G. Feingold, J.A. Ogren, J.L. Jimenez, P. DeCarlo, and A. Nenes, Prediction of cloud condensation nucleus number concentration using measurements of aerosol size distributions and composition and light scattering enhancement due to humidity, *J. Geophys. Res.*, **112**, D10S32, doi:10.1029/2006JD007426, 2007.

Evan, A.T., J. Dunion, J.A. Foley, A.K. Heidinger, and C.S. Velden, New evidence for a relationship between Atlantic tropical cyclone activity and African dust outbreaks, *Geophys. Res. Lett.*, **33**, L19813–19810. 11029/12006GL026408, 2006.

Facchini, M.C., M. Mircea, S. Fuzzi, and R.J. Charlson, Cloud albedo enhancement by surface-active organic solutes in growing droplets, *Nature*, **401**, 257–259, 1999.

Facchini, M.C., S. Decesari, M. Mircea, S. Fuzzi, and G. Loglio, Surface tension of atmospheric wet aerosol and cloud/fog droplets in relation to their organic carbon content and chemical composition, *Atmos. Environ.*, **34**, 4853–4857, 2000.

Fankhauser, J.C., Estimates of thunderstorm precipitation efficiency from field measurements in LLOP, *Mon. Wea. Rev.*, **116**, 663–684, 1988.

Farley, R.D., and C.S. Chen, A detailed microphysical simulation of hygroscopic seeding on the warm process, *J. Appl. Meteor.*, **14**, 718–733, 1975.

Federer, B., A. Waldvogel, W. Schmid, H.H. Schiesser, F. Hampel, M. Schweingruber, W. Stahel, J. Bader, J.F. Mezeix, N. Doras, G. D'Aubigny, G. DerMegreditchian, and D. Vento, Main results of Grossversuch IV, *J. Climate Appl. Meteor.*, **25**, 917–957, 1986.

Feichter, J., E. Roeckner, U. Lohmann, and B. Liepert, Nonlinear aspects of the climate response to greenhouse gas and aerosol forcing, *J. Clim.*, **17**(12), 2384–2398, 2004.

Feingold, G., and P.Y. Chuang, Analysis of the influence of film-forming compounds on droplet growth: Implications for cloud microphysical processes and climate, *J. Atmos. Sci.*, **59**, 2006–2018, 2002.

Feingold, G., and S.M. Kreidenweis, Does heterogeneous processing of aerosol increase the number of cloud droplets? *J. Geophys. Res.*, **105**, 24,351–24,361, 2000.

Feingold, G., and S.M. Kreidenweis, Cloud processing of aerosol as modeled by a large eddy simulation with coupled microphysics and aqueous chemistry, *J. Geophys. Res.*, **107**, 4687, doi:10.1029/2002JD002054, 2002.

Feingold, G., B. Stevens, W.R. Cotton, and R.L. Walko, An explicit cloud microphysics/LES model designed to simulate the Twomey effect, *Atmos. Res.*, **33**, 207–233, 1994.

Feingold, G., S.M. Kreidenweis, B. Stevens, and W.R. Cotton, Numerical simulation of stratocumulus processing of cloud condensation nuclei through collision-coalescence, *J. Geophys. Res.*, **101**, 21,391–21,402, 1996.

Feingold, G., R. Boers, B. Stevens, and W.R. Cotton, A modeling study of the effect of drizzle on cloud optical depth and susceptibility, *J. Geophys. Res.*, **102**, 13,527–13,534, 1997.

Feingold, G., R.L. Walko, B. Stevens, and W.R. Cotton, Simulations of marine stratocumulus using a new microphysical parameterization scheme, *Atmos. Res.*, **47–48**, 505–528, 1998.

Feingold, G., W.R. Cotton, S.M. Kreidenweis, and J.T. Davis, The impact of giant cloud condensation nuclei on drizzle formation in stratocumulus: Implications for cloud radiative properties, *J. Atmos. Sci.*, **56**, 4100–4117, 1999.

Feingold, G., L.A. Remer, J. Ramaprasad, and Y.J. Kaufman, Analysis of smoke impact on clouds in Brazilian biomass burning regions: An extension of Twomey's approach, *J. Geophys. Res.*, **106**, D19, 22,907–22,922, 2001.

Feingold, G., L.E. Wynn, D.E. Veron, and M. Previdi, First measurements of Twomey indirect effect using ground-based remote sensors, *Geophys. Res. Lett.*, **30**, 1287, 2003.

Feingold, G., H. Jiang, and J.Y. Harrington, On smoke suppression of clouds in Amazonia, *Geophys. Res. Lett.*, **32**(2), L02804, doi:10.1029/2004GL021369, 2005.

Feng, M., P. Hacker, R. Lukas, R. Weller, and S.P. Anderson, Upper ocean heat and salt balances in the western equatorial Pacific in response to the intraseasonal oscillation during TOGA COARE, *J. Clim.*, **12**, 2409–2427, 2000.

Ferek, R.J., D.A. Hegg, and P.V. Hobbs, Measurements of ship-induced tracks in clouds off the Washington coast, *J. Geophys. Res.*, **103**, 23,199–23,206, 1998.
Ferrare, R.A., E.V. Browell, J.W. Hair, S. Ismail, D.D. Turner, M. Clayton, C.F. Butler, V.G. Brackett, M.A. Fenn, A. Notari, S.A. Kooi, M. Chin, S. Guibert, M. Schulz, C. Chuang, M. Krol, S.E. Bauer, X. Liu, G. Myhre, Ø. Seland, D. Fillmore, S. Ghan, S. Gong, P. Ginoux, and T. Takemura, The Vertical Distribution of Aerosols: Lidar Measurements vs. Model Simulations, paper presented at 23rd International Laser Radar Conference, 24–28 July 2006, Nara, Japan, 2006.
Ferriday, J.G., and S.K. Avery, Passive microwave remote sensing of rainfall with SSM/I: Algorithm development and implementation, *J. Appl. Meteor.*, **33**, 1587–1596, 1994.
Ferrier, B.S., A double-moment multiple-phase four-class bulk ice scheme. Part I: Description, *J. Atmos. Sci.*, **51**, 249–280, 1994.
Field, P.R., R.J. Cotton, K. Noone, P. Glantz, P.H. Kaye, E. Hirst, R.S. Greenaway, C. Jost, R. Gabriel, T. Reiner, M. Andreae, C.P.R. Saunders, A. Archer, and T. Choularton, Ice nucleation in orographic wave clouds: Measurements made during INTACC, *Q. J. Roy. Meteor. Soc.*, **127**, 1493–1512, 2001.
Field, P.R., R.J. Hogan, P.R.A. Brown, A. Illingworth, T.W. Choularton, and R.J. Cotton, Parameterization of ice-particle size distributions for mid-latitude stratiform cloud, *Q. J. Roy. Meteor., Soc.*, **131**, 1997–2017, 2005.
Field, P.R., O. Mohler, P. Connolly, M. Kramer, R. Cotton, A.J. Heymsfield, H. Saathoff, and M. Schnaiter, Some ice nucleation characteristics of Asian and Saharan desert dust, *Atmos. Chem. Phys.*, **6**, 2991–3006, 2006
Findeisen, W., Die kolloidmeteorologisch Vorgänge bei der Niederschlagsbildung, *Met. Z.*, **55**, 121–132, 1938.
Fisch, G., J. Tota, L.A.T. Machado, M.A.F. Silva Dias, R.F. Da F. Lyra, C.A. Nobre, A.J. Dolman, and J.H.C. Gash, The convective boundary layer over pasture and forest in Amazonia, *Theor. Appl. Climatol.*, **78**, 47–59, 2004.
Fiser, O., Schoenhuber M, and Pesice, P. First results of DSD measurement by videodistrometer in the Czech Republic in 1998–1999, *Studia Geophysica et Geodetica*, **46**, 3, 485–505, 2002.
Fitzgerald, J.W., Dependence of supersaturation spectrum of CCN on aerosol size distribution and composition, *J. Atmos. Sci.*, **30**, 628–634, 1973.
Fitzgerald, J.W., Effect of aerosol composition on cloud droplet size distribution: A numerical study, *J. Atmos. Sci.*, **31**, 1358–1367, 1974.
Fitzgerald, J.M., Marine aerosols: A review, *Atmos. Environ.*, **25A**, 533–545, 1991.
Fletcher, N.H., *The Phys. of Rainclouds*, 242 pp., Cambridge University Press, Cambridge, UK, 1962.
Flossmann, A.I., Interaction of aerosol particles and clouds, *J. Atmos. Sci.*, **55**, 879–887, 1998.
Flossman, A.I., and H.R. Pruppacher, A theoretical study of the wet removal of atmospheric pollutants. Part III: The uptake, redistribution, and deposition of $(NH_4)_2SO_4$ particles by a convective cloud using a two-dimensional cloud dynamics model, *J. Atmos. Sci.*, **45**, 1857–1871, 1988.
Flossman, A.I., W.D. Hall, and H.R. Pruppacher, A theoretical study of the wet removal of atmospheric pollutants. Part I: The redistribution of aerosol particlescaptured through nucleation and impaction scavenging by cloud droplets, *J. Atmos. Sci.*, **42**, 582–606, 1985.
Flueck, J.A., W.L. Woodley, R.W. Burpee, and D.O. Stram, Comments on "FACE rainfall results: Seeding effect or natural variability?" *J. Appl. Meteor.*, **20**, 98–107, 1981.
Folland, C.K., Numerical models of the rain gauge exposure problem, field experiments and an improved collector design, *Q. J. Roy. Meteor. Soc.*, **114**, 1485–1516, 1988.
Foote, G.B., A study of hail growth utilizing observed storm conditions, *J. Climate Appl. Meteor.*, **23**, 84–101, 1984.
Formenti, P., W. Elbert, W. Maenhaut, J. Haywood, and M.O. Andreae, Chemical composition of mineral dust during the Saharan Dust Experiment (SHADE) airborne campaign

in the Cape Verde region, September 2000, *J. Geophys. Res.*, **108**, 8576, doi:10.1029/2002JD002408, 2003.

Franklin, B., Meteorological imaginations and conjectures, *Mem. Manchester Lit. and Phil. Soc.*, **2**, 374–381, 1789.

Fraser, R.S., Y.J. Kaufman, and R.L. Mahoney, Satellite measurements of aerosol mass and transport, *Atmos. Environ.*, **18**, 2577–2584, 1984.

Freud, E., D. Rosenfeld, M.O. Andreae, A.A. Costa, and P. Artaxo, Observed robust relations between CCN and vertical evolution of cloud drop size distributions in deep convective clouds, *Atmos. Chem. Phys. Discuss.*, **5**, 10,155–10,195, 2005.

Fridlind, A.M., A.S. Ackerman, E.J. Jensen, A.J. Heymsfield, M.R. Poellot, D.E. Stevens, D. Wang, L.M. Miloshevich, D. Baumgardner, R.P. Lawson, J.C. Wilson, R.C. Flagan, J.H. Seinfeld, H.H. Jonsson, T.M. VanReken, V. Varutbangkul, and T.A. Rissman, Evidence for the predominance of mid-tropospheric aerosols as subtropical anvil cloud nuclei, *Science*, **304**, 718–722, 2004.

Fromm, M., A. Tupper, Rosenfeld, D., R. Servranckx, and R. McRae, Violent pyro-convective storm devastates Australia's capital and pollutes stratosphere. *Geophys. Res. Let.*, **33**, L05815, doi:10.1029/2005GL025161, 2006.

Fuzzi, S., M.O. Andreae, B.J. Huebert, M. Kulmala, T.C. Bond, M. Boy, S.J. Doherty, A. Guenther, M. Kanakidou, K. Kawamura, V.M. Kerminen, U. Lohmann, L.M. Russell, and U. Pöschl, Critical assessment of the current state of scientific knowledge, terminology, and research needs concerning the role of organic aerosols in the atmosphere, climate, and global change, *Atmos. Chem. Phys.*, **6**, 2017–2038, 2006.

Fuzzi, S., S. Decesari, M.C. Facchini, F. Cavalli, L. Emblico, M. Mircea, M.O. Andreae, I. Trebs, A. Hoffer, P. Guyon, P. Artaxo, L.V. Rizzo, L.L. Lara, T. Pauliquevis, W. Maenhaut, N. Raes, X. Chi, O.L. Mayol-Bracero, L.L. Soto-García, M. Claeys, I. Kourtchev, J. Rissler, E. Swietlicki, E. Tagliavini, G. Schkolnik, A.H. Falkovich, Y. Rudich, G. Fisch, and L.V. Gatti. Overview of the inorganic and organic composition of size-segregated aerosol in Rondônia, Brazil, from the biomass burning period to the onset of the wet season, *J. Geophy. Res.*, **112**, D01201, doi:10.1029/2005JD006741, 2007.

Gabric, A.J., R. Simo, R.A. Cropp, A.C. Hirst, and J. Dachs, Modeling estimates of the global emission of dimethylsulfide under enhanced greenhouse conditions, *Global Biogeochem. Cycles.*, **18**, GB2014, doi:10.1029/2003GB002183, 2004.

Gagin, A., The ice phase in winter continental cumulus clouds, *J. Atmos. Sci.*, **32**, 1604–1614, 1975.

Gagin, A., Evaluation of static and dynamic seeding concepts through analyses of Israeli II experiment and FACE-2 experiments, in *Rainfall Enhancement – A Scientific Challenge, AMS Meteorol. Monogr.*, **43**, 63–70, *Amer. Meteorol. Soc.*, Boston, MA, 1986.

Gagin, A., and K.R. Gabriel, Analysis of recording rain gauge data for the Israeli II experiment. Part I: Effects of cloud seeding on the components of daily rainfall, *J. Clim. Appl. Meteor.*, **26**, 913–926, 1987.

Gagin, A., and J. Neumann, Rain stimulation and cloud physics in Israel, in *Weather and Climate Modification*, edited by W.N. Hess, pp. 454–494, John Wiley and Sons, New York, 1974.

Gagin, A., and J. Neumann, The second Israeli randomized cloud seeding experiment: Evaluation of results, *J. Appl. Meteor.*, **20**, 1301–1311, 1981.

Gagosian, R.B., O.C. Zafiriou, E.T. Peltzer, and J.B. Alford, Lipids in aerosols from the tropical North Pacific – Temporal variability, *J. Geophys. Res. -Oceans and Atmospheres.*, **87**, 1133–1144, 1982.

Gallagher, M.W., K.M. Beswick, T.W. Choularton, J. Duyzer, H. Westrate, and P. Hummelshøj, Measurements of aerosol fluxes to speulderforest using a micrometeorological technique, *Atmos. Environ.*, **31**, 359–373, 1997.

Gallagher, M.W., E. Nemitz, J.R. Dorsey, D. Fowler, M.A. Sutton, M. Flynn, and J. Duyzer, Measurements and parameterizations of small aerosol deposition velocities to grassland, arable crops, and forest: Influence of surface roughness length on deposition, *J. Geophys. Res. Atmos.*, **107**, 4154, doi:10.1029/2001JD000817, 2002.

Gao, S., D.A. Hegg, G. Frick, P.F. Caffrey, L. Pasternack, C. Cantrell, W. Sullivan, J. Ambrusko, T. Albrechcinski, and T.W. Kirchstetter, Experimental and modeling studies of secondary organic aerosol formation and some applications to the marine boundary layer: *J. Geophys. Res.*, **106**, 27,619–27,634, 2001.

Gao, S., D.A. Hegg, P.V. Hobbs, T.W. Kirchstetter, B.I. Magi, and M. Sadilek, Water-soluble organic components in aerosols associated with savanna fires in southern Africa: Identification, evolution, and distribution, *J. Geophys. Res.*, **108**, 8491, doi:10.1029/2002/JD002324, 2003.

Gao, S., N.L. Ng, M. Keywood, V. Varutbangkul, R. Bahreini, A. Nenes, J.W. He, K.Y. Yoo, J.L. Beauchamp, R.P. Hodyss, R.C. Flagan, and J.H. Seinfeld, Particle phase acidity and oligomer formation in secondary organic aerosol, *Environ. Sci. Technol.*, **38**, 6582–6589, 2004.

Gard, E.E., M.J. Kleeman, D.S. Gross, L.S. Hughes, J.O. Allen, B.D. Morrical, D.P. Fergenson, T. Dienes, M.E. Galli, R.J. Johnson, G.R. Cass, and K.A. Prather, Direct observation of heterogeneous chemistry in the atmosphere, *Science*, **279**, 1184–1187, 1998.

Garrett, T.J., and P.V. Hobbs, Long-range transport of continental aerosols over the Atlantic Ocean and their effects on cloud droplet size distributions, *J. Atmos. Sci.*, **52**, 2977–2984, 1995.

Garrett, T.J., P.V. Hobbs, and L.F. Radke, High Aitken nucleus concentrations above cloud tops in the Arctic, *J. Atmos. Sci.*, **59**, 779–783, 2002.

Garrett, T.J., C. Zhao, X. Dong, G.G. Mace, and P.V. Hobbs, Effects of varying aerosol regimes on low-level Arctic stratus, *Geophys. Res. Lett.*, **31**, L17105, doi:10.1029/2004GL019928, 2004.

Garstang, M., R. Bruintjes, R. Serafin, H. Orville, B. Boe, W. Cotton and J. Warburton, Weather Modification: Finding common ground, *Bull. Amer. Meteor. Soc.*, **86**, 647–655, 2005.

GAW, Proceedings of WMO/GAW Aerosol measurement procedures guidelines and recommendations, GAW Report #153, available at http://www.wmo.ch/web/arep/gaw/gawreports.html, 2003.

GAW, Proceedings of WMO/GAW experts workshop on a global surface-based network for long term observations of column aerosol optical properties, Davos, Switzerland, 8–10 March 2004 GAW Report #162 available at http://www.wmo.ch/web/arep/gaw/gawreports.html, 2004

Gayet, J.F., F.Auriol, A. Minikin, J. Ström, M. Seifert, R. Krejci, A. Petzold, G. Febvre, and U.Schumann, Quantitative measurement of the microphysical and optical properties of cirrus clouds with four different in situ probes: Evidence of small ice crystals. *Geophys. Res. Lett.*, **29**, 2230–2233, 2002.

GCOS-107, Systematic Observation requirements for satellite-based products for climate, *Global Climate Observing Systems*, WMO/TD No. 1338, 17–26, 2006.

Gelbard, F., Y. Tambour, and J.H. Seinfeld, Sectional representations for simulating aerosol dynamics, *J. Colloid Interface Sci.*, **76**, 541–556, 1980.

Gelencsér, A., *Carbonaceous aerosol*. Dordrecht, Netherlands, Springer, 350 pp. 350, 2004.

Gelencsér, A., A. Hoffer, Z. Krivácsy, G. Kiss, A. Molnár, and E. Mészáros, On the possible origin of humic matter in fine continental aerosol, *J. Geophys. Res.*, **107**, 4137, doi:10.1029/2001JD001299, 2002.

Generoso, S., F.M. Breon, Y. Balkanski, O. Boucher, and M. Schulz, Improving the seasonal cycle and interannual variations of biomass burning aerosol sources, *Atmos. Chem. Phys.*, **3**, 1211–1222, 2003.

Geogdzhayev, I.V., M.I. Mishchenko, E.I. Terez, G.A. Terez, and G.K. Gushchin, Regional advanced very high resolution radiometer-derived climatology of aerosol optical thickness and size, *J. Geophys. Res.* **110**, D23, Art. No. D23205 DEC 3, 2005.

Georgi, H.W., and E. Kleinjung, Relations between the chemical composition of atmospheric aerosol particles and the concentration of natural ice nuclei, *J. Rech. Atmos.*, **3**, 145–156, 1968.

Gérard, B., J.L. Deuzé, M. Herman, Y.J. Kaufman, P. Lallart, C. Oudard, L.A. Remer, B. Roger, B. Six, and D. Tanré, Comparisons between POLDER 2 and MODIS/Terra aerosol retrievals over ocean, *J. Geophys. Res.*, **110**, D24211, doi:10.1029/2005JD006218, 2005.

Gerber, H., Microphysics of marine stratocumulus clouds with two drizzle modes, *J. Atmos. Sci.*, **53**, 1649–1662, 1996.

Gerber, H., B.G. Arends, and A.S. Ackerman, New microphysics sensor for aircraft use, *Atmos. Res.*, **31**, 235–252, 1994.

Gerber, H., Y. Takano, J.G. Timothy, and P.V. Hobbs, Nephelometer Measurements of the Asymmetry Parameter, Volume Extinction Coefficient, and Backscatter Ratio in Arctic Clouds, *J. Atmos. Sci.*, **57**, 3021–3034, 2000.

Germann, U., and J. Joss, Operational Measurement of Precipitation in Mountainous Terrain, in *Weather Radar: Principles and Advanced Applications*, edited by P. Meischner, pp. 52–77, Springer-Verlag, Berlin, 2004.

Germann, U., G. Galli, M. Boscacci, and M. Bolliger, Radar precipitation measurement in a mountainous region, *Q. J. Roy. Meteor. Soc.*, **132**, 1669–1692, 2006.

Ghan, A.J., L.R. Leung, R.C. Easter, and H. Abdul-Razzak, Prediction of droplet number in a general circulation model, *J. Geophys. Res.*, **102**, 21,777–21,794, 1997.

Ghan, S.J., R.C. Easter, E.G. Chapman, H. Abdul-Razzak, Y. Zhang, L.R. Leung, N.S. Laulainen, R.D. Saylor, and R.A. Zaveri, A physically based estimate of radiative forcing by anthropogenic sulfate aerosol, *J. Geophys. Res.*, **106**, 5279–5293, 2001.

Giebl, H., A. Berner, G. Reischl, H. Puxbaum, A. Kasper-Giebl, and R. Hitzenberger, CCN activation of oxalic and malonic acid test aerosols with the University of Vienna cloud condensation nuclei counter, *J. Aerosol Sci.*, **33**, 1623–1634, 2002.

Gill, P.S., T.E. Graedel, and C.J. Weschler, Organic films on atmospheric aerosol-particles, fog droplets, cloud droplets, raindrops, and snowflakes, *Rev. Geophys.*, **21**, 903–920, 1983.

Ginoux, P., M. Chin, I. Tegen, J. Prospero, B.N. Holben, O. Dubovik, and S.J. Lin, Sources and distributions of dust aerosols simulated with the GOCART model, *J. Geophys. Res.*, **106**, 20,255–220,274, 2001.

Giorgi, F., A particle dry deposition parameterization scheme for use in tracer transport models, *J. Geophys. Res.*, **91**, 9794–9806, 1986.

Girard, E., and J.A. Curry, Simulation of arctic low-level clouds observed during the FIRE Arctic Clouds Experiment using a new bulk microphysics scheme, *J. Geophys. Res. Atmospheres.*, **106**, 15,139–15,154, 2001.

Givati, A., and D. Rosenfeld, Quantifying precipitation suppression due to air pollution, *J. Appl. Meteor.*, **43**, 1038–1056, 2004.

Givati, A., and D. Rosenfeld, Separation between cloud-seeding and air-pollution effects, *J. Appl. Meteor.*, **44**, 1298–1315, 2005.

Gjertsen, U., M. Salek, and D.B. Michelson, Gauge adjustment of radar-based precipitation estimates in Europe, *Proc. Third European Conf. on Radar in Meteor. and Hydrol.*, Visby, Sweden, Copernicus GmbH, 7–11, 2004.

Gokhale, N.R., and J.D. Spengler, Freezing of freely suspended water droops by contact nucleation, *J. Appl. Meteor.*, **11**, 157–160, 1972.

Golaz, J.C., V.E. Larson, and W.R. Cotton, A PDF-based model for boundary layer clouds. Part I: Method and model description, *J. Atmos. Sci.*, **59**, 3540–3551, 2002a.

Golaz, J.C., V.E. Larson, and W.R. Cotton, A PDF-based model for boundary layer clouds. Part II: Model results, *J. Atmos. Sci.*, **59**, 3552–3571, 2002b.

Goldsmith, P., J. Goster, and C. Hume, The ice phase in clouds, Preprints, International Conference on Cloud Physics, Boulder, CO, Amer. Meteor. Soc., Boston, 163–167, 1976.

Goloub P, D. Tanré, J.L. Deuzé, M. Herman, A. Marchand, and F.M. Bréon, Validation of the first algorithm applied for deriving the aerosol properties over the ocean using the POLDER/ADEOS measurements, *IEEE T. on Geosci. Remote.*, **37**, 1586–1596, 1999.

Gong, S.L., A parameterization of sea-salt aerosol source function for sub- and super-micron particles, *Global Biogeochem. Cycles*, **17**, 1097, doi:10.1029/2003GB002079, 2003.

Gong, S.L., and L.A. Barrie, Simulating the impact of sea salt on global nss sulphate aerosols, *J. Geophys. Res.*, **108**, 4516, doi:1029/2002JD003181, 2003.

Gong, S.L., L.A. Barrie, and M. Lazare, Canadian Aerosol Module (CAM): A size-segregated simulation of atmospheric aerosol processes for climate and air quality models. 2. Global sea-salt aerosol and its budgets, *J. Geophys. Res.*, **107**, 4779, doi:10.1029/2001JD002004, 2002.

Gong, S.L., L.A. Barrie, J.P. Blanchet, K.V. Salzen, U. Lohmann, G. Lesins, L. Spacek, L.M. Zhang, E. Girard, H. Lin, R. Leaitch, H. Leighton, P. Chylek, and P. Huang, Canadian Aerosol Module: A size-segregated simulation of atmospheric aerosol processes for climate and air quality models 1. Module development, *J. Geophys. Res.*, **108**, 4007, doi:4010.1029/2001JD002002, 2003a.

Gong, S.L., X.Y. Zhang, T.L. Zhao, I.G. McKendry, D.A. Jaffe, and N.M. Lu, Characterization of Soil Dust Distributions In China And Its Transport During ACE-ASIA 2. Model Simulation and Validation, *J. Geophys. Res.*, **108**, 4262, doi:4210.1029/2002JD002633, 2003b.

Gong, S.L., X.Y. Zhang, T.L. Zhao, X.B. Zhang, L.A. Barrie, I.G. McKendry, and C.S. Zhao, A simulated climatology of asian dust aerosol and its trans-pacific transport. 2: Interannual variability and climate connections, *J. Climate*, **19**, 104–122, 2006.

Goodison, B.E., Accuracy of Canadian snow gauge measurements, *J. Appl. Meteor.*, 17, 1542–1548, 1978.

Goodman, A., G. Underwood, and V. Grassian, A laboratory study of the heterogeneous reaction of nitric acid on calcium carbonate particles, *J. Geophys. Res.*, **105**, 29,053–29,064, 2000.

Gossard, E.E., R.G. Strauch, and R.R. Rogers, Evolution of dropsize distributions in liquid precipitation observed by ground-based Doppler radar, *J. Atmos. Ocean. Tech.*, **7**, 815–828, 1990.

Grabowski, W.W., Indirect impact of atmospheric aerosols in idealized simulations of convective-radiative quasi-equilibrium, *J. Clim.*, 19, 4664–4682, 2006

Grabowski, W.W., X. Wu, and M.W. Moncrieff, Cloud resolving modeling of tropical cloud systems during Phase III of GATE. Part III: Effects of cloud microphysics, *J. Atmos. Sci.*, **56**, 2384–2402, 1999.

Graf, H.F., J. Feichter, and B. Langmann, Volcanic sulfur emissions: Estimates of source strength and its contribution to the global sulfate distribution, *J. Geophys. Res.*, **102**, 10,727–10,738, 1997.

Graham, B., O.L. Mayol-Bracero, P. Guyon, G.C. Roberts, S. Decesari, M.C. Facchini, P. Artaxo, W. Maenhaut, P. Köll, and M.O. Andreae, Water-soluble organic compounds in biomass burning aerosols over Amazonia: 1. Characterization by NMR and GC-MS, *J. Geophys. Res.*, **107**, 8047, doi:10.1029/2001JD000336, 2002.

Graham, B., P. Guyon, W. Maenhaut, P.E. Taylor, M. Ebert, S. Matthias-Maser, O. Mayol-Bracero, R.H.M. Godoi, P. Artaxo, F. Meixner, M.A.L. Moura, C.H.E.D. Rocha, R. van Grieken, M.M. Glovsky, R.C. Flagan, and M.O. Andreae, Composition and diurnal variability of the natural Amazonian aerosol, *J. Geophys. Res.*, **108**, 4765, doi:10.1029/2003JD004049, 2003a.

Graham, B., P. Guyon, P. Taylor, P. Artaxo, W. Maenhaut, M.M. Glovsky, R.C. Flagan, and M.O. Andreae, Organic compounds present in the natural Amazonian aerosol: Characterization by gas chromatography–mass spectrometry, *J. Geophys. Res.*, **108**, 4766, doi:10.1029/2003JD003990, 2003b.

Grant, L.O., and R.D. Elliott, The cloud seeding temperature window, *J. Appl. Meteor.*, **13**, 355–363, 1974.

Grant, L.O., and A.M. Kahan, Weather modification for augmenting orographic precipitation, in *Weather and Climate Modification*, edited by W.N. Hess, pp. 282–317, John Wiley and Sons, New York, 1974.

Grant, L.O., and P.W. Mielke, Jr., A randomized cloud seeding experiment at Climax, Colorado 1960–1965, *Proc. 5th Berkeley Symp. on Mathematical Statistics and Probability*, Volume V, pp. 115–131, University of California Press, Berkeley, Calif., 1967.

Gras, J.L., Cloud condensation nuclei over the southern-ocean, *Geophys. Res. Lett*, **17**, 1565–1567, 1990.

Grassl, H, Albedo reduction and radiative heating of clouds by absorbing aerosol particles, *Contribution to Atmos. Phys.*, Oxford, **48**, 199–210, 1975.

Greenfield, S., Rain scavenging of radioactive particulate matter from the atmosphere, *J. Meteor.*, **14**, 115–125, 1957.

Griffin, R.J., D.R. Cocker, III, J.H. Seinfeld, and D. Dabdub, Estimate of global atmospheric organic aerosols from oxidation of biogenic hydrocarbons, *Geophys. Res. Lett.*, **26**, 2721–2724, 1999.

Griffith, D.A., M.E. Solak, and D.P. Yorty, Is air pollution impacting winter orographic precipitation in Utah? Weather modification association, *J. Wea. Modif.*, **37**, 14–20, 2005.

Grini, A., G. Myhre, J.K. Sundet, and I.S.A. Isaksen, Modeling the annual cycle of sea salt in the global 3D model Oslo CTM2: Concentrations, fluxes, and radiative impact, *J. Clim.*, **15**, 1717–1730, 2002.

Groisman, P.V., and D.R. Legates, The accuracy of United States precipitation data, *Bull. Amer. Meteor. Soc.*, **75**, 215–227, 1994.

Groisman, P.V., V.V. Koknaeva, T.A. Belokrylova, and T.R. Karl, Overcoming biases in precipitation measurement: A history of the USSR experience, *Bull. Amer. Meteor. Soc.*, **72**, 1725–1733, 1991.

Guan, H., S.G. Cober, G.A. Isaac, A. Tremblay and A. Methot, Comparison of three cloud forecast schemes with in-situ aircraft measurements. *Wea. Forecasting*, **17**, 1226–1235, 2002.

Guazzotti, S.A., K.R. Coffee, and K.A. Prather, Continuous measurements of size-resolved particle chemistry during INDOEX-Intensive Field Phase 99, *J. Geophys. Res.*, **106**, 28,607–28,627, 2001.

Guelle, W., M. Schulz, Y. Balkanski, and F. Dentener, Influence of the source formulation on modeling the atmospheric global distribution of sea salt aerosol, *J. Geophys. Res.*, **106**, 27,509–27,524, 2001.

Guibert, S., J.R. Snider, and J.L. Brenguier, Aerosol Activation in Marine Stratocumulus Clouds Part-I: Measurement Validation for a Closure Study PACE Topical Issue, *J. Geophys. Res.*, **108**, D15, 8628 10.1029, 2003.

Gultepe, I., G.A. Isaac, W.R. Leaitch, and C.M. Banic, Parameterizations of marine stratus microphysics based on in situ observations: Implications for GCMs, *J. Clim.*, **9**, 345–357, 1996.

Gultepe, I., G.A. Isaac, S.G. Cober, Ice crystal number concentration versus temperature for climate studies, Internat, *J. Climatol.*, **21**, 1281–1302, 2001.

Gunn R. and B.B. Phillips, An experimental investigation of the effect of air pollution on the initiation of rain, *J. Meteor.*, **14**, 272–280, 1957.

Guyon, P., B. Graham, J. Beck, O. Boucher, E. Gerasopoulos, O.L. Mayol-Bracero, G.C. Roberts, P. Artaxo, and M.O. Andreae, Physical properties and concentration of aerosol particles over the Amazon tropical forest during background and biomass burning conditions, *Atmos. Chem. Phys.*, **3**, 951–967, 2003a.

Guyon, P., B. Graham, G.C. Roberts, O.L. Mayol-Bracero, W. Maenhaut, P. Artaxo, and M.O. Andreae, In-canopy gradients, composition, sources, and optical properties of aerosol over the Amazon forest, *J. Geophys. Res.*, **108**, 4591, doi:10.1029/2003JD003465, 2003b.

Guyon, P., G.P. Frank, M. Welling, D. Chand, P. Artaxo, L. Rizzo, G. Nishioka, O. Kolle, H. Fritsch, M.A.F. Silva Dias, L.V. Gatti, A.M. Cordova, and M.O. Andreae, Airborne

measurements of trace gases and aerosol particle emissions from biomass burning in Amazonia, *Atmos. Chem. Phys.*, **5**, 2989–3002, 2005.
Guzzi, R., and R. Rizzi, The effect of radiative exchange on the growth of a population of droplets, *Contrib. Atmos. Phys.*, **53**, 351–365, 1980.
Gysel, M., E. Weingartner, and U. Baltensperger, Hygroscopicity of aerosol particles at low temperatures. 2. Theoretical and experimental hygroscopic properties of laboratory generated aerosols, *Environ. Sci. Technol.*, **36**, 63–68, 2002.
Habib, E., W.F. Krajewski, and A. Kruger, Sampling errors of tipping bucket rain gauge measurements, *J. Hydrol. Engineering*, **6**, 159–166, 2001.
Haddad, Z.S., J.P. Meagher, R.F. Adler, E.A. Smith, E. Im, and S. Durden, Global variability of precipitation according to the Tropical Rainfall Measuring Mission, *J. Geophys. Res.*, **109**, D17103, doi:10.1029/2004JD004607, 2004.
Hagen, D.E., J. Podzimek, and M.B. Trueblood, Upper-tropospheric aerosol sampled during project FIRE IFO II, *J. Atmos. Sci.*, **52**, 4196–4209, 1995.
Hallett, J., and S.C. Mossop, Production of secondary ice crystals during the riming process, *Nature*, **249**, 26–28, 1974.
Hallett, J., J.G. Hudson, and C.F. Rogers, Characterization of combustion aerosols for haze and cloud formation, *Aerosol Sci. Tech.*, **10**, 70–83, 1989.
Han, Q., W.B. Rossow, and A.A. Lacis, Near-global survey of effective droplet radii in liquid water clouds using ISCCP data, *J. Clim.*, **7**, 465–497, 1994.
Han, Q., W.B. Rossow, J. Chou, and R.M. Welch, Global survey of the relationship of cloud albedo and liquid water path with droplet size using ISCCP, *J. Clim.*, **11**, 1516–1528, 1998.
Han, Z., H. Ueda, K. Matsuda, R. Zhang, K. Arao, Y. Kanai, and H. Hasome, Model study on particle size segregation and deposition during Asian dust events in March 2002, *J. Geophys. Res. D: Atmos.*, **109**, D19205, 19,201–19,222, 2004.
Hansen, A.R., and K.D. Beyer, Experimentally determined thermochemical properties of the malonic acid/water system: Implications for atmospheric aerosols, *J. Phys. Chem. A.*, **108**, 3457–3466, 2004.
Hansen, J., M. Sato, and R. Ruedy, Radiative forcing and climate response, *J. Geophys. Res.*, **102**, 6831–6864, 1997.
Hao, W.M., and M.H. Liu, Spatial and temporal distribution of tropical biomass burning, *Global Biogeochem. Cycles*, **8**, 495–503, 1994.
Hao, W.H., M.H. Liu, M. Lorenzini, K.D. Singh, and D.E. Ward, Spatial distribution of tropical biomass burning in 1990 with 1°x1° resolution, in *Biomass Burning and Global Change*, edited by J.S. Levine, MIT Press, Cambridge, MA, 1996.
Harrington, J.Y., G. Feingold, and W.R. Cotton, Radiative impacts on the growth of a population of drops within simulated summertime arctic stratus, *J. Atmos. Sci.*, **57**, 766–785, 2000.
Harris, D.H., and E. Foufoula-Georgiou, Subgrid variability and stochastic downscaling of modeled clouds: Effects on radiative transfer calculations for rainfall retrieval, *J. Geophys. Res.*, **106**, 10,349–10,362, 2001.
Harsvardhan, S.E. Schwartz, C.M. Benkovitz, and G. Guo, Aerosol influence on cloud microphysics examined by satellite measurements and chemical transport modeling, *J. Atmos. Sci.*, **59**, 714–725, 2002.
Hartmann, D.L., *Global Physical Climatology*, Academic Press, San Diego, 408 pp., 1994.
Hartmann, D.L., M.E. Ockert-Bell, and M.L. Michelson, The effect of cloud type on the earth's energy balance: Global analysis, *J. Clim.*, **5**, 1281–1304, 1992.
Hasse, L., M. Grossklaus, K. Uhlig, and P. Timm, A ship rain gauge for use in high wind speeds, *J. Ocean. Atmos. Tech.*, **15**, 380–386, 1998.
Hatzianastassiou, N., W. Wobrock, and A.I. Flossmann, The effect of cloud-processing of aerosol particles on clouds and radiation, *Tellus*, **50B**, 478–490, 1998.
Haywood, J.M, P. Francis, O. Dubovik, M. Glew, and B. Holben, Comparison of aerosol size distributions, radiative properties, and optical depths determined by aircraft observations

and Sun photometers during SAFARI 2000, *J. Geophys. Res.*, **108**, 8471, doi:10.1029/2002JD002250, 2003a.

Haywood, J.M., S.R. Osborne, P.N. Francis, A. Keil, P. Formenti, M.O. Andreae, and P.H. Kaye, The mean physical and optical properties of regional haze dominated by biomass burning aerosol measured from the C-130 aircraft during SAFARI 2000, *J. Geophys. Res.*, **108**, 8473, doi:10.1029/2002JD002226, 2003b.

Heard, D., *Analytical Techniques for Atmospheric Measurement*, Publisher: Blackwell Publishing, ISBN:1405123575, 2006.

Heald, C.L., D.J. Jacob, R.J. Park, L.M. Russell, B.J. Huebert, J.H. Seinfeld, H. Liao, and R.J. Weber, A large organic aerosol source in the free troposphere missing from current models, *Geophys. Res. Lett.*, **32**, L18809, doi:10.1029/2005GL023831, 2005.

Hegg, D.A., The influence of liquid-phase oxidation of SO_2 in the troposphere, *J. Geophys. Res.*, **99**, 3773–3779, 1985.

Hegg, D.A., and P.V. Hobbs, Cloud water chemistry and the production of sulfates in clouds, *Atmos. Environ.*, **15**, 1597–1604, 1981.

Hegg, D.A., and P.V. Hobbs, Cloud condensation nuclei in the marine atmosphere, in *Nucleation and atmospheric aerosols*, edited by N. Eukuta, and P.E. Wagner, pp. 181–192, A. Deepak Publishing, Hampton, VA, 1992.

Hegg, D.A., and T.V. Larson, The effects of microphysical parameterization on model predictions of sulfate production in clouds, *Tellus*, **42B**, 272–284, 1990.

Hegg, D.A., P.V. Hobbs, and L.F. Radke, Observations of the modification of cloud condensation nuclei in wave clouds, *J. Rech. Atmos.*, **14**, 217–222, 1980.

Hegg, D.A., R.J. Ferek, P.V. Hobbs, and L.F. Radke, Dimethyl sulfide and cloud condensation nucleus correlations in the northeast Pacific Ocean, *J. Geophys. Res.*, **96**, 13,189–13,191, 1991a.

Hegg, D.A., L.F. Radke, and P.V. Hobbs, Measurements of Aitken nuclei and cloud condensation nuclei in the marine atmosphere and their relation to the DMS-cloud-climate hypothesis, *J. Geophys. Res.*, **96**, 18,727–18,733, 1991b.

Hegg, D.A., P.F. Yuen, and T.V. Larson, Modeling the effects of heterogeneous cloud chemistry on the marine particle size distribution, *J. Geophys. Res.*, **97**, 12,927–12,933, 1991c.

Hegg, D.A., R.J. Ferek, and P.V. Hobbs, Cloud condensation nuclei over the Arctic Ocean in early spring, *J. Appl. Meteor.*, **34**, 2076–2082, 1995.

Hegg, D.A., R. Majeed, P.F. Yuen, M.B. Baker, and T.V. Larson, The impacts of SO_2 oxidation in cloud drops and in haze particles on aerosol light scattering and CCN activity, *Geophys. Res. Lett.*, **23**, 2613–2616, 1996.

Hegg, D.A., S. Gao, W. Hoppel, G. Frick, P. Caffrey, W.R. Leaitch, N. Shantz, J. Ambrusko, and T. Albrechcinski, Laboratory studies of the efficiency of selected organic aerosols as CCN, *Atmos. Res.*, **58**, 155–166, 2001.

Heintzenberg, J., Fine particles in the global troposphere: A review, *Tellus*, **41B**, 149–160, 1989.

Heintzenberg, J., and D.C. Covert, Size distribution and chemical composition of marine aerosols: A compilation and review, *Tellus*, **52B**, 1104–1122, 2000.

Heintzenberg, J., K. Okada, and J. Strom, On the composition of non-volatile material in upper tropospheric aerosols and cirrus crystals, *Atmos. Res.*, **41**, 81–88, 1996.

Heisler, S.L., S.K. Friedlan, and R.B. Husar, Relationship of smog aerosol size and chemical element distributions to source characteristics, *Atmos. Environ.*, **7**, 633–649, 1973.

Henning, S., E. Weingartner, M. Schwikowski, H.W. Gaggeler, R. Gehrig, K.P. Hinz, A. Trimborn, B. Spengler, and U. Baltensperger, Seasonal variation of water-soluble ions of the aerosol at the high-alpine site Jungfraujoch (3580 m asl), *J. Geophys. Res.*, **108**, 4030, doi:10.1029/2002JD002439, 2003.

Hering, S., A. Eldering, and J.H. Seinfeld, Bimodal character of accumulation mode aerosol mass distributions in Southern California, *Atmos. Environ.*, **31**, 1–11, 1997.

Herman, M., J.L. Deuzé, A. Marchand, B. Roger, and P. Lallart, Aerosol Remote Sensing from POLDER/ADEOS over the Ocean: Improved Retrieval using Non-Spherical Particle Model, *J. Geophys. Res. Atmos.*, **110**, D10, Art. No. D10S02 MAR 9, 2005.

Herman, J.R., P.K. Bhartia, O. Torres, C. Hsu, C. Seftor, and E. Celarier, Global distribution of UV-absorbing aerosol from Nimbus-7/TOMS data, *J. Geophys. Res.*, **102**, 16,911–16,922, 1997a.

Herman, M., J.L. Deuzé, C. Devaux, P. Goloub, F.M. Bréonm, and D. Tanré, Remote sensing of aerosols over land surfaces including polarization measurements and application to POLDER measurements, *J. Geophys. Res.*, **102**, 17,039–17,049, 1997b.

Heymsfield, A.J., and R.M. Sabin, Cirrus crystal nucleation by homogeneous freezing of solution droplets, *J. Atmos. Sci.*, **46**, 2252–2264, 1989.

Heymsfield, A.J., A.P. Jameson, and H.W. Frank, Hail growth mechanisms in a Colorado storm. Part II: Hail formation processes, *J. Atmos. Sci.*, **37**, 1779–1807, 1980.

Hindman, E.E, II., The nature of aerosol particles from a paper mill and their effects on clouds and precipitation, *Ph.D. dissertation*, University of Washington, Seattle, 242 pp., 1975.

Hindman, E.E. II., Water droplet fogs formed from pyrotechnically generated condensation nuclei, *J. Wea. Modif.*, **10**, 77–96, 1978.

Hindman, E.E. II., M. Tag, B.A. Silverman, and P.V. Hobbs, Cloud condensation nuclei from a paper mill. Part II: Calculated effects on rainfall, *J. Appl. Meteor.*, **16**, 753–755, 1977a.

Hindman, E.E, II., P.V. Hobbs, and L.F. Radke, Cloud condensation nucleus size distributions and their effects on cloud droplet size distributions, *J. Atmos. Sci.*, **34**, 951–955, 1977b.

Hinds, W.C., *Aerosol Technology: Properties, Behavior, and Measurement of Airborne Particles*, John Wiley & Sons, New York, 1982.

Hinds, W.C., *Aerosol Technology: Properties, Behavior, and Measurement of Airborne Particles*, edited by Wiley-Interscience, ISBN:0471194107, 1999.

Hirst, E., P.H. Kaye, R.S. Greenaway, P.R. Field and D.W. Johnson, Discrimination of micrometre-sized ice and super-cooled droplets in mixed-phase cloud, *Atmos. Environ.*, **35**, 33–47, 2001.

Hobbs, P.V., Ice multiplication in clouds, *J. Atmos. Sci.*, **26**, 315–318, 1969.

Hobbs, P.V., Simultaneous airborne measurements of cloud condensation nuclei and sodium-containing particles over the ocean, *Q. J. R. met. Soc.*, **97**, 263–271, 1971.

Hobbs, P.V., *Ice Physics*, p. 837, Clarendon Press, Oxford, Clarendon, 1974.

Hobbs, P.V., The nature of winter clouds and precipitation in the Cascade Mountains and their modification by artificial seeding. Part I: Natural conditions, *J. Appl. Meteor.*, **14**, 783–804, 1975a.

Hobbs, P.V., The nature of winter clouds and precipitation in the Cascade Mountains and their modification by artificial seeding. Part III: Case studies of the effects of seeding, *J. Appl. Meteor.*, **14**, 819–858, 1975b.

Hobbs, P.V., Aerosol-cloud interactions, in *Aerosol-Cloud-Clim. Interactions*, edited by P.V. Hobbs, pp. 33 73, Academic Press, San Diego, 1993.

Hobbs, R.I., and W.A. Cooper, Field evidence supporting qauntitative predictions of secondary ice production rates, *J. Atmos. Sci.*, **44**, 1071–1082, 1987

Hobbs, P.V., and J.D. Locatelli, Ice nuclei from a natural forest fire, *J. Appl. Meteor.*, **8**, 833–834, 1969.

Hobbs, P.V., and L.F. Radke, Cloud condensation nuclei from a simulated forest fire, *Science*, **163**, 279–280, 1969.

Hobbs, P.V., and A.L. Rangno, Comments on the climax randomized cloud seeding experiments, *J. Appl. Meteor.*, **18**, 1233–1237, 1979.

Hobbs, P.V., and A.L. Rangno, Ice particle concentrations in clouds, *J. Atmos. Sci.*, **42**, 2523–2549, 1985.

Hobbs, P.V., and A.L. Rangno, Rapid development of high ice particle concentrations in small polar maritime cumuliform clouds, *J. Atmos. Sci.*, **47**, 2710–2722, 1990.

Hobbs, P.V., and A.L. Rangno, Microstructures of low and middle-level clouds over the Beaufort Sea, *Q. J. Roy. Meteor. Soc.*, **124**, 2035–2071, 1998.

Hobbs, P.V., L.F. Radke and E. Shumway, Cloud condensation nuclei from industrial sources and their apparent effect on precipitation in Washington State, *J. Atmos. Sci.*, **27**, 81–89, 1970.

Hobbs, P.V., T.J. Matejka, P.H. Herzegh, J.D. Locatelli, and R.A. Houze, The mesoscale and microscale structure and organization of clouds and precipitation in midlatitude cycles. I: A case of a cold front, *J. Atmos. Sci.*, **37**, 568–596, 1980.

Hobbs, P.V., D.A. Bowdle, and L.F. Radke, Particles in the lower troposphere over the High Plains of the United States. I: Size distributions, elemental compositions and morphologies, *J. Clim. Appl. Meteor.*, **24**, 1344–1356, 1985a.

Hobbs, P.V., D.A. Bowdle, and L.F. Radke, Particles in the lower troposphere over the high plains of the United States. II: Cloud condensation nuclei, *J. Clim. Appl. Meteor.*, **24**, 1370–1376, 1985b.

Hobbs, P.V., T.J. Garrett, R.J. Ferek, S.R. Strader, D.A. Hegg, G.M. Frick, W.A. Hoppel, R.F. Gasparovic, L.M. Russell, D.W. Johnson, C. O'Dowd, P.A. Durkee, K.E. Nielsen, and G. Innis, Emissions from ships with respect to their effects on clouds, *J. Atmos. Sci.*, **57**, 2570–2590, 2000.

Hoelzemann, J.J., M.G. Schultz, G.P. Brasseur, C. Granier, and M. Simon, Global Wildland Fire Emission Model (GWEM): Evaluating the use of global area burnt satellite data, *J. Geophys. Res.*, **109**, D14S04, doi:10.1029/2003JD003666, 2004.

Hoff, R.M., A. Vandermeer, and L. Spacek, LITE/NARCM retrievals and implications for future missions, in *Advances in Laser Remote Sensing*, edited by A. Dabas, C. Loth and J. Pelon, pp. 27–30, Editions de L'Ecole polytechnique, 91128 Palaiseux Cedex, France, 2001.

Hofschreuder, P., F.G. Römer, N.F.M. Van Leeuwen, and B.G. Arends, Deposition of aerosol on Speulder Forest: Accumulation experiments, *Atmos. Environ.*, **31**, 351–357, 1997.

Holben, B.N., T.F. Eck, I. Slutsker, D. Tanré, J.P. Buis, A. Setzer, E. Vermote, J.A. Reagan, Y.J. Kaufman, T. Nakajima, F. Lavenu, I. Jankowiak and A. Smirnov, AERONET – A federated instrument network and data archive for aerosol characterization, *Remote Sens. Environ.*, **66**, 1–16, 1998.

Holben, B.N., D. Tanré, A. Smirnov, T.F. Eck, I. Slutsker, N. Abuhassan, W.W. Newcomb, J.S. Schafer, B. Chatenet, F. Lavenu, Y.J. Kaufman, J.V. Castle, A. Setzer, B. Markham, D. Clark, R. Frouin, R. Halthore, A. Karneli, N.T. O'Neill, C. Pietras, R.T. Pinker, K. Voss, and G. Zibordi, An emerging ground-based aerosol climatology: Aerosol optical depth from AERONET, *J. Geophys. Res*, **106**, 12,067–12,097, 2001.

Hopke, P.K., Y. Xie, T. Raunemaa, S. Biegalski, S. Landsberger, W. Maenhaut, **P**. Artaxo, and D. Cohen, Characterization of the Gent Stacked Filter Unit PM_{10} sampler, *Aerosol Sci. and Tech.*, **27**, 726–735, 1997.

Hoppel, W.A., Nucleation in the MSA-water vapour system, *Atmos. Environ.*, **21**, 2703–2709, 1987.

Hoppel, W.A., J.E. Dinger, and R.E. Ruskin, Vertical profiles of CCN at various geographical locations, *J. Atmos. Sci.*, **30**, 1410–1420, 1973.

Hoppel, W.A., G.M. Frick, and R.E. Larson, Effect of nonprecipitating clouds on the aerosol size distribution in the marine boundary layer, *Geophys. Res. Lett.*, **13**, 125–128, 1986.

Hoppel, W.A., J.W. Fitzgerald, G.M. Frick, R.E. Larson, and E.J. Mack, Aerosol size distributions and optical properties found in the marine boundary layer over the Atlantic Ocean, *J. Geophys. Res.*, **95**, 3659–3686, 1990.

Hoppel, W.A., G.M. Frick, J.W. Fitzgerald, and B.J. Wattle, A cloud chamber study of the effect that nonprecipitating water clouds have on the aerosol size distribution, *Aerosol Sci. and Tech.*, **20**, 1–30, 1994.

Hoppel, W.A., J.W. Fitzgerald, G. Frick, P. Caffrey, L. Pasternack, D. Hegg, S. Gao, R. Leaitch, N. Shantz, C. Cantrell, T. Albrechcinski, I. Ambrusko, and W. Sullivan, Particle formation and growth from ozonolysis of alpha-pinene, *J. Geophys. Res.*, **106**, 27,603–27,618, 2001.

Hori, M., S. Ohta, N. Murao, and S. Yamagata, Activation capability of water soluble organic substances as CCN, *J. Aerosol Sci.*, **34**, 419–448, 2003.

Houghton, J.T., Y. Ding, D.J. Griggs, M. Noguer, P.J. van der Linden, X. Dai, K. Maskell, and C.A. Johnson, (Eds.), *Climate Change 2001: The Scientific Basis. Contribution of Working Group I to the Third Assessment Report of the Intergovernmental Panel on Climate Change*, Cambridge, UK, and New York, NY, Cambridge University Press, USA, 881 p., 2001.

Hounslow, M.J., R.L. Ryall, and V.R. Marshall, A discretized population balance for nucleation, growth and aggregation, *AIChE J*, **34**, 1821–1832, 1988.

Houze, Jr., R.A., *Cloud Dynamics*, Academic Press, San Diego, 573 pp., 1993.

Houze, Jr., R.A., W. Schmid, RG. Fovell, and H.H. Shiesser, Hailstorms in Switzerland: Left movers, right movers, and false hooks, *Mon. Wea. Rev.*, **121**, 3345–3370, 1993.

Houze, Jr., R.A., S. Brodzik, C. Schumacher, S.E. Yuter, and C.R. Williams, Uncertainties in oceanic radar rain maps at Kwajalein and implications for satellite validation, *J. Appl. Meteor.*, **43**, 1114–1132, 2004.

Howell, W.E., The growth of cloud drops in uniformly cooled air, *J. Meteor.*, **6**, 134–149, 1949.

Hsu, N.C., J.R. Herman, O. Torres, B.N. Holben, Tanre, and T.F. Eck, Comparisons of the TOMS aerosol index and the sun photometer aerosol optical thickness: Results and applications, *J. Geophys. Res.*, **104**, 6269–6279, 1999.

Hudson, J.G., Effects of CCN Concentrations on stratus Clouds, *J. Atmos. Sci.*, **40**, 480–486, 1983.

Hudson, J.G., An instantaneous CCN spectrometer, *Atmos. and Ocean. Tech.*, **6**, 1055–1065, 1989

Hudson, J.G., and X. Da, Volatility and size of cloud condensation nuclei, *J. Geophys. Res.*, **101**, 4435–4442, 1996.

Hudson, J.G., and Y.H. Xie, Vertical distributions of cloud condensation nuclei spectra over the summertime northeast Pacific and Atlantic Oceans, *J. Geophys. Res.*, **104**, 30,219–30,229, 1999.

Hudson, J.G., and S.S. Yum, Droplet spectral broadening in marine stratus, *J. Atmos. Sci.*, **54**, 2642–2654, 1997.

Hudson, J.G. and S.S. Yum, Cloud condensation nuclei spectral and polluted and clean clouds over the Indian Ocean, *J. Geophys. Res.*, **107**, 8022, doi:10.1029/2001JD000829, 2002.

Hudson, J.G., J. Hallett, and C.F. Rogers, Field and laboratory measurements of cloud-forming properties of combustion aerosols, *J. Geophys. Res.*, **96**, 10,847–10,859, 1991.

Huebert, B.J., T. Bates, P.B. Russell, G. Shi, Y.J. Kim, K. Kawamura, G. Carmichael, and T. Nakajima, An overview of ACE-Asia: Strateggies for quantifying the relationships between Asian aerosols and their climatic impacts, *J. Geophys Res.*, **108**, 8633, doi:1029/2003JD003550, 2003.

Huff, F.A., and W.L. Shipp, Spatial correlations of storm, monthly, and seasonal precipitation, *J. App. Meteor.*, **8**, 542–550, 1969.

Huffman, P.J., Supersaturation spectra of AgI and natural ice nuclei, *J. Appl. Met.*, **12**, 6, 1080–1082, 1973a.

Huffman, P.J., Comments on "Numerical Estimates of Humidity in a Membrane-Filter Ice Nucleus Chamber, " **J. Appl. Meteor.**, **12**, 236–236, 1973b.

Huffman, P.J. and G. Vali, The effect of vapour depletion on ice nucleus measurements with membrane filters, *J. Appl. Meteor.*, **12**, 1018–1024, 1973.

Huffman, G.J., R.F. Adler, P. Arkin, A. Chang, R. Ferraro, A. Gruber, J. Janowiak, A. McNab, B. Rudolf, and U. Schneider, The Global Precipitation Climatology Project (GPCP) combined precipitation data set, *Bull. Amer. Meteor. Soc.*, **78**, 5–20, 1997.

Huffman, G.J., R.F. Adler, M.M. Morrissey, D.T. Bolvin, S. Curtis, R. Joyce, B. McGavock, and J. Susskind, Global precipitation at one-degree daily resolution from multisatellite observations, *J. Hydrometeor.*, **2**, 36–50, 2001.

Huffman, G.J., R.F. Adler, E.F. Stocker, D.T. Bolvin, and E.J. Nelkin, Analysis of TRMM 3-hourly multi-satellite precipitation estimates computed in both real and post-real time, Preprints, 12th Conf. on Sat. Meteor. and Oceanog., Long Beach, CA, *Amer. Meteor. Soc.*, P4.**11**, 6 pp., 2003.

Hughes, L.S., J.O. Allen, M.J. Kleeman, R.J. Johnson, G.R. Cass, D.S. Gross, E.E. Gard, M.E. Galli, B.D. Morrical, D.P. Fergenson, T. Dienes, C.A. Noble, P.J. Silva, and K.A. Prather, Size and composition distribution of atmospheric particles in southern California, *Environ. Sci. Technol.*, **33**, 3506–3515, 1999.

Husar, R.B., J. Prospero, and L.L. Stowe, Characterization of tropospheric aerosols over the oceans with the NOAA AVHRR optical thickness operational product, *J. Geophys. Res.*, **102**, 16,889–16,909, 1997.

Husar, R.B., D.M. Tratt, B.A. Schichtel, S.R. Falke, F. Li, D. Jaffe, S. Gassó, T. Gill, N.S. Laulainen, F. Lu, M.C. Reheis, Y. Chun, D. Westphal, B.N. Holben, C. Gueymard, I. McKendry, N. Kuring, G.C. Feldman, C. McClain, R.J. Frouin, J. Merrill, D. DuBois, F. Vignola, T. Murayama, S. Nickovic, W.E. Wilson, K. Sassen, N. Sugimoto, and W.C. Malm, Asian dust evens of April 1998, *J. Geophys. Res.*, **106**, 18,317–18,330, 2001.

Iguchi, T., T. Kozu, R. Meneghini, J. Awaka, and K. Okamoto, Rain-profiling algorithm for TRMM precipitation radar, *J. Appl. Meteor.*, **39**, 2038–2052, 2000.

Illingworth, A., Improved precipitation rates and data quality using polarimetric measurements, in *Weather Radar: Principles and Advanced Applications*, edited by P. Meischner, pp. 130–166, Springer-Verlag, Berlin, 2004.

IPCC, *Climate Change 2001: The Scientific Sasis, Contribution of working Group I to the Third Assessment Report of the Intergovernmental Panel on Climate Change*, Cambridge University Press (http://www.ipcc.ch/activity/tar.htm), 2001.

Iribarne, J.V., and R.G. de Pena, The influence of particle concentration on the evolution of hailstones, *Nubila.*, **5**, 7–30, 1962.

Isaac, G.A., J.W. Strapp, and R.S. Schemenaur, Summer Cumulus Cloud Seeding Experiments near Yellowknife and Thunder Bay, Canada, *J. Applied Met.*, **21**, 1266–1285, 1982.

Israelevich, P.L., Z. Levin, J.H. Joseph, and E. Ganor, Desert aerosol transport in the Mediterranean region as inferred from the TOMS aerosol index, *J. Geophys. Res.*, **107**, 4572, doi:10.1029/2001JD002011, 2002.

Ito, A., and J.E. Penner, Global estimates of biomass burning emissions based on satellite imagery for the year 2000, *J. Geophys. Res.*, **109**, D14S05, doi:10.1029/2003JD004423, 2004.

Ito, A., and J.E. Penner, Historical emissions of carbonaceous aerosols from biomass and fossil fuel burning for the period 1870–2000, *Global Biogeochem. Cycles*, GB2028, doi:10.1029/2004GB002374, 2005.

Jacob, D.J., J.A. Logan, and P.P. Murti, Effect of rising Asian emissions on surface ozone in the United States, *Geophys. Res. Lett.*, **26**, 2175–2178, 1999.

Jacobson, M.Z., Studying the effects of calcium and magnesium on size-distributed nitrate and ammonium with EQUISOLV II. *Atmos. Environ.*, **33**, 3635–3649, 1999.

Jacobson, M.Z., Strong radiative heating due to the mixing state of black carbon in atmospheric aerosols, *Nature*, **409**, 695–697, 2001.

Jacobson, M.Z., Development of mixed-phase clouds from multiple aerosol size distributions and the effect of the clouds on aerosol removal, *J. Geophys. Res.*, **108**, 4245, doi:10.1029/2002JD002691, 2003.

Jacobson, M.Z., Effects of externally-through-internally-mixed soot inclusions within clouds and precipitation on global climate, *J. Phys. Chem. A.*, **110**, 6860–6873, 2006.

Jacobson, M.Z., and J.H. Seinfeld, Evolution of nanoparticle size and mixing state near the point of emission, *Atmos. Environ.*, **38**, 1839–1850, 2004.

Jaeger, L., Monatskarten des Niederschlags für die ganze Erde, Bericht des Deutschen Wetterdienstes, Volume 139, Offenbach a.M., 33 pp. and plates, 1976.

Jaenicke, R., Tropospheric Aerosols, in *Aerosol-Cloud-Climate Interactions*, edited by P.V. Hobbs, pp. 1–27, Academic Press, San Diego, 1993.

Jaenicke, R., Abundance of cellular material and proteins in the atmosphere, *Science*, **308**, 73–73, 2005.

Jaffe, D., T. Anderson, D. Covert, R. Kotchenruther, B. Trost, J. Danielson, W. Simpson, T. Berntsen, S. Karlsdottir, D. Blake, J. Harris, G. Carmichael, and I. Uno, Transport of Asian air pollution to North America, *Geophys. Res. Letts.*, **26**, 711–714, 1999.

Jaffe, D., S. Tamura, and J. Harris, Seasonal cycle, composition and sources of background fine particles along the west coast of the U.S., *Atmos. Environ.*, **39**, 297–306, 2005.

Jang, M., and R.M. Kamens, Atmospheric secondary aerosol formation by heterogeneous reactions of aldehydes in the presence of a sulfuric acid aerosol catalyst, *Environ. Sci. and Tech.*, **35**, 4758–4766, 2001a.

Jang, M., and R.M. Kamens, Characterization of secondary aerosol from the photooxidation of toluene in the presence of NOx and 1-propene, *Environ. Sci. and Tech.*, **35**, 3626–3639, 2001b.

Jang, M., N.M. Czoschke, S. Lee, and R.M. Kamens, Heterogeneous atmospheric aerosol production by acid-catalyzed particle-phase reactions, *Science*, **298**, 814–817, 2002.

Jang, M., B. Carroll, B. Chandramouli, and M.K. Richard, Particle growth by acid-catalyzed heterogeneous reactions of organic carbonyls on preexisting aerosols, *Environ. Sci. and Technol.*, **37**, 3828–3837, 2003.

Jankowiak, I., and D. Tanré, Satellite climatology of Saharan dust outbreaks: Method and preliminary results, *J. Clim.*, **5**, 646–656, 1992.

Janowiak, J.E., Tropical rainfall: A comparison of satellite-derived rainfall estimates with model precipitation forecasts, climatologies, and observations, *Mon. Wea. Rev.*, **120**, 448–462, 1992.

Janowiak, J.E., and P. Xie, CAMS-OPI: A global satellite-rain gauge merged product for real-time precipitation monitoring applications, *J. Clim.*, **12**, 3335–3342, 1999.

Jennings, S.G., Wet processes affecting atmospheric aerosols, Chapter 14, in *Atmospheric Particles*, edited by R.M. Harrison, and R. Van Grieken, pp. 476–507, Wiley, New York, 1998.

Jiang, H., and G. Feingold, Effect of aerosol on warm convective clouds: Aerosol-cloud-surface flux feedbacks in a new coupled large eddy model, *J. Geophys. Res.*, **111**, D01202, doi:10.1029/2005JD006138, 2006.

Jiang, H., G. Feingold, and W.R. Cotton, Simulations of aerosol-cloud-dynamical feedbacks resulting from entrainment of aerosol into the marine boundary layer during the Atlantic Stratocumulus Transition Experiment, *J. Geophys. Res.*, **107**, 4813, doi:10.1029/2001JD001502, 2002.

Jickells, T.D., Z.S. An, K.K. Andersen, A.R. Baker, G. Bergametti, N. Brooks, P.W.B.J.J. Cao, R.A. Duce, K.A. Hunter, H. Kawahata, N. Kubilay, J. LaRoche, P.S. Liss, N. Mahowald, J.M. Prospero, A.J. Ridgwell, L. Tegen, and R. Torres, Global iron connections between desert dust, ocean biogeochemistry, and climate, *Science*, **308**, 67–71, 2005.

Jimenez, J.L., J.T. Jayne, Q. Shi, C.E. Kolb, D.R. Worsnop, I. Yourshaw, J.H. Seinfeld, R.C. Flagan, X.F. Zhang, K.A. Smith, J.W. Morris, and P. Davidovits, Ambient aerosol sampling using the Aerodyne Aerosol Mass Spectrometer, *J. Geophys. Res.*, **108**, 8424, doi:10.1029/2001JD000660, 2003.

Jin, M., J.M. Shepherd, and M.D. King, Urban aerosols and their variations with clouds and rainfall: A case study for New York and Houston, *J. Geophys. Res.*, **110**, D10S20, doi:10.1029/2004JD005081, 2005.

Jirak, I.L., W.R. Cotton, Effect of air pollution on precipitation along the front range of the rocky mountains, *J. Appl. Meteor. Climate.*, **45**, 236–246, 2006.

Jiusto, J.E., Maritime concentration of condensation nuclei, *J. Rech. Atmos.*, **2**, 245–250, 1966.

Johansson, S.A.E., J.L. Campbell, *PIXE: A Novel Technique for Trace Element Analysis*, Wiley, New York, 1988.

John, W., S.M. Wall, J.L. Ondo, and W. Winklmayr, Modes in the size distributions of atmospheric inorganic aerosol, *Atmosp. Environ. Part a-General Topics*, **24**, 2349–2359, 1990.

Johns, T.C., C.F. Durman, H.T. Banks, M.J. Roberts, A.J. McLaren, J.K. Ridley, C.A. Senior, K.D. Williams, A. Jones, G.J. Rickard, S. Cusack, W.J. Ingram, M. Crucifix, D.M.H. Sexton, M.M. Joshi, B.W. Dong, H. Spencer, R.S.R. Hill, J.M. Gregory, A.B. Keen, A.K. Pardaens, J.A. Lowe, A. Bodas-Salcedo, S. Stark and Y. Searl, The new Hadley Centre climate model HadGEM1: Evaluation of coupled simulations, *J. Clim.* **19**, 1327–1353, 2006.

Johnson, D.B., Ultragiant urban aerosol particles, *Science*, **194**, 941–942, 1976.

Johnson, D.B., The role of coalescence nuclei in warm rain initiation, *Ph.D. dissertation*, University of Chicago, 119 pp., 1979.

Johnson, D.B., The role of giant and ultragiant aerosol particles in warm rain initiation, *J. Atmos. Sci.*, **39**, 448–460, 1982.

Johnson, D.B., On the relative efficiency of coalescence and riming, *J. Atmos. Sci.*, **44**, 1672–1680, 1987.

Johnson, B.T., Large-eddy simulations of the semidirect aerosol effect in shallow cumulus regimes, J. Geophys. Res., 110, D14206, doi:10.1029/2004JD005601, 2005.

Johnson, R.H., and P.E. Ciesielski, Rainfall and radiative heating rates from TOGA COARE atmospheric budgets, *J. Atmos. Sci.*, **57**, 1497–1514, 2000.

Johnson, D.W., S.R. Osborne, R. Wood, K. Suhre, P.K. Quinn, T. Bates, M.O. Andreae, K.J. Noone, P. Glantz, B. Bandy, J. Rudolph, and C. O'Dowd, Observations of the evolution of the aerosol, cloud and boundary-layer characteristics during the 1st ACE-2 Lagrangian experiment, *Tellus*, **52B**, 348–374, 2000.

Johnson, B.T., K.P. Shine and P.M. Forster, The semi-direct aerosol effect: Impact of absorbing aerosols on marine stratocumulus, *Q. J. Roy. Meteor. Soc.*, **30**, 1407–1422, 2004.

Jordan, C.E., J.E. Dibb, B.E. Anderson, and H.E. Fuelberg, Uptake of nitrate and sulfate on dust aerosols during TRACE-P, *J. Geophys. Res.*, **108**, 8817, doi:8810.1029/2002JD003101, 2003.

Joss, J. and R. Dyer, Large errors involved in deducing drop-size distributions from Doppler radar data due to vertical air motion, Preprints, 15th Radar Meteorology Conference, Champaign-Urbana, Illinois, *Amer. Meteor. Soc.*, 179–180, 1972.

Joss, J., and E.G. Gori, The parameterization of raindrop size distributions. *Riv. Ital. Geofis.*, **3**, 275–283, 1976.

Joss, J., and A. Waldvogel, Ein Spectrograph fur Niedersclagstropfen mit automatisher Auswertung (A spectrograph for the automatic analysis of raindrops), *Pure Appl. Geophys.*, **68**, 240–246, 1967.

Joss, J., and A. Waldvogel, Raindrop size distribution and sampling size errors, *J. Atmos. Sci.*, **26**, 566–569, 1969.

Joss, J., B. Schädler, G. Galli, R. Cavalli, M. Boscacci, E. Held, G. Della Bruna, G. Kappenberger, V. Nespor, and R. Spiess, *Operational Use of Radar for Precipitation Measurements in Switzerland*, vdf Hochschulverlag AG an der ETH Zürich, 108 pp., 1998.

Jung, C.H., Y.P. Kim, and K.W. Lee, A moment model for simulating raindrop scavenging of aerosols, *Journal of Aerosol Science*, **34**, 1217–1233, 2003.

Junge, C.E., Die Rolle der Aerosole und der gasförmigen Beimengungen der Luft im Spurenstoffhaushalt der Troposphäre, *Tellus*, **5**, 1–26, 1953.

Junge, C.E., *Air Chemistry and Radioactivity*, Academic Press, New York, 382 p., 1963.

Junge, C.E., (Ed.), *Fate of Pollutants in the Air and Water Environments*, 7–26 pp., J. Wiley, New York, 1977.

Junker, C., and C. Liousse, A global emission inventory of carbonaceous aerosol from historic records of fossil fuel and biofuel consumption for the period 1860–1997, *Atmos. Chem. Phys. Discuss.*, **4**, 4897–4927, 2006.

Jylhä, K., Relationship between the scavenging coefficient for pollutants in precipitation and the radar reflectivity factor. Part II: Applications, *J. Appl. Meteorol.*, **38**, 1435–1447, 1999.

Kahn, R., P. Banerjee, and D. McDonald, The sensitivity of multiangle imaging to natural mixtures of aerosols over ocean, *J. Geophys. Res.*, **106**, 18,219–18,238, 2001.

Kahn, R., B. Gaitley, J. Martonchik, D. Diner, K. Crean, and B. Holben, MISR global aerosol optical depth validation based on two years of coincident AERONET observations, *J. Geophys. Res.*, **110**, D10, D10S0410.1029/2004JD004706, 2005.

Kahn, R.A., W.-H. Li, C. Moroney, D.J. Diner, J.V. Martonchik, and E. Fishbein, Aerosol source plume physical characteristics from space-based multiangle imaging, *J. Geophys. Res.*, **112**, D11205, doi:10.1029/2006JD007647, 2007.

Kahnert, M., and L. Tarrasón, *Transboundary Particulate Matter in Europe*, Status report 4/2003, EMEP Report, 2003.

Kajii, Y., S. Kato, D.G. Streets, N.Y. Tsai, A. Shvidenko, S. Nilsson, I. McCallum, N.P. Minko, N. Abushenko, D. Altyntsev, and T.V. Khodzer, Boreal forest fires in Siberia in 1998: Estimation of area burned and emissions of pollutants by advanced very high resolution radiometer satellite data, *J. Geophys. Res.*, **107**, 4745, doi:10.1029/2001JD001078, 2002.

Kalashnikova, O.V., R. Kahn, I.N. Sokolik, and W.H. Li, The ability of multi-angle remote sensing observations to identify and distinguish mineral dust types: Optical models and retrievals of optically thick plumes, *J. Geophys. Res.*, **110**, D18, D18S1410.1029/2004JD004550, 2005

Kalashnikova, O.V., R. Kahn, Ability of multiangle remote sensing observations to identify and distinguish mineral dust types: 2. Sensitivity over dark water, *J. Geophys. Res.*, **111**, D11207, doi:10.1029/2005JD006756, 2006.

Kalberer, M., D. Paulsen, M. Sax, M. Steinbacher, J. Dommen, A.S.H. Prevot, R. Fisseha, E. Weingartner, V. Frankevich, R. Zenobi, and U. Baltensperger, Identification of polymers as major components of atmospheric organic aerosols, *Science*, **303**, 1659–1662, 2004.

Kanakidou, M., K. Tsigaridis, F.J. Dentener, and P.J. Crutzen, Human-activity-enhanced formation of organic aerosols by biogenic hydrocarbon oxidation, *J. Geophys. Res.*, **105**, 9243–9254, 2000.

Kanakidou, M., J.H. Seinfeld, S.N. Pandis, I. Barnes, F.J. Dentener, M.C. Facchini, R. Van Dingenen, B. Ervens, A. Nenes, C.J. Nielsen, E. Swietlicki, J.P. Putaud, Y. Balkanski, S. Fuzzi, J. Horth, G.K. Moortgat, R. Winterhalter, C.E.L. Myhre, K. Tsigaridis, E. Vignati, E.G. Stephanou, and J. Wilson, Organic aerosol and global climate modelling: A review, *Atmos. Chem. Phys.*, 5, 1053–1123, 2005.

Kapustin, V.N., A.D. Clarke, Y. Shinozuka, S. Howell, V. Brekhovskikh, T. Nakajima, A. Higurashi, On the determination of a cloud condensation nuclei from satellite: Challenges and possibilities, *J. Geophys. Res.*, **111**, D04202, doi:10.1029/2004JD005527, 2006.

Karcher, B., T. Peter, U.M. Biermann, and U. Schumann, The initial composition of jet condensation trails, *J. Atmos. Sci.*, **53**, 3066–3083, 1996.

Kavouras, I., N. Mihalopoulos, E.G. Stephanou, Formation of atmospheric particles from organic acids produced by forests. *Nature*, **395**, 683–686, 1998.

Kaufman, Y.J. and R.S. Fraser, The effect of smoke particles on clouds and climate forcing, *Science*, **277**, 1636–1639, 1997.

Kaufman, Y.J. and I Koren, Smoke and Pollution Aerosol Effect on Cloud Cover, *Science*, **313**, 655–658, 2006.

Kaufman, Y.J. and T. Nakajima, Effects of Amazon smoke on cloud microphysics and albedo – Analysis from satellite imagery, *J. Appl. Meteor.*, **32**, 729–744, 1993.

Kaufman, Y.J., D. Tanre, L. Remer, E. Vermote, A. Chu, and B. Holben, Operational remote sensing of tropospheric aerosol over land from EOS moderate resolution imaging spectro-radiometer, *J. Geophys. Res.*, **102**, 17,051–17,067, 1997.

Kaufman, Y.J., D. Tanré, O. Dubovik, A. Karnieli, and L.A. Remer, Absorption of sunlight by dust as inferred from satellite and ground-based remote sensing, *Geophys. Res. Lett.*, **28**, 1479–1483, 2001.

Kaufman, Y.J., D. Tanre, and O. Boucher, A satellite view of aerosols in the climate system. Review. *Nature*, **419**, 215–223, 2002a.

Kaufman, Y.J., J.V. Martins, L.A. Remer, M.R. Schoeberl, and M.A. Yamasoe, Satellite retrieval of aerosol absorption over the oceans using sunglint, *Geophys. Res. Lett.*, **29**, 1298, doi:10.1029/2002GL015403, 2002b.

Kaufman, Y.J., J.M. Haywood, P.V. Hobbs, W. Hart, R. Kleidman, and B. Schmid, Remote sensing of vertical distributions of smoke aerosol off the coast of Africa, *Geophys. Res. Lett.*, **30**, 16, 1831, dio:10.1029/2003GL017068, 2003.

Kaufman, Y.J., I. Koren, L.A. Remer, D. Tanré, P. Ginoux, and S. Fan, Dust transport and deposition observed from the Terra-Moderate Resolution Imaging Spectroradiometer (MODIS) spacecraft over the Atlantic Ocean, *J. Geophys. Res.*, **110**, D10S12, doi:10.1029/2003JD004436, 2005a.

Kaufman, Y.J., O. Boucher, D. Tanré, M. Chin, L.A. Remer, and T. Takemura, Aerosol anthropogenic component estimated from satellite data, **Geoph. Res. Lett.**, **32**, L17804, doi:10.1029/2005GL023125, 2005b.

Kaufman, Y.J., I. Koren, L.A. Remer, D. Rosenfeld, and Y. Rudich, The effect of smoke, dust, and pollution aerosol on shallow cloud development over the Atlantic Ocean, *Proc. Nat. Acad. Sci.*, **102**, 11,207–11,212, 2005c

Kerkweg, A., S. Wurzler, T. Reisin, and A. Bott, On the cloud processing of aerosol particles: An entraining air-parcel model with two dimensional spectral cloud microphysics and a new formulation of the collection kernel, *Q. J. Roy. Meteor. Soc.*, **129**, 1–18, 2003.

Kerminen, V.M., Relative roles of secondary sulfate and organics in atmospheric cloud condensation nuclei production, *J. Geophys. Res.*, **106**, 17,321–17,333, 2001.

Kerminen, V.M., A. Virkkula, R. Hillamo, A.S. Wexler, and M. Kulmala, Secondary organics and atmospheric cloud condensation nuclei production, *J. Geophys. Res.*, **105**, 9255–9264, 2000.

Kerr, R.A., Cloud seeding: One success in 35 years, *Science*, **217**, 519–522, 1982.

Kessler, E., *On the Distribution and Continuity of Water Substance in Atmospheric Circulation, Meteorol. Monogr.*, **10**, 84 pp., American Meteorological Society, Boston, MA, 1969.

Khain, A.P., and M.B. Pinsky, Turbulence effects on the collision kernel. II: Increase of the swept volume of colliding drops, *Q. J.Roy. Meteor. Soc.*, **123**, 1543–1560, 1997.

Khain, A.P., A. Pokrovsky, and I. Sednev, Some effects of cloud-aerosol interaction on cloud microphysics structure and precipitation formation: Numerical experiments with a spectral microphysics cloud ensemble model, *Atmos. Res.*, **52**, 195–220, 1999.

Khain, A.P., M. Ovtchinnikov, M. Pinsky, A. Pokrovsky, and H. Krugliak, Notes on the state-of-the-art numerical modeling of cloud microphysics, *Atmos. Res.*, **55**, 159–224, 2000.

Khain, A.P., D. Rosenfeld, and A. Pokrovsky, Simulating convective clouds with sustained supercooled liquid water down to –37.5̊C using a spectral microphysics model, *Geophys. Res. Lett.*, **28**, 3887–3890, 2001.

Khain, A.P., A. Pokrovsky, M. Pinsky, A. Seifert, and V. Phillips, Simulation of effects of atmospheric aerosols on deep turbulent convective clouds using a spectral microphysics mixed-phase cumulus cloud model. Part I: Model description and possible applications, *J. Atmos. Sci.*, **61**, 2963–2982, 2004.

Leck, C., M. Norman, E.K. Bigg, and R. Hillamo, Chemical composition and sources of the high Arctic aerosol relevant for fog and cloud formation, *J. Geophys. Res.*, **107**, 4135, doi:10.1029/2001JD001463, 2002.

Leck, C., M. Tjernström, P. Matrai, E. Swietlicki, and K. Bigg, Can marine micro-organisms influence melting of the Arctic pack ice? *Eos Trans. AGU*, **85**, 25–36, 2004.

Legates, D.R., and C.J. Willmott, Mean seasonal and spatial variability in gauge-corrected, global precipitation, *Int. J. Climatol.*, **10**, 111–127, 1990.

Legrand, M., and C. N'doumé, Satellite detection of dust using the IR imagery of Meteosat: 1. Infrared difference dust index, *J. Geophys. Res.*, **106**, 18,251–18,274, 2001.

Lelieveld, J., G.J. Roelofsf, L. Ganzeveld, J. Feichter, and H. Rodhe, Terrestrial sources and distribution of atmospheric sulfur, *Phil. Trans. Roy. Soc. Lond.*, **B352**, 149–158, 1997.

Lelieveld, J., H. Berresheim, S. Borrmann, P.J. Crutzen, F.J. Dentener, H. Fischer, J. Feichter, P.J. Flatau, J. Heland, R. Holzinger, R. Korrmann, M.G. Lawrence, Z. Levin, K.M. Markowicz, N. Mihalopoulos, A. Minikin, V. Ramanathan, M. deReus, G.J. Roelofs, H.A. Scheeren, J. Sciare, H. Schlager, M. Schultz, P. Siegmund, B. Steil, E.G. Stephanou, P. Stier, M. Traub, C. Warneke, J. Williams, and H. Ziereis, Global air pollution crossroads over the Mediterranean, *Science*, **298**, 794–799, 2002.

Lensky, I.M., D. Rosenfeld, The time-space exchangeability of satellite retrieved relations between cloud top temperature and particle effective radius, *Atmos. Chem. Phys.*, **6**, 2887–2894, 2006.

Leroy, D., M. Monier, W. Wobrock, and A.I. Flossmann, A numerical study of the effects of the aerosol particle spectrum on the development of the ice phase and precipitation formation, *Atmos. Res.*, **80**, 15–45, 2006.

Levin, Z., and S.A. Yankofsky, Contact versus immersion freezing of freely suspended droplets by bacterial ice nuclei, *J. Clim. and Appl. Meteor.*, **22**, 1964–1966, 1983.

Levin, Z., M. Neiburger and L. Rodriguez, Jr., Experimental evaluation of collection efficiencies and coalescence efficiencies of cloud drops, *J. Atmos. Sci.*, **30**, 944–946, 1973.

Levin, Z., S.A. Yankofsky, D. Pardess, and N. Magal, Possible application of bacterial condensation freezing to artificial rainfall enhancement, *J. Clim. Appl. Meteor.*, **26**, 1188–1197, 1987.

Levin, Z., E. Ganor, and V. Gladstein, The effects of desert particles coated with sulfate on rain formation in the eastern Mediterranean, *J. Appl. Meteor.*, **35**, 1511–1523, 1996.

Levin, Z., A. Teller, E. Ganor and Y. Yin, On the interactions of mineral dust, sea salt particles and clouds – A Measurement and modeling study from the MEIDEX campaign, *J. Geophys. Res.*, **110**, D20202, doi:10.1029/2005JD005810, 2005.

Levkov, L., Congélation de gouttes d'eau au contact particules de CuS, *J. de Rech. Atmos.*, **5**, 133–136, 1971.

Lewis, E.R., and S.E. Schwartz, *Sea Salt Aerosol Production: Mechanisms, Methods, Measurements and Models – A Critical Review*, 413 p., American Geophysical Union, Washington, DC, 2004.

Li, Q.B., D.J. Jacob, I. Bey, P.I. Palmer, B.N. Duncan, B.D. Field, R.V. Martin, A.M. Fiore, R.M. Yantosca, D.D. Parrish, P.G. Simmonds, and S.J. Oltmans, Transatlantic transport of pollution and its effects on surface ozone in Europe and North America, *J. Geophys. Res.*, **107**, 4166, doi:10.1029/2001JD001422, 2002.

Li, L., G.M. Heymsfield, P.E. Racette, L. Tian, and E. Zenker, A 94-GHz cloud radar system on a NASA high altitude ER-2 aircraft, *J. Atmos. Ocean. Tech.*, **21**, 1378–1388, 2004.

Liepert, B.G., J. Feichter, U. Lohmann, and E. Roeckner, Can aerosols spin down the water cycle in a warmer and moister world, *Geophys. Res. Lett.*, **31**, L06207, doi:10.1029/2003GL019060, 2004.

Liggio, J., S.M. Li, and R. McLaren, Heterogeneous reactions of glyoxal on particulate matter: Identification of acetals and sulfate esters, *Environ. Sci. Tech.*, **39**, 1532–1541, 2005a.

Liggio, J., S.M. Li, and R. McLaren, Reactive uptake of glyoxyl on aerosols, *J. Geophys. Res.*, **110**, D10304, doi:10.1029/2004JD005113, 2005b.

Lihavainen, H., V.M. Kerminen, M. Komppula, J. Hatakka, V. Aaltonen, M. Kulmala, and Y. Viisanen, Production of “potential” cloud condensation nuclei associated with atmospheric new-particle formation in northern Finland, *J. Geophys. Res.*, **108**, 4782, doi:10.1029/2003JD003887, 2003.

Li-Jones, X., and J.M. Prospero, Variations in the size distribution of non-sea-salt sulfate aerosol in the marine boundary layer at Barbados: Impact of African dust, *J. Geophys. Res.*, **103**, 16,073–16,084, 1998.

Limbeck, A., M. Kulmala, and H. Puxbaum, Secondary organic aerosol formation in the atmosphere via heterogeneous reaction of gaseous isoprene on acidic particles, *Geophys. Res. Lett.*, **30**, 1996, doi:10.1029/2003GL017738, 2003.

Lin, J.C., T. Matsui, R.A. Pielke Sr. and C. Kummerow, Effects of biomass-burning-derived aerosols on precipitation and clouds in the Amazon Basin: A satellite-based empirical study, *J. Geophys. Res.*, **111**, D19204, doi:10.1029/2005JD006884, 2006.

Liousse, C., J.E. Penner, C. Chuang, J.J. Walton, H. Eddleman, and H. Cachier, A global three-dimensional model study of carbonaceous aerosols, *J. Geophys. Res.*, **101**, 19,411–19,432, 1996a.

Liousse, C., J.E. Penner, J.J. Walton, H. Eddleman, C. Chuang, and H. Cachier, Modeling biomass burning aerosols, in *Biomass Burning and Global Change*, edited by J.S. Levine, pp. 492–508, MIT Press, Cambridge, MA, 1996b.

Liousse, C., M.O. Andreae, P. Artaxo, P. Barbosa, H. Cachier, J.M. Grégoire, P. Hobbs, D. Lavoué, F. Mouillot, J. Penner, M. Scholes, and M.G. Schultz, Deriving global quantitative estimates for spatial and temporal distributions of biomass burning emissions, in *Emissions of Atmospheric Trace Compounds*, edited by C. Granier, P. Artaxo, and C.E. Reeves, pp. 71–113, Kluwer, Dordrecht, 2004.

List, R., A linear radar reflectivity-rainrate relationship for steady tropical rain, *J. Atmos. Sci.*, **45**, 3564–3572, 1988.

List, R., Review and implications of WMO and IUGG recommendations concerning the effect of aerosol pollution and biomass burning on precipitation, *Proc. 14th Intern. Conf. Clouds and Precipitation*, Bologna, Italy, **1**, 139–141, 2004.

Liu, G., and J.A. Curry, Retrieval of precipitation from satellite microwave measurement using both emission and scattering, *J. Geophys. Res.*, **97**, 9959–9974, 1992.

Liu, Z., I. Matsui, N. Sugimoto, High-spectral-resolution lidar using an iodine absorption filter for atmospheric measurements, *Opt. Eng.*, **38**, 1661–1670, 1999.

Liu, P.S.K., W.R. Leaitch, C.M. Banic, and S.M. Li, Aerosol observations at Chebogue Point during the 1993 North Atlantic Regional Experiment: Relationships among cloud condensation nuclei, size distribution, and chemistry, *J. Geophys. Res.*, **101**, 28,971–28,990, 1996.

Liu, Q., Y.L., Kogan, D.K. Lilly, D.W. Johnson, G.E. Innis, P.A. Durkee and K.E. Nielsen, Modeling of Ship effluent transport and its sensitivity to boundary layer structure, *J. Atmos. Sci.*, **57**, 2779–2791, 2000.

Liu, X.H., J.E. Penner, and M. Herzog, Model description, evaluation, and interactions between sulfate and nonsulfate aerosols, *J. Geophys. Res.*, **110**, 2005a.

Liu, Y., R. Fu, and R. Dickinson, Smoke aerosols altering South American monsoon, *Bull. Amer. Meteor. Soc.*, **86**, 1062–1063, 2005b.

Löffler-Mang, M., and J. Joss, An optical disdrometer for measuring size and velocity of hydrometeors, *J. Atmos. Ocean. Tech.*, **17**, 130–139, 2000.

Lohmann, U., A glaciation indirect aerosol effect caused by soot aerosols, *Geophys. Res. Lett.*, **29**, 1052, doi:10.1029/2001GL014357, 2002.

Lohmann, U., Can anthropogenic aerosols decrease the snowfall rate? *J. Atmos. Sci.*, **61**, 2457–2468, 2004.

Lohmann, U. and K. Diehl, Sensitivity studies of the importance of dust ice nuclei for the indirect aerosol effect on stratiform mixed-phase clouds, *J. Atmos. Sci.*, **62**, 968–982, 2006.

Lohmann, U. and J. Feichter, Can the direct and semi-direct aerosol effect compete with the indirect effect on a global scale? *Geophys. Res. Lett.*, **28**, 159–161, 2001.

Lohmann, U. and J. Feichter, Global indirect aerosol effects: A review, *Atmos. Chem. Phys.* **5**, 715–737, 2005.

Lohmann, U. and G. Lesins, Stronger constraints on the anthropogenic indirect aerosol effect, *Science*, **298**, 1012–1016, 2002.

Lohmann, U., J. Feichter, C.C. Chuang, and J.E. Penner, Prediction of the number of cloud droplets in the ECHAM GCM, *J. Geophys. Res.*, **104**, 9169–9198, 1999.

Lohmann, U., W.R. Leaitch, L. Barrie, K. Law, Y. Yi, D. Bergmann, C. Bridgeman, M. Chin, J. Christensen, R. Easter, J. Feichter, A. Jeuken, E. Kjellstrom, D. Koch, P. Rasch, and G.J. Roelofs, Vertical distributions of sulfur species simulated by large scale atmospheric models in COSAM: Comparison with observations, *Tellus B.*, **53**, 646–672, 2001.

Lohmann, U., J. Zhang, and J. Pi, Sensitivity studies of the effect of increased aerosol concentrations and snow crystal shape on the snowfall rate in the Arctic – Art. No. 4341, *J. Geophy. Res. Atmos.*, **108**, 4341, 2003.

Lohmann, U., K. Broekhuizen, R. Leaitch, N. Shantz, and J. Abbatt, How efficient is cloud droplet formation of organic aerosols? *Geophys. Res. Lett.*, **31**, L05108, doi:10.1029/2003GL018999, 2004.

Lu, M.L., and J.H. Seinfeld, Study of the aerosol indirect effect by LES of marine stratocumulus, *J. Atmos. Sci.*, **62**, 3909–3932, 2005

Luo, C., N.M. Mahowald, and J. del Corral, Sensitivity study of meteorological parameters on mineral aerosol mobilization, transport, and distribution, *J. Geophys. Res.*, **108**, 4447, doi:10.1029/2003JD003483, 2003.

Luria, M., C.C. Van Valin, J.N. Galloway, W.C. Keene, D.L. Wellman, H. Sievering, and J.F. Boatman, The relationship between dimethyl sulfide and particulate sulfate in the mid-Atlantic Ocean atmosphere, *Atmos. Environ.*, **23**, 139–147, 1989.

Lynn, B., A.P. Khain, J. Dudhia, D. Rosenfeld, A. Pokrovsky, and A. Seifert, Spectral (bin) microphysics coupled with a Mesoscale Model (MM5). Part 1: Model description and first results, *Mon. Wea. Rev.*, **133**, 44–58, 2005a.

Lynn, B., A.P. Khain, J. Dudhia, D. Rosenfeld, A. Pokrovsky, and A. Seifert, Spectral (bin) microphysics coupled with a Mesoscale Model (MM5). Part 2: Simulation of a CaPe rain event with squall line, *Mon. Wea. Rev.*, **133**, 59–71, 2005b.

Lynn, B., A.P. Khain, D. Rosenfeld, and W.L. Woodley, Effects of aerosols on precipitation from orographic clouds, *J. Geophys. Res.*, **112**, DIO225, doi:10.1029/2006JD007537, 2007.

Maenhaut, W., J. Cafmeyer, S. Dubtsov, and X.G. Chi, Detailed mass size distributions of elements and species, and aerosol chemical mass closure during fall 1999 at Gent, Belgium, *Nuclear Instruments and Methods in Physics Research Section B-Beam Interactions With Materials and Atoms*, **189**, 238–242, 2002.

Magono, C., and C.W. Lee, The vertical structure of snow clouds as revealed by "snow crystal sondes," *J. Meteor. Soc. Japan*, **51**, 176–190, 1973.

Mahowald, N.M., and C. Luo, A less dusty future? *Geophys. Res. Lett.*, **30**, 1903, doi:10.1029/2003GL017880, 2003.

Mahowald, N.M., C.S. Zender, C. Luo, D. Savoie, O. Torres, and J. del Corral, Understanding the 30-year Barbados desert dust record, *J. Geophys. Res.*, **107**, 4561, doi:10.1029/2002JD002097, 2002.

Malm, W.C., and S.M. Kreidenweis, The effects of models of aerosol hygroscopicity on the apportionment of extinction, *Atmos. Environ.*, **31**, 1965–1976, 1997.

Malm, W.C., L.T. Sisler, D. Kaufman, R.A. Eldred, and T.A. Cahill, Spatial and seasonal trends in particle concentrations and optical extinction in the United States, *J. Geophys. Res.*, **99**, 1347–1370, 1994.

Malm, W.C., B.A. Schichtel, R.B. Ames, and K.A. Gebhart, A 10-year spatial and temporal trend of sulfate across the United States – Art. No. 4627, *J. Geophy. Res. Atmos.*, **107**, 4627, 2002.

Malm, W.C., B.A. Schichtel, M.L. Pitchford, L.L. Ashbaugh, and R.A. Eldred, Spatial and monthly trends in speciated fine particle concentration in the United States, *J. Geophys. Res.*, **109**, D03306, doi:10.1029/2003JD003739, 2004.

Manton, M.J. and G.P. Ayers, On the number concentration of aerosols in towns, *Boundary Layer Meteorology.*, **22**, 171–181, 1982.

Manton, M.J., and W.R. Cotton, Parameterization of the atmospheric surface layer, *J. Atmos. Sci.*, **34**, 331–334, 1977.

Marcolli, C., B.P. Luo, and T. Peter, Mixing of the organic aerosol fractions: Liquids as the thermodynamically stable phases, *J. Phys. Chem. A.*, **108**, 2216–2224, 2004.

Maria, S.F., L.M. Russell, M.K. Gilles, and S.C.B. Myneni, Organic aerosol growth mechanisms and their climate-forcing implications, *Science*, **306**, 1921–1924, 2004.

Maring, H., D.L. Savoie, M.A. Izaguirre, L. Custals, and J.S. Reid, Mineral dust aerosol size distribution change during atmospheric transport, *J. Geophys. Res.*, **108**, 8592, doi:10.1029/2002JD002536, 2003.

Market, P., S. Allen, R. Scofield, R. Kuligowski, and A Gruber, Precipitation efficiency of warm season midwestern mesoscale convective systems, *Wea. Forecasting.*, **18**, 1273–1285, 2003.

Marple, V.A., K.L. Rubow and S.M. Behm, A Microorifice Uniform Deposit Impactor (MOUDI): Description, Calibration and Use, *Aerosol Sci. and Tech.*, **14**, 434–446, 1991.

Marshak, A., Y. Knyazikhin, K. Evans, and W. Wiscombe, The "RED versus NIR" plane to retrieve broken-cloud optical depth from ground-based measurements, *J. Atmos. Sci.*, **61**, 1911–1925, 2004.

Mårtensson, E.M., E.D. Nilsson, G. de Leeuw, L.H. Cohen, and H.C. Hansson, Laboratory simulations and parameterization of the primary marine aerosol production, *J. Geophys. Res.*, **108**, 4297, doi:10.1029/2002JD002263, 2003.

Martin, G.M., D.W. Johnson, and A. Spice, The measurement and parameterization of effective radius of droplets in warm stratocumulus clouds, *J. Atmos. Sci.*, **51**, 1823–1842, 1994.

Martins, J.V., D. Tanré, L.A. Remer, Y.J. Kaufman, S. Mattoo, and R. Levy, MODIS Cloud screening for remote sensing of aerosol over oceans using spatial variability, *Geophys. Res. Lett.*, **29**, 8009, doi:10.1029/2001GL013252, 2002.

Martonchik, J.V., D.J. Diner, R. Kahn, M.M. Verstraete, B. Pinty, H.R. Gordon, and T.P. Ackerman, Techniques for the Retrieval of aerosol properties over land ocean using multiangle data, *IEEE Trans. Geosci. Remote. Sens.*, **36**, 1212–1227, 1998.

Martonchik, J.V., D.J. Diner, K.A. Crean, and M.A. Bull, Regional aerosol retrieval results from MISR, *IEEE Trans. Geosci. Remote Sens.*, **40**, 1520–1531, 2002.

Martonchik, J.V., D.J. Diner, R.A. Kahn, B.J. Gaitley, and B.N. Holben, Comparison of MISR and AERONET aerosol optical depths over desert sites, *Geophys. Res. Let.*, **31**, L16102, doi:10.1029/2004GL019807, 2004.

Marwitz, J.D., Precipitation efficiency of thunderstorms on the high plains, *J. Rech. Atmos.*, **6**, 367–370, 1972.

Mason, B.J., (Ed.), *The Physics of Coulds*, Second Edition, Clarendon Press, Oxford, 1971.

Mason, B.J., A review of 3 long-term cloud-seeding experiments, *Meteor. Mag.*, **109**, 335–344, 1980.

Mason, B.J., Personal reflections on 35 years of cloud seeding, *Contemp. Phys.*, **23**, 311–327, 1982.

Mason, B.J. and J. Maybank, The fragmentation and electrification of freezing water drops, *Q. J. Roy. Meteor. Soc.* **86**, 176–186, 1960.

Massie, S.T., O. Torres, and S.J. Smith, Total Ozone Mapping Spectrometer (TOMS) observations of increases in Asian aerosol in winter from 1979 to 2000, *J. Geophys. Res.*, **109**, D18211, doi:10.1029/2004JD004620, 2004.

Mather, G.K., Coalescence enhancement in large multicell storms caused by the emissions from a Kraft paper mill, *J. Appl. Meteor.*, **30**, 1134–1146, 1991.

Mather, G.K., D.E. Terblanche, F.E. Steffens, and L. Fletcher, Results of the South African cloud seeding experiments using hygroscopic flares, *J. Appl. Meteorol.*, **36**, 1433–1447, 1997.

Matsui, T., H. Masunaga, R. Pielke, Sr., and W.K. Tao, Impact of aerosols and atmospheric thermodynamics on cloud properties within climate system, *Geophys. Res. Lett.*, **31**, L06109, doi:10.1029/2003GL019287, 2004.

Matsumoto, K., H. Tanaka, I. Nagao, and Y. Ishizaka, Contribution of particulate sulfate and organic carbon to cloud condensation nuclei in the marine atmosphere, *Geophys. Res. Lett.*, **24**, 655–658, 1997.

Matthias, V., D. Balis, J. Bosenberg, R. Eixmann, M. Iarlori, L. Komguem, I. Mattis, A. Papayannis, G. Pappalardo, M.R. Perrone, and X. Wang, Vertical aerosol distribution over Europe: Statistical analysis of Raman lidar data from 10 European Aerosol Research Lidar Network (EARLINET) stations – Art. No. D18201, *J. Geophys. Res. Atmos.*, **109**, 18201, 2004.

Matthias-Maser, S., and R. Jaenicke, The size distribution of primary biological aerosol particles with radii >0.2 μm in an urban-rural influenced region, *Atmos. Res.*, **39**, 279–286, 1995.

Matthias-Maser, S., J. Brinkmann, and W. Schneider, The size distribution of marine atmospheric aerosol with regard to primary biological aerosol particles over the South Atlantic Ocean, *Atmos. Environ.*, **33**, 3569–3575, 1999.

Matthias-Maser, S., V. Obolkin, T. Khodzer, and R. Jaenicke, Seasonal variation of primary biological aerosol particles in the remote continental region of Lake Baikal/Siberia, *Atmos. Environ.*, **34**, 3805–3811, 2000.

Mayol-Bracero, O.L., O. Rosario, C.E. Corrigan, R. Morales, I. Torres, and V. Perez, Chemical characterization of submicron organic aerosols in the tropical trade winds of the Caribbean using gas chromatography/mass spectrometry, *Atmos. Environ.*, **35**, 1735–1745, 2001.

Mayol-Bracero, O.L., R. Gabriel, M.O. Andreae, T.W. Kirchstetter, T. Novakov, J. Ogren, P. Sheridan, and D.G. Streets, Carbonaceous aerosols over the Indian Ocean during the Indian Ocean Experiment (INDOEX): Chemical characterization, optical properties and probable sources, *J. Geophys. Res.*, **107**, 8030, doi:10.1029/2000JD000039, 2002a.

Mayol-Bracero, O.L., P. Guyon, B. Graham, G. Roberts, M.O. Andreae, S. Decesari, M. . Facchini, S. Fuzzi, and P. Artaxo, Water-soluble organic compounds in biomass burning aerosols over Amazonia: 2. Apportionment of the chemical composition and importance of the polyacidic fraction, *J. Geophys. Res.*, **107**, 8091, doi:10.1029/2001JD000522, 2002b.

Mazurek, M., M.C. Masonjones, H.D. Masonjones, L.G. Salmon, G.R. Cass, K.A. Hallock, and M. Leach, Visibility-reducing organic aerosols in the vicinity of Grand Canyon National Park: Properties observed by high resolution gas chromatography, *J. Geophys. Res.*, **102**, 3779–3793, 1997.

McArthur, B.L.J., D.H. Halliwell, O.J. Neibergall, N.T. O'Neill, J.R. Slusser, and C. Wehrli, Field comparison of network sunphotometers, *J. Geophys. Res.*, **108**, D19, 4596, dio:10.1029/2002JD002964, 2003.

McCormick, M.P., D.M. Winker, E.V. Browell, J.A. Coakley, C.S. Gardner, R.M. Hoff, G.S. Kent, S.H. Melfi, R.T. Menzies, C.M.R. Platt, D.A. Randall, and J.A. Reagan, Scientific investigations planned for the Lidar In-space Technology Experiment (LITE), *Bull. Amer. Meteorol. Soc.*, **74**, 205–214, 1993.

McCormick, M.P., L.W. Thomason, and C.R. Trepte, Atmospheric effects of the Mount Pinatubo eruption, *Nature*, **373**, 399–404, 1995.

McFiggans, G., P. Artaxo, U. Baltensperger, H. Coe, M.C. Facchini, G. Feingold, S. Fuzzi, M. Gysel, A. Laaksonen, U. Lohmann, T.F. Mentel, D.M. Murphy, C.D. O'Dowd, J.R. Snider, E. Weingartner, The effect of physical and chemical aerosol properties on warm cloud droplet activation, *Atmos. Chem. Phys.*, **6**, 2593–2649, 2006.

McGraw, R., Description of aerosol dynamics by the quadrature method of moments. *Aerosol Sci. and Tech.*, **27**, 255–265, 1997.

McInnes, L., D. Covert, and B. Baker, The number of sea-salt, sulfate, and carbonaceous particles in the marine atmosphere: EM measurements consistent with the ambient size distribution, *Tellus*, **49B**, 300–313, 1997.

McMurry, P.H., Review of atmospheric aerosol measurements, *Atmos. Environ.*, **34**, 1959–1999, 2000.

McMurry, P.H., and M.R. Stolzenburg, On the sensitivity of particle-size to relativehumidity for Los Angeles aerosols. *Atmos. Environ.*, **23**, 497–507, 1989.

McNaughton, C.S., A.D. Clarke, S.G. Howell, K.G.M. II, V. Brekhovskikh, R.J. Weber, D.A. Orsini, D.S. Covert, G. Buzorius, F.J. Brechtel, G.R. Carmichael, Y. Tang, F.L. Eisele, R.L. Mauldin, A.R. Bandy, D.C. Thornton, and B. Blomquist, Spatial distribution and size evolution of particles in Asian outflow: Significance of primary and secondary aerosols during ACE-Asia and TRACE-P, *J. Geophys. Res.*, **109**, D19S06, doi:10.1029/2003JD003528, 2004.

Medina, J., and A. Nenes, Effects of film-forming compounds on the growth of giant cloud condensation nuclei: Implications for cloud microphysics and the aerosol indirect effect, *J. Geophys. Res.*, **109**, D20207, doi:10.1029/2004JD004666, 2004.

Meischner, P., (Ed.), *Weather Radar: Principles and Advanced Applications*, 337 pp., Springer-Verlag, Berlin, ISBN:3-540-00328-2, 2004.

Melfi, S.H., J.D. Spinhirne, S.H. Chou, and P.S. Palm., Lidar observation of vertically organized convection in the planetary boundary layer over the ocean, *J. Climate Appl. Meteor.*, **24**, 806–821, 1985.

Meneghini, R., and T. Kozu, *Spaceborne Weather Radar*, 197 pp., Artech House, Boston, MA, 1990.

Menon, S., and A.D. Del Genio, Evaluating the impacts of carbonaceous aerosols on clouds and climate, in *Human-induced Climate Change: An Interdisciplinary Assessment*, edited by M.E. Schlesinger, H.S. Kheshgi, J. Smith, F.C. de la Chesnaye, J.M. Reilly, T. Wilson, and C. Kolstad, Cambridge University Press, UK, 2007.

Menon, S., and L. Rotstayn, The radiative influence of aerosol effects on liquid-phase cumulus and stratiform clouds based on sensitivity studies with two climate models, *Clim. Dyn.*, **27**, 345–356, 2006.

Menon, S., J. Hansen, L. Nazarenko, and Y. Luo, Climate effects of black carbon aerosol in China and India, *Science*, **297**, 2250–2253, 2002.

Menon, S., J.L. Brenguier, O. Boucher, P. Davison, A.D. Del Genio, J. Feichter, S.Ghan, S. Guibert, X.H. Liu, U. Lohmann, H. Pawlowska, J.E. Penner, J. Quaas, D.L. Roberts, L. Schuller, and J. Snider, Evaluating aerosol/cloud/radiation process parameterizations with single-column models and Second Aerosol Characterization Experiment (ACE-2) cloudy column observations, *J. Geophys. Res.*, **108**, 4762, doi:10.1029/2003JD003902, 2003.

Mesinger, F. and N. Mesinger, Has hail suppression in Eastern Yugoslavia led to a reduction in the frequency of hail? *J. Appl. Met.*, **31**, 104–111, 1992.

Messer, H., A. Zinevich, and P. Alpert, Environmental monitoring by wireless communication networks, *Science*, **312**, 713, 2006.

Methven, J., S.R. Arnold, A. Stohl, M.J. Evans, M. Avery, K. Law, A.C. Lewis, P.S. Monks, D. Parrish, C. Reeves, H. Schlager, E. Atlas, D. Blake, H. Coe, R.C. Cohen, J. Crosier, F. Flocke, J.S. Holloway, J.R. Hopkins, G. Hübler, J.D. Lee,

R. Purvis, B. Rappenglück, T.B. Ryerson, G.W. Sachse, H. Singh, N. Watson, L. Whalley, and P. Williams, Establishing Lagrangian connections between observations within air masses crossing the Atlantic during the ICARTT experiment, *J. Geophys. Res.*, **111**, D23S62, doi:10.1029/2006JD007540, 2006.

Metzger, S., F. Dentener, M. Krol, A. Jeuken, and J. Lelieveld, Gas/aerosol partitioning. 2. Global modeling results, *J. Geophys. Res.*, **107**, 4313, doi:10.1029/2001JD001103, 2002a.

Metzger, S., F. Dentener, S. Pandis, and J. Lelieveld, Gas/aerosol partitioning: 1. A computationally efficient model, *J. Geophys. Res.*, **107**, 4312, doi:10.1029/2001JD001102, 2002b.

Meyers, M.P., P.J. DeMott, and W.R. Cotton, New primary ice nucleation parameterizations in an explicit cloud model, *J. Appl. Meteor.*, **31**, 708–721, 1992.

Meyers, M.P., R.L. Walko, J.Y. Harrington, and W.R. Cotton, New RAMS cloud microphysics prameterization. Part II: The two-moment scheme, *Atmos. Res.*, **45**, 3–39, 1997.

Michalsky, J.J., J.A. Schlemmer, W.E. Berkheiser, J.L. Berndt, and L.C. Harrison, Multiyear measurements of aerosol optical depth in the atmospheric radiation measurement and quantitative links program, *J. Geophys. Res.*, **106**, 12,099–12,107, 2001.

Michel, C., C. Liousse, J.M. Grégoire, K. Tansey, G.R. Carmichael, and J.H. Woo, Biomass burning emission inventory from burnt area data given by the SPOT-VEGETATION system in the frame of TRACE-P and ACE-Asia campaigns, *J. Geophys. Res.*, **110**, D09304, doi:10.1029/2004JD005461, 2005.

Middlebrook, A.M., D.M. Murphy, and D.S. Thomson, Observations of organic material in individual marine particles at Cape Grim during the First Aerosol Characterization Experiment (ACE 1), *J. Geophys. Res.*, **103**, 16,475–16,483, 1998.

Mielke, Jr., P.W., L.O. Grant, and C.F. Chappell, Elevation and spatial variation effects of wintertime orographic cloud seeding, *J. Appl. Meteorol.*, **9**, 476–488, 1970.

Mielke, Jr., P.W., L.O. Grant, and C.F. Chappell, An independent replication of the Climax wintertime orographic cloud seeding experiment, *J. Appl. Meteor.*, **10**, 1198–1212, 1971.

Mielke, Jr., P.W., G.W. Brier, L.O. Grant, G.J. Mulvey, and P.N. Rosenweig, A statistical reanalysis of the replicated Climax I and II wintertime orographic cloud seeding experiments, *J. Appl. Meteor.*, **20**, 643–659, 1981.

Mielke, Jr., P.W., K. Berry, A.S. Dennis, P.L. Smith, J.R. Miller, Jr., and B.A. Silverman, HIPLEX-1: Statistical evaluation, *J. Appl. Meteor.*, **23**, 513–522, 1984.

Mikhailov, E., S. Vlasenko, R. Niessner, and U. Pöschl, Interaction of aerosol particles composed of protein and salts with water vapor: Hygroscopic growth and microstructural rearrangement, *Atmos. Chem. Phys.*, **4**, 323–350, 2004.

Milbrandt, J., and M.K. Yau, A multi-moment bulk microphysics parameterization. Part I: Analysis of the role of the shape parameter, *J. Atmos. Sci.*, **62**, 3051–3064, 2005a.

Milbrandt, J., and M.K. Yau, A multimoment bulk microphysics parameterization. Part II: A proposed three-moment closure and scheme description, *J. Atmos. Sci.*, **62**, 3065–3081, 2005b.

Miller, R.L., I. Tegen, and J. Perlwitz, Surface radiative forcing by soil dust aerosols and the hydrologic cycle, *J. Geophys. Res.*, **109**, D04203, doi:10.1029/2003JD004085, 2004.

Min, Q., and L.C. Harrison, Cloud properties derived from surface MFRSR measurements and comparison with GEOS results at the ARM SGP site, *Geophys. Res. Lett.*, **23**, 1641–1644, 1996.

Ming, Yi, V. Ramaswamy, L.J. Donner, V.T.J. Phillips, S.A. Klein, P.A. Ginoux and L.W. Horowitz: Modeling the interactions between aerosols and liquid water clouds with a self- consistent cloud scheme in a general circulation model, *J. Atmos. Sci.*, **64**, 1189–1209, 2007.

Mircea, M., M.C. Facchini, S. Decesari, F. Cavalli, L. Emblico, S. Fuzzi, A. Vestin, J. Rissler, E. Swietlicki, G. Frank, M.O. Andreae, W. Maenhaut, Y. Rudich, and P. Artaxo, Importance of the organic aerosol fraction for modeling aerosol hygroscopic growth and activation: A case study in the Amazon Basin, *Atmos. Chem. Phys.*, **5**, 3111–3126, 2005.

Mishchenko, M.I., L.D. Travis, W.B. Rossow, B. Cairns, B.E. Carlson, and Q. Han, Retrieving CCN column density from single-channel measurements of reflected sunlight over the ocean: A sensitivity study, *Geophys. Res. Lett.*, **24**, 2655–2658, 1997.

Mishchenko, M.I., I.V. Geogdzhayev, L. Liu, J.A. Ogren, A.A. Lacis, W.B. Rossow, J.W. Hovenier, H. Volten, and O. Muñoz, Aerosol retrievals from AVHRR radiances: Effects of particle nonsphericity and absorption and an updated long-term global climatology of aerosol properties, *J. Quant. Spectrosc. Radiat. Transfer.*, **79**, 953–972, 2003.

Mochida, M., Y. Kitamori, K. Kawamura, Y. Nojiri, and K. Suzuki, Fatty acids in the marine atmosphere: Factors governing their concentrations and evaluation of organic films on sea-salt particles, *J. Geophys. Res.*, **107**, 4325, doi:10.1029/2001JD001278, 2002.

Mohler, O., P.R. Field, P. Connolly, S. Benz, H. Saatho, M. Schnaiter, R. Wagner, R. Cotton, M. Kramer, A. Mangold, and A.J. Heymsfield, Efficiency of the deposition mode ice nucleation on mineral dust particles. *Atmos. Chem. Phys. Discuss.*, **6**, 1539–1577, 2006.

Molina, M.J., A.V. Ivanov, S. Trakhtenberg, and L.T. Molina, Atmospheric evolution of organic aerosol, *Geophys. Res. Lett.*, **31**, L22104, doi:10.1029/2004GL020910, 2004.

Molnar, A., E. Meszaros, H.C. Hansson, H. Karlsson, A. Gelencser, G.Y. Kiss, and Z. Krivacsy, The importance of organic and elemental carbon in the fine atmospheric aerosol particles, *Atmos. Environ.*, **33**, 2745–2750, 1999.

Mooney, M.L. and G.W. Lunn. The Area of Maximum Effect Resulting from the Lake Almanor Randomized Cloud Seeding Experiment, *J. Appl. Meteor.*, **8**(1), 68–74, 1969.

Moore, K.G., A.D. Clarke, V.N. Kapustin, and S.G. Howell, Long-range transport of continental plumes over the Pacific Basin: Aerosol physiochemistry and optical properties during PEM-Tropics A and B, *J. Geophys. Res.*, **108**, 8236, doi:10.1029/2001JD001451, 2003.

Mordy, W.A., Computations of the growth by condensation of a population of cloud droplets, *Tellus*, **11**, 16–44, 1959.

Morin, E., W.F. Krajewski, D.C. Goodrich, X. Gao, and S. Sorooshian, Estimating rainfall intensities from weather radar data: The scale-dependency problem, *J. Hydrometeor.*, **4**, 782–797, 2003.

Mossop, S.C., Concentrations of Ice Crystals in Clouds, *Bull. of Amer. Meteor. Soc.*, **51**, 474–480, 1970.

Mossop, S.C., and J. Hallett, Ice crystal concentration in cumulus clouds: Influence of the drop spectrum, *Science*, **186**, 632–634, 1974.

Mossop, S.C. and A. Ono, Measurements of ice crystal concentration in clouds, *J. Atmos. Sci.*, **26**, 130–137, 1969.

Mossop, S.C., A. Ono, and K.J. Heffernan, Studies of ice crystals in natural clouds, *J. Atmos. Res.*, **1**, 44–64, 1967.

Mossop, S.C., R.E. Ruskin, and J.K. Hefferman, Glaciation of a cumulus at –4°C, *J. Atmos. Sci.*, **25**, 889–899, 1968.

Mossop, S.C., R.E. Cottis and B.M. Bartlett, Ice crystal concentrations in cumulus and stratocumulus clouds, *Q. J. Roy. Meteor. Soc.*, **98**, 105–123, 1972.

Moulin, C., C.E. Lambert, F. Dulac and U. Dayan, Control of atmospheric export of dust from North Africa by the North Atlantic oscillation, *Nature*, **387**, 691–694, 1997.

Murakami, M., K. Kikuchi, and C. Magono, Experiments on aerosol scavenging by natural snow crystals. Part I: Collection efficiency of uncharged snow crystals for micron and sub-micron particles, *J. Meteor. Soc. of Japan.*, **63**, 119–128, 1985.

Murphy, D.M., and D.S. Thomson, Chemical composition of single aerosol particles at Idaho Hill: Negative Ion Measurements, *J. Geophys. Res.*, **102**, 6341–6352, 1997.

Murphy, D.M., J.R. Anderson, P.K. Quinn, L.M. McInnes, F.J. Brechtel, S.M. Kreidenweis, A.M. Middlebrook, M. Posfai, D.S. Thomson, and P.R. Buseck, Influence of sea-salt on aerosol radiative properties in the Southern Ocean marine boundary layer, *Nature*, **392**, 62–65, 1998a.

Murphy, D.M., D.S. Thomson, A.M. Middlebrook, and M.E. Schein, In situ single-particle characterization at Cape Grim, *J. Geophys. Res.*, **103**, 16,485–16,491, 1998b.

Murty, A.S.R., A.M. Selvam, P.C.S. Devara, K. Krishna, R.N. Chatterjee, B.K. Mukherjee, L.T. Kemani, G.A. Momin, R.S. Reddy, S.K. Sharma, D.B. Jadhav, R. Vijayakumar, P.E. Raj, G.K. Manohar, S.S. Kandalgaonkar, S.K. Paul, A.G. Pillai, S.S. Parasnis, C.P. Kulkarni, A.L. Londhe, C.S. Bhosale, S.B. Morwal, P.D. Safai, J.M. Pathan, K. Indira, M.S. Naik, P.S.P. Rao, P. Sikka, K.K. Dani, M.K. Kulkarni, H.K. Trimbake, P.N. Sharma, R.K. Kapoor, and M.I.R. Tinmaker, Eleven-year warm cloud seeding experiment in Maharashtra state, India, *J. Weather Modif.*, 32, 10–20, 2000.

Nadal, F., and F.M. Bréon, Parameterization of surface polarized reflectance derived from POLDER spaceborne measurements, *IEEE Trans. Geosci. Remote Sens.*, **37**, 1709–1718, 1999.

Nakajima, T., and M.D. King, Determination of the optical thickness and effective particle radius of clouds from reflected solar radiation measurements. Part I: Theory, *J. Atmos. Sci.*, **47**, 1878–1893, 1990.

Nakajima, T., A. Higurashi, K. Kawamoto and J. Penner, A possible correlation between satellite-derived cloud and aerosol microphysical parameters, *Geophys. Res. Lett.* **28**, 1171–1174, 2001.

Nelson, R.T., and N.R. Gokhale, Concentration of giant particles below cloud bases, Reprint, *Proceedings of the First National Conf. on Weather Mod.*, April 28 to May 1, 1968.

Nenes, A., and J.H. Seinfeld, Parameterization of cloud droplet formation in global climate models, *J. Geophys. Res. D: Atmos.*, **108**, 4415, doi:10.1029/2002JD002911, 2003.

Nenes, A., S.N. Pandis, and C. Pilinis, ISORROPIA: A new thermodynamic equilibrium model for multiphase multicomponent inorganic aerosols, *Aquat. Geochem.*, **4**, 123–152, 1998.

Nenes, A., Chuang, P.Y, Flagan, R., and Seinfeld, J.H, A Theoretical Analysis of Cloud Condensation Nucleus (CCN) Instruments, *J.Geophys. Res.*, **106**, D4, 3449–3474, 2001.

Nenes, A., R.J. Charlson, M.C. Facchini, M. Kulmala, A. Laaksonen, and J.H. Seinfeld, Can chemical effects on cloud droplet number rival the first indirect effect? *Geophys. Res., Lett.*, **29**, 1848, doi:10.1029/2002GL015295, 2002.

Nešpor, V., and B. Sevruk, Estimation of wind-induced error of rainfall gauge measurements using a numerical simulation, *J. Atmos. Ocean. Tech.*, **16**, 450–464, 1999.

Neusüß, C., M. Pelzing, A. Plewka, and H. Herrmann, A new analytical approach for size-resolved speciation of organic compounds in atmospheric aerosol particles: Methods and first results, *J. Geophys. Res.*, **105**, 4513–4527, 2000.

Neusüß, C., H. Wex, W. Birmili, A. Wiedensohler, C. Koziar, B. Busch, E. Bruggemann, T. Gnauk, M. Ebert, and D.S. Covert, Characterization and parameterization of atmospheric particle number-, mass-, and chemical-size distributions in central Europe during LACE 98 and MINT, *J. Geophys. Res.*, **107**, 8127, doi:10.1029/2001JD000514, 2002.

Nho-Kim, E.Y., M. Michou, and V.H. Peuch, Parameterization of size-dependent particle dry deposition velocities for global modeling, *Atmos. Environ.*, **38**, 1933–1942, 2004.

Nicholson, K.W., The dry deposition of small particles: A review of experimental measurements, *Atmos. Environ.*, **22**, 2653–2666, 1988.

Nickerson, E.C., FACE rainfall results: Seeding effect or natural variability? *J. Appl. Meteor.*, **18**, 1097–1105, 1979.

Nickerson, E.C., Reply – The FACE-1 seeding effect revisited, *J. Appl. Meteor.*, **20**, 108–114, 1981.

Nickerson, E.C., E. Richard, R. Rosset, and D.R. Smith, The numerical simulation of clouds, rain, and airflow over the Vosges and Black Forest Mountains: A meso-beta model with parameterized microphysics, *Mon. Wea. Rev.*, **114**, 398–414, 1986.

Nilsson, E.D., Ü. Rannik, E. Swietlicki, C. Leck, P.P. Aalto, J. Zhou, and M. Norman, Turbulent aerosol fluxes over the Arctic Ocean 2. Wind-driven sources from the sea, *J. Geophys. Res.*, **106**, 32,139–32,154, 2001.

Nissen, R., R. List, D. Hudak, and G.M. McFarquhar, Constant raindrop fall speed profiles derived from Doppler radar data analysis for steady nonconvective rain, *J. Atmos. Sci.*, **62**, 220–230, 2005.

Nober, F., H.F. Graf and D. Rosenfeld, Sensitivity of the global circulation to the suppression of precipitation by anthropogenic aerosols, *Global Planet Change.*, **37**, 57–80, 2003.

Noble, C.A., and K.A. Prather, Real-time measurement of correlated size and composition profiles of individual atmospehric aerosol particles, *Environ. Sci. Tech.*, **30**, 2667–2680, 1996.

Noone, K., U. Baltensperger, A. Flossmann, S. Fuzzi, H. Hass, E. Nemitz, J.P. Putaud, H. Puxbaum, U. Schurath, K. Tørseth, and H.t. Brink, Tropospheric aerosols and clouds, in *Towards Cleaner Air for Europe – Science, Tools and Application. Part 1: Results from the EUROTRAC-2 Synthesis and Integration Project*, edited by P.M. Midgley, P.J.H. Builtjes, R.M.H.D. Fowler, C.N. Hewitt, N. Moussiopoulos, K. Noone, K. Thrseth and A. Volz-Thomas, Margraf Verlag, Weikersheim, 2003.

Norris, J.R., Low cloud type over the ocean from surface observations. Part II: Geographical and seasonal variations, *J. Clim.*, **11**, 383–403, 1998.

Novakov, T., and C.E. Corrigan, Cloud condensation nucleus activity of the organic component of biomass smoke particles, *Geophys. Res. Lett.*, **23**, 2141–2144, 1996.

Novakov, T., and J.E. Penner, Large contribution of organic aerosols to cloud-condensation-nuclei concentrations, *Nature*, **365**, 823–826, 1993.

Novakov, T., C.E. Corrigan, J.E. Penner, C.C. Chuang, O. Rosario, and O.L.M. Bracero, Organic aerosols in the Caribbean trade winds: A natural source? *J. Geophys. Res.*, **102**, 21,307–21,313, 1997.

Noziere, B., and D.D. Riemer, The chemical processing of gas-phase carbonyl compounds by sulfuric acid aerosols: 2, 4-pentandione, *Atmos. Environ.*, **37**, 841–851, 2003.

NRC (National Research Council) of the National Academy of Sciences, *Weather and Climate Modification: Progress and Problems*, 258 pp., Government Printing Office, Washington, DC, 1973.

NRC (National Research Council) of the National Academy of Sciences, *Critical Issues in Weather Modification Research*, Board on Atmospheric Sciences and Climate, Division on Earth and Life Studies, The National Academy Press, 123 pp., 2003.

O'Conner, E.J., R.J. Hogan, and A.J. Illingworth, Retrieving stratocumulus drizzle parameters using Doppler radar and lidar, *J. Appl. Meteor.*, **44**, 14–27, 2005.

Ochs, H.T., K.V. Beard, R.R. Czys, N.F. Laird, D.E. Schaufelberger, and D.J. Holdridge, Collisions between small precipitation drops. Part I: Laboratory measurements of bounce, coalescence and temporary coalescence, *J. Atmos. Sci.*, **52**, 2258–2275, 1995.

O'Dowd, C.D., and M.H. Smith, Physicochemical properties of aerosols over the Northeast Atlantic: Evidence for wind-speed-related submicron sea-salt aerosol production, *J. Geophys. Res.*, **98**, 1137–1149, 1993.

O'Dowd, C.D., M.H. Smith, I.E. Consterdine, and J.A. Lowe, Marine aerosol, sea-salt, and the marine sulphur cycle: A short review, *Atmos. Environ.*, **31**, 73–80, 1997.

O'Dowd, C.D., P. Aalto, K. Hameri, M. Kulmala, and T. Hoffmann, Aerosol formation – Atmospheric particles from organic vapours, *Nature*, **416**, 497–498, 2002.

O'Dowd, C.D., M.C. Facchini, F. Cavalli, D. Ceburnis, M. Mircea, S. Decesari, S. Fuzzi, Y.J. Yoon, and J.P. Putaud, Biogenically driven organic contribution to marine aerosol, *Nature*, **431**, 676–680, 2004.

Ogren, J.A. A systematic approach to in situ observations of aerosol properties, in *Aerosol Forcing of Climate*, edited by R.J. Charlson and J. Heintzenberg, pp. 215–226, John Wiley & Sons, Ltd., Chichester; 1995.

Okamoto, H., T. Nishizawa, S. Sato, A. Kamei, H. Kuroiwa, H. Kumagai, M. Yasui, N. Sugimoto, I. Matsui, and A. Shimizu, Study of clouds by shipborne radar and lidar measurements during R-V Mirai MR01-K02 cruise, in Proc. EarthCARE workshop (second international workshop on space-borne cloud profiling radar) CRL/ARS, Report 02-02, pp. 155–163, 2002.

Oke, T.R., *Boundary Layer Climates*, Cambridge University Press, Cambridge, 435 p., 1987.
Oki, T., The global water cycle, in *Global Energy and Water Cycles*, edited by K.A. Browning and R.J. Gurney, 292 pp., Cambridge University Press, Cambridge, UK, 1999.
Olson, T.M., and M.R. Hoffmann, Hydroxyalkylsulfonate formation: Its role as a sulphur (IV) reservoir in atmospheric water droplets, *Atmos. Environ.*, **23**, 985–997, 1989.
Olson, W.S., P. Bauer, N.F. Viltard, D.E. Johnson, W.K. Tao, R. Meneghini, and L. Liao, A melting-layer model for passive/active microwave remote sensing applications. Part I: Model formulation and comparison with observations, *J. Appl. Meteor.*, **40**, 1145–1163, 2001.
Omar, A.H., and C.S. Gardner, Observations by the lidar In-space Technology Experiment (LITE) of high altitude clouds over the Equator in regions exhibiting extremely cold temperatures, *J. Geophys. Res.*, **106**, 1227, 2001.
Oraltay, R.G., J. Hallett, Evaporation and melting of ice crystals: A laboratory study, *Atmos. Res.*, **24**, 169–189, 1989.
Paeth, H. and J. Feichter, Greenhouse-gas versus aerosol forcing and African climate response, *Clim Dyn.*, **26**, 35–54, 2006.
Paluch, I.R., and D.H. Lenschow, Stratiform cloud formation in the marine boundary layer, *J. Atmos. Sci.*, **48**, 2141–2158, 1991.
Pandis, S.N., L.M. Russell, and J.H. Seinfeld, The relationship between DMS flux and CCN concentration in remote marine regions, *J. Geophys. Res.*, **99**, 16,945–16,957, 1994.
Pankow, J.F., Review and Comparative analysis of the theories on partitioning between the gas and aerosol particulate phases in the atmosphere, *Atmos. Environ.*, **21**, 2275–2283, 1987.
Pankow, J.F., An absorption model of the gas/aerosol partitioning involved in the formation of secondary organic aerosol, *Atmos. Environ.*, **28**, 189–193, 1994a.
Pankow, J.F., An absortion model of gas/particle partitioning of organic compounds in the atmosphere, *Atmos. Environ.*, **28**, 185–188, 1994b.
Pankow, J.F., and T.F. Bidleman, Effects of temperature, TSP and percent non-exchangeable material in determining the gas-particle partitioning of organic compounds, *Atmos. Environ.*, **25A**, 2241–2249, 1991.
Park, S.H., and K.W. Lee, Condensational growth of polydisperse aerosol for the entire particle size range, *Aerosol Sci. Tech.*, **33**, 222–227, 2000.
Parol, F., J.C. Buriez, C. Vanbauce, P. Couvert, G. Seze, P. Goloub, and S. Cheinet, First results of the POLDER "Earth Radiation Budget and Clouds" operational algorithm, IEEE Trans. Geosci., *Remote Sens.*, **37**, 1597–1612, 1999.
Patashnick, H., G. Rupprecht, J.L. Ambs, M.B. Meyer, Development of a reference standard for particulate matter mass in ambient air, *Aerosol Sci. and Tech.*, **34**, 42–45, 2001.
Pawlowska, H., and J.L. Brenguier, Microphysical properties of stratocumulus clouds during ACE-2, *Tellus*, **52B**, 867–886, 2000.
Pawlowska, H., and J.L. Brenguier, An Observational Study of Drizzle Formation in Stratocumulus Clouds during ACE-2 for GCM parameterizations PACE Topical Issue, *J. Geophys. Res.*, **108**, D19, 8587 10.1029, 2003.
Peng, C., M.N. Chan, and C.K. Chan, The hygroscopic properties of dicarboxylic and multifunctional acids: Measurements and UNIFAC predictions, *Environ. Sci. Tech.*, **35**, 4495–4501, 2001a.
Peng, C., A.H.L. Chow, and C.K. Chan, Hygroscopic study of glucose, citric acid, and sorbitol using an electrodynamic balance: Comparison with UNIFAC predictions, *Aerosol Sci. and Tech.*, **35**, 753–758, 2001b.
Peng, Y., U. Lohmann, and W.R. Leaitch, Importance of vertical velocity variations in the cloud droplet nucleating process of marine stratus clouds, *J. Geophys. Res.*, **110**, D21213, doi:10.1029/2004JD004922, 2005.
Penner, J.E., M.O. Andreae, H. Annegarn, L. Barrie, J. Feichter, D. Hegg, A. Jayaraman, R. Leaitch, D. Murphy, J. Nganga, and G. Pitari, Aerosols, their direct and indirect effects, in *Climate Change 2001: The Scientific Basis. Contribution of Working Group I to*

the Third Assessment Report of the Intergovernmental Panel on Climate Change, edited by J.T. Houghton, Y. Ding, D.J. Griggs, M. Noguer, P.J. van der Linden, X. Dai, K. Maskell, and C.A. Johnson, pp. 289–348, Cambridge University Press, Cambridge, UK, and New York, NY, USA, 2001.

Penner, J.E., J. Quaas, T. Storelvmo, T. Takemura, O. Bouche, H. Guo, A. Kirkevåg, J.E. Kristjánsson, and Ø. Seland, Model intercomparison of indirect aerosol effects, *Atmos. Chem. Phys.*, 6, 3391–3405, 2006.

Perry, K.D., T.A. Cahill, R.A. Eldred, D.D. Dutcher, and T.E. Gill,. Long-range transport of North African dust to the eastern United States, *J. Geophys. Res.*, **102**, 11,225–211,238, 1997.

Peters, G., B. Fischer, and T. Andersson, Rain observations with a vertically looking Micro Rain Radar (MRR), *Boreal Environ. Res.*, 7, 353–362, 2002.

Petters, M.D., and S.M. Kreidenweis, A single parameter representation of hygroscopic growth and cloud condensation nucleus activity, *Atmos. Chem. Phys. Discuss.*, **6**, 8435–8456, 2006.

Petzold, A., and M. Schönlinner, Multi-angle absorption photometry – A new method for the measurement of aerosol light absorption and atmospheric black carbon, *J. Aerosol Sci.*, **35**, 421–441, 2004.

Petzold, A., J. Strom, S. Ohlsson, and F.P. Schroder, Elemental composition and morphology of ice-crystal residual particles in cirrus clouds and contrails, *Atmos. Res.*, **49**, 21–34, 1998.

Petzold, A., M. Fiebig, H. Flentje, A. Keil, U. Leiterer, F. Schroder, A. Stifter, M. Wendisch, and P. Wendling, Vertical variability of aerosol properties observed at a continental site during the Lindenberg Aerosol Characterization Experiment (LACE 98), *J. Geophys. Res.*, **107**, 8128, doi:10.1029/2001JD001043, 2002.

Petzold, A, M. Gysel, X. Vancassel, R. Hitzenberger, H. Puxbaum, S. Vrochticky, E.,-Weingartner, U. Baltensperger, and P. Mirabel, On the effects of organic matter and sulphur-containing compounds on the CCN activation of combustion particles, *Atmos. Chem. Phys.*, **5**, 3187–3203, 2005.

Pielke, R.A., W.R. Cotton, R.L. Walko, C.J. Tremback, W.A. Lyons, L.D. Grasso, M.E. Nicholls, M.D. Moran, D.A. Wesley, T.J. Lee, J.H. Copeland: A comprehensive meteorological modeling system – RAMS, *Meteorol. Atmos. Phys.*, **49**, 69–91, 1992.

Piironen, A.K., and E.W. Eloranta, Convective boundary layer mean depths, cloud base altitudes, cloud top altitudes, cloud coverages, and cloud shadows obtained from Volume Imaging Lidar data, *J. Geophys. Res.*, **100**, 25,569–25,576, 1995.

Pilinis, C., and J.H. Seinfeld, Continued development of a general equilibrium model for inorganic multicomponent atmospheric aerosols, *Atmos. Environ.*, **21**, 2453–2466, 1987.

Pilinis, C., K.P. Capaldo, A. Nenes, and S.N. Pandis, MADM – A new multicomponent aerosol dynamics model, *Aerosol Sci. and Tech.*, **32**, 482–502, 2000.

Pincus, R., and S.A. Klein, Unresolved spatial variability and microphysical process rates in large-scale models, *J. Geophys. Res.*, **105**, 27,059–27,065, 2000.

Pinsky, M.B., and A.P. Khain, Formation of inhomogeneity in drop concentration induced by the inertia of drops falling in a turbulent flow, and the influence of the inhomogeneity on the drop-spectrum broadening, *Q. J. Roy. Meteorol. Soc.*, **123**, 165–186, 1997a.

Pinsky, M.B., and A.P. Khain, Turbulence effects on droplet growth and size distributions in clouds – A review, *J. Aerosol Sci.*, **28**, 1127–1214, 1997b.

Pirjola, L., C.D. O'Dowd, I.M. Brooks, and M. Kulmala, Can new particle formation occur in the clean marine boundary layer? *J. Geophys. Res.*, **105**, 26,531–26,546, 2000.

Pitter, R.L., and H.R. Pruppacher, A wind tunnel investigation of freezing of small water drops falling at terminal velocity in air, *Q. J. Roy. Meteor. Soc.*, **99**, 540–550, 1973.

Platnick, S., M.D. King, S.A. Ackerman, W.P. Menzel, B.A. Baum, J.C. Riédi, and R.A. Frey, The MODIS cloud products: Algorithms and examples from Terra, *IEEE Trans. Geosci. Remote Sens.*, **41**, 459–473, 2003.

Popovitcheva, O.B., M.E. Trukhin, N.M. Persiantseva, and N.K. Shonija, Water adsorption on aircraft-combustor soot under young plume conditions, *Atmos. Environ.*, **35**, 1673–1676, 2001.

Porch, W.M., C.Y.J. Kao, and R.G. Kelley, Jr., Ship trails and ship induced cloud dynamics, *Atmos. Environ.*, **24A**, 1051–1059, 1990.

Posfai, M., J. Li, J.R. Anderson, and P.R. Buseck, Aerosol bacteria over the Southern Ocean during ACE-1, *Atmos. Res.*, **66**, 231–240, 2003.

Powell, K.A., C.R. Trepte, and G.S. Kent, Observations of Saharan Dust by LITE, paper presented at Advances in Atmospheric Remote Sensing with Lidar, Selected Papers of the 18th ILRC, Berlin, 1996.

Prather, K.A., T. Nordmeyer, and Kimberly Salt, Real-time characterization of individual aerosol particles using time-of-flight mass spectrometry, *Anal. Chem.*, **66**, 1403–1407, 1994

Prenni, A.J., P.J. DeMott, S.M. Kreidenweis, D.E. Sherman, L.M. Russell, and Y. Ming, The effects of low molecular weight dicarboxylic acids on cloud formation, *J. Phys. Chem. A.*, **105**, 11,240–11,248, 2001.

Prenni, A.J., P.J. DeMott, and S.M. Kreidenweis, Water uptake of internally mixed particles containing ammonium sulfate and dicarboxylic acids, *Atmos. Environ.*, **37**, 4243–4251, 2003.

Prenni, A.J., J.Y. Harrington, M. Tjernström, P.J. DeMott, A. Avramov, C.N. Long, S.M. Kreidenweis, P.Q. Olsson, and J. Verlinde, Can Ice-Nucleating Aerosols Affect Arctic Seasonal Climate? *Bull. Amer. Meteor. Soc.*, **88**, 541–550, 2007.

Procopio, A.S., P. Artaxo, Y.J. Kaufman, L.A. Remer, J.S. Schafer, and B.N. Holben, Multiyear analysis of amazonian biomass burning smoke radiative forcing of climate, *Geophys. Res. Lett.*, **31**, L03108, doi:10.1029/2003GL018646, 2004.

Prospero, J.M., Long-term measurements of the transport of African mineral dust to the southeastern United States: Implications for regional air quality, *J. Geophys. Res. D: Atmos.*, **104**, 15,917–15,927, 1999.

Prospero, J.M., P. Ginoux, O. Torres, S.E. Nicholson, and T.E. Gill, Environmental characterization of global sources of atmospheric soil dust identified with the Nimbus 7 Total Ozone Mapping Spectrometer (TOMS) absorbing aerosol product, *Rev. Geophys.*, **40**, 1002, doi:10.1029/2000RG000095, 2002.

Pruppacher, H.R., and J.D. Klett, *Microphysics of Clouds and Precipitation*, pp. 954, Reidel, Dordrecht, 1997.

Putaud, J.P., R. van Dingenen, M. Mangoni, A. Virkkula, F. Raes, H. Maring, J.M. Prospero, E. Swietlicki, O.H. Berg, H. Hillamo, and T. Mäkelä, Chemical mass closure and origin assessment of the submicron aerosol in the marine boundary layer and the free troposphere at Tenerife during ACE-2, *Tellus*, **52B**, 141–168, 2000.

Putaud, J.P., R.V. Dingenen, U. Baltensperger, E. Brüggemann, A. Charron, M.C. Facchini, S. Decesari, S. Fuzzi, R. Gehrig, H.C. Hansson, R.M. Harrison, A.M. Jones, P. Laj, G. Lorbeer, W. Maenhaut, N. Mihalopoulos, K. Müller, F. Palmgren, X. Querol, S. Rodriguez, J. Schneider, G. Spindler, H.T. Brink, P. Tunved, K. Tørseth, B. Wehner, E. Weingartner, A. Wiedensohler, P. Wåhlin, and F. Raes, A European aerosol phenomenology. Physical and chemical characteristics of particulate matter at kerbside, urban, rural and background sites in Europe, EUR 20411 European Commission, Joint Research Centre, European Commission, EUR 20411 EN, 2003.

Putaud, J.P., F. Raes, R. Van Dingenen, E. Brüggemann, M.C. Facchini, S. Decesari, S. Fuzzi, R. Gehrig, C. Hüglin, P. Laj, G. Lorbeer, W.Maenhaut, N. Mihalopoulos, K. Müller, X. Querol, S. Rodriguez, J. Schneider, G. Spindler, H. ten Brink, K. Tørseth and A. Wiedensohler, European aerosol phenomenology. 2: Chemical characteristics of particulate matter at kerbside, urban, rural and background sites in Europe, *Atmos. Environ.*, **38**, 2579–2595, 2004.

Quaas, J., O. Boucher, and F.M. Bréon, Aerosol indirect effects in POLDER satellite data and the Laboratoire de Météorologie Dynamique-Zoom (LMDZ) general circulation model, *J. Geophys. Res.*, **109**, doi:10.1029/2003JD004317, 2004.

Quaas, J., O. Boucher, and U. Lohmann, Constraining the total aerosol indirect effect in the LMDZ and ECHAM4 GCMs using MODIS satellite data, *Atmos. Chem. Phys.*, **6**, 947–955, 2006.

Quinn, P.K., S.F. Marshall, T.S. Bates, D.S. Covert, and V.N. Kapustin, Comparison of measured and calculated aerosol properties relevant to the direct radiative forcing of tropospheric sulfate aerosol on climate, *J. Geophys. Res.*, **100**, 8977–8991, 1995.

Quinn, P.K., D.J. Coffman, V.N. Kapustin, T.S. Bates, and D.S. Covert, Aerosol optical properties in the marine boundary layer during the First Aerosol Characterization Experiment (ACE 1) and the underlying chemical and physical aerosol properties, *J. Geophys. Res.*, **103**, 16,547–16,563, 1998.

Radke, L.F., and P.V. Hobbs, An automatic cloud condensation nuclei counter, *J. Appl. Meteor.*, **8**, 105–109, 1969.

Radke, L.F., and P.V. Hobbs, Cloud condensation nuclei on the Atlantic seaboard of the United States, *Science*, **193**, 999–1002, 1976.

Radke, L.F., P.V. Hobbs, and M.W. Eltgroth, Scavenging of aerosol particles by precipitation., *J. App. Meteor.*, **19**, 715–722, 1980.

Radke, L.F., D.A. Hegg, J.H. Lyons, C.A. Brock, P.V. Hobbs, R. Weiss, and R. Rasmussen, Airborne measurements on smokes from biomass burning, in *Aerosols and Climate*, edited by P.V. Hobbs, and M.P. McCormick, pp. 411–422, A. Deepak Publishing, Hampton, VA, 1988.

Radke, L.F., J.A. Coakley Jr., and M.D. King, Direct and remote sensing observations of the effects of ship tracks on clouds, *Science*, **246**, 1146–1149, 1989.

Radke, L.F., D.A. Hegg, P.V. Hobbs, J.D. Nance, J.H. Lyons, K.K. Laursen, R.E. Weiss, P.J. Riggan, and D.E. Ward, Particulate and trace gas emissions from large biomass fires in North America, in *Global Biomass Burning: Atmospheric, Climatic and Biospheric Implications*, edited by J.S. Levine, pp. 209–224, MIT Press, Cambridge, MA, 1991.

Raemdonck, H., W. Maenhaut, and M.O. Andreae, Chemistry of marine aerosol over the tropical equatorial Pacific, *J. Geophys. Res.*, **91**, 8623–8636, 1986.

Raes, F., T. Bates, F. McGovern, and M. van Liedekerke, The 2nd Aerosol Characterization Experiment (ACE-2): General overview and main results, *Tellus*, **52B**, 111–125, 2000.

Ramanathan, V., P.J. Crutzen, J. Lelieveld, A.P. Mitra, D. Althausen, J. Anderson, M.O. Andreae, W. Cantrell, G.R. Cass, C.E. Chung, A.D. Clarke, J.A. Coakley, W.D. Collins, W.C. Conant, F. Dulac, J. Heintzenberg, A.J. Heymsfield, B. Holben, S. Howell, J. Hudson, A. Jayaraman, J.T. Kiehl, T.N. Krishnamurti, D. Lubin, G. McFarquhar, T. Novakov, J.A. Ogren, J Heintzenberg, and R.J. Charlson, In-situ sampling of clouds with a droplet to aerosol converter, *Geophys. Res. Lett.*, **12**, 121–124, 1985.

Ramanathan, V., P.J. Crutzen, J. Lelieveld, A.P. Mitra, D. Althausen, J. Anderson, M.O. Andreae, W. Cantrell, G.R. Cass, C.E. Chung, A.D. Clarke, J.A. Coakley, W.D. Collins, W.C. Conant, F. Dulac, J. Heintzenberg, A.J. Heymsfield, B. Holben, S. Howell, J. Hudson, A. Jayaraman, J.T. Kiehl, T.N. Krishnamurti, D. Lubin, G. McFarquhar, T. Novakov, J.A. Ogren, I.A. Podgorny, K. Prather, K. Priestley, J.M. Prospero, P.K. Quinn, K. Rajeev, P. Rasch, S. Rupert, R. Sadourny, S.K. Satheesh, G.E. Shaw, P. Sheridan, and F.P.J. Valero, Indian Ocean Experiment: An integrated analysis of the climate forcing and effects of the great Indo-Asian haze, *J. Geophys. Res.*, **106**, 28,371–28,398, 2001a.

Ramanathan, V., P. Crutzen, J. Kiehl, and D. Rosenfeld, Aerosols, climate, and the hydrological cycle, *Science*, **294**, 2119–2124, 2001b.

Ramanathan, V., C. Chung, D. Kim, T. Bettge, L. Buja, J.T. Kiehl, W.M. Washington, Q. Fu, D.R. Sikka, and M. Wild, Atmospheric brown clouds: Impacts on South Asian climate and hydrological cycle, *Proc. Nat. Acad. Sci.*, **102**, 5326–5333, 2005.

Randall, D., M. Khairoutdinov, A. Arakawa, and W. Grabowski, Breaking the cloud parameterization deadlock, *Bull. Amer. Meteorol. Soc.*, **84**, 1547–1564, 2003.

Rangno, A.L., and P.V. Hobbs, A re-evaluation of the Climax cloud seeding experiments using NOAA published data, *J. Clim. Appl. Meteor.*, **26**, 757–762, 1987.

Rangno, A.L., and P.V. Hobbs, Ice particle concentrations and precipitation development in small polar maritime continental clouds, *Q. J. Roy. Meteor. Soc.*, **117**, 207–241, 1991.

Rangno, A.L., and P.V. Hobbs, Further analyses of the Climax cloud-seeding experiments, *J. Appl. Meteor.*, **32**, 1837–1847, 1993.

Rangno, A.L., and P.V. Hobbs, Ice particle concentrations and precipitation development in small continental cumuliform clouds, *Q. J. Roy. Meteor. Soc.*, **120**, 573–601, 1994.

Rangno, A.L., and P.V. Hobbs, Reply to Woodley, *J. Appl. Meteor.*, **36**, 253–254, 1997a.

Rangno, A.L., and P.V. Hobbs, Reply to Ben-Zvi, *J. Appl. Meteor.*, **36**, 257–259, 1997b

Rangno, A.L., and P.V. Hobbs, Reply to Rosenfeld, *J. Appl. Meteor.*, **36**, 272–276, 1997c.

Rangno, A.L., and P.V. Hobbs, Reply to Dennis and Orville, *J. Appl. Meteor.*, **36**, 279–279, 1997d.

Rangno, A.L., and P.V. Hobbs, *Comprehensive Reply to Rosenfeld*, Cloud and Aerosol Research Group, Dept. Atmospheric Sciences, University of Washington, Seattle, 25 pp., 1997e.

Rangno, A.L. and P.V. Hobbs, Microstructures and precipitation development in cumulus and small cumulonimbus clouds over the warm pool of the tropical Pacific Ocean, *Q. J. Roy. Meteor. Soc.*, **131**, 639–673, 2005.

Rasch, P.J., J. Feichter, K. Law, N. Mahowald, J. Penner, C. Benkovitz, C. Genthon, C. Giannakopoulos, P. Kasibhatla, D. Koch, H. Levy, T. Maki, M. Prather, D.L. Roberts, G.J. Roelofs, D. Stevenson, Z. Stockwell, S. Taguchi, M. Kritz, M. Chipperfield, D. Baldocchi, P. McMurry, L. Barrie, Y. Balkansi, R. Chatfield, E. Kjellstrom, M. Lawrence, H.N. Lee, J. Lelieveld, K.J. Noone, J. Seinfeld, G. Stenchikov, S. Schwartz, C. Walcek, and D. Williamson, A comparison of scavenging and deposition processes in global models: Results from the WCRP Cambridge Workshop of 1995, *Tellus B.*, **52**, 1025–1056, 2000.

Rasmussen, R.M., J. Hallett, R. Purcell, J. Cole, and M. Tryhane, The hotplate snowgauge. Abstracts, *11th Conference on Cloud Physics*, American Meteorological Society, 3–7 June 2002 in Ogden, UT (CD) P1.6, 2002a.

Rasmussen, R.M., I. Geresdi, G. Thompson, K, Manning, and E. Karplus, Freezing drizzle formation in stably stratified layer clouds: The role of radiative cooling of cloud droplets, cloud condensation nuclei, and ice initiation, *J. Atmos. Sci*, **59**, 837–860, 2002b.

Rasool, S.I., (Ed.), *Chemistry of the Lower Atmosphere*, Plenum Press, New York, 1973.

Rau, J.A., and M.A.K. Khalil, Anthropogenic contributions to the carbonaceous content of aerosols over the Pacific Ocean, *Atmos. Environ.*, **27A**, 1297–1307, 1993.

Ravishankara, A.R., and C.A. Longfellow, Reactions on tropospheric condensed matter, *Phys. Chem. Chem. Phys.*, **1**, 5433–5441, 1999.

Raymond, T.M., and S.N. Pandis, Cloud activation of single-component organic aerosol particles, *J. Geophys. Res.*, **107**, 4787, doi:10.1029/2002JD002159, 2002.

Raymond, T.M., and S.N. Pandis, Formation of cloud droplets by multicomponent organic particles, *J. Geophys. Res.*, **108**, 4469, doi:10.1029/2003JD003503, 2003.

Reddy, M.S., and C. Venkataraman, Inventory of aerosol and sulphur dioxide emissions from India. Part II – Biomass combustion, *Atmos. Environ.*, **36**, 699–712, 2002.

Reid, J.S., and P.V. Hobbs, Physical and optical properties of young smoke from individual biomass fires in Brazil, *J. Geophys. Res.*, **103**, 32,013–32,030, 1998.

Reid, J.S., P.V. Hobbs, R.J. Ferek, D.R. Blake, J.V. Martins, M.R. Dunlap, and C. Liousse, Physical, chemical, and optical properties of regional hazes dominated by smoke in Brazil, *J. Geophys. Res.*, **103**, 32,059–32,080, 1998.

Reid, J.S., H.H. Jonsson, H.B. Maring, A. Smirnov, D.L. Savoie, S.S. Cliff, E.A. Reid, J.M. Livingston, M.M. Meier, O. Dubovik, and S.C. Tsay, Comparison of size and morphological measurements of coarse mode dust particles from Africa, *J. Geophys. Res.*, **108**, 8593, doi:10.1029/2002JD002485, 2003.

Reid, J.S., R. Koppmann, T.F. Eck, and D.P. Eleuterio, A review of biomass burning emissions. Part II: Intensive physical properties of biomass burning particles, *Atmos. Chem. Phys.*, **5**, 799–825, 2005.

Reisin, T., Z. Levin, and S. Tzivion, Rain production in convective clouds as simulated in an axisymmetric model with detailed microphysics. Part I: Description of the model, *J. Atmos. Sci.*, **53**, 497–519, 1996a.

Reisin, T., Z. Levin, and S. Tzivion, Rain production in convective clouds as simulated in an axisymmetric model with detailed microphysics. Part II: Effects of varying drops and ice nucleation, *J. Atmos. Sci.*, **53**, 1815–1837, 1996b.

Reisin, T., S. Tzivion, and Z. Levin, Seeding convective clouds with ice nuclei or hygroscopic particles: A numerical study using a model with detailed microphysics, *J. Appl. Meteor.*, **35**, 1416–1434, 1996c.

Reisner, J., R.M. Rasmussen, and R.T. Bruintjes, Explicit forecasting of supercooled water in winter storms using the MM5 mesoscale model, *Q. J. Roy. Meteorol. Soc.*, **124**, 1071–1107, 1998.

Remer, L.A, D. Tanré, Y.J. Kaufman, C. Ichoku, S. Mattoo, R. Levy, D.A. Chu, B. Holben, O. Dubovik, A. Smirnov, J.V. Martins,5 R.R. Li, and Z. Ahmad. Validation of MODIS aerosol retrieval over ocean, *Geophys. Res. Lett.*, **29**, 8008, doi:10.1029/2001GL013204, 2002a.

Remer, L.A., Y.J. Kaufman, Z. Levin, and S. Ghan, Model assessment of the ability of MODIS to measure top of atmosphere direct radiative forcing from smoke aerosols, *J. Atmos. Sci.*, **59**, 657–667, 2002b.

Remer, L A., Y.J. Kaufman, D.Tanré, S. Mattoo, D.A. Chu, J.V. Martins, R.R. Li, C. Ichoku, R.C. Levy, R.G. Kleidman, T.F. Eck, E. Vermote, and B.N. Holben, The MODIS aerosol algorithm, products and validation, *J. Atmos. Sci.*, **62**, 947–972, 2005.

Respondek, P.S., A.I. Flossmann, R.R. Alheit, and H.R. Pruppacher, A theoretical study of the wet removal of atmospheric pollutants. Part V: The uptake, redistribution and deposition of $(NH_4)_2SO_4$ by a convective cloud containing ice, *J. Atmos. Sci.*, **52**, 2121–2132, 1995.

Reynolds, O., On the manner in which raindrops (snowflakes) and hailstones are formed, *Proc. Manchester Lit. Phil. Soc.*, **16**, 23–33, 1877. (Also, **17**, 15–24, 1878; **6**, 48–60, and 161–170, 1879; also *Nature*, **15**, 163–165, 1877 and **17**, 207–209, 1878.)

Reynolds, D.W., A report on winter snowpack-augmentation, *Bull. Amer. Meteor. Soc.*, **69**, 1290–1300, 1988.

Reynolds, D.W., and A.S. Dennis, A review of the Sierra Cooperative Pilot Project, *Bull. Amer. Meteor. Soc.*, **67**, 513–523, 1986.

Richards, F., and P.A. Arkin, On the relationship between satellite-observed cloud cover and precipitation, *Mon. Wea. Rev.*, **109**, 1081–1093, 1981.

Riédi, J., M. Doutriaux-Boucher, P. Goloub, and P. Couvert, Global distribution of cloud top phase from POLDER/ADEOS I, *Geophys. Res. Lett.*, **27**, 1707–1710, 2000.

Riemer, N., H. Vogel, and B. Vogel, Soot aging time scales in polluted regions during day and night, *Atmos. Chem. Phys.*, **4**, 1885–1893, 2004.

Rinehart, R., *Radar for Meteorologists*, Fourth Edition, p. 482, Rinehart Publication, Columbia, MO, USA, ISBN:0-9658002-1-0, 2004.

Rissler, J., E. Swietlicki, J. Zhou, G. Roberts, M.O. Andreae, L.V. Gatti, and P. Artaxo, Physical properties of the sub-micrometer aerosol over the Amazon rain forest during the wet-to-dry season transition – Comparison of modeled and measured CCN concentrations, *Atmos. Chem. Phys.*, **4**, 2119–2143, 2004.

Rissler, J., A. Vestin, E. Swietlicki, G. Fisch, J. Zhou, P. Artaxo, and M.O. Andreae, Size distribution and hygroscopic properties of aerosol particles from dry-season biomass burning in Amazonia, *Atmos. Chem. Phys.*, **6**, 471–491, 2006.

Rivera-Carpio, C.A., C.E. Corrigan, T. Novakov, J.E. Penner, C.F. Rogers, and J.C. Chow, Derivation of contributions of sulfate and carbonaceous aerosols to cloud condensation nuclei from mass size distributions, *J. Geophys. Res.*, **101**, 19,483–19,493, 1996.

Roach, W.T., On the effect of radiative exchange on the growth by condensation of a cloud or fog droplet, *Q. J. Roy. Meteorol. Soc.*, **102**, 361–372, 1976.

Roberts, P., and J. Hallett, A laboratory study of the ice nucleating properties of some mineral particulates, *Q. J. Roy. Meteorol. Soc.*, **94**, 25–34, 1968.

Roberts, G.C., and A. Nenes, A continuous-flow streamwise thermal-gradient CCN chamber for atmospheric measurements, *Aerosol Sci. Tech.*, **39**, 206–221, 2005.

Roberts, G.C., M.O. Andreae, J. Zhou, and P. Artaxo, Cloud condensation nuclei in the Amazon Basin: "Marine" conditions over a continent? *Geophys. Res. Lett.*, **28**, 2807–2810, 2001.

Roberts, G.C., P. Artaxo, J. Zhou, E. Swietlicki, and M.O. Andreae, Sensitivity of CCN spectra on chemical and physical properties of aerosol: A case study from the Amazon Basin, *J. Geophys. Res.*, **107**, 8070, doi:10.1029/2001JD000583, 2002.

Rodriguez, M.A., and D. Dabdub, IMAGES-SCAPE2: A modeling study of size- and chemically resolved aerosol thermodynamics in a global chemical transport model, *J. Geophys. Res.*, **109**, D02203, doi:10.1029/2003JD003639, 2004.

Roeckner, E., P. Stier, J. Feichter, S. Kloster, M. Esch, and I. Fischer-Bruns, Impact of carbonaceous aerosol emissions on regional climate change, *Clim. Dynamics*, doi:10.1007/s00382-006-0147-3, 2006.

Roelofs, G.J., Drop size dependent sulfate distribution in a growing cloud, *J. Atmos. Chem.*, **14**, 109–118, 1992.

Roelofs, G.J., A cloud chemistry sensitivity study and comparison of explicit and bulk cloud model performance, *Atmos. Environ.*, **27A**, 2255–2264, 1993.

Roelofs, G.J., P. Kasibhatla, L. Barrie, D. Bergmann, C. Bridgeman, M. Chin, J. Christensen, R. Easter, J. Feichter, A. Jeuken, E. Kjellstrom, D. Koch, C. Land, U. Lohmann, and P. Rasch, Analysis of regional budgets of sulfur species modeled for the COSAM exercise, *Tellus*, **53B**, 673–694, 2001.

Rogers, D.C., Measurements of natural ice nuclei with a continuous flow diffusion chamber, *Atmos. Res.*, **29**, 209–228, 1993.

Rogers, R.R., and M.K. Yau, *A Short Course in Cloud Physics*, 304 pp., Pergamon, Tarrytown, N.Y., 1989.

Rogers, C.F., J.G. Hudson, B. Zielinska, R.L. Tanner, J. Hallett, and J.G. Watson, Cloud condensation nuclei from biomass burning, in *Global Biomass Burning: Atmospheric, Climatic and Biospheric Implications*, edited by J.S. Levine, pp. 431–438, MIT Press, Cambridge, MA, 1991.

Rogers, D.C., P.J. DeMott, S.M. Kreidenweis, Y. Chen, Measurements of ice nucleating aerosols during SUCCESS, *Geophys. Res. Letters*, **25**, 9, 1383–1386, 1998.

Rogers, D.C., P.J. DeMott, and S.M. Kreidenweis, Airborne measurements of tropospheric ice-nucleating aerosol particles in the Arctic spring, *J. Geophys. Res.*, **106**, 15,053–15,063, 2001.

Rokicki, M.L., and K.C. Young, The initiation of precipitation in updrafts, *J. Appl. Meteor.*, **17**, 745–754, 1978.

Rosenfeld, D., Comment on "A new look at the Israeli cloud seeding experiments," *J. Appl. Meteor.*, **36**, 260–271, 1997.

Rosenfeld, D., TRMM observed first direct evidence of smoke from forest fires inhibiting rainfall, *Geophys. Res. Lett.*, **26**, 3105–3108, 1999.

Rosenfeld, D., Suppression of rain and snow by urban and industrial air pollution, *Science*, **287**, 1793–1796, 2000.

Rosenfeld, D., and G. Feingold, Explanation of discrepancies among satellite observations of the aerosol indirect effects, *Geophys. Res. Lett.*, **30**, 1766, doi:10.1029/2003GL017684, 2003.

Rosenfeld, D., and A. Givati, Evidence of orographic precipitation suppression by air pollution induced aerosols in the Western US, *J. Appl. Meteor. and Climat.*, **45**, 893–911, 2006.

Rosenfeld, D., and G. Gutman, Retrieving microphysical properties near the tops of potential rain clouds by multispectral analysis of AVHRR data, *Atmos. Res.*, **34**, 259–283, 1994.

Rosenfeld, D., and I.M. Lensky, Satellite-based insights into precipitation formation processes in continental and maritime convective clouds, *Bull. Amer. Meteor. Soc.*, **79**, 2457–2476, 1998.

Rosenfeld, D., and W.L. Woodley, Effects of cloud seeding in west Texas, *J. Appl. Meteor.*, **28**, 1050–1080, 1989.

Rosenfeld, D., and W.L. Woodley, Effects of cloud seeding in west Texas: Additional results and new insights, *J. Appl. Meteor.*, **32**, 1848–1866, 1993.

Rosenfeld, D., and W.L. Woodley, Deep convective clouds withs sustaine highly supercooled liquid water until −37.5°C, *Nature*, **405**, 440–442, 2000.

Rosenfeld, D., Y. Rudich, and R. Lahav, Desert dust suppressing precipitation: A possible desertification feedback loop, *Proc. Nat. Acad. Sci.*, **98**, 5975–5980, 2001.

Rosenfeld, D., R. Lahav, A.P. Khain, and M. Pinsky, The role of sea spray in cleansing air pollution over ocean via cloud processes, *Science*, **297**, 1667–1670, 2002.

Rosenfeld, D., I.M. Lensky, J. Peterson, A. Gingis, Potential impacts of air pollution aerosols on precipitation in Australia, *Clean Air and Environ. Quality*, **40**(2), 43–49, 2006a.

Rosenfeld, D., W.L. Woodley, T.W. Krauss, V. Makitov, Aircraft microphysical documentation from cloud base to anvils of hailstorm feeder clouds in argentina, *J. Appl. Meteor.*, **45**, 1261–1281, 2006b.

Ross, K.E., S.J. Piketh, R.T. Bruintjes, R.P. Burger, R.J. Swap, and H.J. Annegarn, Spatial and seasonal variations in CCN distribution and the aerosol-CCN relationship over southern Africa, *J. Geophys. Res.*, **108**, 8481, doi:10.1029/2002JD002384, 2003.

Rossow, W.B., and R.A. Schiffer, ISCCP cloud data products, *Bull. Amer. Meteor. Soc.*, **72**, 2–20, 1991.

Rossow, W.B., A.W. Walker, and L.C. Gardner, Comparisons of ISCCP and other cloud amounts, *J. Clim.*, **6**, 2394–2418, 1993.

Rotstayn, L.D., and Y.G. Liu, A smaller global estimate of the second indirect aerosol effect, *Geophys. Res. Lett.*, **32**, L05708, doi:10.1029/2004GL021922, 2005.

Rotstayn, L.D., and U. Lohmann, Tropical rainfall trends and the indirect aerosol effect, *J. Clim.*, **15**, 2103–2116, 2002.

Rotstayn, L.D., B.F. Ryan, and J.E. Penner, Precipitation changes in a GCM resulting from the indirect effects of anthropogenic aerosols, *Geophys. Res. Lett.*, **27**, 3045–3048, 2000.

Rudich, Y., A. Sagi, D. Rosenfeld, Influence of the Kuwait oil fires plume (1991) on the microphysical development of clouds, *J. Geophys. Res.*, **108**, 4478, doi:10.1029/2003JD003472, 2003.

Rudich, Y., O. Khersonsky, and D. Rosenfeld, Treating clouds with a grain of salt, *Geophys. Res. Lett.*, **29**, 2060, doi:10.1029/2002GL016055, 2002.

Rudolf, B., H. Hauschild, W. Rueth, and U. Schneider, Terrestrial precipitation analysis: operational method and required density of point measurements, in *Global Precipitations and Climate Change*, edited by M. Desbois and G. Desalmond, pp. 173–186, NATO ASI Series I, 26, Springer, Verlag, 1994.

Ruijgrok, W., C.I. Davidson, and K.W. Nicholson, Dry deposition of particles: Implications and recommendations for mapping of deposition over Europe, *Tellus*, **47B**, 587–601, 1995.

Russell, L.M., Aerosol organic-mass-to-organic-carbon ratio measurements, *Environ. Sci. Technol.*, **37**, 2982–2987, 2003.

Ryan, B.F., and W.D. King. A critical review of the Australian experience in cloud seeding, *Bull. Amer. Meteor. Soc.* **78**, 239–354, 1997.

Saleeby, S.M., and W.R. Cotton, A large-droplet mode and prognostic number concentration of cloud droplets in the Colorado State University Regional Atmospheric Modeling System (RAMS). Part I: Module descriptions and supercell test simulations, *J. Appl. Meteorol.*, **43**, 182–195, 2004.

Saleeby, S.M., and W.R. Cotton, A large-droplet mode and prognostic number concentration of cloud droplets in the Colorado State University Regional Atmospheric Modeling System (RAMS). Part II: Sensitivity to a Colorado winter snowfall event, *J. Appl. Meteorol.*, **44**, 1912–1929, 2005.

Sassen, K., Dusty ice clouds over Alaska, *Nature*, **434**, 456, 2005.

Satake, S., I. Uno, T. Takemura, G.R. Carmichael, Y. Tang, D. Streets, N. Sugimoto, A. Shimizu, M. Uematsu, J.S. Han, and S. Ohta, Characteristics of Asian aerosol transport simulated with a regional-scale chemical transport model during the ACE-Asia observation, *J. Geophys. Res. D: Atmos.*, **109**, D19S22, 11–16, 2004.

Sax, R.I., S.A. Changnon, L.O. Grant, W.F. Hitchfield, P.V. Hobbs, A.M. Kahan, and J.S. Simpson, Weather modification: Where are we now and where are we going? An editorial overview, *J. Appl. Meteor.*, **14**, 652–672, 1975.

Saxena, P., and L.M. Hildemann, Water-soluble organics in atmospheric particles: A critical review of the literature and application of thermodynamics to identify candidate compounds, *J. Atmos. Chem.*, **24**, 57–109, 1996.

Saxena, P., L.M. Hildemann, P.H. McMurry, and J.H. Seinfeld, Organics alter hygroscopic behavior of atmospheric particles, *J. Geophys. Res.*, **100**, 18,755–718,770, 1995.

Schaap, M., H. Van Der Gon, F.J. Dentener, A.J.H. Visschedijk, M. Van Loon, H.M. ten Brink, J.P. Putaud, B. Guillaume, C. Liousse, and P.J.H. Builtjes, Anthropogenic black carbon and fine aerosol distribution over Europe, *J. Geophys. Res.*, **109**, D18207, doi:10.1029/2003JD004330, 2004.

Schemenauer, R.S., and A.A. Tsonis, Comments on "Physical Interpretation of Results from the HIPLEX-1 Experiment," *J. Appl. Meteor.*, **24**, 1269–1274, 1985.

Schery, S.D., and S. Whittlestone, Evidence of high deposition of ultrafine particles at Mauna Loa Observatory, *Atmos. Environ.*, **29**, 3319–3324, 1995.

Schmid, O., P. Artaxo, W.P. Arnott, D. Chand, L.V. Gatti, G.P. Frank, A. Hoffer, M. Schnaiter, and M.O. Andreae, Spectral light absorption by ambient aerosols influenced by biomass burning in the Amazon Basin. I: Comparison and field calibration of absorption measurement techniques, *Atmos. Chem. Phys.*, **6**, 3443–3462, 2006.

Schnell, R.C. and G. Vali, Biogenic ice nuclei. Part I: Terrestrial and marine sources, *J. Atmos . Sci.*, **33**, 1554–1564, 1976.

Schüller, L., J.L. Brenguier, and H. Pawlowska, Retrieval of Microphysical, Geometrical and Radiative Properties of Marine Stratocumulus from Remote Sensing, PACE Topical Issue, *J. Geophys. Res.*, **108**, D15, 8631 10.1029, 2003.

Schüller, L., R. Bennartz, J.L. Brenguier, and J. Fischer: An algorithm for the retrieval of droplet number concentration and geometrical thickness of stratiform marine boundary layer clouds applied to MODIS Radiometric Observations, *J. Appl. Meteor.*, **44**, 28–38, 2005.

Schulz, M., G. de Leeuw, and Y. Balkanski, Sea-salt aerosol source functions and emissions, in *Emissions of Atmospheric Trace Compounds*, edited by C. Granier, P. Artaxo, and C.E. Reeves, pp. 333–359, Kluwer, Dordrecht, 2004.

Schumann, T., Large discrepancies between theoretical and field-determined scavenging coefficients, *J. Aerosol Sci.*, **20**, 1159–1162, 1989

Schwartz, S.E., F. Arnold, J.P. Blanchet, P.A. Durkee, D.J. Hofmann, W.A. Hoppel, M.D. King, A.A. Lacis, T. Nakajima, J.A. Ogren, and O.B. Toon, Group report: Connections between aerosol properties and forcing of climate, in *Aerosol Forcing of Climate*, edited by R.J. Charlson and J. Heintzenberg, pp. 251–280, John Wiley and Sons, Chichester; 1995.

Schwartz, S.E., Harsvardhan, and C.M. Benkovitz, Influence of anthropogenic aerosol on cloud optical depth and albedo shown by satellite measurements and chemical transport modeling, *Proc. Natl. Acad. Sci. USA*, **99**, 1784–1789, 2002.

Scott, W.D., and P.V. Hobbs, The formation of sulfate in water droplets, *J. Atmos. Sci.*, **24**, 54–57, 1967.

Scott, B.C., and P.V. Hobbs, A theoretical study of the evolution of mixed-phase cumulus clouds, *J. Atmos. Sci.*, 34, 812–826, 1977.

Segal, Y., A.P. Khain, M. Pinsky and D. Rosenfeld, Effects of hygroscopic seeding on raindrop formation as seen from simulations using a 2000-bin spectral cloud parcel model, *Atmos. Res.*, **71**, 3–34, 2004

Sehmel, G.A., Particle and gas dry deposition: A review, *Atmos. Environ. Part A: General Topics*, **14**, 983–1011, 1980.

Seifert, A. and K.D. Beheng, A double-moment parameterization for simulating autoconversion, accretion and self collection, *Atmos. Res.* **59–60**, 265–281, 2001.

Seifert, A. and K.D. Beheng, A two-moment cloud microphysics parameterization for mixed-phase clouds. Part I: Model Description, *Meteorol. and Atmos. Phys.*, **92**, 45–66, 2006a.

Seifert, A. and K.D. Beheng, A two-moment cloud microphysics parameterization for mixed-phase clouds. Part II: Maritime vs. continental deep convective storms, *Meteorol. and Atmos. Phys.*, **92**, 67–82, 2006b.

Seifert, A., A.P. Khain, A. Pokrovsky and K. Beheng, A comparison of spectral bin and two-moment bulk mixed-phase microphysics, *Atmos. Res*, **80**, 44–66, 2006.

Seinfeld, J.H., and S.N. Pandis, *Atmospheric Chemistry and Physics: From Air Pollution to Climate Change*, 1326 p., John Wiley, New York, 1998.

Seinfeld, J.H., and J.F. Pankow, Organic atmospheric particulate material, *Annu. Rev. Phy. Chem.*, **54**, 121–140, 2003.

Seinfeld, J.H., G.R. Carmichael, R. Arimoto, W.C. Conant, F.J. Brechtel, T.S. Bates, T.A. Cahill, A.D. Clarke, S.J. Doherty, P.J. Flatau, B.J. Huebert, J. Kim, K.M. Markowicz, P.K. Quinn, L.M. Russell, P.B. Russell, A. Shimizu, Y. Shinozuka, Y.T.C.H. Song, I. Uno, A.M. Vogelmann, R.J. Weber, J.H. Woo, and X.Y. Zhang, ACE-ASIA: Regional Climatic and Atmospheric Chemical Effects of Asian Dust and Pollution, *Bull. Amer. Meteor. Soc.*, **85**, 367–380, 2004a.

Seinfeld, J.H., R.A. Kahn, T.L. Anderson, R.J. Charlson, R. Davies, D.J. Diner, J.A. Ogren, S.E. Schwartz, and B.A. Wielicki, Scientific objectives, measurement needs, and challenges motivating the PARAGON aerosol initiative, *Bull. Amer. Meteor. Soc.*, **85**, 1501–1508, 2004b.

Sekiguchi, M., T. Nakajima, K. Suzuki, K. Kawamoto, A. Higurashi, D. Rosenfeld, I. Sano, and S. Mukai, A study of the direct and indirect effects of aerosols using global satellite data sets of aerosol and cloud parameters, *J. Geophys. Res.*, **108**, 4699, doi:10.1029/2002JD003359, 2003

Sevruk, B., *Methods of Correction for Systematic Error in Point Precipitation Measurement for Operational Use*, Operational Hydrology Report No. 21, WMO – No. 589, 35 pp., Secretariat of the WMO, Geneva, 1982.

Shantz, N.C., W.R. Leaitch, and P.F. Caffrey, Effect of organics of low solubility on the growth rate of cloud droplets, *J. Geophys. Res.*, **108**, 4168, doi:10.1029/2002JD002540, 2003.

Shaw, R.A., and D. Lamb, Experimental determination of the thermal accommodation and condensation coefficients of water, *J. Chem. Phys.*, **111**, 10,659–10,663, 1999.

Shaw, R.A., W.C. Reade, L.R. Collins, and J. Verlinde, Preferential concentration of cloud droplets by turbulence: Effects on the early evolution of cumulus cloud droplet spectra, *J. Atmos. Sci.*, **55**, 1965–1976, 1998.

Shimizu, A., N. Sugimoto, and I. Matsui, Seasonal and Inter-Annual Variations of Vertical Aerosol Distribution Observed in Thailand, Presented at the 23rd International Laser Radar Conference, July 2006 Nara, Japan (23ILRC, ISBN:4-9902916-0-3), 801–804, 2006.

Shinozuka, Y., A.D. Clarke, S.G. Howell, V.N. Kapustin, and B.J. Huebert, Sea-salt vertical profiles over the Southern and tropical Pacific oceans: Microphysics, optical properties, spatial variability, and variations with wind speed, *J. Geophys. Res.*, **109**, D24201, doi:10.1029/2004JD004975, 2004.

Short, D.A., P.A. Kucera, B.S. Ferrier, J.C. Gerlach, S.A. Rutledge, and O.W. Thiele, Shipboard radar rainfall patterns within the TOGA COARE IFA, *Bull. Amer. Meteor. Soc.*, **78**, 2817–2836, 1997.

Shrestha, A.B., C.P. Wake, and J.E. Dibb, Chemical composition of aerosol and snow in the high Himalaya during the summer monsoon season, *Atmos. Environ.*, **31**, 2815–2826, 1997.

Shulman, M.L., M.C. Jacobson, R.J. Charlson, R.E. Synovec, and T.E. Young, Dissolution behavior and surface tension effects of organic compounds in nucleating cloud droplets, *Geophys. Res. Lett.*, **23**, 277–280, 1996.

Sicre, M.A., J.C. Marty, and A. Saliot, N-alkanes, fatty-acid esters, and fatty-acid salts in size fractionated aerosols collected over the Mediterranean Sea, *J. Geophys. Res.*, **95**, 3649–3657, 1990.

Sievering, H., The dry deposition of small particles: A review of experimental measurement, *Atmos. Environ.*, **23**, 2863–2864, 1989.

Silverman, B.A., Static mode seeding of summer cumuli-a review, in *Rainfall Enhancement – A Scientific Challenge, AMS Meteor. Monogr.*, **21**, 7–24, Amer. Meteor. Soc., Boston, MA, 1986.

Silverman, B.A., An independent statistical reevaluation of the South African hygroscopic flare seeding experiment, *J. Appl. Meteor.*, **39**, 1373–1378, 2000.

Silverman, B.A., A critical assessment of glaciogenic seeding of convective clouds for rain enhancement, *Bull. Amer. Meteor. Soc.*, **82**, 903–924, 2001.

Silverman, B.A., A critical assessment of hygroscopic seeding of convective clouds for rainfall enhancement, *Bull. Amer. Meteor. Soc.*, **84**, 1219–1230, 2003.

Silverman, B.A., and W. Sukarnjanasat, Results of the Thailand warm-cloud hygroscopic particle seeding experiment, *J. Appl. Meteor.*, **39**, 1160–1175, 2000.

Simoneit, B.R.T., J.N. Cardoso, and N. Robinson, An assessment of the origin and composition of higher molecular weight organic matter in aerosols over Amazonia, *Chemosphere*, **21**, 1285–1301, 1990.

Simpson, J., Comment on "Field experimentation in weather modification," *J. Amer. Statist. Assoc.*, **74**, 95–97, 1979.

Simpson, J., and W.L. Woodley, Florida area cumulus experiments 1970–1973 rainfall results, *Appl. Meteor.*, **14**, 734–744, 1975.

Simpson, J., N.E. Westcott, R.J. Clerman, and R.A. Pielke, On cumulus mergers. *Arch. Meteor. Geophy. Bioklim.*, Series A, **29**, 1–40, 1980.

Singh, N., N.A. Sontakke, H.N. Singh, and A.K. Pandey, Recent trend in spatiotemporal variation of rainfall over India-an investigation into basin-scale rainfall fluctuations, in *Regional Hydrological Impacts of Climatic Change-Hydroclimatic Variability*, edited by S. Franks, S.T. Wagener, E. Bøgh, H.V. Gupta, L. Bastidas, C. Nobre, and C.de Oliveira Galvão, IAHS Publ., **296**, 273–282, 2005.

Sinha, P., P.V. Hobbs, R.J. Yokelson, D.R. Blake, S. Gao, and T.W. Kirchstetter, Distributions of trace gases and aerosols during the dry biomass burning season in southern Africa, *J. Geophys. Res.*, **108**, 4536, doi:10.1029/2003JD003691, 2003.

Sirois, A., and L.A. Barrie, Arctic lower tropospheric aerosol trends and composition at Alert, Canada: 1980–1995, *J. Geophys. Res.*, **104**, 11,599–11,618, 1999.

Slinn, W.G.N., Precipitation scavenging, in *Atmospheric Science and Power Production*, edited by D. Randerson, pp. 466–532, Document DOE/TIC-27601, Technical Information Center, Office of Scientific and Technical Information, U.S. Department of Energy, 1984.

Smith, P.L., A.S. Dennis, B.A. Silverman, A.B. Super, E.W. Holroyd, W.A. Cooper, P.W. Mielke, K.J. Berry, H.D. Orville, and J.R. Miller, HIPLEX-1: Experimental design and response variables, *J. Clim. Appl. Meteor.*, **23**, 497–512, 1984.

Smith, D.M., M.S. Akhter, J.A. Jassim, C.A. Sergides, W.F. Welch, and A.R. Chughtai, Studies of the structure and reactivity of soot, *Aerosol Sci. Tech.*, **10**, 311–325, 1989.

Smith, L.R. Johnson, D.L. Priegnitz, B.A. Boe, and P.J. Mielke, Jr., An exploratory analysis of crop hail insurance data for evidence of cloud seeding effects in North Dakota, *J. Appl. Meteor.*, **36**, 463–73, 1997.

Smith, E.A, J.E. Lamm, R. Adler, J. Alishouse, K. Aonashi, E. Barrett, P. Bauer, W. Berg, A. Chang, R. Ferraro, J. Ferriday, S. Goodman, N. Grody, C. Kidd, D. Kniveton, C. Kummerow, G. Liu, F. Marano, A. Mugnai, W. Olson, G. Petty, A. Shibata, R. Spencer, F. Wentz, T. Wilheit, and E. Zipser, Results of the WetNet PIP-2 project, *J. Atmos. Sci.*, **55**, 1483–1536, 1998.

Smith, S.J., H. Pitcher, and T.M.L. Wigley, Global and regional anthropogenic sulfur dioxide emissions, *Global Planet. Change*, **29**, 99–119, 2001.

Smith, R.B., Q. Jiang, M.G. Fearon, P. Tabary, M. Dorninger, J.D. Doyle, and R. Beniot, Orographic precipitation and air mass transformation: An Alpine example, *Q. J. Roy. Meteor. Soc.*, **129**, 433–454, 2003.

Smith, R.B., I. Barstad, L. Bonneau, Orographic precipitation and Oregon's climate transition, *J. Atmos. Sci.*, **62(1)**, 177–191, 2005.

Smith, R.B. and J.P. Evans, Orographic precipitation and water vapour fractionation over the southern Andes, *J. Hydrometeor.*, **8**, 3–19, 2007.

Snider, J.R., and J.L. Brenguier, Cloud condensation nuclei and cloud droplet measurements during ACE-2, *Tellus*, **52B**, 828–842, 2000.

Snider, J.R., S. Guibert, and J.L. Brenguier, Aerosol Activation in Marine Stratocumulus Clouds, Part-II: Köhler and Parcel Theory Closure Studies, PACE Topical Issue, *J. Geophys. Res.*, **108**, D15, 8629 10.1029, 2003.

Spencer, R.W., D.W. Martin, B. Hinton and J.A. Weinman, Satellite microwave radiances coorelated with radar rain rates over land, *Nature*, **304**, 141–143, 1983.

Squires, P., The microstructure and colloidal stability of warm clouds. I. The relation between structure and stability, *Tellus*, **10**, 256–271, 1958.

Squires, P., and T. Twomey, The relation between cloud drop numbers and the spectrum of cloud nuclei, in *Physics of Precipitation, Monograph, No. 5, Amer. Geophys.*, Union., Washington, DC, 211–219, 1961.

Squires, P., and S. Twomey, A comparison of cloud nucleus measurements over central North America and the Caribbean Sea, *J. Atmos. Sci.*, **23**, 401–404, 1966.

Steiner, M., J.A. Smith, S.J. Burges, C.V. Alonso, and R.W. Darden, Effect of bias adjustment and rain gauge data quality control on radar rainfall estimation, *Water Resour. Res.*, **35**, 2487–2503, 1999.

Steiner, M., T.L. Bell, Y. Zhang, and E.F. Wood, Comparison of two methods for estimating the sampling-related uncertainty of satellite rainfall averages based on a large radar data set, *J. Clim.*, **16**, 3759–3778, 2003.

Stephens, G.L., Radiation profiles in extended water clouds. II: Parameterization schemes, *J. Atmos. Sci.*, **35**, 2123–2132, 1978.

Stephens, G.L., D.G. Vane, R.J. Boain, G.G. Mace, K. Sassen, Z. Wang, A.J. Illingworth, E.J. O'Connor, W.B. Rossow, S.L. Durden, S.D. Miller, R.T. Austin, A. Benedetti, C. Mitrescu, and the CloudSat Science Team, The Cloud-SAT mission and the A-Train: A new dimension of space-based observations of clouds and precipitation, *Bull. Amer. Meteorol. Soc.*, **83**, 1771–1790, 2002.

Stevens, B., G. Feingold, W.R. Cotton, and R.L. Walko, Elements of the microphysical structure of numerically simulated stratocumulus, *J. Atmos. Sci.*, **53**, 980–1006, 1996.

Stevens, B., W.R. Cotton, and G. Feingold, A critique of one- and two-dimensional models of boundary layer clouds with a binned representations of drop microphysics, *Atmos. Res.*, **47–48**, 529–553, 1998.

Stevens, B., D.H. Lenschow, G. Vali, H. Gerber, A. Bandy, B. Blomquist, J.L. Brenguier, C.S. Bretherton, F. Burnet, T. Campos, S. Chai, I. Faloona, D. Friesen, S. Haimov,

K. Laursen, D.K. Lilly, S. Loehrer, S.P. Malinowski, B. Morley, M.D. Petters, Dynamics and chemistry of marine stratocumulus -DYCOMS-II, 2003, *Bull. Amer. Meteorol. Soc.*, **84**, 579–593, 2003.

Stevens, B., G. Vali, K. Comstock, M.C. van Zanten, P.H. Austin, C.S. Bretherton and D.H. Lenschow, Pockets of Open Cells (POCs) and Drizzle in Marine Stratocumulus, *Bull. Amer. Meteorol. Soc.*, **86**, 51–57, 2005.

Stewart, R.E., Precipitation types in winter storms, *Pure and Appl. Geophys.*, **123**, 597–609, 1985.

Stier, P., J. Feichter, S. Kinne, S. Kloster, E. Vignati, J. Wilson, L. Ganzeveld, I. Tegen, M. Werner, Y. Balkanski, M. Schulz, O. Boucher, A. Minikin, and A. Petzold, The aerosol-climate model ECHAM5-HAM, *Atmos. Chem. Phys.*, **5**, 1125–1156, 2005.

Stier, P., J. Feichter, S. Kloster, E. Vignati, and J. Wilson, Emission-induced nonlinearities in the global aerosol system: Results from the ECHAM5-HAM aerosol-climate model, *J. Clim.*, **19**, 3845–3862, 2006.

Stith, J.L., L.F. Radke, and P.V. Hobbs, Particle emissions and the production of ozone and nitrogen oxides from the burning of forest slash, *Atmos. Environ.*, **15**, 73–82, 1981.

Stith, J.L., D.A. Burrows, P.J. DeMott, Initiation of ice: Comparison of numerical model results with observations of ice development in a cumulu cloud, *Atmos. Environ.*, **32**, 13–30, 1994.

Storelvmo, T., J.E. Kristjansson, S.J. Ghan, A. Kirkevåg, Ø. Seland and T. Iversen, Predicting cloud droplet number concentration in CAM-Oslo, *J. Geophys. Res.*, **111**, D24208, doi:10.1029/2005JD006300, 2006.

Strangeways, I.C., Back to basics: The "met. enclosure": Part 2(b)–Rain gauges, their errors, *Weather.*, **51**, 298–303, 1996.

Strapp, J.W., J. Oldenburg, R. Ide, L. Lilie, S. Bacic, Z. Vukovic, M. Oleskiw, D. Miller, E. Emery and G. Leone, Wind tunnel measurements of the response of hot-wire liquid water content instruments to large droplets, *J. Atmos. Oceanic Tech.*,. **20**, 791–806, 2003.

Strawbridge, K.B., and R.M. Hoff, LITE validation experiment along California's coast: Preliminary results, *Geophy. Res. Lett.*, **23**, 73–76, 1996.

Strawbridge, K.B., and B.J. Snyder, Planetary boundary layer height determination during Pacific 2001 using the advantage of a scanning lidar instrument, *Atmos Environ.*, **38**, 5861–5871, 2004.

Streets, D.G., and S.T. Waldhoff, Present and future emissions of air pollutants in China: SO_2, NO_x, and CO, *Atmos. Environ.*, **34**, 363–374, 2000.

Streets, D.G., N.Y. Tsai, H. Akimoto, and K. Oka, Sulfur dioxide emissions in Asia in the period 1985–1997, *Atmos. Environ.*, **34**, 4413–4424, 2000.

Streets, D.G., S. Gupta, S.T. Waldhoff, M.Q. Wang, T.C. Bond, and B. Yiyun, Black carbon emissions in China, *Atmos. Environ.*, **35**, 4281–4296, 2001.

Streets, D.G., K.F. Yarber, J.H. Woo, and G.R. Carmichael, Biomass burning in Asia: Annual and seasonal estimates and atmospheric emissions, *Global Biogeochem. Cycles*, **17**, 1099, doi:10.1029/2003GB002040, 2003.

Streets, D.G., T.C. Bond, T. Lee, and C. Jang, On the future of carbonaceous aerosol emissions, *J. Geophys. Res.*, **109**, D24212, doi:10.1029/2004JD004902, 2004.

Strom, J., and S. Ohlsson, In situ measurements of enhanced crystal number densities in cirrus clouds caused by aircraft exhaust, *J. Geophys. Res.*, **103**, 11,355–11,361, 1998.

Ström, J., B. Strauss, T. Anderson, F. Schröder, J. Heintzenberg, and P. Wendling, In situ observations of the microphysical properties of young cirrus clouds, *J. Atmos. Sci.*,. **54**, 2542–2553, 1997.

Sullivan, R.C., and K.A. Prather, Recent advances in our understanding of atmospheric chemistry and climate made possible by on-line aerosol analysis instrumentation, *Anal. Chem.*, **77**, 3861–3885, 2005.

Sun, J.M., and P.A. Ariya, Atmospheric organic and bio-aerosols as Cloud Condensation Nuclei (CCN): A review, *Atmos. Environ.*, **40**, 795–820, 2006.

Super, A.B., Further exploratory analysis of the bridger range winter cloud seeding experiment, *J. Appl. Meteor.*, **25**(12),1926–1933, 1986.

Super, A.B., and B.A. Boe, Microphysical effects of wintertime cloud seeding with silver iodide over the Rocky Mountains. Part III: Observations over the Grand Mesa, Colorado, *J. Appl. Meteor.*, **27**, 1166–1182, 1988.

Super, A.B., and J.A. Heimbach. Evaluation of the bridger range winter cloud seeding experiment using control gages, *J. Appl. Meteor.*, **22**(12),1989–2011, 1983.

Super, A.B., and J.A. Heimbach. Microphysical effects of wintertime cloud seeding with silver iodide over the rocky mountains. Part II: Observations over the Bridger Range, Montana, *J. Appl. Meteor.*, **27**, 1152–1165, 1988.

Super, A.B., B.A. Boe, and E.W. Holroyd, Microphysical effects of wintertime cloud seeding with silver iodide over the Rocky Mountains. Part I: Experimental design and instrumentation, *J. Appl. Meteor.*, **27**, 1145–1151, 1988.

Svenningsson, B., J. Rissler, E. Swietlicki, M.M., M. Bilde, M.C. Facchini, S. Decesari, S. Fuzzi, J. Zhou, J. Monster, and T. Rosenorn, Hygroscopic growth and critical supersaturations for mixed aerosol particles of inorganic and organic compounds of atmospheric relevance, *Atmos. Chem. Phys.*, **6**, 1937–1952, 2006.

Sverdrup, G.M., K.T. Whitby, and W.E. Clark, Characterization of California aerosols. 2: Aerosol size distribution measurements in Mojave Desert, *Atmos. Environ.*, **9**, 483–494, 1975.

Swap, R.J., H.J. Annegarn, J.T. Suttles, M.D. King, S. Platnick, J.L. Privette, and R.J. Scholes, Africa burning: A thematic analysis of the Southern African Regional Science Initiative (SAFARI 2000), *J. Geophys. Res.*, **108**, 8465, doi:8410.1029/2003JD003747, 2003.

Szidat, S., T.M. Jenk, H.A. Synal, M. Kalberer, L. Wacker, I. Hajdas, A. Kasper-Giebl, and U. Baltensperger, Contributions of fossil fuel, biomass burning, and biogenic emissions to carbonaceous aerosols in Zas traced by ^{14}**C, J. Geophys. Res.**, **111**, D07206, doi:10.1029/2005JD006590, 2006.

Szyrmer, W., and I. Zawadzki, Biogenic and anthropogenic sources of ice-forming nuclei: A review, *Bull. Am. Meteorol. Soc.*, **78**, 209–228, 1997.

Tabazadeh, A., R.J. Yokelson, H.B. Singh, P.V. Hobbs, J.H. Crawford, and L.T. Iraci, Heterogeneous chemistry involving methanol in tropospheric clouds, *Geophys. Res. Lett.*, **31**, L06114, doi:10.1029/2003GL018775, 2004.

Takahashi, T., Hail in an axisymmetric cloud model, *J. Atmos. Sci.*, **33**, 1579–1601, 1976.

Takemura, T., T. Nozawa, S. Emori, T.Y. Nakajima, and T. Nakajima, Simulation of climate response to aerosol direct and indirect effects with aerosol transport-radiation model, *J. Geophys. Res.*, **110**, D02202, doi:10.1029/2004JD00502, 2005.

Talbot, R.W., M.O. Andreae, T.W. Andreae, and R.C. Harriss, Regional aerosol chemistry of the Amazon Basin during the dry season, *J. Geophys. Res.*, **93**, 1499–1508, 1988.

Talbot, R.W., M.O. Andreae, H. Berresheim, P. Artaxo, M. Garstang, R.C. Harriss, K.M. Beecher, and S.M. Li, Aerosol chemistry during the wet season in Central Amazonia: The influence of long-range transport, *J. Geophys. Res.*, **95**, 16,955–16,969, 1990.

Tang, I.N., Thermodynamic and optical properties of mixed-salt aerosols of atmospheric importance, *J. Geophys. Res.*, **102**, 1883–1893, 1997.

Tang, I.N., and H.R. Munkelwitz, Water activities, densities, and refractive indices of aqueous sulfates and sodium nitrate droplets of atmospheric importance, *J. Geophys. Res.*, **99**, 18,801–18,808, 1994.

Tanré, D., Y.J. Kaufman, M. Herman, and S. Mattoo, Remote sensing of aerosol over oceans from EOS-MODIS, *J. Geophys. Res.*, **102**, 16,971–16,988, 1997.

Taylor, G.R., Sulfate production and deposition in midlatitude continental cumulus clouds. Part II: Chemistry model formulation and sensitivity analysis, *J. Atmos. Sci.*, **46**, 1991–2007, 1989.

Taylor, J.P., and A. McHaffie, Measurements of cloud susceptibility, *J. Atmos. Sci.*, **51**, 1298–1306, 1994.

Tegen, I., Modeling the mineral dust aerosol cycle in the climate system, *Quaternary Sci. Rev.*, **22**, 1821–1834, 2003.

Tegen, I., P. Hollrig, M. Chin, I. Fung, D. Jacob, and J. Penner, Contribution of different aerosol species to the global aerosol extinction optical thickness: Estimates from model results, *J. Geophys. Res.*, **102**, 23,895–23,915, 1997.

Tegen, I., S.P. Harrison, K. Kohfeld, I.C. Prentice, M. Coe, and M. Heimann, Impact of vegetation and preferential source areas on global dust aerosol: Results from a model study, *J. Geophys. Res.*, **107**, 4576, doi:10.1029/2001JD000963, 2002.

Tegen, I., M. Werner, S.P. Harrison, and K.E. Kohfeld, Relative importance of climate and land use in determining present and future global soil dust emission, *Geophys. Res. Lett.*, **31**, L05105, doi:10.1029/2003GL019216, 2004.

Telford, J.W., A new aspect of coalescence theory, *J. Meteorol.*, **12**, 436–444, 1955.

Telford, J.W., and S.K. Chai, A new aspect of condensation theory, *Pure and App. Geophys.*, **118**, 720–742, 1980.

Telford, J.W., T.S. Keck, and S.K. Chai, Entrainment at cloud tops and the droplet spectra, *J. Atmos. Sci.*, **41**, 3170–3179, 1984.

Teller, A. and Z. Levin, The effects of aerosols on precipitation and dimensions of subtropical clouds: A sensitivity study using a numerical cloud model, *Atmos. Chem. and Phys.*, **6**, 67–80, 2006.

Ten Brink, H., W. Maenhaut, R. Hitzenberger, T. Gnauk, G. Spindler, A. Even, X.G. Chi, H. Bauer, H. Puxbaum, J.P. Putaud, J. Tursic, and A. Berner, INTERCOMP2000: The comparability of methods in use in Europe for measuring the carbon content of aerosol, *Atmos. Environ.*, **38**, 6507–6519, 2004.

Tervahattu, H., K. Hartonen, V.M. Kerminen, K. Kupiainen, P. Aarnio, T. Koskentalo, A.F. Tuck, and V. Vaida, New evidence of an organic layer on marine aerosols, *J. Geophys. Res.*, **107**, 4053, doi:10.1029/2000JD000282, 2002a.

Tervahattu, H., J. Juhanoja, and K. Kupiainen, Identification of an organic coating on marine aerosol particles by TOF-SIMS, *J. Geophys. Res.*, **107**, 4319, doi:10.1029/2001JD001403, 2002b.

Tesmer, J.R., and T.T. Wilheit, An improved microwave radiative transfer model for tropical oceanic precipitation, *J. Atmos. Sci.*, **55**, 1674–1688, 1998.

Textor, C., H.F. Graf, C. Timmreck, and A. Robock, Emissions from volcanoes, in *Emissions of Atmospheric Trace Compounds*, edited by C. Granier, P. Artaxo, and C.E. Reeves, Kluwer, Dordrecht, 2004.

Textor, C., M. Schulz, S. Guibert, S. Kinne, Y. Balkanski, S. Bauer, T. Berntsen, T. Berglen, O. Boucher, M. Chin, F. Dentener, T. Diehl, R. Easter, H. Feichter, D. Fillmore, S. Ghan, P. Ginoux, S. Gong, A. Grini, J. Hendricks, L. Horowitz, P. Huang, I. Isaksen, I. Iversen, S. Kloster, D. Koch, A. Kirkevåg, J.E. Kristjansson, M. Krol, A. Lauer, J.F. Lamarque, X. Liu, V. Montanaro, G. Myhre, J. Penner, G. Pitari, S. Reddy, Ø. Seland, P. Stier, T. Takemura, and X. Tie, Analysis and quantification of the diversities of aerosol life cycles within AeroCom, *Atmos. Chem. Phys.*, **6**, 1777–1813, 2006.

Thiessen, A.H., Precipitation averages for large areas, *Mon. Wea. Rev.*, **39**, 1082–1084, 1911.

Thomson, A.D., and R. List, Raindrop spectra and updraft determination by combining Doppler radar and disdrometer, *J. Atmos. Ocean. Tech.*, **13**, 465–476, 1996.

Tobias, H.J., and P.J. Ziemann, Thermal desorption mass spectrometric analysis of organic aerosol formed from reactions of 1-Tetradecene and O3 in the presence of alcohols and carboxylic acids, *Environ. Sci. and Tech.*, **34**, 2105–2115, 2000.

Tompkins, A.M., C. Cardinali, J.J. Morcrette, and M. Rodwell, Influence of aerosol climatology on forecasts of the African Easterly Jet, *Geophy. Res. Lett.*, **32**, L10801, doi:10810.11029/12004GL022189, 2005.

Toon, O.B., R.P. Turco, D. Westphal, R. Malone, and M. Liu, A multidimensional model for aerosols: Description of computational analogs, *J. Atmos. Sci.*, **45**, 2123–2144, 1988.

Torres, O., P.K. Bhartia, J.R. Herman, A. Sinyuk, P. Ginoux, and B. Holben, A Long-term record of aerosol optical depth from TOMS observations and comparison to AERONET measurements, *J. Atmos. Sci.*, **59**, 398–413, 2002.

Tørseth, K., *Transboundary Particulate Matter in Europe*, Status Report 4/2004, EMEP Report., 2004.

Tørseth, K., L. Tarrasn, and M. Amann, *Transboundary Particulate Matter in Europe*, in EMEP Report 5/2002, 2002.

Trochkine, D., Y. Iwasaka, A. Matsuki, M. Yamada, Y.S. Kim, T. Nagatani, D. Zhang, G.Y. Shi, and Z. Shen, Mineral aerosol particles collected in Dunhuang, China, and their comparison with chemically modified particles collected over Japan, *J. Geophys. Res.*, **108**, 8642, doi:10.1029/2002JD003268, 2003.

Tsai, F., T.H. Liu, S.C. Liu, T.Y. Chen, T.L. Anderson, and S.J. Masonis, Model simulation and analysis of coarse and fine particle distributions during ACE-Asia, *J. Geophys. Res.*, **109**, D19S20, doi:10.1029/2003JD003665, 2004.

Tsigaridis, K., and M. Kanakidou, Global modelling of secondary organic aerosol in the troposphere: A sensitivity analysis, *Atmos. Chem. Phys.*, **3**, 1849–1869, 2003.

Tsigaridis, K., J. Lathière, M. Kanakidou, and D.A. Hauglustaine, Naturally driven variability in the global secondary organic aerosol over a decade, *Atmos. Chem. Phys.*, **5**, 1891–1904, 2005.

Tukey, J.W., L.V. Jones, and D.R. Brillinger, *The Management of Weather Resources, Volume I, Proposals for a National Policy and Program*, Report of the Statistical Task Force to the Weather Modification Advisory Board, Government Printing Office, 118 pp., 1978a.

Tukey, J.W., D.R. Brillinger, and L.V. Jones, *Report of the Statistical Task Force to the Weather Modification Advisory Board*, Volume II, U.S. Government Printing Office, pE-3, 1978b.

Turekian, V.C., S.A. Macko, and W.C. Keene, Concentrations, isotopic compositions, and sources of size-resolved, particulate organic carbon and oxalate in near-surface marine air at Bermuda during spring, *J. Geophys. Res.*, **108**, 4157, doi:10.1029/2002JD002053, 2003.

Turk, F.J., E.E. Ebert, H.J. Oh, B.J. Sohn, V. Levizzani, E.A. Smith, and R.R. Ferraro, Validation of an operational global precipitation analysis at short time scales, 12th *AMS Conf. Satellite Meteor. and Ocean.*, 9–13 February, Long Beach, CA, 2003.

Turpin, B.J., and H.J. Lim, Species contributions to PM2.5 mass concentrations: Revisiting common assumptions for estimating organic mass, *Aerosol Sci. and Tech.*, **35**, 602–610, 2001.

Turpin, B.J., P. Saxena, and E. Andrews, Measuring and simulating particulate organics in the atmosphere: Problems and prospects, *Atmos. Environ.*, **34**, 2983–3013, 2000.

Tustison, B., D. Harris, and E. Foufoula-Georgiou, Scale issues in verification of precipitation forecasts, *J. Geophys. Res.*, **106**, 11,775–11,784, 2001.

Twohy, C.H. and D. Rogers, Airflow and water-drop trajectories at instrument sampling points around the beechcraft king air and lockheed electra, *J. Atmos. and Oceanic Tech.*, **10**, 4, pp. 566–578, 1993

Twohy, C.H., A.J. Schanot, and W.A. Cooper, Measurement of condensed water content in liquid and ice clouds using an airborne counterflow virtual impactor, *J. Atmos. Oceanic Tech.*, **14**, 197–202, 1997.

Twohy, C.H., J.W. Strapp, and M. Wendisch, Performance of a counterflow virtual impactor in the NASA icing research tunnel, *J. Atmos. Oceanic Tech.*, **20**, 781–790, 2003.

Twomey, S., On the nature and origin of natural cloud nuclei, *Bull. Obs. de Puy de Dome*, **1**, 1–5, 1960.

Twomey, S., On the composition of cloud nuclei in the northeastern United States, *J. de Rech. Atmos.*, **3**, 281–285, 1968.

Twomey, S., Composition of cloud nuclei, *J. Atmos. Sci.*, **28**, 377–381, 1971.

Twomey, S., The influence of pollution on the shortwave albedo of clouds, *J. Atmos. Sci.*, **34**, 1149–1152, 1977.

Twomey, S., Aerosols, clouds and radiation, *Atmos. Environ.*, *25A*, 2435–2442, 1991.

Twomey, S., and T. Cocks, Remote sensing of cloud parameters from spectral reflectance measurements in the near-infrared, *Beitr. Phys. Atmos.*, **62**, 172–179, 1989.

Twomey, S., and K.A. Davidson, Automatic observations of cloud nucleus concentration, *J. Atmos. Sci.*, **27**, 1056–1059, 1970.

Twomey, S., and K.A. Davidson, Automatic observations of cloud nuclei, September 1969–August 1970, *J. Atmos. Sci.*, **28**, 1295–1296, 1972.

Twomey, S., and J. Warner, Comparison of measurements of cloud droplets and cloud nuclei, *J. Atmos. Sci.*, **24**, 702–703, 1967.

Twomey, S. and T.A. Wojciechowski, Observations of the geographical variation of cloud nuclei, *J. Atmos. Sci.*, **26**, 648–651, 1969.

Twomey, S., Davidson, K.A. and Seton, K.J., Results of five years' observations of cloud nucleus concentration at Roberson, New South Wales', *J. Atmos. Sci.*, **35**, 650–656, 1978

Tzivion, S., G. Feingold and Z. Levin, An efficient numerical solution to the stochastic collection equation, *J. Atmos. Sci.*, 44, 3139–3149, 1987.

U.S. CLIVAR Pan American Implementation Panel, U.S. CLIVAR Pan American Research: A Scientific Prospectus and Implementation Plan, U.S. CLIVAR Office, Washington, DC, 58 pp., 2002.

Uno, I., Z. Wang, M. Chiba, Y.S. Chun, S.L. Gong, Y. Hara, E. Jung, S.S. Lee, M. Liu, M. Mikami, S. Music, S. Nickovic, S. Satake, Y. Shao, Z. Song, N. Sugimoto, T. Tanaka, and D.L. Westphal, Dust Model Intercomparison (DMIP) study over Asia: Overview, *J. Geophys. Res.*, **111**, D12213, doi:12210.11029/12005JD006575, 2006.

Usher, C.R., C.A. Cleveland Jr., D.R. Strongin, and M.A. Schoonen, Reactions on Mineral Dust, *Chem. Rev.*, **103**, 4883–4939, 2003.

Vaillancourt, P.A., M.K. Yau, P. Bartello, and W.W. Grabowski, Microscopic approach to cloud droplet growth by condensation. Part II: Turbulence, clustering, and condensational growth, *J. Atmos. Sci.*, **59**, 3421–3435, 2002.

Vali, G., Atmospheric ice nucleation – A review, *J. Rech. Atmos.*, **19**, 105–115, 1985.

Vali, G., M. Christensen, R.W. Fresch, E.L. Galyan, L.R. Maki, and R.C. Schnell, Biogenic ice nuclei. II: Bacterial sources, *J.Atmos. Sci.*, **33**, 1565–1570, 1976.

van den Heever, S., and W.R. Cotton, Urban aerosol impacts on downwind convective storms, *J. Appl. Meteor. Climat.*, **46**, 828–850, 2007.

van den Heever, S.C., G.G. Carrio, W.R. Cotton, P.J. DeMott, and A.J. Prenni, Impacts of nucleating aerosol on Florida convection. Part I: Mesoscale Simulations, *J. Atmos. Sci.*, **63**, 1752–1775, 2006.

Van der Werf, G.R., J.T. Randerson, J. Collatz, and L. Giglio, Carbon emissions from fires in tropical and subtropical ecosystems, *Global Change Biology.*, **9**, 547–562, 2003.

Van Dingenen, R., F. Raes, J.P. Putaud, U. Baltensperger, A. Charron, M.C. Facchini, S. Decesari, S. Fuzzi, R. Gehrig, H.G. Hansson, R.M. Harrisonc, C.H. Uglin, A.M. Jones, P. Laj, G. Lorbeer, W. Maenhaut, F. Palmgren, X. Querol, S. Rodriguez, J. Schneider, H. ten Brink, P. Tunved, K. Torseth, B. Wehner, E. Weingartner, A. Wiedensohler, and P. Wahlin, A European aerosol phenomenology-1: physical characteristics of particulate matter at kerbside, urban, rural and background sites in Europe, *Atmos. Environ.*, **38**, 2561–2577, 2004.

Van Grieken, R., C. Xhoffer, L. Wouters, P. Artaxo, Micro-analysis techniques for the characterization of individual environmental particles, *Anal. Sci.*, **7**, 1117–1122, 1991.

van Zanten, M.C., Stevens, B., Vali, G., and Lenschow, D.H., Observations of drizzle in nocturnal marine stratocumulus, *J. Atmos. Sci.*, **62**, 88–106, 2005.

Vanbauce, C., J.C. Buriez, F. Parol, B. Bonnel, G. Seze G, and P. Couvert, Apparent pressure derived from ADEOS-POLDER observations in the oxygen A-band over ocean, *Geophys. Res. Lett.*, **25**, 3159–3162, 1998.

Vanbauce, C., B. Cadet, and R.T. Marchand, Comparison of POLDER apparent and corrected oxygen pressure to ARM/MMCR cloud boundary pressures, *Geophys. Res. Lett.*, **30**, 1212, 2003.

VanCuren, R.A., and T.A. Cahill, Asian aerosols in North America: Frequency and concentration of fine dust, *J. Geophy. Res. Atmos.*, **107**, 4804, doi:10.1029/2002JD002204, 2002.

VanReken, T.M., T.A. Rissman, G.C. Roberts, V. Varutbangkul, H.H. Jonsson, R.C. Flagan, and J.H. Seinfeld, Toward Aerosol/Cloud Condensation Nuclei (CCN) closure during CRYSTAL-FACE, *J. Geophys. Res.*, **108**, 4633, doi:10.1029/2003JD003582, 2003.

Vardiman, L., The generation of secondary ice particles in cloud crystal-crystal collisions, *J. Atmos. Sci.*, **35**, 2168–2180, 1978.

Varutbangkul, V., F.J., Brechtel, R., Bahreini, N.L., Ng, M.D., Keywood, J.H., Kroll, R.C., Flagan, J.H., Seinfeld, A., Lee, and A.H., Goldstein, Hygroscopicity of secondary organic aerosols formed by oxidation of cycloalkenes, monoterpenes, sesquiterpenes, and related compounds, *Atmos. Chem. Phys.*, **6**, 2367–2388, 2006.

Veefkind, J.P., G. de Leeuw, and P.A. Durkee, Retrieval of aerosol optical depth over land using two-angle view satellite radiometry during TARFOX, *Geophys. Res. Lett.*, **25**, 3135–3138, 1998.

Velders, G.J.M., Heijboer, L.C, Kelder, H., The simulation of the transport of aircraft emissions by a 3-dimensional global-model, *Annales Geophysicae-Atmospheres Hydrospheres and Space Sci.*, **12**(5), 385–393, 1994

Verlinde, J., P.J. Flatau, and W.R. Cotton, Analytical solutions to the collection growth equation: Comparison with approximate methods and application to cloud microphysics parameterization schemes, *J. Atmos. Sci.*, **47**, 2871–2880, 1990.

Vermote, E. and Y.J. Kaufman, Absolute calibration of AVHRR visible and near infrared channels using ocean and cloud views, *Int. J. Rem. Sens.*, **16**, 2317–2340, 1995.

Vet, R.J., J.R. Brook, T.F. Dann, and J. Dion, *The Nature of PM2.5 Mass, Composition and Precursors in Canada*, Meteorogical Service of Canada, Toronto, 2001.

Vignati, E., G. de Leeuw, and R. Berkowicz, Modeling coastal aerosol transport and effects of surf-produced aerosols on processes in the marine atmospheric boundary layer, *J. Geophys. Res.*, **106**, 20,225–20,238, 2001.

Vignati, E., J. Wilson, and P. Stier, M7: An efficient size-resolved aerosol microphysics module for large-scale aerosol transport models, *J. Geophys. Res. Atmos.*, **109**, 22202, doi:10.1029/2003JD004485, 2004.

Volken, M., and T. Schumann, A critical review of below-cloud aerosol scavenging results on Mt. Rigi, *Water, Air, and Soil Pollution.*, **68**, 15–28, 1993.

von Blohn, N., S.K. Mitra, K. Diehl, and S. Borrmann, The ice nucleating ability of pollen. Part III: New laboratory studies in immersion and contact freezing modes including more pollen types, *Atmos. Res.*, **78**, 182–189, 2005.

Walko, R.L, W.R. Cotton, J.L. Harrington, and M.P. Meyers, New RAMS cloud microphysics parameterization. Part I: The single-moment scheme, *Atmos. Res.*, **38**, 29–62, 1995.

Wallace, J.M. and P.V. Hobbs, *Atmospheric Science: An Introductory Science*, Second Edition, 504 pp., Academic Press, Burlington, 2006.

Wang, P.K., Moisture plumes above thunderstorm anvils and their contributions to cross-tropopause transport of water vapour in midlatitudes, *J. Geophys. Res.*, **108**, 4194, doi:10.1029/2002JD002581, 2003.

Wang, C., A modeling study on the climate impacts of black carbon aerosols, *J. Geophys. Res.*, **109**, D03106, doi:10.1029/2003JD004084, 2004.

Wang J.Y. and B. Geerts, Identifying drizzle within marine stratus with W-band radar reflectivity profiles, *Atmos. Res.*, **69**, 1–27, 2003.

Wang, T., K.B. Earnshaw, and R.S. Lawrence, Path-averaged measurements of rain rate and raindrop size distributions using a fast-response optical sensor, *J. Appl. Meteor.*, **18**, 654–660, 1979.

Wang, P.H., P. Minnis, M.P. McCormick, G.S. Kent, and K.M. Skeens, A 6-year climatology of cloud occurrence frequency from stratospheric aerosol and gas experiment II observations (1985–1990), *J. Geophys. Res.*, **101**, 29,407–29,429, 1996.

Wang, S., Q. Wang, and G. Feingold, Turbulence, condensation and liquid water transport in numerically simulated nonprecipitating stratocumulus clouds, *J. Atmos. Sci.*, **60**, 262–278, 2003.

Warner, J., A reduction of rain associated with smoke from sugar-cane fires – An inadvertent weather modification, *J. App. Meteor.*, **7**, 247–251, 1968.

Warner, J., Smoke from sugar-cane fires and rainfall, in Proceedings, International Conference on Weather Modification, *Canberra, Amer. Meteor. Soc.*, 191–192, 1971.

Warner, J., and S. Twomey, The production of cloud nuclei by cane fires and the effect on cloud droplet concentration, *J. Atmos. Sci.*, **24**, 704–706, 1967.

Warren, S.G., C.J. Hahn, J. London, R.M. Chervine and R.L. Jenne, Global distribution of total cloud cover and cloud type amounts over land, *NCAR Technical Note*, NCAR/TN-273 + STR, 29 pp., 1986a.

Warren, S.G., C.J. Hahn, J. London, R.M. Chervine and R.L. Jenne, Global distribution of total cloud cover and cloud type amounts over ocean, *NCAR Technical Note*, NCAR/TN-317 + STR, 42 pp., 1986b.

Weber, R.J., P.H. McMurry, L. Mauldin, D.J. Tanner, F.L. Eisele, F.J. Brechtel, S.M. Kreidenweis, G.L. Kok, R.D. Schillawski, and D. Baumgardner, A study of new particle formation and growth involving biogenic and trace gas species measured during ACE 1, *J. Geophys. Res.*, **103**, 16,385–16,396, 1998.

Webster, P.J., and R. Lukas, TOGA COARE: The coupled ocean-atmosphere response experiment, *Bull. Amer. Meteor. Soc.*, **73**, 1377–1416, 1992.

Wegener, A., *Thermodynamik der Atmosphäre*, J.A. Barth, Leipzig, 1911.

Wehrli, C., Calibration of filter radiometers for determination of atmospheric optical depth, *Metrologia*, **37**, 419–422, 2000.

Weingartner, E., H. Burtscher, and U. Baltensperger, Hygroscopic properties of carbon and diesel soot particles, *Atmos. Environ.*, **31**, 2311–2327, 1997.

Weingartner, E., S. Nyeki, and U. Baltensperger, Seasonal and diurnal variation of aerosol size distributions ($10<D<750$ nm) at a high-alpine site (Jungfraujoch 3580 m asl), *J. Geophys. Res.*, **104**, 26,809–26,820, 1999.

Weingartner, E., M. Gysel, and U. Baltensperger, Hygroscopicity of aerosol particles at low temperatures. 1. New low-temperature H-TDMA instrument: Setup and first applications, *Environ. Sci. Technol.*, **36**, 55–62, 2002.

Welch, H.E., D.C.G. Muir, B.N. Billeck, W.L. Lockhart, G.J. Brunskill, H.J. Kling, M.P. Olson, and R.M. Lemoine, Brown snow: A long range transport event in the Canadian Arctic, *Environ. Sci. Technol.*, **25**, 280–286, 1991.

Weller, R.A., F. Bradley, and R. Lukas, The interface or air-sea flux component of the TOGA coupled ocean-atmosphere response experiment and its impact on subsequent air-sea interaction studies, *J. Atmos. Ocean. Tech.*, **21**, 223–257, 2004.

Wendisch, M., T.J. Garrett, and J.W. Strapp, Wind tunnel tests of the airborne PVM-100A response to large droplets, *J. Atmos. Oceanic Tech.*, **19**, 1577–1584, 2002

Werner, M., I. Tegen, S.P. Harrison, K.E. Kohfeld, I.C. Prentice, Y. Balkanski, H. Rodhe, and C. Roelandt, Seasonal and interannual variability of the mineral dust cycle under present and glacial climate conditions, *J. Geophys. Res.*, **107**, 4744, doi:10.1029/2002JD002365, 2002.

Wesely, M.L., and B.B. Hicks, A review of the current status of knowledge on dry deposition, *Atmos. Environ.*, **34**, 2261–2282, 2000.

Wetzel, M.A, L.L. Stowe, Satellite-observed patterns in stratus microphysics, aerosol optical thickness, and shortwave radiative forcing, *J. Geophys. Res. Atmos.* **104**, D24, 31,287–31,299, 1999.

Wexler, A.S., and J.H. Seinfeld, Second-generation inorganic aerosol model, *Atmos. Environ.*, **25A**, 2731–2748, 1991.

Weymouth, G., G.A. Mills, D. Jones, E.E. Ebert, and M.J. Manton, A continental-scale daily rainfall analysis system, *Aust. Met. Mag.*, **48**, 169–179, 1999.

Whelpdale, D.M., and R. List, The coalescence process of raindrop growth, *J. Geophys. Res.*, **76**, 2836–2856, 1971.

Whitby, K.T., Physical characteristics of sulfur aerosols, *Atmos. Environ.*, **12**, 135–159, 1978.

Whitby, E.R., and P.H. McMurry, Modal aerosol dynamics modeling, *Aerosol Sci. Tech.*, **27**, 673–688, 1997.

Whitby, K.T., B.Y.H. Liu, and R.B. Husar, Aerosol size distribution of Los Angeles smog, *J. Colloid and Interface Sci.*, **39**, 177–204, 1972.

Whitby, K.T., W.E. Clark, V.A. Marple, G.M. Sverdrup, G.J. Sem, K. Willeke, B.Y.H. Liu, and D.Y.H. Pui, Characterization of California aerosols. 1: Size distributions of freeway aerosol, *Atmos. Environ.*, **9**, 463–482, 1975.

Whiteman, D.N., Examination of the traditional Raman lidar technique. I: Evaluating the temperature-dependent lidar equations, *Appl. Opt.*, **42**, 2571–2593, 2003.

Wiedensohler, A., and D.S. Covert, Number concentrations and size distributions of atmospheric aerosol under baseline condition at Cape Grim, *J. Aerosol Sci.*, **27**, S99–S100, 1996.

Wieland, W., Die Wasserdampfkondensation an naturlichen Aerosol bei geingen Übersaffigungen, *Z. Angew. Math. Phys.*, **7**, 428–436, 1956.

Wilheit, T.T., Some comments on passive microwave measurement of rain, *Bull. Amer. Meteor. Soc.*, **67**, 1226–1232, 1986.

Williams, C.R., Simultaneous ambient air motion and raindrop size distributions retrieved from UHF vertical incident profiler observations, *Radio Sci.*, **37**, doi:10.1029/2000RS002603, 2002.

Williams, D.J., Carras, J.N., Milne, J.W. and Heggie, A.C., The oxidation and long-range transport of sulphur dioxide in a remote region, *Atmos. Environ.*, **15**, 2255–2262, 1981.

Williams, K.D., A. Jones, D.L. Roberts, C.A. Senior, and M.J. Woodage, The response of the climate system to the indirect effects of anthropogenic sulfate aerosol, *Clim. Dynamics.*, **17**, 845–856, 2001.

Williams, E., D. Rosenfeld, N. Madden, J. Gerlach, N. Gears, L. Atkinson, N. Dunnemann, G. Frostrom, M. Antonio, B. Biazon, R. Camargo, H. Franca, A. Gomes, M. Lima, R. Machado, S. Manhaes, L. Nachtigall, H. Piva, W. Quintiliano, L. Machado, P. Artaxo, G. Roberts, N. Renno, R. Blakeslee, J. Bailey, D. Boccippio, A. Betts, D. Wolff, B. Roy, J. Halverson, T. Rickenbach, J. Fuentes, and E. Avelino, Contrasting convective regimes over the Amazon: Implications for cloud electrification, *J. Geophys. Res.*, **107**, 8082, doi:10.1029/2001JD000380, 2002.

Winker, D.M., and C.R. Trepte, Laminar cirrus observed near the tropical tropopause by LITE, *Geophys. Res. Lett.*, **25**, 3351–3354, 1998.

Winker, D.M., J. Pelon, and M.P. McCormick, The CALIPSO mission: Spaceborne lidar for observation of aerosols and clouds, paper presented at SPIE Asia-Pacific Sym., *On Rem. Sen. of the Atmosphere, Environment and Space*, Hangzhou, China, 23–27 October 2002.

Winker, D.M., R.H. Couch, and M.P. McCormick, The CALIPSO mission: Spaceborne lidar for observation of aerosols and clouds, *Proc. SPIE*, **4893**, 1–11, 2003.

Winker, D.M., Hunt, W.H. and Hostetler, C.A., Status and Performance of the CALIOP Lidar, *Proc. SPIE*, **5575**, 8–15, 2004.

Wise, M.E., J.D. Surratt, D.B. Curtis, J.E. Shilling, and M.A. Tolbert, Hygroscopic growth of ammonium sulfate/dicarboxylic acids, *J. Geophys. Res.*, **108**, 4638, doi:10.1029/2003JD003775, 2003.

WMO, *World Meteorological Organization, Report of the WMO workshop on hygroscopic seeding*, WMP Report No. 35, World Meteor. Org., WMO/TD No. 1006, Geneva, Switzerland, 68 pp., 2000.

WMO, *GAW Aerosol Measurement Procedures, Guidelines and Recommendations*, GAW Report No. 153, Geneva, 2003

WMO, *WMO/GAW Experts Workshop on a Global Surface Based Network for Long Term Observations of Column Aerosol Optical Properties*, edited by U. Baltensperger, L. Barrie, C. Wehrli, GAW Report No. 162, Geneva, 2005.

Wolf, M.E., and G.M. Hidy, Aerosols and climate: Anthropogenic emissions and trends for 50 years, *J. Geophys. Res.*, **102**, 11,113–11,121, 1997.

Wood, R., Drizzle in stratiform boundary layer clouds. Part I: Vertical and horizontal structure, *J. Atmos. Sci.*, **62**, 3011–3033, 2005.

Wood, R., Cancellation of aerosol indirect effects in marine stratocumulus through cloud thinning, *J. Atmos. Sci.*, **64**, 2657–2669, 2007.

Woodcock, A.H., Salt nuclei in marine air as a function of altitude and wind force, *J. Meteor.*, **10**, 362–371, 1953.

Woodcock, A.H., and R.H. Jones, Rainfall trends in Hawaii, *J. Appl. Meteor.*, **9**, 690–695, 1970.

Woodcock, A.H., R.A. Duce, and J.L. Moyers, Salt particles and raindrops in Hawaii, *J. Atmos. Sci.*, **28**, 1252–1257, 1971.

Woodley, W.L., Comments on "A new look at the Israeli cloud seeding experiments," *J. Appl. Meteor.*, **36**, 250–252, 1997.

Woodley, W.L., B. Jordan, A. Barnston, J. Simpson, R. Biondini, and J.A. Flueck, Rainfall results of the Florida area cumulus experiment, 1970–1976, *J. Appl. Meteor.*, **21**, 139–164, 1982.

Woodley, W.L., A. Barnston, J.A. Flueck, and R. Biondini, The Florida Area Cumulus Experiment's second phase (FACE-2). Part II: Replicated and confirmatory analyses, *J. Appl. Meteor.*, **22**, 1529–1540, 1983.

Woodley, W.L., D. Rosenfeld, and B.A. Silverman. Results of on-top glaciogenic cloud seeding in Thailand. Part I: The demonstration experiment, *J. Appl. Meteor.*, **42**, 920–938, 2003a.

Woodley, W.L., D. Rosenfeld, and B.A. Silverman. Results of on-top glaciogenic cloud seeding in Thailand. Part II: Exploratory analyses, *J. Appl. Meteor.*, **42**, 939–951, 2003b.

Wurzler, S., T.G. Reisin, and Z. Levin, Modification of mineral dust particles by cloud processing and subsequent effects on drop size distributions, *J. Geophys. Res.*, **105**, 4501–4512, 2000.

Xie, P., and P.A. Arkin, Global precipitation: A 17-year monthly analysis based on gauge observations, satellite estimates, and numerical model outputs, *Bull. Amer. Meteor. Soc.*, **78**, 2539–2558, 1997.

Xue, H., and G. Feingold, Large eddy simulations of trade-wind cumuli: Investigation of aerosol indirect effects, *J. Atmos. Sci.*, 1605–1622, 2006.

Yang, D., B.E. Goodison, J.R. Metcalfe, V.S. Golubev, R. Bates, T. Pangburn, and C.L. Hanson, Accuracy of NWS 8" Standard nonrecording precipitation gauge: Results and Application of WMO Intercomparison, *J. Oceanic Atmos. Tech.*, **15**, 54–68, 1998.

Yankofsky, S., Z. Levin, T. Bertold, and N. Sandlerman, Some basic characteristics of bacterial freezing nuclei, *J. Appl. Meteor.*, **20**, 1013–1019, 1981.

Yevich, R., and J.A. Logan, An assessment of biofuel use and burning of agricultural waste in the developing world, *Global Biogeochem. Cycles*, **17**, 1095, doi:10.1029/2002GB001952, 2003.

Yienger, J.J., M. Galanter, T.A. Holloway, M.J. Phandnis, S.K. Guttikunda, G.R. Carmichael, W.J. Moxim, and H.L. Levi II, The episodic nature of air pollution transport from Asia to North America, *J. Geophys. Res.*, **105**, 26,931–926,945, 2000.

Yin, Y., Z. Levin, T.G. Reisin, and S. Tzivion, The effects of giant condensation nuclei on the development of precipitation in convective clouds – A numerical study, *Atmos. Res.*, **53**, 91–116, 2000a.

Yin, Y., Z. Levin, T.G. Reisin and S. Tzivion: Seeding convective clouds with hygroscopic flares: Numerical simulations using a cloud model with detailed microphysics, *J. Appl. Meteor.*, **39**, 1460–1472, 2000b.

Yin, Y., Z. Levin, T.G. Reisin, and S. Tzivion. On the response of radar-derived properties to hygroscopic flare seeding, *J. Appl. Meteor.*, **40**, 1654–1661, 2001.

Yin, Y., S. Wurzler, Z. Levin, and T.G. Reisin, Interactions of mineral dust particles and clouds: Effects on precipitation and cloud optical properties, *J. Geophys. Res.*, **107**, 4724, doi:10.1029/2001JD001544, 2002.

Yin, Y., K.S. Carslaw, and G. Feingold, Vertical transport and processing of aerosols in a mixed-phase convective cloud and the feedback on cloud development, *Q. J. Roy. Meteorol. Soc.*, **131**(605), 221–245, 2005.

Young, K.C., A numerical examination of some hail suppression concepts, in *Hail: A review of hail science and hail suppression*, edited by G.B. Foote and C.A. Knight, *Meteorol. Monogr.*, **16**(38), 195–214, 1977.

Young, K.C., *Microphysical Processes in Clouds*, Oxford University Press, New York, 335–336, 1993.

Young, K.C., Weather modification – A theoretician's viewpoint, *Bull. Amer. Meteor. Soc.*, **77**, 2701–2710, 1996.

Yuter, S.E., Precipitation radar, in *Encycl. Atmos. Sci.*, edited by J. Holton, J. Pyle, and J. Curry, pp. 1833–1852, Academic Press, London, UK, 2002.

Yuter, S.E., and R.A. Houze, Jr., The natural variability of precipitating clouds over the western Pacific warm pool, *Q. J. Roy. Met. Soc.*, **124**, 53–99, 1998.

Yuter, S.E., and W.S. Parker, Rain measurement on ship revisited: The 1997 PACS TEPPS cruise, *J. Appl. Meteor.*, **40**, 1003–1018, 2001.

Yuter, S.E., R.A. Houze, Jr., E.A. Smith, T.T. Wilheit, and E. Zipser, Physical characterization of tropical oceanic convection observed in KWAJEX, *J. Appl. Meteor.*, **44**, 385–415, 2005.

Yuter, S.E., D. Kingsmill, L.B. Nance, and M.Löffler-Mang, Observations of precipitation size and fall speed characteristics within coexisting rain and wet snow, *J. Appl. Meteor. Climatology*, **45**, 1450–1464, 2006.

Zappoli, S., A. Andracchio, S. Fuzzi, M.C. Facchini, A. Gelencser, G. Kiss, Z. Krivacsy, A. Molnar, E. Meszaros, H.C. Hansson, K. Rosman, and Y. Zebühr, Inorganic, organic and macromolecular components of fine aerosol in different areas of Europe in relation to their water solubility, *Atmos. Environ.*, **33**, 2733–2743, 1999.

Zawadzki, I., and A. Bellon, Error statistics of VPR correction in stratiform precipitation, Preprints, 31st Conf. Radar Meteor., Seattle, Wash., *Amer. Meteor. Soc.*, 225–228, 2003.

Zender, C.S., H.S. Bian, and D. Newman, Mineral Dust Entrainment and Deposition (DEAD) model: Description and 1990s dust climatology, *J. Geophys. Res.*, **108**, 4416, doi:10.1029/2002JD002775, 2003.

Zender, C.S., R.L. Miller, and I. Tegen, Quantifying mineral dust mass budgets: Terminology, constraints, and current estimates, *Eos Trans. AGU*, **85**, 509–512, 2004.

Zhai, P., X. Zhang, H. Wan, and Z. Pan, Trends in Total Precipitation and Frequency of Daily Precipitation Extremes over China, *J. Climate*, **18**, 1096–1108, 2005.

Zhang, L., S.L. Gong, J. Padro, and L. Barrie, A size-segregated particle dry deposition scheme for an atmospheric aerosol module, *Atmos. Environ.*, **35**, 549–560, 2001.

Zhang, Q., C.O. Stanier, M.R. Canagaratna, J.T. Jayne, D.R. Worsnop, S.N. Pandis, and J.L. Jimenez, Insights into the chemistry of new particle formation and growth events in Pittsburgh based on aerosol mass spectrometry, *Environ. Sci. Technol.*, **38**, 4797–4809, 2004a.

Zhang, Y.H., X.L. Zhu, S. Slanina, M. Shao, L.M. Zeng, M. Hu, M. Bergin, and L. Salmon, Aerosol pollution in some Chinese cities (IUPAC Technical Report), *Pure and Applied Chemistry*, **76**, 1227–1239, 2004b.

Zhang, L., D.V. Michelangeli, and P.A. Taylor, Numerical studies of aerosol scavenging by low-level, warm stratiform clouds and precipitation, *Atmos. Environ.*, **38**, 4653–4665, 2004c.

Zhang, X.Y., Y.Q. Wang, D. Wang, S.L. Gong, R. Arimoto, L.J. Mao, and J.Li, Characterization and sources of regional-scale transported carbonaceous and dust aerosols from different pathways in costal and sandy land areas of China, *J. Geophys. Res.*, **110**, D15301, doi:10.1029/2004JD005457, 2005a.

Zhang, J., U. Lohmann, and P. Stier, A microphysical parameterization for convective clouds in the ECHAM5 Climate Model: 1. Single column model results evaluated at the Oklahoma ARM site, *J. Geophys. Res.*, **110**, D15S07, doi:10.1029/2004JD005128, 2005b.

Zhao, T.L., S.L. Gong, X.Y. Zhang, and I.G. McKendry, Modelled size-segregated wet and dry deposition budgets of soil dust aerosol during ACE-Asia, 2001: Implications for trans-pacific transport, *J. Geophys. Res.*, **108**, 8665, doi:8610.1029/2002JD003363, 2003.

Zhao, T.L., S.L. Gong, X.Y. Zhang, and I.G. McKendry, Hydrophilic properties of aged soot, *Geophys. Res. Lett.*, **32**, L01807, doi:01810.01029/02004GL021496, 2005.

Zhao, T.L., S.L. Gong, X.Y. Zhang, J.P. Blanchet, I.G. McKendry, and Z.J. Zhou, A simulated climatology of asian dust aerosol and its trans-pacific transport. 1: Mean climate and validation, *J. Clim.*, **19**, 88–103, 2006.

Zhou, J.C., E. Swietlicki, O.H. Berg, P.P. Aalto, K. Hameri, E.D. Nilsson, and C. Leck, Hygroscopic properties of aerosol particles over the central Arctic Ocean during summer, *J. Geophys. Res.*, **106**, 32,111–32,123, 2001.

Zuberi, B., K.S. Johnson, G.K. Aleks, L.T. Molina, and A. Laskin, Hydrophilic properties of aged soot, *Geophys. Res. Lett.*, **32**, L01807, doi:10.1029/2004GL021496, 2005.

Zufall, M.J., and C.I. Davidson, Dry deposition of particles, *Atmospheric Particles*, 425–473, 1998.

Index

MIX
Papier aus verantwortungsvollen Quellen
Paper from responsible sources
FSC® C105338

If you have any concerns about our products,
you can contact us on
ProductSafety@springernature.com

In case Publisher is established outside the EU,
the EU authorized representative is:
Springer Nature Customer Service Center GmbH
Europaplatz 3, 69115 Heidelberg, Germany

Printed by Libri Plureos GmbH
in Hamburg, Germany